Bifurcation Phenomena in Mathematical Physics and Related Topics

NATO ADVANCED STUDY INSTITUTES SERIES

Proceedings of the Advanced Study Institute Programme, which aims at the dissemination of advanced knowledge and the formation of contacts among scientists from different countries

The series is published by an international board of publishers in conjunction with NATO Scientific Affairs Division

A	Life Sciences	Plenum Publishing Corporation
B	Physics	London and New York
C	Mathematical and Physical Sciences	D. Reidel Publishing Company Dordrecht, Boston and London
D	Behavioural and Social Sciences	Sijthoff & Noordhoff International Publishers
E	Applied Sciences	Alphen aan den Rijn and Germantown U.S.A.

Series C – Mathematical and Physical Sciences

Volume 54 – Bifurcation Phenomena in Mathematical Physics and Related Topics

Bifurcation Phenomena in Mathematical Physics and Related Topics

Proceedings of the NATO Advanced Study Institute held at Cargèse, Corsica, France, June 24-July 7, 1979

edited by

CLAUDE BARDOS

Université de Paris-Nord, Départment de Mathematiques, Villetaneuse, France

and

DANIEL BESSIS

CEN, CEA-Saclay, Gif-sur-Yvette, France

D. Reidel Publishing Company

Dordrecht : Holland / Boston : U.S.A. / London : England

Published in cooperation with NATO Scientific Affairs Division

Library of Congress Cataloging in Publication Data

CIP

NATO Advanced Study Institute, Cargese, Corsica, 1979.
Bifurcation phenomena in mathematical physics and related topics.

(NATO advanced study institutes series: Series C, Mathematical and physical sciences ; v. 54)
1. Bifurcation theory–Congresses. 2. Mathematical physics–Congresses. I. Bardos, Claude, 1940. II. Bessis, D., 1933. III. North atlantic treaty organization. Division of scientific affairs. IV. Title. V. Series.
QC20.7.B54N37 1979 530'515353 80-11588
ISBN-13: 978-94-009-9006-7 e-ISBN-13: 978-94-009-9004-3
DOI: 10.1007/978-94-009-9004-3

Published by D. Reidel Publishing Company
P. O. Box 17, 3300 AA Dordrecht, Holland

Sold and distributed in the U.S.A. and Canada
by Kluwer Boston Inc., Lincoln Building,
160 Old Derby Street, Hingham, MA 02043, U.S.A.

In all other countries, sold and distributed
by Kluwer Academic Publishers Group,
P.O. Box 322, 3300 AH Dordrecht, Holland

D. Reidel Publishing Company is a member of the Kluwer Group

Softcover reprint of the hardcover 1st edition 1980

TABLE OF CONTENTS

FOREWORD

One of the main ideas in organizing the Summer Institute of Cargese on "Bifurcation Phenomena in Mathematical Physics and Related Topics" was to bring together Physicists and Mathematicians working on the properties arising from the non linearity of the phenomena and of the models that are used for their description.

Among these properties the existence of bifurcations is one of the most interesting, and we had a general survey of the mathematical tools used in this field. This survey was done by M. Crandall and P. Rabinowitz and the notes enclosed in these proceedings were written by E. Buzano and C. Canuto. Another mathematical approach, using Morse Theory was given by J. Smoller reporting on a joint work with C. Conley. An example of a direct application was given by M. Ghil.

For physicists the theory of bifurcation is closely related to critical phenomena and this was explained in a series of talks given by J.P. Eckmann, G. Baker and M. Fisher. Some related ideas can be found in the talk given by T.T. Wu on a joint work with Barry Mc Coy on quantum field theory. The description of these phenomena leads to the use of Padé approximants (it is explained for instance in the lectures of J. Nuttall) and then to some problems in drop hot moment problems. (cf. the lecture of D. Bessis).

The technique of Padé approximants is also related to explicit constructions of solutions by analytic methods. One example of these constructions was done by M. Barnsley for the solutions of the Boltzmann equation.

Non linearity in partial differential equations leads to two very active fields, one of which is the theory of the mappings and the existence of attractors ; in this field we had lectures given by B. Derrida, Y. Pomeau and some applications given by S. Aubry. The other is the theory of solitons and R. Bullough presented a survey of this topic. Of course this theory is closely related with algebraic properties of the equations as was explained in talks given by D. and G. Chudnovsky and by F. Calogero.

The theory of non linear Partial Differential Equations found a challenging range of applications in the Yang–Mills theory. A survey was given by R. Stora and some mathematical approaches to the problem, including regularity of solutions and existence of solitary waves were described by H. Berestycki and P.L. Lions, B. Gidas, R. Glassey and W. Strauss.

Finally we had a set of lectures on some fundamental questions of the nonlinear theory of Partial Differential Equations. This included a study of sharp estimates on Sobolev spaces with some applications to nonlinear Schrodinger equation (lecture by H. Brezis), and classical Navier– Stokes equations. A lecture on the structure of the steady state solutions and the invariant set for these equations was given by R. Temam (on a joint work with C. Foias and on some joint works with J.C. Saut).

The Navier–Stokes equations are used to describe flows of incompressible fluids. In the case of compressible fluids the non linearity leads to the appearance of shocks. We had some lectures (not published in these proceedings) given by R. Di Perna on the theory of shock waves. In the case of a single equation a very precise estimate can be obtained on the asymptotic behaviour of the solution and this was the goal of the lectures of C. Dafermos.

The study of Nematic Crystals by J.P. Dias and the Hartree– Fock equation (lecture given by B. Giraud) are two other examples of the case of non linear analysis in physical problems.

Professor J.L. Lions has constantly supervised the organization of the meeting, and we have also received much valuable advice and help from Professor R. Dautray; we want to express to them our deep gratitude.

The organizers would like to thank the Scientific Committee of NATO for a grant which made this Institute possible and especially Dr. Mario di Lullo for his help and the constant interest that he has shown in this course. We also thank the National Science Foundation of the U.S.A. for travel grants. Thanks are due to several French agencies that have given their support, the Commissariat à l'Energie Atomique, the D.G.R.S.T. and the D.R.E.T., and especially to Mr. Heugas Scienfific Director Assistant, for his continuous help.

We had the opportunity to use the facility of the Institute of Scientific Studies of Cargèse, and we are happy to thank the director Prof. M. Levy and in particular Marie-France Hanseler.

At the end of the Meeting a "Mechoui" was organized by the participants. The excellent cooking and the very warm atmosphere of the evening were due to Professors S. Benachour, J. Della Dora and H. Ghidouche. To them we express the gratitude of everyone.

Finally we have to express our warmest thanks to Madeleine Porneuf and Francine Lefevre, who by their constant work and enthusiasm made possible the success of this Institute and of the Publication of the Proceedings. We hope that these publications by Reidel Publishing Company will help the increasing collaboration between mathematicians and physicists.

C. Bardos and D. Bessis

I

BIFURCATION IN MATHEMATICS AND APPLICATIONS

MATHEMATICAL THEORY OF BIFURCATION

Michael G. CRANDALL and Paul H. RABINOWITZ

Mathematics Research Center, Madison, Wisconsin

Notes by E. Buzano and C. Canuto

0. INTRODUCTION

One of the first studies of bifurcation is due to Euler (1744), who treated the buckling of a column subjected to axial compression (the so-called Elastica). He considered the following boundary-value problem

$$\begin{cases} \theta'' + \lambda \sin \theta = 0 \\ \theta'(0) = \theta'(1) = 0 \end{cases} \tag{0.1}$$

where θ is the angle between the tangent to the column and the real axis, λ is the thrust applied and 1 is the length of the column:

fig. 1

Obviously $\theta \equiv 0$ is a solution of (0.1) for any real λ. This means that the column remains straight in equilibrium. But it is not difficult to see that as λ exceeds π^2 a new family of solutions appears and the column buckles.

We can plot the state of the system (as given e.g. by $\theta(0)$) versus λ:

C. Bardos and D. Bessis (eds.), Bifurcation Phenomena in Mathematical Physics and Related Topics, 3–46.

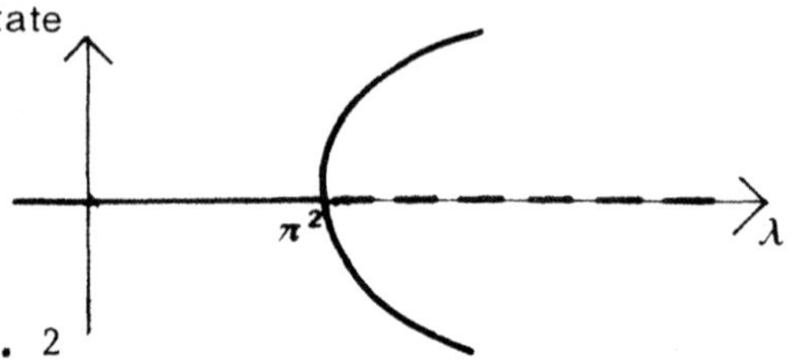

fig. 2

Of course one may consider the buckling of many other elastic structures as plates, shells, etc. ... See for example BERGER [11], BERGER and FIFE [13], FRIEDRICHS and STOKER [37] and ANTMAN [8].

These phenomena can be settled into an abstract formulation as follows. Let us consider two Banach spaces X and Y and a mapping $G: X \to Y$. Suppose there exists a curve Z of zeroes of G. An interior point $w \in Z$ is called a <u>bifurcation point for G with respect to Z</u> if every neighbourhood of w in X contains solutions of $G(x) = 0$ not on Z (see fig. 3). Usually $X \subset \mathbb{R} \times \tilde{X}$ and $Z = \mathbb{R} \times \{0\}$ (see fig. 4).

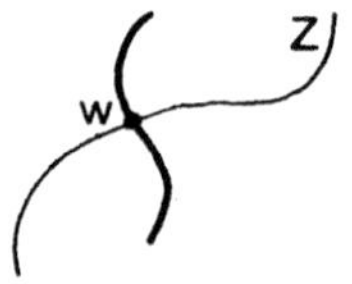

fig. 3

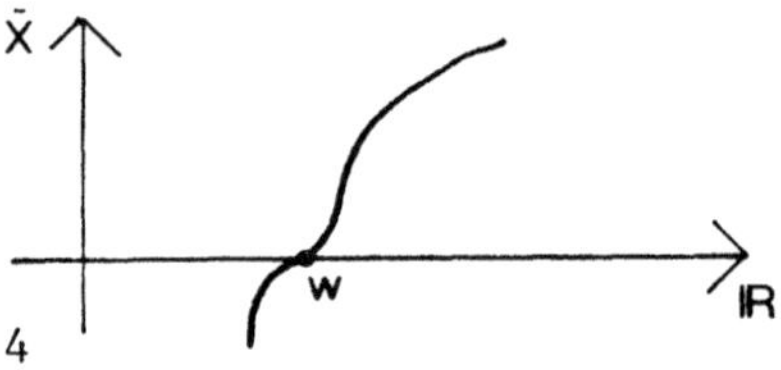

fig. 4

Let us point out that here the solutions have the form $(\lambda, \tilde{x})$ where λ is a (bifurcation) parameter usually with a physical meaning like temperature or Reynold's number or the magnitude of a force, etc. ... The elements of Z are called the <u>trivial zeroes</u> of G. We are interested in the study of families of nontrivial solutions which bifurcate from Z. The importance of such a study is due to the great number of applications as shown by the following examples.

a) <u>Fission Theory of Binary Stars</u>. This theory attempts to explain a binary system of stars as the evolution of a rotating self gravitating fluid mass undergoing contraction. The first models considered were ellipsoids of uniform density. Therefore the problem was reduced to find the three semiaxes a_1, a_2, a_3 of the ellipsoid or the density, mass and angular momentum. In 1742 Mac-

Laurin showed that oblate spheroids ($a_1 = a_2$, $a_3 < a_1$) are figures of equilibrium. In 1834 Jacobi found that when $a_3/a_1 \sim 0.6$ a new family of ellipsoids bifurcates from the MacLaurin spheroids.

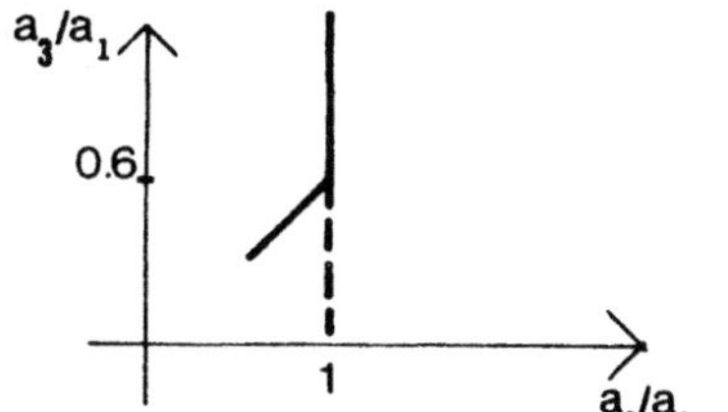

fig. 5

In 1883 Thompson and Tait noted that far enough along the Jacobi branch the figure of the stars becomes unstable with respect to a deformation tending to thin it in the middle and thicken it at the ends. This led Poincaré to conjecture in 1885 the appearing of a sequence of "pear-shaped" figures bifurcating from the Jacobi ellipsoids as a cause of the fission of the star. Unfortunately the new branch was proved to be unstable and the theory turned out to be not very convincing.

Further developments in the relativistic framework have been carried out during the last years. (See e.g. LEBOWITZ [63]).

b) Thermal Convection - The Bénard Problem. A viscous incompressible fluid is contained between two infinite horizontal conducting plates. The two plates are mantained at different temperatures T_0 and T_1, the lower being warmer. Mathematically the system is described by the Navier-Stokes equations coupled with the heat-conduction equation. For a small enough gradient $T = T_0 - T_1$ the fluid is at rest and there is heat conduction only. When T exceeds a critical value a steady-state motion appears which is called convective current and heat is transferred both by conduction and by convection. Further increasing of T can give secondary bifurcation. See CHANDRASEKAR [17], KIRCHGÄSSNER and KIELHÖFER [58] and SATTINGER [87].

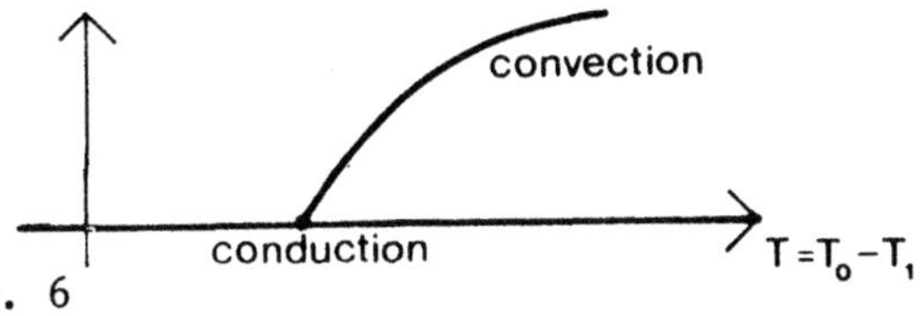

fig. 6

c) Rotating fluids - The Taylor Problem. A viscous incompressible fluid lies between two vertical coaxial cylinders, of which the inner rotates with constant angular velocity ω, while the outer is at rest. The fluid's motion is described by the Navier-Stokes equations in cylindrical coordinates; for any value of ω, they admit a stationary solution which describes a motion consisting of circular orbits with velocity depending on the distance from the inner cylinder. This solution is called Couette flow and remains stable for ω smaller than a critical value at which the fluid breaks up into horizontal bands called Taylor vortices and a new fluid motion periodic in the vertical direction is superimposed on the Couette flow (see fig. 8). When ω is increased still further the Taylor vortices lose their stability and are replaced by a new flow (sometimes called wavy vortices), which is periodic both in the vertical direction and in time. This is an example of bifurcation of time-independent into time-periodic solutions. See CHANDRASEKAR [17] , KIRCHGÄSSNER and KIELHÖFER [58] and SATTINGER [87] .

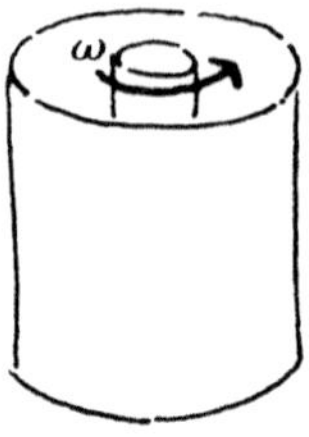

fig. 7

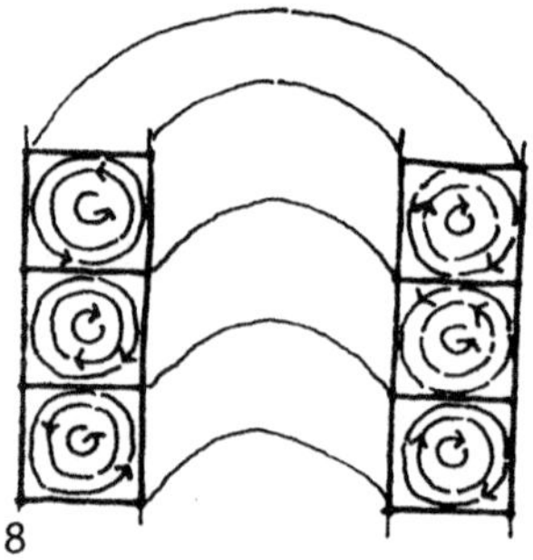

fig. 8

d) Reaction - Diffusion Problems. Other equations which exibit bifurcation of steady-state solutions into time-independent or time-periodic solutions are the so-called reaction-diffusion equations:

$$\frac{\partial u}{\partial t} = D\Delta u + f(\mu,u)$$

where u is the vector of the chemical concentrations of an n-component system. D is a diagonal matrix, Δ in the Laplace operator in the spatial coordinates describing the diffusion process, while f describes all reactions and interations and μ is a real parameter. Among the chemical problems that can be modelized by these equations let us mention here the Belusov-Zhabotinski reaction consisting in the oxidation of malonic acid by bromate in the presence of cerium ions, which in one of the few examples of substained oscillations in chemical systems. Even the behaviour of certain biological systems can be described by equations of the reaction-diffusion type: for instance pulse solutions of the celebrated Hodgking-Huxley equations succesfully model electric signals propagating along a nerve axon. Excellent references are FIFE [33] and [34] . An interesting approach to the stability of steady-state solutions of reaction-diffusion equations is given in Smoller's lecture in this volume.

e) Bifurcation into an Invariant Torus of the Navier-Stokes Equations. Direct and secondary bifurcation of steady-state and time-periodic solutions of the general Navier-Stokes equations have been studied by G. Iooss. The bifurcated solutions lie on an invariant torus. For more details refer to the book by IOOSS [49].

f) Imperfect Bifurcation. Going back to the Elastica it is interesting to compare the "exact" bifurcation diagram with the one obtained after experiments. The latter does not present bifurcation any longer (see fig. 9). This feature can be explained by perturbing the "exact" problem by suitable parameters which take into account physical or geometrical imperfections. See MATKOWSKY and REISS [68] and GOLUBITSKY and SCHAEFFER [38].

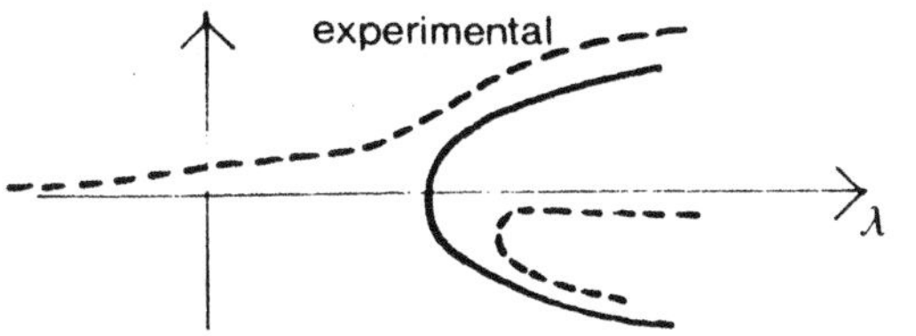

fig. 9

What are the sort of questions one studies in the theory of bifurcation? The main ones are the following:

1. Given a curve of zeroes Z of our operator equation $F(\lambda,x)=0$, what are necessary and sufficient conditions for a point on Z to be a bifurcation point ?
2. What is the structure of the set of zeroes of F near a bifurcation point ?
3. What are effective procedures for calculating this set ?
4. Can one continue this set into the large ? Does local knowledge of bifurcation have global implications for the zero set of F ?
5. For problems in which $F(\lambda,x)=0$ is the equilibrium form of a corresponding evolution equation, what can one say about the stability of the various solutions obtained ?

We will provide some answers to this questions in this series of lectures. The topics to be covered are the following:

1. Bifurcation from Simple Eigenvalue
2. The Hopf Bifurcation
3. Linearized Stability

4. The Liapunov-Schmidt Procedure
5. Global Results
6. The Variational Case
7. Periodic Solutions of Hamiltonian Systems.

Points 1 to 4 are covered by Prof.M.G.Crandall; they are mainly concerned with analytical and constructive procedures (essentially the Implicit Function Theorem) which lead to a local analysis of the bifurcation point. In the second part of these lectures, Prof. P.H. Rabinowitz first employs topological degree arguments to give global bifurcation theorems. Then the variational case is studied by reducing the problem to a finite-dimension one. Lastly a survey of recent results on periodic solutions of Hamiltonian Systems is given.

A large number of general references, both on theoretical methods and applications is available. See for example KELLER and ANTMAN [57], SATTINGER [87], IOOSS [44] and [49] and RABINOWITZ [79] and [80].

1. BIFURCATION FROM SIMPLE EIGENVALUE

Let us consider the following evolution problem

$$\begin{cases} \dfrac{dw}{dt} = G(\lambda, w) \\ w(0) = w_0 \end{cases} \tag{1.1}$$

where t ranges over $[0,+\infty)$ and let us suppose that there exist λ_0, w_0 such that $G(\lambda_0, w_0) = 0$; then $w(t) \equiv w_0$ solves (1.1) for $\lambda = \lambda_0$ and is called an equilibrium (or stationary) solution. Equilibria can be regarded as the "simplest" solutions of (1.1). We assume that we know a family of stationary solutions depending on λ, i.e. a mapping $\lambda \mapsto \varphi(\lambda)$ such that $G(\lambda, \varphi(\lambda)) = 0$ for any λ in a suitable set. Then by the change of variable $u + \varphi(\lambda) = w$, the problem (1.1) becomes:

$$\begin{cases} \dfrac{du}{dt} = F(\lambda, u) \\ u(0) = u_0 \end{cases} \tag{1.2}$$

where $F(\lambda,u) = G(\lambda, u + \varphi(\lambda))$, and $F(\lambda,0) \equiv 0$ (0 is called the trivial solution).

We are interested in finding other equilibria for (1.1) near

$w = \varphi(\lambda)$ i.e. nontrivial equilibria for (1.2) bifurcating from the trivial line $\{(\lambda,0)\}$.

First let us consider the finite-dimensional case: assume that $F : \mathbb{R} \times \mathbb{R}^n \to \mathbb{R}^n$ and satisfies $F(\lambda,0) \equiv 0$.

Definition 1.1. $(\lambda_0,0)$ is a bifurcation point for F if any neighbourhood of $(\lambda_0,0)$ contains nontrivial solutions of the equation $F(\lambda,u) = 0$.

This definition is appropriate for the cases we have in mind; for a suitable definition in generality see MARSDEN [66].

We look for necessary and sufficient conditions to have bifurcation. The main tool of our analysis will be the Implicit Function Theorem. We recall here the finite dimensional version of it.

Let $\mathcal{O}$ be an open set in $\mathbb{R} \times \mathbb{R}^n$ and $(\lambda_0,u_0) \in \mathcal{O}$; assume we have $F \in C^1(\mathcal{O},\mathbb{R}^n)$ such that $F(\lambda_0,u_0) = 0$. Denote by $F_u(\lambda,u) = \{\frac{\partial F_j}{\partial u_i}(\lambda,u)\}_{1 \leqslant i,j \leqslant n}$ the Jacobian matrix of F at (λ,u) with respect to a fixed basis in $\mathbb{R}^n$, so that F may be written as

$$F(\lambda,u+\hat{u}) = F(\lambda,u) + F_u(\lambda,u)\hat{u} + o(\|\hat{u}\|)$$

in a neighbourhood of u ($\|.\|$ denotes any norm in $\mathbb{R}^n$).

THEOREM 1.1 (Implicit Function Theorem). If $F_u(\lambda_0,u_0)$ is non-singular, there exists a curve $u = u(\lambda)$ defined on a neighbourhood L of λ_0 such that $u(\lambda_0) = u_0$ and $F(\lambda,u(\lambda)) \equiv 0$. Moreover any solution of $F(\lambda,u) = 0$ in a neighbourhood of (λ_0,u_0) is of the form $(\lambda,u(\lambda))$. Finally if $F \in C^k(\mathcal{O},\mathbb{R}^n)$, then $u \in C^k(L,\mathbb{R}^n)$, $k \geqslant 1$. ◄

The theorem can be proved in a constructive way, using a contraction argument which yields $u(\lambda)$ and all its existing derivatives as limit of well-defined sequences of approximations: see for instance the proof given by CRANDALL [20] in the infinite dimensional case.

The Implicit Function Theorem easily gives a good necessary condition for bifurcation from the trivial branch.

THEOREM 1.2 If $F_u(\lambda_0,0)$ is non-singular, then $(\lambda_0,0)$ is not a bifurcation point.

Proof. According to Theorem 1.1, $u(\lambda) \equiv 0$ is the unique curve of solutions near $(\lambda_0,0)$. ◄

If $F_u(\lambda_0,0)$ is singular, $(\lambda_0,0)$ may or may not be a bifurcation point. Before exhibiting some examples, let us express $F(\lambda,u)$ by Taylor's formula near $(\lambda_0,0)$; if $F \in C^2(\mathbb{R} \times \mathbb{R}^n,\mathbb{R}^n)$ we have:

$$F(\lambda,u) = L_0u + (\lambda - \lambda_0)L_1u + p(\lambda,u)$$

where $L_0 = F_u(\lambda_0,0)$ and $L_1 = F_{\lambda u}(\lambda_0,0)$ are nxn matrices and $p \in C^2(\mathbb{R} \times \mathbb{R}^n, \mathbb{R}^n)$ satisfies $p(\lambda,0) \equiv 0$, $p_u(\lambda_0,0) = 0$ and $p_{\lambda u}(\lambda_0,0) = 0$.

Example 1.1 Set $u = (x,y)^T$ and define $F : \mathbb{R} \times \mathbb{R}^2 \to \mathbb{R}^2$ by

$$F(\lambda,u) = \begin{pmatrix}1 & 0\\ 0 & 1\end{pmatrix}\begin{pmatrix}x\\ y\end{pmatrix} + (\lambda-\lambda_0)\begin{pmatrix}a & b\\ 0 & 0\end{pmatrix}\begin{pmatrix}x\\ y\end{pmatrix} + \begin{pmatrix}0\\ x^2+y^2\end{pmatrix}.$$

Then $F_u(\lambda_0,0) = \begin{pmatrix}1 & 0\\ 0 & 0\end{pmatrix}$ has zero as an algebraically simple eigenvalue, however $F(\lambda,u) = 0$ implies $x^2+y^2 = 0$, so no bifurcation can occur.

Note that $\begin{pmatrix}1 & 0\\ 0 & 0\end{pmatrix} + (\lambda - \lambda_0)\begin{pmatrix}a & b\\ 0 & 0\end{pmatrix}$ (the linear part in u for F) is singular for every value of λ; but even singularity at $\lambda - \lambda_0$ only is not sufficient for bifurcation, as the following example shows.

Example 1.2 Take

$$F(\lambda,u) = (\lambda - \lambda_0)\begin{pmatrix}1 & 0\\ 0 & 1\end{pmatrix}\begin{pmatrix}x\\ y\end{pmatrix} + \begin{pmatrix}-y^3\\ x^3\end{pmatrix}$$

$F(\lambda,u) = 0$ yields $x^4+y^4 = 0$, so again $u = 0$. In this case $\det(L_0 + (\lambda - \lambda_0)L_1) = (\lambda - \lambda_0)^2$ has $\lambda = \lambda_0$ as a double zero.

But if $\det(L_0 + (\lambda - \lambda_0)L_1)$ has a zero of odd multiplicity at $\lambda = \lambda_0$, then we obtain a positive result.

THEOREM 1.3 *If $\det(L_0 + (\lambda - \lambda_0)L_1)$ changes sign at λ_0, then $(\lambda_0,0)$ is a bifurcation point.* ◄

This theorem is derived by means of Brower's topological degree, in a non-constructive way. Actually it is the finite-dimensional version of the subsequent theorem 5.1 due to Krasnoselskij. If we add the assumption that 0 is a geometrically simple eigenvalue for $F_u(\lambda_0,0)$, we can derive precise information about the local structure of the bifurcating branch and moreover we can obtain it by a constructive procedure: we are considering the so-called "Bifurcation at a Simple Eigenvalue".

THEOREM 1.4 (CRANDALL and RABINOWITZ [21]) *Assume that $N(L_0) = \operatorname{span}\{u_0\}$ and $L_1u_0 \notin R(L_0)$ ($N(L_0)$ and $R(L_0)$ being the null space and the range of L_0 respectively). Let Z be such that $\mathbb{R}^n = \operatorname{span}\{u_0\} \oplus Z$. Then there exists a C^1-curve $(\lambda,\psi) : (-\delta,\delta) \to \mathbb{R} \times Z$ defined on some interval $(-\delta,\delta)$ such that*

$$\begin{cases} \lambda(0) = \lambda_0 \\ \psi(0) = 0 \\ F(\lambda(s), s(u_0 + \psi(s))) = 0 \qquad \forall s \in (-\delta, \delta). \end{cases}$$

Moreover in a neighbourhood of $(\lambda_0, 0)$ any zero of F either lies on this curve or is a trivial solution.

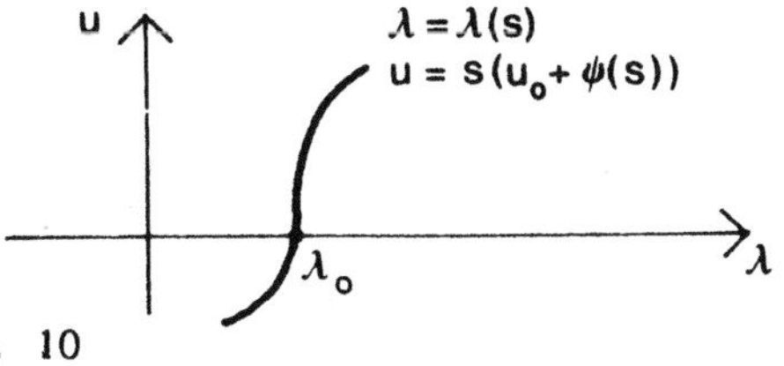

fig. 10

REMARK 1.1. Under the assumption $N(L_0) = \text{span}\{u_0\}$, condition $L_1 u_0 \notin R(L_0)$ is equivalent to the simplicity of the root $\mu = 0$ for $\mu \to \det(L_0 + \mu L_1)$, hence in particular $\det(L_0 + (\lambda - \lambda_0)L_1)$ changes sign at $\lambda = \lambda_0$. Indeed, $L_1 u_0 \notin R(L_0)$ iff $\tilde{u}_0{}^T L_1 u_0 \neq 0$, where $\tilde{u}_0$ spans the null space of $L_0{}^T$. Denoting by L_{ij} the cofactor of L_0 obtained by deleting the i-row and the j-column, there exists a cofactor L_{rs} different from zero since $\dim N(L_0) = 1$ and one easily checks that u_0 and $\tilde{u}_0$ are proportional to $(L_{r1}, \dots, L_{rn})^T$ and $(L_{1s}, \dots, L_{ns})^T$ respectively. Identities $L_{is} L_{rj} = L_{ij} L_{rs}$ hold for any i,j,r,s since $\det L_0 = 0$, hence $\tilde{u}_0{}^T L_1 u_0 = L_{rs} \cdot \sum_{i,j} l_{ij} L_{ij}$, where $L_1 = \{l_{ij}\}_{1 \leq i,j \leq n}$. Finally $\sum_{i,j} l_{ij} L_{ij}$ equals $\frac{\partial}{\partial \mu} \det(L_0 + \mu L_1)|_{\mu=0}$ (see KOPELL and HOWARD [60]).

Sketch of Proof of Theorem 1.4. (Existence) Define

$$\mathcal{F}(s,\lambda,z) = \begin{cases} s^{-1} F(\lambda, s(u_0+z)) = L_0(u_0+z) + (\lambda - \lambda_0) L_1(u_0+z) + \\ \quad + s^{-1} p(\lambda, s(u_0+z)) & \text{if } s \neq 0 \\ F_u(\lambda, 0)(u_0+z) & \text{if } s = 0 \end{cases}$$

and check that $\mathcal{F} \in C^1(\mathbb{R} \times \mathbb{R} \times Z, \mathbb{R}^n)$. Since $\mathcal{F}(0,\lambda_0,0) = 0$ and $\mathcal{F}_{(\lambda,z)}(0,\lambda_0,0)(\hat{\lambda},\hat{z}) = L_0 \hat{z} + \hat{\lambda} L_1 u_0$ is non-singular being onto by the hypotheses, we can uniquely solve $\mathcal{F}(s,\lambda,z) = 0$ for $\lambda = \lambda(s)$, $z = \psi(s)$ near $(0,\lambda_0,0)$ by the Implicit Function Theorem.

(Uniqueness) Since $\mathcal{F}_{(\lambda,z)}(0,\lambda_0,0)$ is an isomorphism, one proves by continuity arguments that for every $\epsilon > 0$ there exists a neighbourhood of $(\lambda_0, 0)$ such that any solution $(\lambda, u) = (\lambda, su_0 + w)$ of $F = 0$ (with $w \in Z$) in this neighbourhood satisfies the estimates $|\lambda - \lambda_0| < \epsilon$, $|s| < \epsilon$, $\|w\| < |s|\epsilon$. Then either $u = 0$ or $s \neq 0$; in the latter case $(s,\lambda,s^{-1}u)$ is also zero for F, hence for ϵ small enough it equals $(s,u(s),\psi(s))$ by the uniqueness in the Implicit Function Theorem. ◄

The previous results can be extended to the infinite dimensional case, essentially under the same hypotheses. We only need to replace some concepts of Calculus in R^n with the corresponding ones in Banach spaces. To begin with, we recall the definition of Fréchet derivative. Let U and V be two real or complex Banach spaces, equipped with norms $\|.\|_U$ and $\|.\|_V$ respectively, and let B(U,V) be the Banach space of linear bounded mappings of U into V equipped with the norm $\|L\|_{B(U,V)} = \sup\{\|Lu\|_V | u \in U, \|u\|_U = 1\}$. Let us consider a continuous function $f : \mathcal{O} \subseteq U \to V$, where $\mathcal{O}$ is an open set in U. We say that f is Fréchet-differentiable at a point $u_0 \in \mathcal{O}$ iff there exists $L \in B(U,V)$ such that $\lim_{u \to u_0} \|f(u) - f(u_0) - L(u-u_0)\|_V / \|u-u_0\|_U = 0$. If this occurs L is unique and is called the Fréchet derivative of f at u_0 (being denoted by $f_u(u_0)$). Most of the rules concerning classical differentiation hold for Fréchet differentiation: for more details we refer to CARTAN [15]. In the following $C^k(\mathcal{O},V)$ will denote the space of all functions $f : \mathcal{O} \to V$ k-times continuously Fréchet-differentiable over $\mathcal{O}$.
As for the Implicit Function Theorem, it works as well in infinite dimensional spaces, as stated in the next theorem.

THEOREM 1.5 (Implicit Function Theorem in Banach spaces). Assume $F \in C^0(\mathcal{O},V)$, with $\mathcal{O}$ open set in $\Lambda \times U$, Λ being a topological space, U and V Banach spaces, and suppose $F(\lambda_0,u_0) = 0$ for $(\lambda_0,u_0) \in \mathcal{O}$. Denote by $F_u(\lambda,u)$ the Fréchet derivative at u of the mapping $\hat{u} \to F(\lambda,\hat{u})$ and define $F_u(\lambda,u)$ to be *nonsingular* if it has a bounded inverse, i.e. it is a (topological linear) isomorphism of U onto V. Then if the mapping $(\lambda,u) \to F_u(\lambda,u)$ is defined and continuous near (λ_0,u_0), and if $F_u(\lambda,u_0)$ is nonsingular, there exists a continuous curve $u = u(\lambda)$ defined on a neighbourhood L of λ_0 such that $u(\lambda_0) = u_0$ and $F(\lambda,u(\lambda)) \equiv 0$. Moreover any solution of $F(\lambda,u) = 0$ near (λ_0,u_0) is of the form $(\lambda,u(\lambda))$. Finally if Λ is a Banach space and $F \in C^k(\mathcal{O},V)$ with $k \geqslant 1$, then $u \in C^k(L,U)$. ◄
For a proof see CRANDALL [20].

We are now ready to state the Simple Eigenvalue Theorem in the abstract framework.

THEOREM 1.6 Let U and V be Banach spaces and let $F \in C^2(\mathbb{R} \times U,V)$ be such that $F(\lambda,0) \equiv 0$ for any real λ. Set $L_0 = F_u(\lambda_0,0)$, $L_1 = F_{\lambda u}(\lambda_0,0)$ and suppose that

(i) $N(L_0) = \text{span}\{u_0\}$
(ii) $\text{codim } R(L_0) = \dim V/R(L_0) = 1$
(iii) $L_1 u_0 \notin R(L_0)$

Let Z be any closed subspace of U such that $U = \text{span}\{u_0\} \oplus Z$ (i.e. any $u \in U$ can be uniquely written as $u = \alpha u_0 + z$, $\alpha \in \mathbb{R}$, $z \in Z$). Then the same results as in Theorem 1.4 hold. ◄

REMARK 1.2 In finite dimension, $\dim N(L_0) = 1$ is equivalent to $\text{codim } R(L_0) = 1$; this is not the case in an arbitrary Banach space, whence hypothesis (ii) to get $(\hat{\lambda}, \hat{z}) \mapsto L_0 \hat{z} + \hat{\lambda} L_1 u_0$ to be nonsingular.

REMARK 1.3 More general situations are considered in a paper by VEGA [96].

Example 1.3 If $[a,b]$ is a bounded real interval, denote by $C^k(a,b)$ the Banach space $C^k([a,b],\mathbb{R})$ equipped with the usual maximum norm and by $C_0^k(a,b)$ the subspace $\{u \in C^k(a,b) | u(a) = u(b) = 0\}$. Consider the mapping $F : \mathbb{R} \times C_0^2(a,b) \to C^0(a,b)$ defined as

$$F(\lambda, u) = u'' + \lambda(u + u^2 g(u))$$

for g regular enough. Note that if $g(u)$ is such that $u^2 g(u) + u = \sin u$, $F(\lambda, u) = 0$ is the equation for the Elastica discussed in the Introduction. In any case

$$F_u(\lambda, 0)\hat{u} = \hat{u}'' + \lambda \hat{u} \qquad \text{and} \qquad F_{\lambda u}(\lambda, 0)\hat{u} = \hat{u}, \qquad \forall \hat{u} \in C_0^2(a,b)$$

hence $F_u(\lambda, 0)$ is nonsingular iff $\lambda \neq \ell^2\pi^2$, $\ell = 1, 2, \ldots$; otherwise $N(F_u(\ell^2\pi^2, 0)) = \text{span}\{(\sqrt{2}/\pi) \sin \ell\pi x\}$, $R(F_u(\ell^2\pi^2, 0)) = \{v \in C^0(a,b) | \int_0^1 v(x) \sin \ell\pi x \, dx = 0\}$ and $\sin \ell\pi x \notin R(F_u(\ell^2\pi^2, 0))$. Setting

$$Z = \{u \in C_0^2(a,b) | \int_0^1 u(x) \sin \ell\pi x \, dx = 0\}$$

then all the hypotheses of the above theorem are verified, hence a bifurcation curve occurs at any $\lambda = \ell^2\pi^2$. This curve can be obtained by an iterative algorithm, since the Implicit Function Theorem is constructive. Choosing for simplicity $\ell = 1$, set $u_0(x) = (\sqrt{2}/\pi) \sin \pi x$, then

$$\mathcal{F}(s, \lambda, z) = (u_0 + z)'' + \lambda[(u_0 + z) + (u_0 + z)^2 \cdot g(s(u_0 + z))]$$

is the function $\mathcal{F}$ defined in the proof of Theorem 1.4 and the algorithm employed in the Implicit Function Theorem reads as follows

$$\begin{cases} (\lambda^0, z^0) = (\pi^2, 0) \\ (\lambda^{n+1}, z^{n+1}) = -[\mathcal{F}_{(\lambda,z)}(0, \pi^2, 0)]^{-1} \cdot \mathcal{F}(s, \lambda^n, s^n) + (\lambda^n, z^n) \end{cases}$$

that is, by making explicite the expression on the right:

$$\lambda^{n+1} = \lambda^n - \int_0^1 F(s,\lambda^n,z^n(y))\, u_0(y)\, dy$$

$$z^{n+1} = z^n - \int_0^1 K(x,y)[F(s,\lambda^n,z^n(y)) - (\lambda^n - \lambda^{n+1})u_0(y)]\, dy$$

where for any $f \in R(F_u(\pi^2,0))$, $z(x) = \int_0^1 K(x,y)f(y)dy \in Z$ solves $z'' + \pi^2 z = f$, $z(0) = z(1) = 0$.

2. THE HOPF BIFURCATION

In the previous section, bifurcation of equilibria for equation (1.1) from the trivial branch has been investigated. In this section we are interested in bifurcation of trivial solutions into periodic solutions. Fluid dynamics is a good source of examples of such a phenomenon.

With a small change in notations with respect to section 1, let us consider a map $f \in C^2(\mathbb{R} \times \mathbb{R}^n, \mathbb{R}^n)$ such that $f(\mu,0) \equiv 0$, and look for small periodic solutions - with period $2\pi\rho$ - of the equation

$$\frac{du}{dt} + f(\mu,u) = 0 \tag{2.1}$$

The period being unknown, we can make it explicitly appear in the equation by the change of time scale $t = \tau\rho$ which yields

$$F(\rho,\mu,u) = \frac{du}{d\tau} + \rho f(\mu,u) = 0 \tag{2.2}$$

Hence we look for a curve of 2π-periodic solutions of this problem bifurcating at $\rho = 1$, $\mu = 0$ (the requirement on ρ is a normalization). Since two parameters are involved we need a result which generalizes Theorem 1.5 to the case of arbitrary parameters spaces.

LEMMA 2.1 Let Λ, U, V be Banach spaces; consider a mapping $F \in C^2(\Lambda \times U, V)$ such that $F(\lambda,u) = Lu + B(\lambda-\lambda_0,u) + p(\lambda,u)$, with $L \in B(U,V)$, $B: (\hat{\lambda},\hat{u}) \mapsto B(\hat{\lambda},\hat{u})$ bilinear and bounded, $p(\lambda,u) \equiv 0$, $p_u(\lambda_0,0) = 0$, $p_{\lambda u}(\lambda_0,0) = 0$. Assume there exist $0 \neq u_0 \in N(L)$ and a closed subspace $Z \subset U$ such that the mapping $\Lambda \times Z \to V$ given by

$$(\hat{\lambda},\hat{z}) \mapsto L\hat{z} + B(\hat{\lambda},u_0) \tag{2.3}$$

is one-to-one and onto. Then the zeroes of F in a neighbourhood of $(\lambda_0,0)$ in $\Lambda \times \mathrm{span}\{u_0\} \times Z$ consist exactly of the trivial solutions $(\lambda,0)$ and of a C^1-curve $s \mapsto (\lambda(s), s(u_0 + \psi(s)))$, where $(\lambda,\psi) : (-\delta,\delta) \to \Lambda \times Z$ and $(\lambda(0),\psi(0)) = (\lambda_0,0)$. ◄

The proof is conceptually the same as for Theorem 1.4; however, note that 0 is no longer required to be a "geometrically simple" eigenvalue for L: the essential condition to obtain the result is that the mapping (2.3) be an isomorphism. Roughly speaking if V is fixed, the "larger" is Λ, the "smaller" must be Z; if Z is same orthogonal complement to N(L), this means that dim N(L) must "grow" proportionally to dim Λ.

We apply the Lemma with $F:\mathbb{R}^2 x C^1_{2\pi}(\mathbb{R},\mathbb{R}^n) \to C^0_{2\pi}(\mathbb{R},\mathbb{R}^n)$ (where the sybscript 2π means restriction to 2π-periodic functions), $\lambda = (\rho,\mu)$, $\lambda_0 = (1,0)$. If $f(\mu,u) = L_0 u + \mu L_1 u + p(\mu,u)$, with L_0, $L_1 \in B(\mathbb{R}^n,\mathbb{R}^n)$ and p satisfying the usual conditions, we write F as:

$$F(\lambda,u) = \frac{du}{d\tau} + L_0 u + (\rho-1)L_0 u + \mu L_1 u + p(\mu,u)$$

and we set $Lu = \frac{du}{d\tau} + L_0 u$ and $B(\lambda-\lambda_0,u) = (\rho-1)L_0 u + \mu L_1 u$.
Now $L : C^1_{2\pi}(\mathbb{R},\mathbb{R}^n) \to C^0_{2\pi}(\mathbb{R},\mathbb{R}^n)$ is singular iff it is not one-to-one, that is iff $\pm ki \in \sigma(L_0)$ (the spectrum of L_0) for at least one integer $k = 1,2,\ldots$ (since L_0 is a real matrix, eigenvalues occur in conjugate pairs).
We make the two following hypotheses, first considered by E.HOPF (see [42]):

(H_1) $\pm$ i are algebraically simple eigenvalues for L_0, and
$\pm ki \notin \sigma(L_0)$ for $n = 0,2,3,\ldots$

Since i is algebraically simple, by spectral perturbation results (see e.g. KATO [56]) we derive the existence of a continuously differentiable curve $\beta:(-\epsilon,\epsilon) \to \mathbb{C}$ such that $\beta(\mu) \in \sigma(L_0+\mu L_1)$ for $|\mu| < \epsilon$ and $\beta(0) = i$. Then we require

(H_2) Re $\beta'(0) \neq 0$.

Under these assumptions, the hypotheses of Lemma 2.1 can be fulfilled:

<u>THEOREM 2.1</u> <u>Take</u> $a \in N(L_0 - iI)$, <u>and set</u> $u_0(\tau) = \mathrm{Re}(e^{-i\tau}a)$, $u_1(\tau) = u_0'(\tau) = \mathrm{Im}(e^{-i\tau}a)$, <u>and</u> $Z = \{z \in C^1_{2\pi}(\mathbb{R},\mathbb{R}^n) \mid \int_0^{2\pi} (z(\tau), u_i(\tau))_{\mathbb{R}^n} d\tau = 0$ for $i = 1,2\}$. <u>Then if</u> (H_1) <u>and</u> (H_2) <u>hold there exists a "unique"</u> C^1<u>-curve of</u> 2π<u>-periodic solutions of (2.2) bifurcating at</u> (1,0,0).

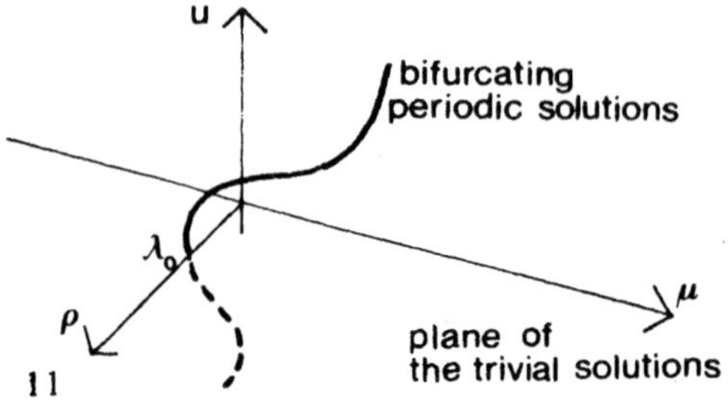

fig. 11

Shetch of Proof. Since $ki \notin \sigma(L_0)$ for $n = 0, \pm 2, \pm 3, \ldots$, it is easily checked that $N(L) = \text{span}\{u_0, u_1\}$. We must prove that the equation

$$\frac{d\hat{z}}{d\tau} + L_0\hat{z} + \hat{\rho}L_0u_0 + \hat{\mu}L_1u_0 = v$$

has a unique solution $(\hat{\rho}, \hat{\mu}, \hat{u}) \in \mathbb{R}^2 \times Z$ for any $v \in C^0_{2\pi}(\mathbb{R}, \mathbb{R}^n)$. By Fredholm Alternative Theorem this is true iff there exists a unique $(\hat{\rho}, \hat{\mu}) \in \mathbb{R}^2$ such that $v - \hat{\rho}L_0u_0 - \hat{\mu}L_1u_0$ is orthogonal to the null space of $L^* = -\frac{d}{d\tau} + L_0^*$, L_0^* being the adjoint of L_0. If $b \in N(L_0^* + iI)$ with $(a,b)_{\mathbb{C}^n} = \pi^{-1}$, define $\varphi_0(\tau) = \mathrm{Re}(e^{-i\tau}b)$, $\varphi_1(\tau) = \varphi_0'(\tau) = \mathrm{Im}(e^{-i\tau}b)$. Then $N(L^*) = \text{span}\{\varphi_0, \varphi_1\}$, and a straightforward computation yields $(u_i, \varphi_j)_{C^0_{2\pi}(\mathbb{R},\mathbb{R}^n)} = \delta_{ij}$ for $i,j = 0,1$. It follows that $\hat{\rho}, \hat{\mu}$ can be uniquely determined iff

$$\det \begin{vmatrix} (L_0u_0, \varphi_0) & (L_1u_0, \varphi_0) \\ (L_0u_0, \varphi_1) & (L_1u_0, \varphi_1) \end{vmatrix} = (L_1u_0, \varphi_0) \neq 0$$

Proof is complete once we show that $(L_1u_0, \varphi_0) = \mathrm{Re}\beta'(0)$. First we find a continuously differentiable curve $\mu \mapsto a(\mu)$ on $(-\epsilon, \epsilon)$ such that $a(\mu)$ is an eigenfunction of $L_0 + \mu L_1$ relative to the eigenvalue $\beta(\mu)$ and $a(0) = a$. Moreover we define $b(\mu)$ to be the eigenfunction of $L_0^* + \mu L_1^*$ for the eigenvalue $\beta(\mu)$ such that $(a(\mu), b(\mu))_{\mathbb{C}^n} = \pi^{-1}$: then $\mu \mapsto b(\mu)$ is continuously differentiable and $b(0) = b$. Differentiating the identity

$$(L_0a(\mu), b(\mu)) + \mu(L_1a(\mu), b(\mu)) = \pi^{-1}\beta(\mu)$$

at $\mu = 0$ we get $(L_1a, b) = \pi^{-1}\beta(0)$ since $(L_0a'(0), b) + (L_0a, b'(0)) = i\frac{d}{d\mu}(a(\mu), b(\mu))|_{\mu=0}$. By integrating $(L_1a, b) = (L_1e^{-i\tau}a, e^{-i\tau}b)$ over $[0, 2\pi]$ and taking the real part of the result we obtain

$$(L_1u_0,\varphi_0) + (L_1u_1,\varphi_1) = 2\,\mathrm{Re}\,\beta'(0)$$

But one easily checks that $(L_1u_0,\varphi_0) = (L_1u_1,\varphi_1)$, and the proof is complete. ◄

The following theorem provides a rather general extension of the Hopf Bifurcation Theorem to the infinite-dimensional case. Much work has been done by many authors in this direction, using different techniques (see e.g. CHAFEE [16], HENRY [40], IOOSS [44]); particular attention has been devoted to bifurcation into time-periodic solutions for the Navier-Stokes equations (see IUDOVICH [51], SATTINGER [86], JOSEPH-SATTINGER [55], IOOSS [43] and the book by MARSDEN and MCCRACKEN [67]).

The result we are going to state involves analytic semigroups and fractional powers of unbounded operators; the reader unfamiliar with these concepts may refer to KATO [56] or FRIEDMAN [36]. We seek periodic solutions of the equation

$$\frac{du}{dt} + f(\mu,u) = \frac{du}{dt} + L_0u + g(\mu,u) = 0 \tag{2.4}$$

Our assumptions are the following: let U be a real Banach space with complexification $U_c = U + iU$. Denote the extension to U_c of any linear operator L on U again by L; then $\lambda\in\sigma(L)$ iff $\bar\lambda\in\sigma(L)$.

(HL_0) $L_0 : D(L_0)\subset U\to U$ is a densely defined linear operator on U such that $-L_0$ is the infinitesimal generator of a strongly continuous semigroup on U, which is a holomorphic semigroup on U_c;
L_0 has compact resolvent;
i is a simple eigenvalue for L_0, and $ni\notin\sigma(L_0)$ for $n = 0,2,3,\ldots$

As a consequence of (HL_0), for any real r such that $r > -\mathrm{Re}\,\lambda$ for all $\lambda\in\sigma(L_0)$, we can define the fractional powers $(L_0+rI)^\alpha$ for $\alpha\geq 0$. Their domains being independent of r, we can set $U_\alpha = D((L_0+rI)^\alpha)$, which are Banach spaces under the norm $\|u\|_\alpha = \|(L_0+rI)^\alpha u\|_U$. Then

(Hg) There exist an $\alpha\in[0,1)$ and a neighbourhood Ω of $(0,0)$ in $\mathbb{R}\times U_\alpha$ such that $g\in C^2(\Omega,U)$. Moreover $g(\mu,0)=0$ if $(\mu,0)\in\Omega$, and $g_u(0,0)=0$.

Finally, let $\beta=\beta(\mu)$ be the continuously differentiable function defined in a neighbourhood of 0 such that $\beta(\mu)\in\sigma(L_0+g_u(\mu,0))$

and $\beta(0) = i$. Then

($H\beta$) $\mathrm{Re}\,\beta'(0) \neq 0$.

Define a function $u \in C^0_{2\pi\rho}(\mathbb{R}, U_\alpha)$ to be a solution of (2.4) if it takes its values in $D(L_0)$, du/dt exists and is continuous with values in U and (2.4) is satisfied pointwise.

THEOREM 2.2 (CRANDALL and RABINOWITZ [24]). *If* (HL_0), (Hg) *and* ($H\beta$) *are verified, there exist a positive* η *and continuously differentiable functions*

$$(\rho, \mu, u) : (-\eta, \eta) \to \mathbb{R}^2 \times C^0(\mathbb{R}, U_\alpha)$$

such that: i) *for* $0 < |s| < \eta$, $u(s)$ *is a* $2\pi\rho(s)$-*periodic solution of* (2.4) *corresponding to* $\mu = \mu(s)$; ii) $\rho(0) = 1$, $\mu(0) = 0$, $u(0) = 0$ *and* $u(s) \neq 0$ *if* $s \neq 0$; iii) *any* $2\pi\rho$-*periodic solution* u *of* (2.4) *with* $|\rho - 1|$, $|\mu|$ *and* $\|u\|$ *in* $C^0_{2\pi\rho}(\mathbb{R}, U_\alpha)$ *small enough is of the above form for some* $|s| < \eta$ *up to a translation of the real line. Moreover if* $g \in C^{k+1}(\Omega, U_\alpha)$, *then the functions* ρ, μ, u *are of class* C^k. ◄

Example 2.1 The problem

$$\begin{cases} u_t = u_{xx} + \mu u + 2v + \text{h.o.t.} \\ v_t = v_{xx} - 2u \qquad\quad + \text{h.o.t.} \\ u = v = 0 \qquad \text{at } x = 0, \pi \end{cases} \tag{2.5}$$

(where h.o.t. stands for terms not affecting the calculations below) has a bifurcating curve of periodic solutions at $\mu_0 = 2$ with period near to $2\pi/\sqrt{3}$. Indeed, split the lower order terms on the right as

$$\left[\begin{pmatrix} \dfrac{\partial^2}{\partial x^2} & 0 \\ 0 & \dfrac{\partial^2}{\partial x^2} \end{pmatrix} + \begin{pmatrix} 2 & 2 \\ 2 & 0 \end{pmatrix}\right]\begin{pmatrix} u \\ v \end{pmatrix} + (\mu - 2)\begin{pmatrix} 1 & 0 \\ 0 & 0 \end{pmatrix}\begin{pmatrix} u \\ v \end{pmatrix} = \tilde{L}_0 \underline{w} + \tilde{\mu}\tilde{L}_1 \underline{w}.$$

with $\underline{w} = (u, v)^T$ and $\tilde{\mu} = \mu - 2$. To solve the eigenvalue problem

$$\tilde{L}_0 \underline{z} + \tilde{\mu}\tilde{L}_1 \underline{z} = \tilde{\beta}\underline{z} \tag{2.6}$$

try the ansatz

$$\underline{z} = \begin{pmatrix} \alpha \\ \beta \end{pmatrix} \sin nx$$

Then $\tilde{\beta} = \tilde{\beta}(\mu)$ is a (simple) eigenvalue for (2.6) iff $2\tilde{\beta} = (\mu - 2n^2) \pm$ $\pm i\sqrt{16 - \mu^2}$; hence $\tilde{L}_0$ has a purely immaginary eigenvalue iff $n = 1$, precisely $\tilde{\beta}(0)=\sqrt{3}i$, and $\mathrm{Re}\,\tilde{\beta}'(0)=\frac{1}{2}\neq 0$. So $L_0=\frac{1}{\sqrt{3}}\tilde{L}_0$ satisfies the hypotheses (HL_0) and $(H\beta)$ of the theorem (take $U=C^0(0,\pi)$ and $D(L_0)=C_0^2(0,\pi)$). After changing the time variable into $\tau=\sqrt{3}t$ and assuming that the higher order terms in (2.5) satisfy (Hg), apply the theorem to get a curve of periodic solutions with period near 2π with respect to τ, that is near $2\pi/\sqrt{3}$ with respect to t.

3. LINEARIZED STABILITY

In this section we investigate the stability properties of the branch of stationary or periodic solutions of the equation

$$\frac{du}{dt}=F(\lambda,u) \qquad (F(\lambda,0)\equiv 0) \tag{3.1}$$

bifurcating from the trivial equilibria.
Heuristically, suppose we are in the following situation: for $\lambda<\lambda_0$ the spectrum $\sigma(F_u(\lambda,0))$ lies entirely in the left half plane, while for $\lambda>\lambda_0$ $\sigma(F_u(\lambda,0))$ intersects the right half plane (i.e. the trivial solution is stable for $\lambda<\lambda_0$ and unstable for $\lambda>\lambda_0$ - see below); then at $\lambda=\lambda_0$ at least one eigenvalue crosses

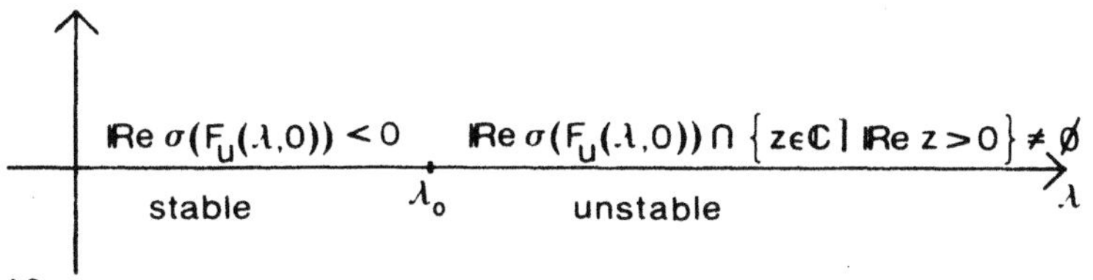

fig. 12

the imaginary axes, that is $\pm iw\in\sigma(F_u(\lambda_0,0))$ for some real w. Under suitable hypotheses, the case $w=0$ yields bifurcation into new equilibria, while the case $w\neq 0$ yields bifurcation into periodic solutions. Stability for both kind of solutions can be discussed in a similar way using linearized stability criteria: Floquet exponents for Hopf bifurcation play the same rôle as eigenvalues of F_u for equilibria. Even the tool of analysis is conceptually the same: a relationship between the "perturbed eigenvalue" along the bifurcating curve and some geometric features of the curve itself. Hence stability essentially depends on the shape of the bifurcating curve.

Before stating the main results, we recall some well known

facts about stability. Let us begin with:

a) Stability of equilibria. Assume $F : \mathbb{R} \times U \to V$, U, V Banach spaces; for any $t > 0$ define $S(t) : U \to U$ as $S(t)u_0 = u(t)$ if u solves (3.1) with initial datum $u(0) = u_0$.

Definition 3.1 An equilibrium solution $u \equiv \bar{u}$ of (3.1) is stable if for any neighbourhood $\mathcal{U}$ of $\bar{u}$ there exists a neighbourhood $\mathcal{V}$ of $\bar{u}$ such that $S(t)\mathcal{V} \subset \mathcal{U}$ for all $t > 0$; $\bar{u}$ is asymptotically stable if in addition $S(t)u_0$ converges to $\bar{u}$ as t tends to ∞, for any $u_0 \in \mathcal{V}$.

The following result is known as Principle of Linearized Stability:

THEOREM 3.1 (LIAPUNOV [64]). Assume $F_u(\lambda, \bar{u})$ has only point spectrum. Then

(i) If $\bar{u}$ is stable, $\sigma(F_u(\lambda,\bar{u})) \subset \{z \in \mathbb{C} \mid \operatorname{Re} z \leqslant 0\}$

(ii) If $\sigma(F_u(\lambda,\bar{u})) \subset \{z \in \mathbb{C} \mid \operatorname{Re} z < 0\}$, $\bar{u}$ is asymptotically stable. ◄

From now on, suppose F satisfies all the hypotheses of Theorem 1.6.

Definition 3.2. Assume $K \in B(U,V)$; $\eta \in \mathbb{C}$ is a K-simple eigenvalue of L_0 if

(i) $N(L_0 - \eta K) = \operatorname{span}\{\hat{u}\}$, $\hat{u} \neq 0$

(ii) $\operatorname{Codim} R(L_0 - \eta K) = 1$

(iii) $K\hat{u} \notin R(L_0 - \eta K)$.

Note that if U=V and η is isolated, then η is an I-simple eigenvalue for L_0 iff η is algebraically simple in the usual sense. According to this definition 0 is a L_1-simple eigenvalue of L_0 with eigenfunction u_0.

LEMMA 3.1. Let $K \in B(U,V)$ and let 0 be a K-simple eigenvalue for L_0. Then any $L \in B(U,V)$ near L_0 has a unique K-simple real eigenvalue $\eta(L)$ near 0 and a unique small vector $u(L) \in N(L_0 - \eta(L)K)$ such that $u(L) - u_0 \in Z$. Moreover $L \mapsto (\eta(L), u(L))$ is an analytic function of L.

Proof (Hint) Define $\mathcal{F} : B(U,V) \times \mathbb{R} \times Z \to V$ as $\mathcal{F}(L,\eta,z) = L(u_0+z) - \eta K(u_0+z)$. Apply the Implicit Function Theorem at $(L_0,0,0)$ since $\mathcal{F}(L_0,0,0) = 0$ and $\mathcal{F}_{(\eta,z)}(L_0,0,0)(\hat{\eta},\hat{z}) = L_0\hat{z} - \hat{\eta}Ku_0$ is non-singular by assumption. ◄

Set $F_u(s) = F_u(\lambda(s), u(s))$. Then by the lemma there exist regular functions $\lambda \mapsto (\gamma(\lambda), w(\lambda)) \in \mathbb{R}\times Z$ and $s \mapsto (k(s), z(s)) \in \mathbb{R}\times Z$ defined in a neighbourhood of λ_0 and 0 respectively, such that

$$F_u(\lambda,0)(u_0+w(\lambda)) = \gamma(\lambda)K(u_0+w(\lambda))$$
$$F_u(s)(u_0+z(s)) = k(s)K(u_0+z(s))$$
$$(\gamma(\lambda_0), w(\lambda_0)) = (k(0), z(0)) = (0,0)$$

The characterization for $k(s)$ (the <u>critical value</u> for $F_u(s)$) is given in the following theorem.

<u>THEOREM 3.2.</u> (CRANDALL and RABINOWITZ [22]). <u>There exists a continuous real-valued function</u> $s \mapsto A(s)$ <u>such that</u>

$$k(s) = s\lambda'(s)A(s)$$

<u>for any</u> s <u>small enough and</u>

$$A(0)Ku_0 + L_1u_0 \in R(L_0)$$

(<u>hence</u> $A(0) \neq 0$). <u>Precisely</u> $A(0) = -\gamma'(\lambda_0)$. ◄

To apply this result in discussing stability, assume $U \subset V$ with continuous injection I; moreover assume that 0 is an algebraically simple eigenvalue for L_0, so that we can take $K = I$. In the situation pictured at the beginning of this section, we have $\gamma(\lambda) < 0$ if $\lambda < \lambda_0$ and $\gamma(\lambda) > 0$ if $\lambda > \lambda_0$, hence $\gamma'(\lambda_0) > 0$. The sign of $k(s)$ depends on the sign of $s\lambda'(s)$. If $s\lambda'(s) > 0$ for $s \neq 0$ <u>(supercritical bifurcation)</u>, $k(s) < 0$ hence the bifurcating curve is entirely stable near $(\lambda_0, 0)$.

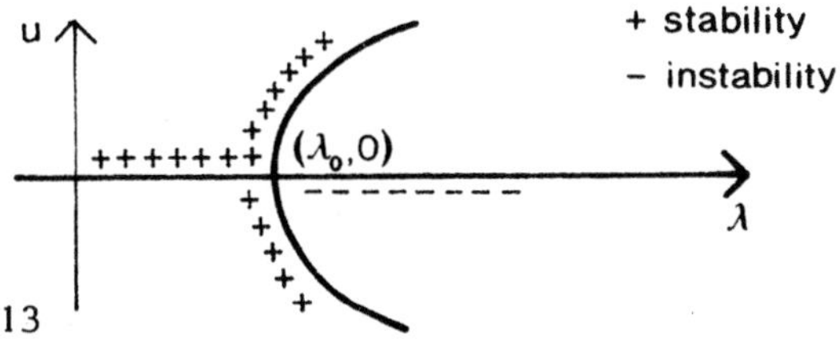

fig. 13

The converse happens when $s\lambda'(s) < 0$ for $s \neq 0$ <u>(subcritical bifurcation)</u>

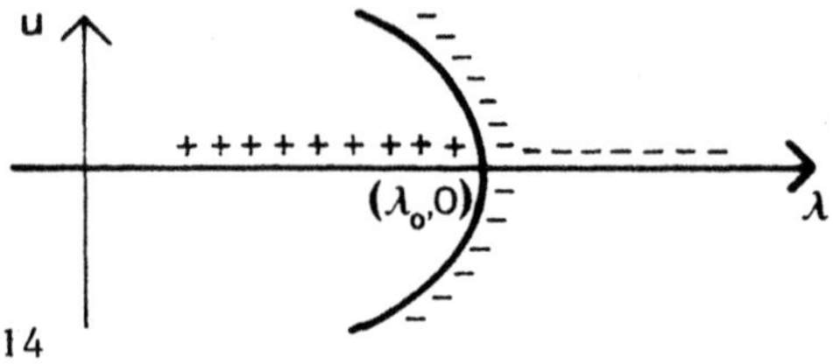

fig. 14

If $\lambda'(s)$ does not change sign at $s = 0$ (transcritical bifurcation) the bifurcation curve is partially stable, partially unstable.

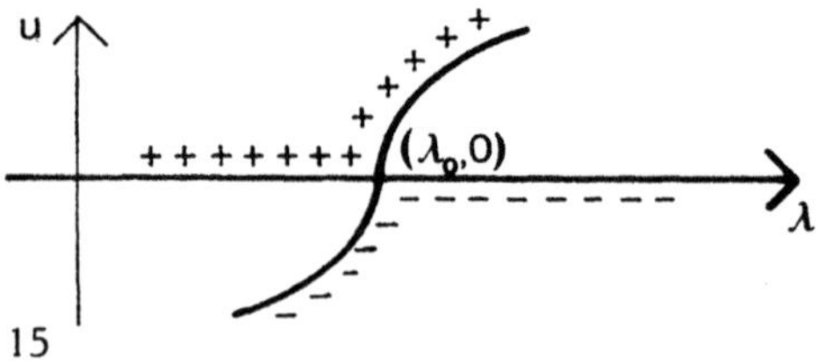

fig. 15

This is what is commonly called the Principle of Exchange of Stability.

b) Stability of Periodic Solutions. For the sake of simplicity, we only consider the finite dimensional case: functional framework and assumptions are now the same as in Theorem 2.1. For the infinite-dimensional case see CRANDALL and RABINOWITZ [24].

For any $t > 0$, let $S(t): \mathbb{R}^n \to \mathbb{R}^n$ be defined by $S(t)u_0 = u(t)$ where u solves

$$\frac{du}{dt} + f(\mu, u) = 0 \tag{3.2}$$

with intial datum $u(0) = u_0$.

Definition 3.3. Let $\overline{u} \in C^1_T(\mathbb{R}, \mathbb{R}^n)$ be a T-periodic solution of (3.2). Then $\overline{u}$ is (orbitally) stable if for any neighbourhood $\mathcal{U} \subset \mathbb{R}^n$ of $\mathcal{O}_{\overline{u}} = \{u(t) | \, 0 \leq t \leq T\}$ there exists a neighbourhood $\mathcal{V} \subset \mathbb{R}^n$ of $\mathcal{O}_{\overline{u}}$ such that $S(t)\mathcal{V} \subset \mathcal{U}$ for all $t > 0$; $\overline{u}$ is asymptotically (orbitally) stable if in addition for any $u_0 \in \mathcal{V}$ $S(t)u_0$ tends to $\mathcal{O}_{\overline{u}}$ as t tends to ∞. (see fig. 16).

In order to establish a principle of linearized stability, first consider the problem

$$\begin{cases} \dfrac{du}{dt} = A(t)u & (3.3) \\ u(0) = u_0 & (3.4) \end{cases}$$

where $A(t)$ is a T-periodic time-dependent linear operator. Define $U(t) : \mathbb{R}^n \to \mathbb{R}^n$ as $U(t)u_0 = u(t)$. Note that $U(t+mT) = U(t)U(T)^m$ since $A(t)$ is T-periodic.

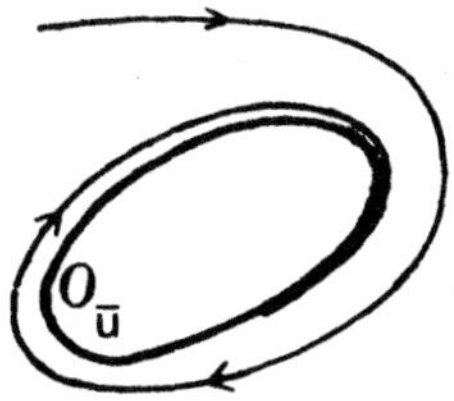

fig. 16

Definition 3.4. Any $\eta \in \sigma(U(T))$ is called a Floquet multiplier for (3.3); any $k \in \mathbb{C}$ such that $\exp(Tk) \in \sigma(U(T))$ is called a Floquet exponent for (3.3) (Note that $0 \notin \sigma(U(T))$.

PROPERTY 3.1. $-k$ is a Floquet exponent for (3.3) iff the problem

$$\frac{dz}{dt} A(t)z = kz \qquad (3.5)$$

has a nontrivial T-periodic solution.

Proof. $\exp(-kT) \in \sigma(U(T))$ iff there exists u_0 such that $U(T)u_0 = \exp(-kT)u_0$, i.e. iff $z(t) = \exp(kt)U(t)u_0$ satisfies (3.5). ◄

About stability recall that $\lim_{m\to\infty} \|U(T)^m\|^{1/m} = \sup\{|\eta| : \eta \in \sigma(U(T))\}$. Then $U(t)u_0 \to 0$ exponentially as $t \to \infty$, for whatever $u_0 \in \mathbb{R}^n$, iff all Floquet exponents have strictly negative real part, while there exists $u_0 \in \mathbb{R}^n$ such that $U(t)u_0$ is unbounded over $(0,+\infty)$ iff some Floquet exponent has strictly positive real part.

Now we can go back to our original problem.

Definition 3.5. Let $\bar{u} \in C^1_T(\mathbb{R},\mathbb{R}^n)$ be a T-periodic solution of (3.2). We define the Floquet multipliers and exponents for $\bar{u}$ to be those relative to the linearized problem

$$\frac{du}{dt} = -f_u(\mu,\bar{u}(t))u.$$

Note that time differentiation of (3.2.) with $u=\bar{u}$ yields

$$\frac{d\dot{\bar{u}}}{dt} + f_u(\mu,\bar{u}(t))\,\dot{\bar{u}}=0 \qquad (\dot{\bar{u}}=\frac{d\bar{u}}{dt})$$

hence 1 (resp.0) is always a Floquet multiplier (resp.exponent) for $\bar{u}$.

THEOREM 3.3. (Principle of Linearized Stability for Periodic Solutions - see e.g. HARTMAN [39] , HENRY [40]). (i) If $\bar{u}$ is stable, then all exponents for $\bar{u}$ have nonpositive real part; (ii) if 0 is a simple exponent (i.e. 1 is an algebraically simple multiplier) and the remaining exponents (i.e. exponents different from $0 \pm in$, $n=0,1,2,\ \ldots$) have strictly negative real part, then $\bar{u}$ is asymptotically stable. ◄

If $(\rho(s),\mu(s),u(s))$ (with $u(s)=s(u_0+\psi(s))$, $\psi(s)\in Z$) is the curve of periodic solutions bifurcating at (1,0,0) given by Theorem 2.1 , the Floquet exponents for u(s) are the numbers -k such that

$$\frac{dw}{dt} + \rho(s)f_u(\mu(s),u(s))\cdot w = kw \tag{3.6}$$

has a nontrivial 2π-periodic solution; since everything is continuous, they are small perturbations (if s is small) of the Floquet exponents corresponding to $s=0$, i.e. those for which the problem

$$\begin{cases} \dfrac{dw}{dt} + L_0 w = kw \\ w(0)=w(2\pi) \end{cases} \tag{3.7}$$

has a nontrivial solution. This is the case iff $k\in\{\lambda\pm in \mid \lambda\in\sigma(L_0),\ n=0,1,2,\ \ldots\}$; the related multipliers are therefore $\{\exp(-2\pi\lambda) \mid \lambda\in\sigma(L_0)\}$. We assumed $\pm i\in\sigma(L_0)$ and hence 0 is an exponent (with double multiplicity). Suppose that $\sigma(L_0)-\{\pm i\}$ lies in the open right half plane so that every exponent for (3.7) except 0 has strictly negative real part. It follows that exponents for (3.6) all lie in the open right half plane, except 0 (which is always an exponent) and possibly a second small exponent k(s), obtained as a perturbation of the second zero eigenvalue of (3.7). According to the Principle of Linearized Stability this is the very exponent responsible for the stability of u(s). The following theorem provides a rigorous definition of k(s), together with an analytic expression of it in terms of the geometric behaviour of the bifurcating curve.

THEOREM 3.4. (CRANDALL and RABINOWITZ [24]). Under the assumptions of Theorem 2.1., there exists a unique continuous curve $s \mapsto (k(s), \eta(s), z(s)) \in \mathbb{R} \times \mathbb{R} \times Z$ defined in a neighbourhood of 0 such that $(k(0), \eta(0), z(0)) = (0,0,0)$ and

$$\frac{d}{d\tau}(u_0+z(s)) + \rho(s) f_u(\mu(s), u(s)) \cdot (u_0+z(s)) =$$

$$= k(s)(u_0+z(s)) + \eta(s)(u_1+\dot{\psi}(s))$$

for s small enough. Moreover there exists a continuous function $s \mapsto A(s)$ defined in a neighbourhood of 0, such that

$$k(s) = s\mu'(s) \cdot A(s) \qquad \text{with} \qquad A(0) = \mathrm{Re}\,\beta'(0)$$

Recalling that $\frac{d}{d\tau}(u_1+\dot{\psi}(s)) = \rho(s) \cdot f_u(\mu(s), u(s))(u_1+\psi(s)) = 0$, we observe that either $k(s) = 0$, and then $u_0+z(s)$ is a (generalized) eigenvector of $\frac{d}{d\tau} + \rho(s) \cdot f_u(\mu(s), u(s))$ for the null eigenvalue which therefore has double multiplicity; or $k(s) \neq 0$ and then

$$\frac{dw(s)}{d\tau} + \rho(s) \cdot f_u(\mu(s), u(s)) \cdot w(s) = k(s) w(s)$$

for $w(s) = u_0+z(s) + k(s)^{-1}\eta(s)(u_1+\dot{\psi}(s))$, i.e. $-k(s)$ is a Floquet exponent. In any case, k(s) is the required perturbation of the second zero eigenvalue for (3.7).

Therefore we can conclude exactly in the same way as for bifurcation of equilibra. Denoting by $\tilde{\beta}(\mu) \in \sigma(f_u(\mu,0))$ the perturbation of the eigenvalue $i \in \sigma(f_u(0,0))$ for μ near 0, one may check that $\tilde{\beta}'(0) = \beta'(0)$. Hence if $\tilde{\beta}(\mu)$ crosses the imaginary axes from right to left as μ crosses the origin (i.e. if $\mathrm{Re}\,\tilde{\beta}(\mu) > 0$ for $\mu < 0$ and

$\mathrm{Re}\,\tilde{\beta}(\mu) < 0$ when $\mu > 0$), we have $\beta'(0) < 0$. In this situation the bifurcating periodic solutions are stable if bifurcation is "supercritical" ($s\mu'(s) > 0$ for $s \neq 0$), unstable if bifurcation is "subcritical" ($s\mu'(0) < 0$ for $s \neq 0$).

4. LIAPUNOV-SCHMIDT PROCEDURE

Liapunov-Schmidt procedure provides a way to reduce the dimension of the space in which one tries to solve an equation of the type

$$F(\lambda,u) = 0$$

near a singular point. The idea is to split the space into two components and project the equation over each of them. Of the two equations so obtained, one is always solvable via the Implicit Function Theorem, the other usually is over a finite dimensional space and therefore can be attacked by several, often sophisticated techniques from geometry and topology.

We present here a fairly general version of the method. We are given two Banach spaces X,Y and a C^1-mapping

$$T_0 : X \to Y$$

such that $T_0(x_0) = 0$, $x_0 \in X$. We want to study nearby problems

$$T(x) = 0$$

with T and x close to T_0 and x_0. To be precise, suppose T_0 belongs to a family $\mathcal{T}$ of C^1-mappings $T : X \to Y$ equipped with a suitable topology, such that the mapping

$$\mathcal{T} \times X \ni (T,x) \longmapsto T(x) \in Y$$

is continuous. Assume

$$X = N(T_0'(x_0)) \oplus N^c$$
$$Y = R(T_0'(x_0)) \oplus R^c$$

where N, N^c, R, R^c are closed subspaces, and define the mapping $F : \mathcal{T} \times N \times N^c \times R^c \to Y$ as

$$F(T,n,n^c,r^c) = T(x_0+n+n^c) + r^c$$

We have $F(T_0,0,0,0) = 0$ and $F_{(n^c,r^c)}(T_0,0,0,0)(\hat{n}^c,\hat{r}^c) = T_0'(x_0)\hat{n}^c+\hat{r}^c$ is nonsingular. Therefore by the Implicit Function Theorem we can solve the equation $F = 0$ for $n^c(T,n)$, $r^c(T,n)$ in a neighbourhood of $(T_0,0)$ in $\tau \times N$. Finally note that $T(x_0+n+n^c(T,n)) = 0$ iff $r^c(T,n) = 0$. $r^c(T,n) = 0$ is called the <u>bifurcation equation</u>. Usually we have $F(\lambda,u) = 0$, with $F \in C^1(\mathbb{R} \times X, Y)$; then we can take $\tau = \{F_\lambda = F(\lambda,.) | \lambda \in \mathbb{R}\}$ equipped with the topology of the real line. If $\dim N(T_0'(x_0)) = n$ and $\dim R^c = m$ the bifurcation equation reduces to a system of m equations in n+1 unknowns.

The reduction of a problem to a finite dimensional one allows to use singularity theory techniques to handle imperfect bifurcation. Indeed one can describe "all the possible imperfections" in terms of the universal unfolding of a singularity. For more details refer to the work by GOLUBITSKY and SCHAEFFER [38] .

5. GLOBAL RESULTS

Let us begin by recalling the concept of <u>compact mapping</u>. Consider a real Banach space E and a mapping $T : E \to E$. We say that T is compact if it is continuous and maps bounded sets into relatively compact sets.

<u>Example 5.1.</u> Consider the differential operator

$$L : C_0^\alpha(0,1) \longrightarrow C^0(0,1), \quad \alpha \geqslant 2,$$

defined as $Lu = -(p(x)u')' + q(x)u$
with $p(x) > 0$ and $q(x) \geqslant 0$ on $[0,1]$. We can invert L in terms of Green's functions expressing the solution of $Lu = f$ with $f \in C^0(0,1)$ by:

$$u = \int_0^1 G(x,y)f(y)dy \equiv T(f)$$

Then

$$\begin{array}{ccc} T: \quad C_0^0 & \longrightarrow & C_0^0 \\ \text{continuous} \downarrow & \nearrow & \text{compact} \\ C_0^\alpha & & \end{array}$$

is compact. Of course the choice of the spaces is not unique: for example also $T : C_0^1 \to C_0^1$ is compact.

Example 5.2. Consider the equation

$$Lu = g(x,u,u'), \qquad 0 \leqslant x \leqslant 1,\ u(0) = u(1) = 0,$$

where g is a continuous function. By inverting L we get a nonlinear compact operator $T : C_0^1 \to C_0^1$:

$$u(x) = \int_0^1 G(x,y)g(y,u(y),u'(y))\,dy \equiv T(u)$$

In general we deal with equations of this kind:

$$0 = L_0 u + \lambda L_1 u + P(\lambda,u) \tag{5.1}$$

with L_0 invertible. Then (5.1) reduces to

$$0 = F(\lambda,u) \equiv u + \lambda L_0^{-1} L_1 u + L_0^{-1} P(\lambda,u) \equiv u + T(\lambda,u) \tag{5.2}$$

If L_0^{-1} is compact and L_1 and $P(\lambda,.)$ are continuous, then T is compact.

Topological degree arguments have been successfully employed to study bifurcation problems like (5.2). The first such result we know is due to KRASNOSELSKIJ [62] :

THEOREM 5.1. Let E be a real Banach space and $F(\lambda,u) = u - \lambda L + H(\lambda,u)$, where L is compact linear and $H : \mathbb{R} \times E \to E$ is compact with $H(\lambda,u) = o(\|u\|)$ at 0 uniformly on bounded λ-intervals. Then if μ is a *characteristic value* of L (i.e. there exists a nontrivial v such that $v = \mu L v$) of *odd* multiplicity, then $(\mu,0)$ is a bifurcation point for F with respect to the line of trivial solutions. ◄

Actually the above hypotheses give information on the global behaviour of the nontrivial solutions of F:

THEOREM 5.2. (RABINOWITZ [76]). Under the above hypotheses let S denote the closure of the set of nontrivial zeroes of F. Then S contains a component $\mathcal{C}$ (i.e. a maximal closed and connected subset) which meets $(\mu,0)$ and either is unbounded in $\mathbb{R} \times E$ or meets $(\hat{\mu},0)$, where $\hat{\mu} \neq \mu$ and $\hat{\mu}$ is a characteristic value of L. (see fig. 17)

REMARK 5.1. If μ is a simple characteristc value of L, then $\mathcal{C}$ consists of two subcontinua that near $(\mu,0)$ meet only at $(\mu,0)$

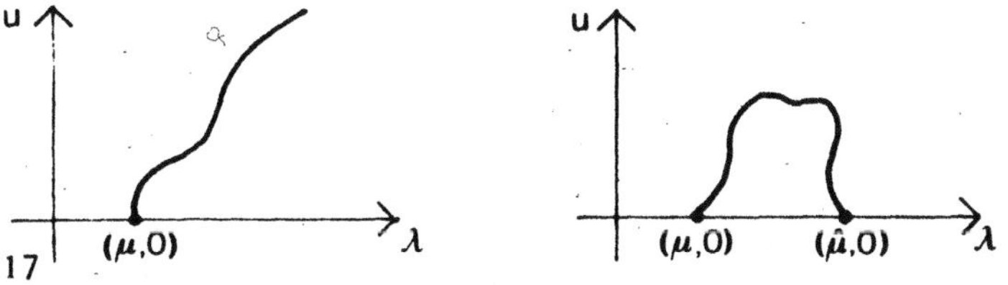

fig. 17

(see fig. 18). If μ has multiplicity greater than 1 it can happen that only one bifurcation branch occurs.

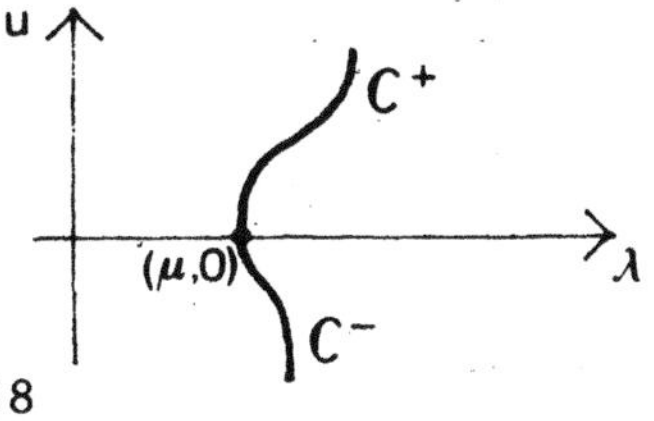

fig. 18

We shall come back to the proof of Theorem 5.2. after discussing the following significant example.

Example 5.3. Consider

$$\mathcal{L}u \equiv -(p(x)u')' + q(x)u = \lambda a(x)u + g(x\ ,u,u'),\quad 0 \leqslant x \leqslant 1,$$
$$u(0) = u(1) = 0$$
$$a(x) > 0 \text{ on } [0,1]\text{, and } g(x,\xi,\eta) = o((\xi^2+\eta^2)^{1/2}) \text{ at } (\xi,\eta) = (0,0).$$

By inverting $\mathcal{L}$ on $E = C_0^1$ the problem reduces to the following equation

$$u = \lambda Lu + H(\lambda,u) \qquad (L = \mathcal{L}^{-1})$$

Now $\mathcal{L}v = \mu\, a\, v$ has an increasing sequence of simple eigenvalues $0 < \lambda_1 < \lambda_2 < \ldots < \lambda_n < \ldots$ with $\lambda_n \to \infty$ as $n \to \infty$. Any eigenfunction v_k corresponding to λ_k has exactly k-1 simple nodal zeroes on (0,1) (at a simple nodal zero the derivative is nonzero). Then we can apply Theorem 5.2 and conclude that a continuum $\mathcal{C}_k$ of nontrivial solutions branches from each eigenvalue λ_k. We want to explore more closely the properties of these continua. Eigenfunctions are normalized as $\|v_k\| = 1$, $v_k'(0) > 0$, $k = 1,2,\ldots$.
First we examine the linear case: if $g \equiv 0$, then a line $\{(\lambda_k, \alpha v_k),\ \alpha \in \mathbb{R}\}$ bifurcates from each eigenvalue λ_k. If $g \not\equiv 0$, we are in the nonlinear case: let us introduce the following continua:

$$\mathcal{S}_k = \{\varphi \in C_0^1 \mid \varphi \text{ has exactly k-1 simple zeroes for } 0 < x < 1 \text{ and all zeroes of } \ \text{ in } [0,1] \text{ are simple}\}$$

Of course $v_k \in S_k$. We have the following:

THEOREM 5.3. *The branch C_k bifurcating from $(\lambda_k, 0)$ is contained in $(\mathbb{R} \times S_k) \cup (\lambda_k, 0)$ and is unbounded.*

Sketch of Proof. The result follows from these two facts:
Fact 1 There exists a neighbourhood $\mathcal{U}$ of $(\lambda_k, 0)$ in $\mathbb{R} \times E$ such that all nontrivial zeroes in $\mathcal{U}$ lie in $\mathbb{R} \times S_k$. Actually for smooth g this follows from Theorem 1.6. In fact the multiplicity of the characteristic value is 1, then near $(\lambda_k, 0)$ the second component of C_k is given by $v(s) = s(v_k + \psi(s))$ with $\psi(s) \to 0$ as $s \to 0$. Therefore $v(s) \in S_k$ because $v_k \in S_k$ and S_k is open.
Fact 2 $(C_k - \{(\lambda_k, 0)\}) \cap \partial(\mathbb{R} \times S_k) = \phi$ because a solution (λ, u) with u having a double zero has u identically zero. ◄

To prove Theorem 5.2 we need the notion of degree due to Leray and Schauder (see e.g. SCHWARTZ [88]).
Let us consider a mapping $f \in C^1(\mathcal{O}, \mathbb{R}^n)$ where $\mathcal{O}$ is an open and bounded subset of $\mathbb{R}^n$. We want to solve the equation $f(x) = b$ with $b \in \mathbb{R}^n$ given and $f \neq b$ on $\partial \mathcal{O}$. Assume that $\mathcal{O} \cap f^{-1}(b)$ is finite and that the Jacobian matrix $f'(x)$ is nonsingular at these points. Then we can define

$$d(f, \mathcal{O}, b) = \sum_{x \in f^{-1}(b) \cap \mathcal{O}} \operatorname{sign} |f'(x)|$$

where $|f'(x)|$ is the determinant of the Jacobian matrix.

Properties of the degree.

(i) $d(I, \mathcal{O}, b) = 1$ *if* $b \in \mathcal{O}$, $d(I, \mathcal{O}, b) = 0$ *if* $b \notin \mathcal{O}$;

(ii) if $f \neq b$ in $\mathcal{O}$ *then* $d(f, \mathcal{O}, b) = 0$;

(iii) (*Additivity*) *if* $\mathcal{O} = \mathcal{O}_1 \cup \mathcal{O}_2$, $b \notin f(\partial \mathcal{O} \cup \overline{\mathcal{O}_1 \cap \mathcal{O}_2})$ *then* $d(f, \mathcal{O}, b) = d(f, \mathcal{O}_1, b) + d(f, \mathcal{O}_2, b)$;

(iv) (*Continuity*) d *is continuous in* $f, \mathcal{O}$ *and* b. Note that since $d(f, \mathcal{O}, b) = d(f - b, \mathcal{O}, 0)$ it sufficies to verify the continuity with respect to f and $\mathcal{O}$.

From the continuity of d it follows that we may define $d(f, \mathcal{O}, b)$ for $f \in C^0(\overline{\mathcal{O}}, \mathbb{R}^n)$ by approximating f by regular functions.

(v) (Homotopy invariance). Let $h \in C^0([0,1] \times \mathcal{O}, \mathbb{R}^n)$ such that $h \neq b$ on $[0,1] \times \partial\mathcal{O}$, then $d(h(t,.),\mathcal{O},b)$ is constant with respect to t.

COROLLARY. The degree only depends on the boundary values of the function f.

Proof. Let $f,g \in C^0(\bar{\mathcal{O}},\mathbb{R}^n)$, $f,g \neq b$ on $\partial\mathcal{O}$, $f = g$ on $\partial\mathcal{O}$; then use the homotopy $h(t,x) = tf(x) + (1-t)g(x)$, $t \in [0,1]$. ◄

(vi) Define $f_\lambda = I - \lambda L$ where L is an nxn matrix. f_λ vanishes at zero and when λ is a characteristic value of L. Then, if B_r in the ball of radius r and center 0, $d(f_\lambda,B_r,0)$ is well defined as long as λ is not a characteristic value. (In fact if $f_\lambda(x) = 0$, $\|x\| = 1$, then $rx \in \partial B_r$ and $f_\lambda(rx) = 0$). Furthermore $d(f_\lambda, B_r,0)$ changes by a factor $(-1)^m$ as λ crosses a characteristic value of multiplicity m. This follows from the fact that $d(f_\lambda,B_r,0) = \text{sign}\det(I-\lambda L)$.

The degree may be extended to functions on a Banach space E. Let $\Phi(u) = u - T(u)$, where $\Phi \in C^0(\Omega,E)$ with Ω bounded and open subset of E and T is compact. If $b \in E$ and $b \notin \Phi(\partial\Omega)$ we may define $d(\Phi,\Omega,b)$ by approximating Φ with mappings over finite-dimensional spaces. The degree so extended enjoys all the properties (i)-(vi) above.

Let us now go back to the proof of Theorem 5.2. Let $\mathcal{C}$ denote the component of $S \cup (\mu,0)$ containing $(\mu,0)$. Suppose the theorem is false. Using the compactness of F, by simple topological arguments one shows there exists a bounded open set $\mathcal{O}$ which contains $\mathcal{C}$ and intersects the segment $(\mu-\delta, \mu+\delta)$ on $u = 0$ and having no nontrivial solutions on $\partial\mathcal{O}$. Let $\mathcal{O}_s = \{u \in E \mid (s,u) \in \mathcal{O}\}$. Consider for each $\lambda \neq \mu$ a positive $r(\lambda)$ such that $(\lambda,0)$ is the only solution of (5.1) on $\{\lambda\} \times B_{r(\lambda)}$ (see fig.19). For $\lambda \neq \mu$ there is no solution of (5.1) in $\{\lambda\} \times \partial(\mathcal{O}_\lambda - \bar{B}_{r(\lambda)})$ and therefore $d(F(\lambda,.), \mathcal{O}_\lambda - \bar{B}_{r(\lambda)},0)$ is well defined. By homotopy invariance $d(F(\lambda,.), \mathcal{O}_\lambda - \bar{B}_{r(\lambda)},0)$ is constant for $\lambda > \mu$. But for large λ (5.1) has no solution on $\mathcal{O}_\lambda - \bar{B}_{r(\lambda)}$ and then

$$d(F(\lambda,.), \mathcal{O}_\lambda - \bar{B}_{r(\lambda)}) = 0 \qquad (5.2)$$

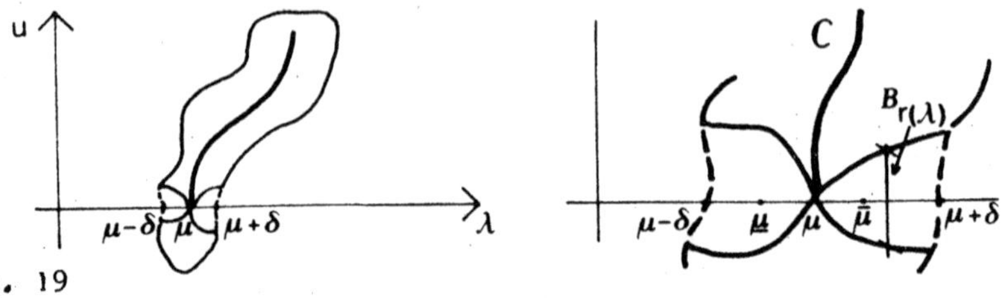

fig. 19

for $\lambda > \mu$. By a similar argument we conclude that (5.2) holds also for $\lambda < \mu$. Again by homotopy invariance

$$d(F(\lambda,.),\mathcal{O}_\lambda,0) = \text{const.} \tag{5.3}$$

for $|\lambda-\mu| < \delta$.

Now suppose $\mu-\delta < \underline{\mu} < \bar{\mu} < \mu+\delta$. We have $\mathcal{O}_{\bar{\mu}} = B_{r(\bar{\mu})} \cup (\mathcal{O}_{\bar{\mu}} - B_{r(\bar{\mu})})$, hence by additivity $d(F(\bar{\mu},.),\mathcal{O}_{\bar{\mu}},0) = d(F(\bar{\mu},.),\ \mathcal{O}_{\bar{\mu}} - \bar{B}_{r(\bar{\mu})},0) +$ $+\ d(F(\bar{\mu},.),B_{r(\bar{\mu})},0) = d(F(\bar{\mu},.),\ B_{r(\bar{\mu})},0)$ by (5.2). Therefore by (5.3) we get

$$d(F(\underline{\mu},.),\ B_{r(\underline{\mu})},0) = d(F(\bar{\mu},.),\ B_{r(\bar{\mu})},0) \tag{5.4}$$

Now define the homotopy $F(\lambda,u,\theta) = u - \lambda Lu + \theta H(\lambda,u)$ with $\theta \in [0,1]$. By choosing $r(\bar{\mu})$ small enough, since $H(\lambda,u) = o(\|u\|)$ at 0, we have $F(\bar{\mu},u,\theta) \neq 0$, for any $(u,\theta) \in \partial B_{r(\bar{\mu})} \times [0,1]$ and therefore $d(F(\bar{\mu},.),\ B_{r(\bar{\mu})},0) = d(I-\bar{\mu}L,\ B_{r(\bar{\mu})},0)$. By employing the same argument for $\underline{\mu}$ and using (5.4) we conclude

$$d(I-\underline{\mu}L,\ B_{r(\underline{\mu})},0) = d(I-\bar{\mu}L,\ B_{r(\bar{\mu})},0).$$

But μ is a characteristic value of odd multiplicity, therefore

$$d(I-\underline{\mu}L,\ B_{r(\underline{\mu})},0) = -\ d(I-\bar{\mu}L,\ B_{r(\bar{\mu})},0) \neq 0$$

Thus we have a contradiction and the proof is complete. ◄

<u>REMARK 5.2.</u> If the map $F(\lambda,u)$ is defined only on a neighbourhood Ω of $(\mu,0)$ then the component $\mathcal{C}$ containing $(\mu,0)$ either meets $\partial\Omega$ or contains $(\hat{\mu},0) \in \Omega$, where $\hat{\mu} \neq \mu$ is a characteristic value.

REMARK 5.3. The compactness requirement can be eliminated to obtain local result. If F is a Fredholm operator of index 0, the Liapunov-Schmidt procedure reduces equation (5.1) to a finite-dimensional one, so that the Brouwer degree can be used to conclude bifurcation for odd multiplicity.

The odd multiplicity requirement can be dropped in situations involving positive operators:

THEOREM 5.4. (DANCER [25] and TURNER [92]). Let K be a cone in E, $L : K \to K$ be compact and linear and $H \in C^0(\mathbb{R}^+ \times K, K)$ be compact with $H(\lambda,u) = o(\|u\|)$ at 0 uniformly on bounded λ-intervals. If L has positive spectral radius and μ is the smallest positive characteristic value of L, then $F(\lambda,u) = u - \lambda(Lu + H(\lambda,u))$ has a component of nontrivial zeroes C that meets $(\mu,0)$ and is unbounded in $\mathbb{R}^+ \times K$. ◄

A global Hopf bifurcation theorem is also available, see ALEXANDER and YORKE [2] .

6. THE VARIATIONAL CASE

Let E be a real Hilbert space and $f \in C^0(E,\mathbb{R})$. Assuming f differentiable on E, we can consider the following mapping

$$u \to f'(u) \in E^*$$

Since $E^* \simeq E$ we get the application

$$u \to f'(u) \in E$$

which is called a potential operator.

Example 6.1. Consider a self-adjoint operator $L : E \to E$ and a differentiable mapping $h : E \to R$. Set

$$f(u) = \tfrac{1}{2}(Lu,u) + h(u)$$

($(.,.)$ is the scalar product in E), then (after the identification $E^* \simeq E$)

$$f'(u) = Lu + h'(u)$$

is a potential operator. We suppose $h(u) = o(\|u\|^2)$ at the origin and set $h'(u) = H(u)$; therefore $f'(0) = 0$.

We are interested in the following equation

$$f'(u) = \lambda u \tag{6.1}$$

On the existence of bifurcation points for (6.1) we have the following result that is an extension of a theorem by KRASNOSELSKIJ [62]:

THEOREM 6.1. (BÖHME [14] and MARINO [65]). *Let* E *be a real Hilbert space,* $f \in C^2(E, \mathbb{R})$ *with* $f'(u) = Lu + H(u)$, L *linear and* $H(u) = o(\|u\|)$ *at* 0. *Suppose* μ *is an isolated eigenvalue for* L *of finite multiplicity, then* $(\mu, 0)$ *is a bifurcation point for*

$$f'(u) = \lambda u \tag{6.1}$$

Moreover (6.1) *possesses at least two distinct one-parameter families of solutions* $(\lambda(r), u(r))$ *having* $\|u(r)\| = r$ *and* $\lambda(r) \to \mu$ *as* $r \to 0$.

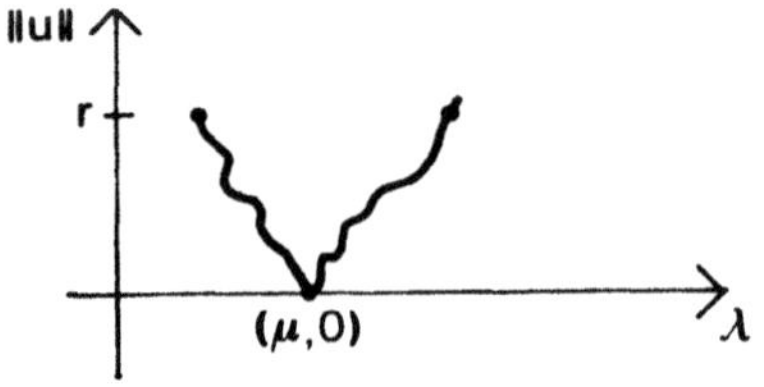

fig. 20

REMARK 6.1. The set of solutions does not need to be a continuum, as shown in an example by BÖHME [14].

REMARK 6.2. If f is even and μ has multiplicity m, then by employing the Ljusternik-Schnirelmann theory it can be shown that (6.1) possesses at least m one-parameter families of pairs of solutions $(\lambda(r), u(r))$ and $(\lambda(r), -u(r))$ having $\|u(r)\| = r$ and $\lambda(r) \to \mu$ as $r \to 0$. See BÖHME [14] and MARINO [65].

Of course it is more natural to let λ *vary, rather than* $\|u\|$. *From this point of view, under the same hypotheses as above there are three possibilities* (FADELL and RABINOWITZ [29]):

(i) $(\mu, 0)$ *is not an isolated solution of* (6.1) *in* $\{\mu\} \times E$:

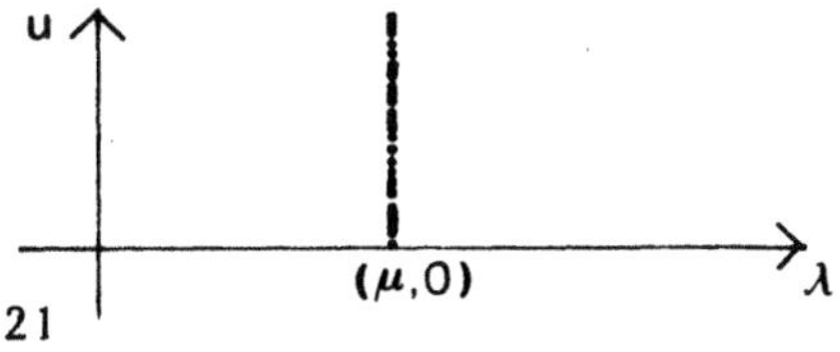

fig. 21

(ii) There exists a one-sided neighbourhood $\mathcal{O}$ of μ in $\mathbb{R}$ such that for any $\lambda \in \mathcal{O}$, (6.1) has at least two nontrivial solutions which converge to 0 as λ tends to μ:

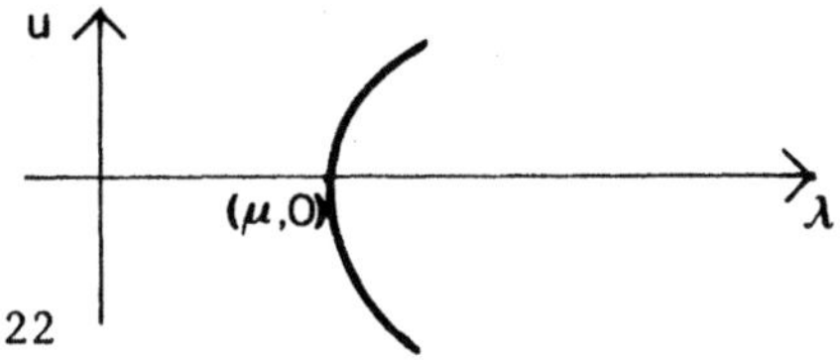

fig. 22

(iii) There exists a two-sided neighbourhood $\mathcal{O}$ of μ in $\mathbb{R}$ such that for any $\lambda \in \mathcal{O}$, (6.1) has at least one nontrivial solution which converges to 0 as λ tends to μ:

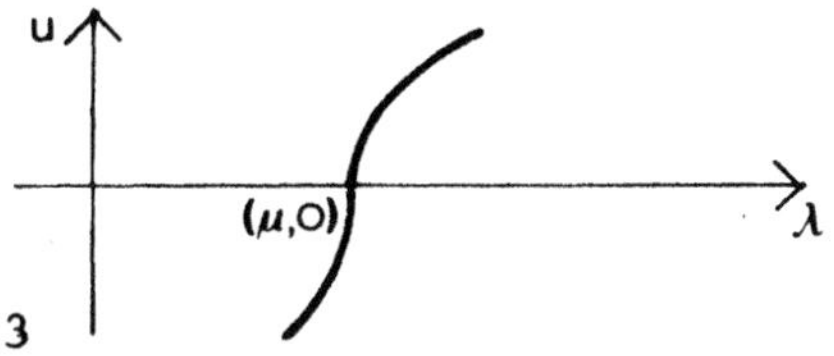

fig.23

FADELL and RABINOWITZ [28] improved these results when f is an even function. Then if μ has multiplicity n, points (i) and (ii) above are replaced by the following unique statement:

(ii)' There exist left and right neighbourhoods $\mathcal{O}_\ell$ and $\mathcal{O}_r$ of μ in R and integers k and $m \geq 0$ such that $k+m \geq n$ and if $\lambda \in \mathcal{O}_\ell$ (resp. $\lambda \in \mathcal{O}_r$) (6.1) possesses at least k (resp. m) pairs of nontrivial solutions which converge to 0 as λ tends to μ.

Sketch of Proof for Theorem 6.1. First we reduce equation (6.1) by the Liapunov-Schmidt procedure. Let P and $P^\perp$ be the orthogonal projections onto $N(L-\mu I) = N$ and $N^\perp$ respectively. We write u =

$= v + w \in N \oplus N^{\perp} = E$. Equation (6.1) splits into

$$Lv + PH(v+w) - \lambda v = 0 \quad \text{or} \quad (\mu - \lambda) v + PH(v+w) = 0 \tag{6.2}$$

and $\quad Lw - \lambda w + P^{\perp} H(v+w) = 0 \qquad\qquad (6.3)$

Set $F(\lambda, v, w) = (L - \lambda I) w + P^{\perp} H(v+w)$ with $F: \mathbb{R} \times N \times N^{\perp} \to N^{\perp}$. We seek solutions of $F(\lambda, v, w) = 0$ near $(\mu, 0, 0)$. We have $F(\mu, 0, 0) = 0$ and $F_w(\mu, 0, 0)\hat{w} = (L - \mu I)\hat{w}$. $L - \mu I$ is an isomorphism of $N^{\perp}$ onto $N^{\perp}$, so $F = 0$ can be solved for w as a function φ of λ and v near $(\mu, 0)$. Substituting $w = \varphi(\lambda, v)$ into (6.3), we have:

$$\varphi(\lambda, v) = -[(L - \mu I)_{|N^{\perp}}]^{-1} P^{\perp} H(v + \varphi(\lambda, v))$$

for λ near μ, and therefore $\varphi(\lambda, v) = o(\|v\|)$ at 0, uniformly in λ. By (6.2) we get

$$\mu - \lambda = (H(v + \varphi(\lambda, v)), v) \, / \, \|v\|^2$$

By the hypotheses on H and the fact that $\varphi(\lambda, v) = o(\|v\|)$ uniformly in λ, the derivative of the right hand side at $(\mu, 0)$ is 0. Therefore we can apply the Implicit Function Theorem 1.5. and get $\lambda = \chi(v)$ near $v = 0$. Of course $\mu = \chi(0)$. Consider now

$$g(v) = f(v + \varphi(\chi(v), v)).$$

It can be shown that critical points of $g(v)$ on

$$\|v\|^2 + \|\varphi(\chi(v), v)\|^2 = r^2, \text{ with } r \geqslant 0, \tag{6.4}$$

satisfy equation (6.1). But since $\varphi = o(\|v\|)$, the manifold defined by (6.4) is almost a sphere. Therefore we have a maximum and a minimum for $g(v)$ on it and the theorem is proved. ◄

Discussion of the three cases. The solutions of $f'(u) = \lambda u$ are the critical points of

$$f(u) - \frac{\lambda}{2} \|u\|^2 .$$

But we know by the Liapunov-Schmidt procedure that these critical points near $(\mu, 0)$ are of the kind $u = v + \varphi(\lambda, v)$ and therefore we have to look for the critical points near $(\mu, 0)$ of:

$$\psi(\lambda, v) = \tfrac{1}{2} \left((L - \lambda I)(v + \varphi(\lambda, v)), \, v + \varphi(\lambda, v)\right) + h(v + \varphi(\lambda, v)) =$$

$$= \frac{\mu-\lambda}{2}\|v\|^2 + \frac{1}{2}((L-\lambda I)\varphi,\varphi) + h(v+\varphi).$$

Now since $\varphi = o(\|v\|)$ at $v = 0$ and by the hypotheses on h, the first term is determinant except at $\lambda=\mu$. To simplify suppose $\dim N(L-\mu I) = 1$. For fixed λ the three cases are:

(i) $v = 0$ is a nonisolated critical point for $\psi(\mu,v)$.

(ii) $v = 0$ is a local maximum (or minimum) for $\psi(\mu,v)$.
If $\mu>\lambda$ the dominant term $(\mu-\lambda)\|v\|^2/2$ has a minimum at the origin and two local maxima near $v = 0$. If $\lambda>\mu$ $(\mu-\lambda)\|v\|^2/2$ has a maximum and the diagram does not change:

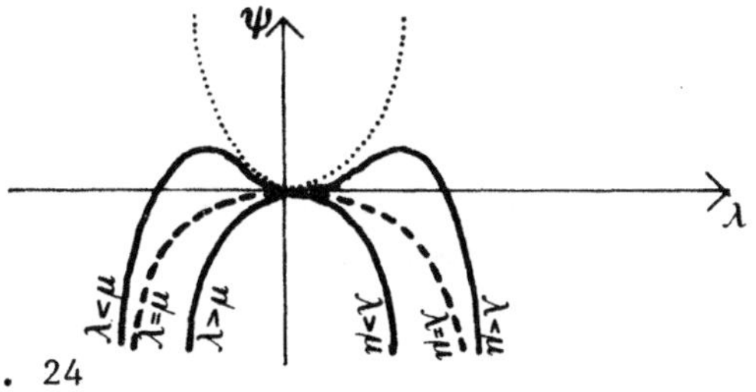

fig. 24

(iii) $v = 0$ is neither a local maximum nor a local minimum for $\psi(\mu,v)$. Then, if $\lambda<\mu$, $\psi(\lambda,v)$ has a minimum at the origin and a maximum nearby. If $\lambda>\mu$ we have a maximum at the origin and a minimum nearby:

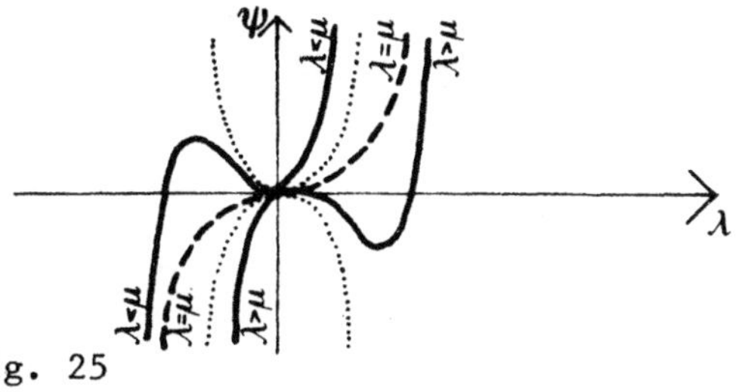

fig. 25

7. PERIODIC SOLUTIONS OF HAMILTONIAN SYSTEMS

Let $p,q \in \mathbb{R}^n$ and $H \in C^2(\mathbb{R}^{2n},\mathbb{R})$, with $H(0,0) = 0$, $H_p(0,0) = 0$ and $H_q(0,0) = 0$. Consider the Hamiltonian system

$$\dot{p} = -H_q \quad , \quad \dot{q} = H_p \tag{7.1}$$

Setting $z = (p,q)^T$ and $J = \begin{pmatrix} 0 & -I \\ I & 0 \end{pmatrix}$, (7.1) can be rewritten as

$$\text{(HS)} \quad \dot{z} = J H_z \tag{7.2}$$

The assumptions on $H_z(0)$ imply that (7.2) possesses the trivial solution $z = 0$. We seek for small nontrivial time-periodic solutions. The first result of this nature is the Liapunov Center Theorem (see e.g. NEMYTSKII and STEPHANOV [70]), which in fact can be deduced from a clever application of the Hopf Bifurcation Theorem:

THEOREM 7.1. (LIAPUNOV). *If $\sigma\{J H_{zz}(0)\} = \{\pm i\xi_1, \pm i\xi_2, \ldots, \pm i\xi_n\}$ where ξ_j is real for any j, and ξ_j/ξ_1 is not an integer for $j \neq 1$, then there exists a one parameter family of small periodic solutions with period near $2\pi/\xi_1$ which bifurcates from $z = 0$.* ◄

The irrationality condition can be dropped:

THEOREM 7.2. (WEINSTEIN [97]). *Suppose $H_{zz}(0)$ is positive definite, then for any $c > 0$ (HS) possesses at least n geometrically distinct periodic solutions on $H^{-1}(c)$ whose periods are near those of the linearized problem.* ◄

Recently Moser gave another proof of Weinstein's result. He stated the Theorem as follows:

THEOREM 7.3. (MOSER [69]). *Suppose $\mathbb{R}^{2n} = E_1 \oplus E_2$ where E_1 and E_2 are invariant subspaces for*

$$\dot{w} = J H_{zz}(0) w. \tag{7.3}$$

If all the solutions of (7.3) with initial data in E_1 have a common period $T > 0$ while no solution of (7.3) in $E_2 - \{0\}$ has period T and if $H_{zz}(0)$ is positive defined on E_1, then for small $c > 0$, (HS) possesses at least $\frac{1}{2}\dim E_1$ distinct periodic solutions on $H^{-1}(c)$ whose periods are near T. ◄

These results are related to the problem of finding the critical points of the action integral:

$$A(z) = \int_0^{2\pi} p\,\dot{q}\,dt$$

subjected to the constraint $\frac{1}{2\pi}\int_0^{2\pi} H(z)\,dt = c$, over a space of 2π-periodic functions. In fact if we look at this variational problem formally, by a Lagrange multipliers argument, we get the Euler equation:

$$J\,\dot{z} = -\lambda H_z(z)$$

which is equivalent to (HS) after stretching the time variable : $t = \lambda\tau$ (note that $J^2 = -I$). Since A is neither bounded from above nor from below on the constraint manifold, direct methods from the calculus of variations do not seem to apply here. However it is possible to use a finite dimensional approximation procedure. Indeed the functional is restricted to a finite-dimensional subspace of $(L^2(S^1))^{2n}$ (S^1 is the unit sphere of $\mathbb{R}^2$). Secondly a minimax argument is employed to obtain a critical value and corresponding nontrivial critical point for the finite-dimensional problem. Finally resorting to suitable estimates of the solutions one can pass to the limit and obtain the following result in the large:

THEOREM 7.4. (RABINOWITZ [83]). Let H be $C^1(\mathbb{R}^{2n},\mathbb{R})$ and suppose that for some $c \neq 0$ $H^{-1}(c)$ is radially diffeomorphic to S^{2n-1} (the unit sphere in $\mathbb{R}^{2n}$). Then $\dot{z} = J H_z$ possesses a periodic solution on $H^{-1}(c)$. ◄

Recently EKELAND and LASRY [27] proved the existence of n geometrically distinct periodic solutions of (HS) over $H^{-1}(c)$ under the assumption that $H^{-1}(c)$ is a C^2-variety of strictly positive curvature, which bounds a convex region and is contained between two spheres whose radii are in the ratio 2/1. Thereby they obtained a partial globalization of Weinstein's Theorem 7.2.

REFERENCES

[1] ABRAHAM, R. and MARSDEN, J.: Foundations of Mechanics, 2nd. ed., Benjamin, New York, 1978.

[2] ALEXANDER, J.C. and YORKE, J.A.: Global Bifurcation of Periodic Orbits, Amer. J. Math., 100 (1978), 263-292.

[3] AMANN, H.: On the Existence of Positive Solutions of Nonlinear Elliptic Boundary-Value Problems, Indiana Univ. Math. J., 21 (1971), 125-146.

[4] AMANN, H.: Ljusternik-Schnilermann Theory and Nonlinear Eigenvalue Problems, Math. Ann., 199 (1972), 55-72.

[5] AMANN, H.: On the Number of Solutions of Nonlinear Equations in Ordered Banach Spaces, J. Funct Anal., 11 (1972), 346-384.

[6] AMANN, H.: Nonlinear Operators in Ordered Banach Spaces and Some Applications to Nonlinear Boundary-Value Problems, Nonlinear Operators and the Calculus of Variations, 1-15, Lecture Notes in Math., vol. 543, Springer, Berlin, 1976.

[7] AMBROSETTI, A. and RABINOWITZ, P.H.: Dual Variational Methods in Critical Points Theory and Applications, J. Fuct. Anal., 14 (1973), 349-381.

[8] ANTMAN, S.S.: Bifurcation Problems for Nonlinearly Elastic Structures, Applications of Bifurcation Theory, P.H. Rabinowitz (ed.), 73-125, Academic Press, New York, 1977.

[9] ARNOLD, V.I.: Lectures on Bifurcation and Versal Families, Russian Math. Surveys, 27 (1972), 54-123.

[10] AUCHMUTY, J.F.G. and NICOLIS, J.: Bifurcation Analysis of Nonlinear Reaction-Diffusion Equations I, Bull. Math. Biol., 37 (1973), 323-365.

[11] BERGER, M.S.: On Von Kármán's Equations and the Buckling of a Thin Elastic Plate I, Comm. Pure Appl. Math., 20 (1967), 687-719.

[12] BERGER, M.S.: Nonlinearity and Functional Analysis, Academic Press, New York, 1977.

[13] BERGER, M.S. and FIFE P.C.: On Von Kármán's Equations and the Buckling of a Thin Elastic Plate II, Comm. Pure Appl. Math., 21 (1968), 227-247.

[14] BÖHME, R.: Die Lösung der Verzweigungsgleichungen für nichlineare Eigenwertproblem, Math. Z., 127 (1972), 105-126.

[15] CARTAN, H.: Differential Calculus, Hermann, Paris; Houghton, Boston, 1971.

[16] CHAFEE, N.: The Bifurcation of one or more Closed Orbits from an Equilibrium Point of an Autonomous Differential Equation, J. Diff. Eq., 4 (1968), 661-679.

[17] CHANDRASEKAR, S.: Hydrodynamic and Hydromagnetic Stability, Oxford Univ. Press, Oxford, 1961.

[18] CHOW, S.N., HALE, J.K. and MALLET-PARET, J.: Application of Generic Bifurcation I and II, Arch. Rat. Mech. Anal., 59 (1975), 159-188 and 62 (1976), 209-235.

[19] COLES, D.: Transition in Circular Couette Flow, J. Fluid Mech., 21 (1965), 385-425.

[20] CRANDALL, M.G.: An Introduction to Constructive Aspects of Bifurcation and the Implicit Function Theorem, Application of Bifurcation Theory, P.H. Rabonowitz (ed.), 1-35, Academic Press, New York, 1977.

[21] CRANDALL, M.G. and RABINOWITZ, P.H.: Bifurcation from Simple Eigenvalues, J. Funct. Anal., 8 (1971), 321-340.

[22] CRANDALL, M.G. and RABINOWITZ, P.H.: Bifurcation, Perturbation of Simple Eigenvalues, and Linearized Stability, Arch. Rat. Mech. Anal., 52 (1973), 161-180.

[23] CRANDALL, M.G. and RABINOWITZ, P.H.: The Principle of Exchange of Stability, Dynamical Systems, A.R. Bednarek and L. Cesari (eds.), Academic Press, 1977.

[24] CRANDALL, M.G. and RABINOWITZ, P.H.: The Hopf Bifurcation Theorem in Infinite Dimensions, Arch. Rat. Mech. Anal., 67 (1977), 53-72.

[25] DANCER, E.N.: Global Solution Branches for Positive Mappings, Arch. Rat. Mech. Anal., 52 (1973), 181-192.

[26] DAVEY, A., DI PRIMA, R.C. and STUART, J.T.: On the Instability of Taylor Vortices, J. Fluid Mech., 31 (1968), 17--52.

[27] EKELAND, I. and LASRY, J.M.: Nombre de solutions périodiques des équations de Hamilton, C.R. Acad. Sc. Paris, 288 Série A (1979), 209-211.

[28] FADELL, E.R. and RABINOWITZ, P.H.: Bifurcation for Odd Potential Operators and an Alternative Topological Index, J. Funct. Anal., 26 (1977), 48-67.

[29] FADELL, E.R. and RABINOWITZ, P.H.: Generalized Cohomological Index Theories for Lie Groups Actions with an Application to Bifurcation Questions for Hamiltonian Systems, Inv. Math., 45 (1978), 139-175.

[30] FIFE, P.C.: The Bénard Problem for Generalized Fluid Dynamical Equations and Remarks on Boussinesq Equations, India-

na Univ. Math. J., 20 (1970), 303-326.

[31] FIFE, P.C.:Branching Phenomena in Fluid Dynamics and Chemical Reaction-Diffusion Theory, Eigenvalues of Nonlinear Problems, G. Prodi (ed.), 23-83, Edizioni Cremonese, Roma, 1974.

[32] FIFE, P.C.: Stationary Patterns for Reaction Diffusion Equations,Nonlinear Diffusion, W.E. Fitzgibbon and H.F. Walker (eds.), Research Notes in Math., Pitman, London, 1977.

[33] FIFE, P.C.: Asymptotic States for Equations of Reaction and Diffusion, Bull. Am. Math. Soc., 84 (1978), 693-726.

[34] FIFE, P.C.: Mathematical Aspects of Reaction-Diffusion Systems, Lecture Notes in Biomath., Vol. 28, Springer, Berlin, 1979.

[35] FIFE, P.C. and JOSEPH, D.: Existence of Convective Solutions of the Generalized Bénard Problem, Arch. Rat. Mech.Anal., 33 (1969), 116-138.

[36] FRIEDMAN, A.: Partial Differential Equations, Holt, Reinehart and Winston, Inc., New York, 1969.

[37] FRIEDRICHS, K.O. and STOKER, J.: The Nonlinear Boundary-Value Problem of the Buckled Plate, Am. J. Math., 63 (1941), 839-888.

[38] GOLUBITSKY, M. and SCHAEFFER, D.: A Theory for Imperfect Bifurcation Via Singularity Theory, Comm. Pure Appl. Math., 32 (1978), 21-98.

[39] HARTMAN, R.: Ordinary Differential Equations, John Wiley, New York, 1964.

[40] HENRY, D.: Geometric Theory of Semilinear Parabolic Equations, University of Kentucky Lecture Notes, 1974.

[41] HERSCHKOWITZ, H. and KAUFMAN, M.: Bifurcation Analysis of Nonlinear Reaction-Diffusion Equations II, Bull. Math. Biol., 37 (1975), 589-636.

[42] HOPF, E.: Abzweigung einer periodischen Lösung von einer stationaren Lösung eines Differentialsystems, Ber. Math.-Phys. Sachsische Academie der Wissenschaften Leiptzig, 94 (1942), 1-22.

[43] IOOSS, G.: Existence et stabilité de la solution périodique sécondaire intervenant dans les problèmes d'evolution du type Navier-Stokes, Arch. Rat. Mech. Anal., 47 (1972), 301-329.

[44] IOOSS, G.: Stabilité et bifurcation, Dept. of Math., Univ. of Paris Sud, Orsay, 1973.

[45] IOOSS, G.: Bifurcation of a Periodic Solution of the Navier-Stokes Equations into an Invariant Torus, Arch. Rat.Mech. Anal., 58 (1975), 35-56.

[46] IOOSS, G.: Secondary Bifurcation of a Steady Solution into an Invariant Torus for Evolution Problems of Navier-Stokes Type, Applications of Methods of Functional Analysis to Problems in Mechanics, 354-365, Lecture Notes in Math., vol. 503, Springer, Berlin, 1976.

[47] IOOSS, G.: Direct Bifurcation of a Steady Solution of the Navier-Stokes Equations into an Invariant Torus, Turbulence and Navier-Stokes Equations, 113-120, Lectures Notes in Math., vol. 565, Springer, Berlin, 1976.

[48] IOOSS, G.: Sur la deuxième bifurcation d'une solution stationnaire de système du type Navier-Stokes, Arch. Rat. Mech. Anal., 64 (1977), 339-369.

[49] IOOSS, G.: Bifurcation of Maps and Applications, North-Holland, Amsterdam, 1979.

[50] IUDOVICH, V.I.: On the Origin of Convection, Prikl. Mat.Mek. (J. Appl. Math. Mech.), 30 (1966), 1193-1199.

[51] IUDOVICH, V.I.: Investigation of Auto-oscillations of a Continuous Medium Occurring at Loss of Stability of a Stationary Mode, Prikl. Mat. Mek. (J. Appl. Math.Mech.), 36 (1972), 450-459.

[52] JOSEPH, D.: Stability of Convection in Containers of Arbitrary Shape, J. Fluid Mech., 47 (1971), 257-282.

[53] JOSEPH, D.: Stability of Fluid Motions I, II, Springer, Berlin, 1976.

[54] JOSEPH, D. and NIELD, D.A.: Stability of Bifurcating Time-periodic and Steady Solutions of Arbitrary Amplitude, Arch. Rat. Mech. Anal.,58 (1975), 369-380.

[55] JOSEPH, D. and SATTINGER D.H.: Bifurcating Time-periodic Solutions and their Stability, Arch. Rat. Mech. Anal., 45, (1972), 79-109.

[56] KATO, T.: Perturbation Theory for Linear Operators, Springer, Berlin, 1966.

[57] KELLER, J.B. and ANTMAN, S. (eds.): Bifurcation Theory and Nonlinear Eigenvalue Problems, Benjamin, New York, 1969.

[58] KIRCHGÄSSNER,K. and KIELHÖFER, H.: Stability and Bifurcation in Fluid Dynamics, Rocky Mountain Math.J., 3 (1973), 275-318.

[59] KIRCHGÄSSNER, K. and SORGER, P.: Stability Analysis of Branching Solutions of the Navier-Stokes Equations, Proc. 12th Int. Cong. Appl. Mech., Stanford Univ., August 1968.

[60] KOPELL, N. and HOWARD, L.N.: Bifurcations under Nongeneric Conditions, Advances in Math., 13 (1974), 274-283.

[61] KRASNOSELSKIJ, M.A.: Positive Solutions of Operator Equations, Nordhoff, Groningen, 1964.

[62] KRASNOSELSKIJ, M.A.: Topological Methods in the Theory of Nonlinear Integral Equations, Pergamon Press, Oxford, 1964.

[63] LEBOWITZ, N.R.: Bifurcation and Stability Problems in Astrophysics, Applications of Bifurcation Theory, P.H. Rabinowitz (ed.), 259-284, Academic Press, New York, 1977.

[64] LIAPUNOV, M.A.: Problème générale de la stabilité du mouvement, Annals of Mathematical Studies, 17, Princeton, 1949.

[65] MARINO, A.: La biforcazione nel caso variazionale, Conf. Sem. Mat. dell'Univ. Bari, 132 (1977).

[66] MARSDEN, J.: Qualitative Methods in Bifurcation Theory, Bull. Am. Math. Soc., 84 (1978), 1125-1148.

[67] MARSDEN, J. and MCCRACKEN, M.: The Hopf Bifurcation and its Applications, Springer, Berlin, 1976.

[68] MATKOWSKY, B.J. and REISS, E.L.: Singular Perturbations of Bifurcations, SIAM J. Appl. Math., 33 (1977), 230-255.

[69] MOSER, J.: Periodic Orbits Near an Equilibrium and a Theorem by Alan Weinstein, Comm. Pure Appl. Math., 29 (1976), 727-747.

[70] NEMYTSKII, V.V. and STEPHANOV, V.V.: Qualitative Theory of Differential Equations, Princeton Univ. Press, Princeton, 1960.

[71] NIREMBERG, L.: Topics in Nonlinear Functional Analysis, Courant Institute Lecture Notes, Courant Institute of Mathematical Sciences, 1974.

[72] PALAIS, R.S.: Critical Point Theory and the Minimax Principle, Proc. Symp. Pure Math., 15, Am. Math. Soc., 185-212, Providence, R.I., 1970.

[73] PRODI, G.(ed.): Eigenvalues of Nonlinear Problems, C.I.M.E., Edizioni Cremonese, Roma, 1973.

[74] PRODI, G. and AMBROSETTI, A.: Analisi Non Lineare, Scuola Normale Superiore, Pisa, 1973.

[75] RABINOWITZ, P.H.: Existence and Nonuniqueness of Rectangular Solutions of the Bénard Problem, Arch. Rat. Mech. Anal., 24 (1968), 32-57.

[76] RABINOWITZ, P.H.: Some Global Results for Nonlinear Eigenvalue Problems, J. Funct. Anal., 7 (1971), 487-513.

[77] RABINOWITZ, P.H.: Some Aspects of Nonlinear Eigenvalue Problems, Rocky Mountain Math. J., 3 (1973), 161-202.

[78] RABINOWITZ, P.H.: Variational Methods for Nonlinear Eigenvalue Problems, Eigenvalue of Nonlinear Problems, G. Prodi (ed.), 141-195, C.I.M.E., Edizioni Cremonese, Roma, 1975.

[79] RABINOWITZ, P.H.: Survey of Bifurcation Theory, Dynamical Systems, An International Symposium, vol. I, L. Cesari, J.K. Hale and J.P. LaSalle (eds.), 83-96, Academic Press, New York, 1976.

[80] RABINOWITZ, P.H.: (ed.) Applications of Bifurcation Theory, Academic Press, New York, 1977.

[81] RABINOWITZ, P.H.: A Bifurcation Theorem for Potential Operators, J. Funct. Anal., 25 (1977), 412-424.

[82] RABINOWITZ, P.H.: Periodic Solutions of Hamiltonian Systems, Comm. Pure Appl. Math., 31 (1978), 157-184.

[83] RABINOWITZ, P.H.: A Variational Method for Finding Periodic Solutions of Differential Equations, Nonlinear Evolution Equations, M.G. Crandall (ed.), 225-251, Academic Press, New York, 1978.

[84] RUELLE, D. and TAKENS, F.: On the Nature of Turbulence, Comm. Math. Phys., 20 (1971), 167-192, 23 (1971), 343-344.

[85] SATHER, D.: Branching of Solutions of Nonlinear Equations in Hilbert Spaces, Rocky Mountain Math. J., 3 (1973), 203-250.

[86] SATTINGER, D.H.: Bifurcation of Periodic Solutions of the Navier-Stokes Equations, Arch. Rat. Mech. Anal., 41(1971), 66-80.

[87] SATTINGER, D.H.: Topics in Stability and Bifurcation Theory, Lecture Notes in Math., vol. 309, Springer, Berlin, 1973.

[88] SCHWARTZ, J.: Nonlinear Functional Analysis, Lecture Notes New York Univ., Gordon and Breach, New York, 1969.

[89] STAKGOLD, I.: Branching of Solutions of Nonlinear Equations, SIAM Review, 13 (1971), 289-332.

[90] STOKER, J.J.: Nonlinear Elasticity, Gordon and Breach, New York, 1968.

[91] THOM, R.: Structural Stability and Morphogenesis, Benjamin, New York, 1972.

[92] TURNER, R.E.L.: Transversality and Cone Maps, Arch. Rat. Mech. Anal., 58 (1975), 151-179.

[93] VAINBERG, M.M.: Variational Methods in the Study of Nonlinear Operators, Holden-Day, San Francisco, 1964.

[94] VAINBERG, M.M. and TRENOGIN, V.A.: The Method of Liapunov-Schmidt in the Theory of Nonlinear Equations and their Further Development, Russian Math. Surveys, 17 (1962), 1-60.

[95] VAINBERG, M.M. and TRENOGIN, V.A.: Theory of Branching of Solutions of Nonlinear Equations, Monographs and Textbooks on Pure and Applied Math., Nordhoff International Publisher, Leyden, 1974.

[96] VEGA, J.M.: A Constructive Approach to the Problem of Bifurcation from Simple Eigenvalues, Applied Mathematics, Technical Report No. 7808, Department of Engeneering Sciences and Applied Math., Northwestern University, Evanston, Ill., 1979.

[97] WEINSTEIN, A.: Normal Modes for Nonlinear Hamiltonian Systems, Inv. Math., 20 (1973), 47-57.

REMARKS ON THE STABILITY OF STEADY-STATE SOLUTIONS OF REACTION-DIFFUSION EQUATIONS

by C. Conley and J. Smoller

§1. Introduction

In this note we shall describe a new approach to the stability problem for steady-state solutions of reaction-diffusion equations. We shall explain our techniques and results by discussing a few examples which model quite well the general theory.

Our approach is topological, and is based on a generalized "Morse theory" as developed in [1]. This theory extends the Morse index to isolated invariant sets in finite or infinite dimensional spaces, and is therefore applicable to stability questions. Thus, we think of our equation as an ordinary differential equation in a function space, with the rest-points of this ordinary differential equation being the steady-state solutions of the given partial differential equation. If the rest point in question is an isolated invariant set, then we can associate to it an index. This index, being a generalization of the usual Morse index, also carries similar information; namely, it tells the dimension of the unstable manifold of the given steady-state solution. Furthermore, this index is "invariant under continuation", so that if the system depends smoothly on parameters, then the index of corresponding isolated invariant sets are the same. This often enables us to compute the index merely by "deforming" the equation to a simpler one. We illustrate this technique in §4. Finally, there is an analogue of the usual "sum" formula for indices; this too is useful for computing indices.

C. Bardos and D. Bessis (eds.), Bifurcation Phenomena in Mathematical Physics and Related Topics, 47–56.

In order to give very precise results, we need a little more; namely, we must be able to count the number of steady-state solutions. This problem is intimately connected with bifurcation theory, and requires a knowledge of the number of critical points of a certain function, the so-called "time map". This is discussed and illustrated in §3.

We wish to point out that one can view our methods as extending to infinite dimensions, a standard technique used in finite dimensions. Namely that of finding all of the rest points of the given differential equation, and then linearizing about these points in order to determine the nature of the flow near these rest points. In our case, linearization is replaced by topological techniques.

§2. The Problem

Consider the reaction-diffusion equation

$$u_t = u_{xx} + f(u) , \quad -L < x < L , \tag{1}$$

together with homogeneous Neumann or Dirichlet boundary conditions:

$$u(\pm L,t) = 0 , \text{ or } u_x(\pm L,t) = 0 , \quad t > 0 ,$$

respectively*. By a steady state solution of (1), we mean a solution V which doesn't depend on t ; ie, V solves the equation

$$u'' + f(u) = 0 , \tag{2}$$

together with the same boundary conditions. We will often refer to such solutions as "rest points" of (1). The problem which we want to discuss is that of the stability of the rest points V , now considered as particular solutions of (1). Here stability is meant in the Lyapunov sense. In other words, if the initial data $u(x,0)$ is close to $V(x)$, we ask whether the solution u of (1) with this data, tends to V as $t \to +\infty$, say in the sense of $L_2(-L,L)$. If this is true, we say that the solution V is stable, and we call V an "attractor" of (1).

---

* More general boundary conditions and certain systems could also be considered, but these offer only technical complications, and we shall not discuss them here.

As discussed in [1], it is possible to assign an "index" to any isolated invariant set of a flow. For the equation (1), we see that the functional

$$\Phi(u) = \int_{-L}^{L}\left(\frac{u_x^2}{2} - F(u)\right)dx \qquad (3)$$

decreases along solutions; ie, the flow is "gradient-like". It follows that every isolated rest point of (1) is also an isolated invariant set, and thus has associated to it, an index. This index takes the form of an equivalence class of pointed topological spaces, the relation being homotopy equivalance.

To be a little more precise, an invariant set S is called <u>isolated</u>, if it is the maximal invariant set in some compact neighborhood N of itself; the neighborhood N is called an isolating neighborhood. The index of S is stated in terms of index pairs $\langle N_1,N_2\rangle$ in N. By an index pair $\langle N_1,N_2\rangle$, we mean that N_1 and N_2 are subsets of N which are compact and positively invariant relative to N, such that $S \subset \text{int } N_1 \setminus N_2$, and in addition, if $x \in N_1$, $t > 0$ and $x \cdot t \notin N$, then $\exists\, t' > 0 \ni: x\cdot[0,t'] \subset N_1$, and $x \cdot t' \in N_2$. For a given isolated invariant set S, it can be shown that index pairs exist, and that the quotient spaces N_1/N_2 are all homotopically equivalent. The homotopy type of N_1/N_2 is, by definition, the <u>index</u> of the isolated invariant set S. It is independent of N, and depends only on S.

If we have a family of equations depending smoothly on a parameter, and if N is an isolating neighborhood for all parameter values in a closed connected subset of parameter space, then the various isolated invariant sets are said to be <u>related by continuation</u>; this relation is also defined to be transitive. It can be proved that the index of an isolated invariant set is <u>stable under continuation</u>; ie, if two isolated invariant sets are related by continuation, then they have the same index, [1]. This powerful result often enables one to caluclate the index of an isolated invariant set without actually constructing index pairs or even isolating neighborhoods; see §4.

§3. Steady-State Solutions

For the equation (2), it is possible to interpret solutions geometrically, in the phase plane. To illustrate, consider the equation (2), together with, say, homogeneous Dirichlet boundary conditions, $u(\pm L) = 0$. If we write (2) as a system in the usual way,

$$u' = v \ , \quad v' = -f(u) \ , \tag{4}$$

then we see that a solution is a "curve" which begins and ends on $u = 0$, and takes "time" 2L to make the journey.

In order to study such solutions, it is convenient to introduce the "time" map, which we shall now do. Thus, for a point p on the (positive) v-axis, we denote by $\alpha(p)$, the corresponding (first) point on the u-axis to which p is carried by the flow, see figure 1.

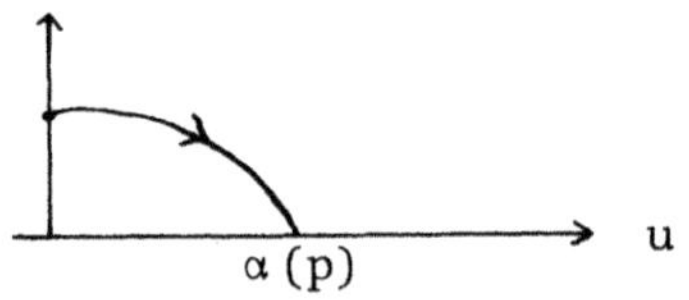

Figure 1

We let $T(p)$ denote the "time" that this orbit takes in going from p to $\alpha(p)$. Now a moment's reflection shows that the graph of T contains all the information on the number of solutions for a given L. In fact, in [3] it was proved that if $(f(u) - u f^1(u)) > 0$ along the orbit (except possibly at the end points), then T is monotone at p, and hence the solution does not bifurcate, for small changes in L.

Now in order to count the number of critical points of T. This is generally a very difficult problem, but for certain interesting functions f, one can actually carry this program out. We shall now discuss an example wherein one can give a rather exact picture of the graph of T. Thus, consider the case where f is the cubic, $f(u) = -u(u-a)(u-b)$, $0 < a < b$, and $\int_0^b f\, du > 0$.

The phase portrait for (3) takes the form of figure 2.

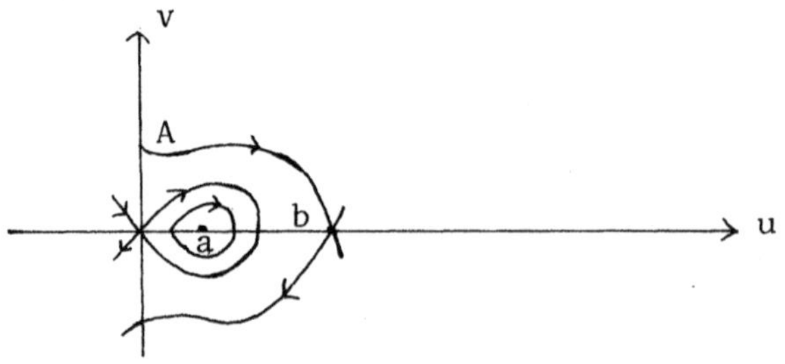

Figure 2

We let A denote the point where $A^2 = 2F(b)$, and we consider the "time" map on $(0,A)$, defined by

$$T(p) = \int_0^{\alpha(p)} [F(\alpha(p)) - F(u)]^{-1/2} \, du ,$$

where F is a primitive of f ; ie $F' = f$. Now it is easy to see that $\alpha'(p) > 0$, so that we can think of T either as a function of α , where $c < \alpha < b$, with $F(c) = F(0)$, or as a function of p , $0 < p < A$.

Note that if α is near 0 or b , then the orbit through α comes near a rest point, so that $T(\alpha)$ is very large. Hence it follows that T must have at least one critical point. In fact*, one can prove the estimate

$$\frac{3}{\alpha} T'(\alpha) + T''(\alpha) > 0 , \text{ if } \alpha > \frac{4\sigma}{9} , \sigma = a + b .$$

It follows at once, from this, that T has exactly one critical point, and so the graph of T takes the form sketched in figure 3.

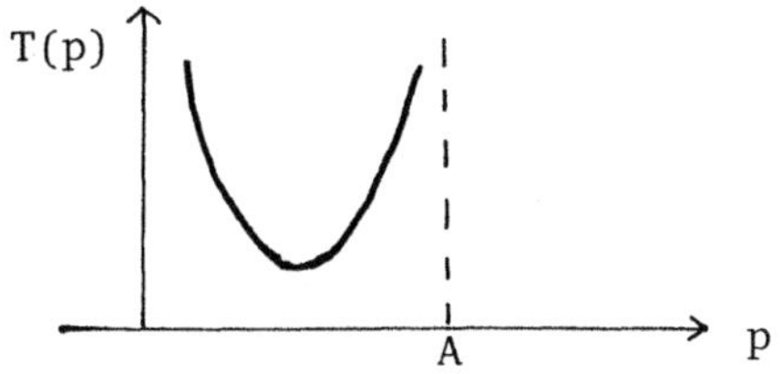

Figure 3

Using this, we see that if L is very small, then (3) has only $u \equiv 0$ as a solution. For a certain critical value L_0 of L , a single solution bifurcates out of 0 , while for $L > L_0$, (2) has exactly three solutions. In the next section, we shall discuss the stability of these solutions.

---

* The results in this section were obtained jointly with A. Wasserman, and will appear in detail elsewhere.

§4. Stability

As we mentioned earlier, every isolated rest point of (1) is an isolated invariant set. In view of the discussion in the last section, it follows that all rest points of (1) are isolated invariant sets. We shall show how to compute the indices of the rest points of each of these solutions. Before doing this however, we must discuss a little of the general theory; see [1].

We consider the equation (1), and let F be any primative of f ; ie $F' = f$. We assume that f is smooth, and that $F(u) \to -\infty$ as $|u| \to \infty$. We also assume that $f(u) \neq 0$ outside a compact set, and that for C sufficiently large, the set of u for which $F(u) \geq -C$ is a convex set. For C^1 functions which satisfy the boundary conditions, we can define the functional Φ , as given in (3). Then for any constant K , and C sufficiently large, the set M of u , such that $F(u) \geq -C$ and $J(u) \geq K$ is positively invariant; ie, if $u(\cdot,t)$ lies in M , then $u(\cdot,T)$ lies in M for all $T > t$. Moreover, it can also be shown that i) M is compact in the uniform topology, ii) each element of M determines a unique solution in M , and iii) the set of solutions whose data is in M , is uniformly equi-continuous in t . If we now append past histories in M , which satisfy the same equi-continuity conditions, we obtain a flow in which the general theory of isolated invariant sets is applicable.*

Having discussed the general framework, we are now in a position to investigate the stability of steady-state solutions of (1). In order to do this, we will calculate the index of the particular solution in question. This latter calculation will be a consequence of the stability of the index under continuation.

Consider first the homogeneous Dirichlet problem for (1), where $f(u) = -u(u-a)(u-b)$, $0 < a < b$, and f has positive area on the interval $0 \leq u \leq b$. As we have seen in §3, if L is sufficiently large, there are exactly three steady-state solutions, 0, u_1, and u_2 . These are depicted in figure 4, where $u_2(0) > u_1(0)$.

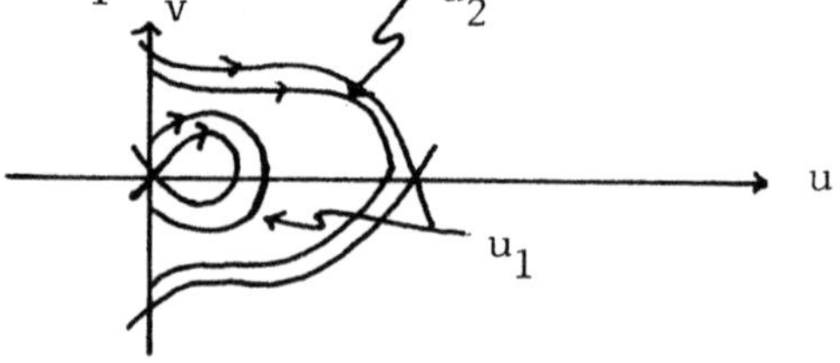

Figure 4

* This will appear in a forthcoming publication.

We claim that 0 and u_2 are stable, while u_1 is not stable. To see this, we fix L, and consider the function f as a "parameter". Thus, to show that u_2 is stable, we deform f to a new function f_1 obtained by "pulling up the valley". We then "straighten out f_1" to a linear function f_2, and finally, we deform f_2 to the function $f_3 = -k\,u$, $k > 0$; see figure 5.

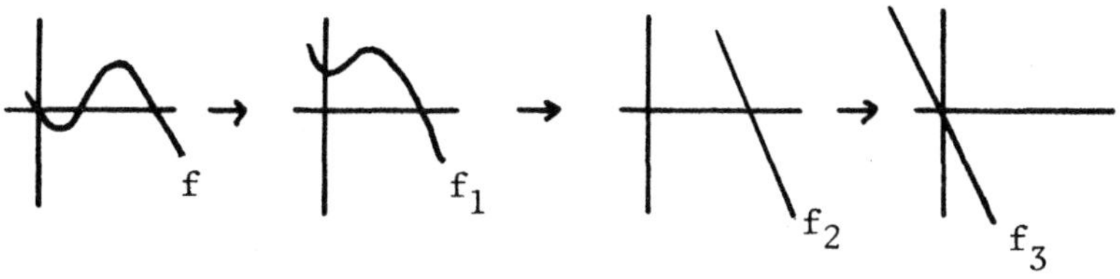

Figure 5

Observe that under this deformation, the solution u_2 of (1) has been continued to the zero solution of the linear equation $u_t = u_{xx} - ku$. For this latter equation, it is well-known that the zero solution is an attractor*, and hence has the index of an attractor, $\bar{1}$, the homotopy type of a pointed zero sphere. It follows from the continuation theorem, that u_2 must have index $\bar{1}$, so that u_2 also must be stable. To see that 0 is stable, we use a similar argument; namely, we could deform f first to a new function g_1, obtained by "pushing down the hill", and then continuing g_1 to the function $-ku$ as before; see Figure 6.

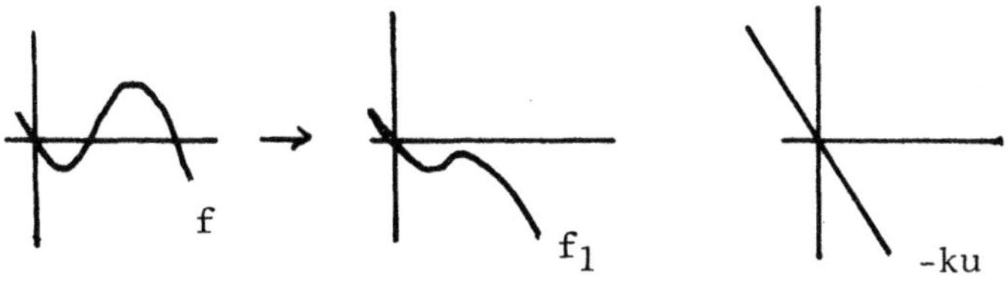

Figure 6

* In fact, it is a global attractor, in the sense that all solutions tend to zero as $t \to +\infty$.

Here we have continued the zero solution of (1) to the zero solution of $u_t = u_{xx} - k u$. It follows as before that 0 is an attractor. Finally, to see that u_1 is unstable, we observe that all solutions of (1), tend asymptotically, as $t \to +\infty$, to the set $0 \leq u \leq b$; ie, this set itself is a (global) attractor for the flow. Hence, it too has index $\bar{1}$; from this it is easy to conclude (using the sum formula) that the index of u_1 cannot be $\bar{1}$, so that u_1 cannot be stable*.

We turn now to a second example which illustrates another aspect of the theory. Consider the system (1) where $f(u) = -u(u-a)(u-b)$, where now $a < 0 < b$, and $\int_a^b f > 0$.

Here we are concerned with the Neumann problem for (1); ie, we consider (1) together with the homogeneous Neumann boundary conditions $u'(\pm L) = 0$. In [2], it is shown that all non-constant solutions are unstable. We would like to illustrate here how our methods also yield this result.

If we again write (2) as a system $u' = v$, $v' = -f(u)$, we see that solutions of the Neumann problem are curves which begin and end on the line $v = 0$, and take "time" $2L$ to make the journey. Now if L is very small, there are only three solutions; namely the constants a, 0 and b. By continuing f to linear functions, as above, we can easily show that the constant solutions $u = a$ and $u = b$ are stable, while the solution $u = 0$ has a one-dimensional unstable manifold; ie $\text{ind}(0) = \Sigma^1$ (the homotopy type of a pointed 1-sphere). Now as L increases, two solutions u_1, u_2, bifurcate out of 0; these are depicted in figure 7. We claim that

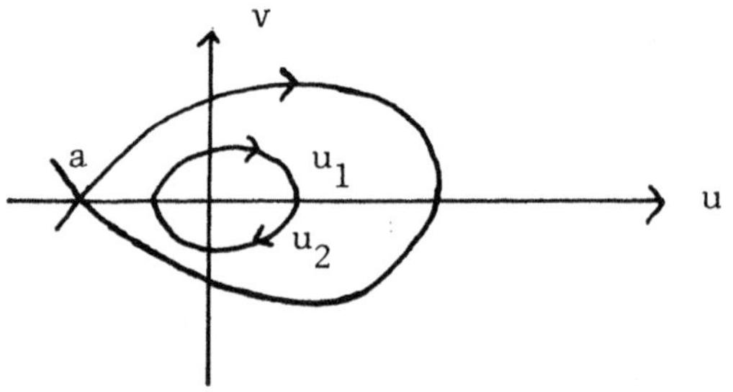

Figure 7

both of these solutions must be unstable. The argument goes as

* In fact, more can be said; namely, it can be shown, via index arguments, that u_1 has index Σ^1, and if u_1 is non-degenerate; it then has a 1-dimensional unstable manifold.

follows. First we note that the total index of the isolated invariant set S, consisting of the three rest points u_1, u_2, 0 and (possibly) any connecting orbits must be Σ^1, the index of 0 for small L. Next, symmetry considerations show that $\text{ind}\,(u_1) = \text{ind}\,(u_2)$. Finally, the index of 0 is now Σ^2, the pointed two-sphere, as one checks by linearization. Now suppose that u_1 and u_2 were both stable solutions. Then $\text{ind}\,(u_1) = \text{ind}\,(u_2) = \bar{1} \equiv \Sigma^0$, the pointed zero sphere. Now since the flow is gradient-like, and there are a finite number of critical points, there is a Morse filtration* of S of the following form

$$M_1 \underset{M_{21}}{\longrightarrow} M_2 \underset{M_{32}}{\longrightarrow} M_3 \quad .$$

This induces the following two exact sequences on homology groups (with, say, real coefficients):

$$H_p(h_2) \to H_p(h_{21}) \to H_p(h_1)$$

$$H_p(h_3) \to H_p(h) \to H_p(h_{21})$$

where $h_i = \text{ind}\,(u_i) = \Sigma^0$ or Σ^2, $h = \text{ind}\,(S) = \Sigma^1$, and

h_{21} is the index of S_{21}, the isolated invariant set in M_{21}. Now it is well-known that $H_p(h_i) = 0$ if p is odd. Thus from the exactness of the first sequence we see that $H_p(h_{21}) = 0$, so that from the second sequence we get $H_p(h) = 0$, if p is odd. However, since $h = \Sigma^1$, it follows that $H_1(h) = \mathbb{R}$; this is the desired contradiction. Similar considerations show that all non-constant solutions of the Neumann problem cannot be stable. These topics will be more fully discussed in a forthcoming paper.

---

* This is defined in [1]; roughly speaking, a Morse filtration of an isolated invariant set S is an increasing sequence of attractor neighborhoods, relative to S, ending with S. When the associated invariant sets are identified as distinct points, one gets a gradient-like quotient flow whose critical points correspond to these points, such that the ordering of these critical points agrees with that induced by the "gradient-function". In our case the M_i's are isolating neighborhoods isolating the critical points u_1, u_2 or 0, in some order, and M_{21} (resp. M_{32}) is an isolating neighborhood containing M_1 and M_2 (resp. M_2 and M_3), but not M_3 (resp. M_1).

References

1. Conley, Charles, Isolated Invariant Sets and the Morse Index, Conf. Board of the Math. Sci., No. 38, A.M.S., Providence, 1978

2. Fife, Paul, Bull. Amer. Math. Soc., 84, 693-726, (1978)

3. Smoller, J., Tromba, A., and A. Wasserman, J. of Nonlinear Anal., to appear

University of Wisconsin
Madison, Wisconsin 53706

University of Michigan
Ann Arbor, Michigan 48109

SUCCESSIVE BIFURCATIONS AND THE ICE-AGE PROBLEM

Michael Ghil

Courant Institute of Mathematical Sciences
New York University, New York, N. Y. 10012, U.S.A.

The inherent nonlinearity of reasonable models for the Earth's climatic system suggests the relevance of bifurcation studies to the problem of ice ages. Exchanges of stability between equilibria, periodic and quasiperiodic solutions are possible explanations of different features in the paleoclimatic record.

We present a number of climate models in which successive bifurcations lead to increasingly complex solution behavior. One class of models is governed by a nonlinear parabolic partial differential equation (PDE) with a time lag τ. For $\tau = 0$, the model's steady-state solutions [1,3] exhibit an S-shaped bifurcation curve with respect to the solar-input parameter. Hopf bifurcation of a stable equilibrium to a stable periodic solution occurs as $\tau > 0$ is increased [1,4]. Given slightly different forms of the nonlinear source term, as τ increases further the periodic solution can be replaced by a quasiperiodic one of a fairly complex structure [1,2]. Characteristics of these different solutions can explain some aspects of the Earth's climate history.

Another class of models is governed by a system of autonomous ordinary differential equations (ODEs). One of these is the space averaged version of the PDE above. Another ODE models explicitly the mechanism responsible for the time lag τ in the previous model. This two-equation system also exhibits Hopf bifurcation; the amplitude, period and phase relationships of the solutions compare favorably with those recorded for glaciation cycles [6,7]. We consider also an augmented system, in which the third equation models the deflection of the Earth's crust under the varying size

C. Bardos and D. Bessis (eds.), Bifurcation Phenomena in Mathematical Physics and Related Topics, 57–58.

of ice sheets [5]. Its critical points correspond to those of the preceding system, but their stability properties differ.

REFERENCES

1. Bhattacharya, K.: 1979 "A Study of Almost-Intransitivity as a Possible Cause of Terrestrial Climate Changes," Ph.D. Thesis, submitted to the Physics Department, Columbia University, New York.

2. Bhattacharya, K., and Ghil, M: 1978, "An Energy-Balance Model with Multiply-Periodic and Quasi-Chaotic Free Oscillations," in: Evolution of Planetary Atmospheres and Climatology of the Earth, Centre National d'Etudes Spatiales, Toulouse, France, pp. 299-310.

3. Ghil, M.: 1976, "Climate Stability for a Sellers-Type Model," J. Atmos. Sci., 33, pp. 3-20, reviewed in Math. Rev. 54, Aug. 1977, No. 4568.

4. Ghil, M., and Bhattacharya, K.: 1979, "An Energy-Balance Model of Glaciation Cycles," to appear in: A Review of Climate Models, W. L. Gates, Ed., GARP Publication Series, World Meteorological Organization/International Council of Scientific Unions, Geneva Switzerland.

5. Ghil, M., Källén, E., Crafoord, C., and Le Treut, H.: 1979, "Simple Climate Models, II: Systems of Nonlinear ODEs," preprint, SIAM Fall Meeting, Denver, Nov. 1979.

6. Källén, E., Crafoord, C., and Ghil, M.: 1978, "Free Oscillations in a Coupled Atmosphere-Hydrosphere-Cryosphere System," in: Evolution of Planetary Atmospheres and Climatology of the Earth, Centre National d'Etudes Spatiales, Toulouse, France, pp. 285-298.

7. Källén, E., Crafoord, C., and Ghil, M.: 1979, "Free Oscillations in a Climate Model with Ice-Sheet Dynamics," J. Atmos. Sci., in press.

II

BIFURCATION IN THEORETICAL PHYSICS AND STATISTICAL MECHANICS ; APPLICATIONS

CRITICAL PHENOMENA IN STATISTICAL MECHANICS — ASPECTS OF RENORMALIZATION GROUP THEORY

Michael E. Fisher

Baker Laboratory, Cornell University
Ithaca, New York 14853

The principle topics covered in the lectures delivered at the School have been reviewed in earlier literature: accordingly, only a summary account with some useful references is presented here.

I. FACTS, MODELS, and PROBLEMS

The crucial experimental facts that pose the basic question for the theory of critical phenomena may be seen most readily in a simple ferromagnet, such as iron or nickel which is cooled in zero magnetic field, $H = 0$, from high temperatures. Initially, the observed magnetization, M, is strictly zero; but at a sharply defined point, the critical temperature, $T = T_c$, a "bifurcation" phenomena occurs and the equilibrium value of the magnetization for $T < T_c$ becomes nonzero: actually any value of M between two extremes, $M_0(T)$, the "spontaneous magnetization", and $-M_0(T)$, may arise depending on precise experimental conditions. Likewise the magnetic susceptibility, $\chi(T) = (\partial M/\partial H)_T$, is finite above T_c but diverges to $+\infty$ as $T \to T_c+$ [1, 2].

The science of thermodynamics establishes that knowledge of a master function of two variables, namely the free energy, $F(T, H)$, which is jointly convex in its arguments, is sufficient to describe the observed singular behavior in the critical region. Equivalently, one can perform a Legendre transform to the new free energy function, $A(T, M) = F + MH$, from which $M = (\partial A/\partial M)_T$.

The classical or standard phenomenological theory of critical behavior has the appearance of great generality. It simply

C. Bardos and D. Bessis (eds.), Bifurcation Phenomena in Mathematical Physics and Related Topics, 61–68.

postulates an expansion of A in powers of M with coefficients analytic in T. Mean field theory and nearly all other approximate theories are equivalent to this in the critical region; so is catastrophe theory. A simple bifurcation analysis together with the imposition of convexity then yields the unambiguous predictions [1, 2]:

$$M_0(T) \approx B|t|^{\beta} \qquad \text{and} \qquad \chi(T) \approx Ct^{-\gamma}, \tag{1}$$

as $t = (T - T_c)/T_c \to 0$ with <u>critical exponents</u> $\beta = \frac{1}{2}$ and $\gamma = 1$.

These predictions, however, are in total disagreement with the experimental facts! Indeed, for most systems exhibiting critical behavior (ferromagnets, fluids, etc.) the most accurate observations may be summarized by (1) but with $\beta \simeq 0.31$ to 0.37 ($\neq \frac{1}{2}$) and $\gamma \simeq 1.22$ to 1.38 ($\neq 1$) [1, 2]. Furthermore, for wide classes of real physical systems these nonclassical critical exponent values appear to be <u>universal</u> and independent of almost all the detailed physico-chemical properties of the system under study. In addition, the equation of state in the critical region is found to obey a <u>scaling law</u>. For a ferromagnet this can be written [1, 2]

$$M \equiv M(T, H) \approx |t|^{\beta} Y(H/|t|^{\beta+\gamma}), \tag{2}$$

as $t, H \to 0$, where the scaling function $Y(z)$ also appears to be universal. The scaling law implies exponent relations such as $\alpha + 2\beta + \gamma = 2$ where α describes the critical divergence of the specific heat. It follows from these observations that the free energy functions $F(T, H)$ <u>and</u> $A(T, M)$ are singular, that is non-analytic, at the critical point in a fundamental way: no Taylor series expansion of any sort can be justified about the critical point.

To understand critical phenomena more deeply one must utilize the science of statistical mechanics. The first step is to choose an appropriate microscopic model of a physical system. This should contain the essential physical features: namely, (i) translational invariance in Euclidian space of d dimensions; and (ii) short range interactions coupling nearby physical degrees of freedom. The most tractable models mathematically, embody a uniform space lattice of N sites, $i = 1, 2, \ldots, N$, with, at each site, a spin variable or degree of freedom, $\vec{s}_i$. In the simplest case (a) one takes just $s_i = \pm 1$ (defining an Ising model); then (b) one also considers vectors $\vec{s}_i$ of n components and fixed length, $|\vec{s}_i| = 1$; finally, to ease the mathematics further and to make contact with field theory one discusses (c) continuous spin vectors $\vec{s}_i \equiv (s_i^{(1)}, s_i^{(2)}, \ldots s_i^{(n)})$ of unbounded length, $-\infty < s_i^{(\mu)} < \infty$.

II. STATISTICAL MECHANICS AND THE RENORMALIZATION GROUP CONCEPT

To implement the statistical mechanical analysis of a microscopic model of, say, a ferromagnet one must supply the reduced Hamiltonian or energy functional $\bar{\mathcal{H}} = -\mathcal{E}(\{s_i\})$. In the most basic models $\bar{\mathcal{H}}$ contains first an external field term $\vec{h}\cdot\Sigma_i^N \vec{s}_i$ with $h = H/k_BT$, and then a nearest neighbor coupling term $K\Sigma_{(i,j)}\vec{s}_i \cdot \vec{s}_j$ with $K = J/k_BT$ (where k_B is Boltzmann's constant and J is the "exchange parameter"). In addition, one must specify an integration or trace operation, Tr_N^s, on the spin variables $\{\vec{s}_i\}$ which sums over all possible values or "configurations". Thence one defines the partition function for a system of N spins by

$$Z_N[\bar{\mathcal{H}}] = \mathrm{Tr}_N^s\left\{\exp[\bar{\mathcal{H}}(\{s_i\})]\right\}. \tag{3}$$

Finally, the fundamental reduced free energy per spin is given by

$$f[\bar{\mathcal{H}}] = -F_N/Nk_BT = N^{-1}\ln Z_N[\bar{\mathcal{H}}]. \tag{4}$$

Knowing $f[\bar{\mathcal{H}}] \equiv f(T, H)$, yields all the critical behavior. Actually, it is also essential to consider $N \gg 1$ and, in fact, to calculate the thermodynamic limit, $N \to \infty$ (since for finite N no sharp critical point can occur).

The basic difficulty in statistical mechanics is evidently the calculation of the trace integrals or sums over an unbounded number, N, of spin variables. The central idea of the renormalization group may be regarded as splitting the task of computing all N sums or integrals into a series of successive "renormalizations" in which, at each stage, only some of the traces are performed [3,4]. The number of remaining spin variables or degrees of freedom at the first stage is reduced from N to

$$N' = N/b^d, \qquad \text{with} \qquad b > 1. \tag{5}$$

Here b is a spatial rescaling factor such that distances $\underset{\sim}{x}$ in the renormalized physical system are rescaled to

$$\underset{\sim}{x}' = \underset{\sim}{x}/b. \tag{6}$$

This serves to preserve the spatial density of spins or degrees of freedom. In fact, however, the Hamiltonian $\bar{\mathcal{H}}$ must also change at each stage: one may write the renormalized or new Hamiltonian formally as

$$\bar{\mathcal{H}}' = \mathbb{R}_b[\bar{\mathcal{H}}]. \tag{7}$$

Repeated iteration of the operation $\mathbb{R}_b$ reduces the number of variables further and generates the renormalization group (which, in general, is only a semi-group).

This idea may be explored via its exact implementation in a simple $d=1$-dimensional chain of Ising spins, $s_i = \pm 1$ $(i=1,2,\ldots)$ [3]. In the so-called "dedecoration" or "decimation" renormalization group, one proceeds by first taking the trace over only every second spin, say over the odd labelled spins s_{2k+1} $(k=0,1,2,\ldots)$ so that $b=2$. Because of the simplicity of the model one then obtains a representation of the renormalization group $\mathbb{R}$ as a closed set of recursion relations which give the new, or renormalized values of temperature and field, T' and H', as explicit functions of the original, "bare" or "unrenormalized" parameters T and H [3]. In addition, the original free energy per spin can be recaptured from the new free energy via the general formula

$$f(T, H) \equiv f[\overline{\mathcal{H}}] = b^{-d} f[\overline{\mathcal{H}}'] \equiv b^{-d} f(T', H'). \tag{8}$$

which follows from the fact that, by construction of $\mathbb{R}$, one has

$$Z_{N'}[\overline{\mathcal{H}}'] \equiv Z_N[\overline{\mathcal{H}}]. \tag{9}$$

This may be termed the "unitarity relation" for renormalization groups.

Examination of the exact recursion relations reveals that they have an unstable fixed point, which will be indicated $\mathcal{H}^* \equiv (T^*, H^*)$, at $(T^*=0,\ H^*=0)$. This in turn can be seen to correspond to the critical point of the Ising chain (which, because of its low dimensionality, can occur only at $T=0$). In terms of the appropriate $d=1$ Ising model variable, $t = \exp(-4K) = \exp(-4J/k_BT)$, the critical behavior in zero field is found from the exact solution (obtained by other means) to be $f(t) \approx \text{const}\ t^{2-\alpha}$ as $t \to 0$ with $2-\alpha = \frac{1}{2}$. However, this singular behavior also can be found without solving the model exactly merely by linearizing the renormalization group recursion relations about the nontrivial fixed point and then calculating the appropriate eigenvalue, $\Lambda_1 = b^{\lambda_1}$. Iteration of the linearized renormalization group and use of the flow equation (8) then leads directly to the relation $2-\alpha = d/\lambda_1$. With the numerical values $d=1$ and $\Lambda_1 = 4$ this gives the correct critical point exponent. But how far can the method be generalized to more realistic higher dimensional models?

III. RENORMALIZATION GROUPS IN THE SPACE OF HAMILTONIANS

A special feature of the one-dimensional Ising chain and its generalizations [3], which allows the exact construction and analysis of the dedecoration or decimation renormalization groups, is that the renormalized Hamiltonian, $\mathcal{H}'$, differs from the original Hamiltonian $\mathcal{H}$, only in that the values of the magnetic field parameter $h = H/k_BT$ and the first neighbor coupling parameter $K = J/k_BT$ are renormalized: no extra couplings or interactions of any sort are introduced into the Hamiltonian. One of the important insights of K. G. Wilson when he developed the full renormalization group approach, is that one must, in general, allow for the renormalized Hamiltonian $\bar{\mathcal{H}}'$ to be much more complex than the initial, bare Hamiltonian $\bar{\mathcal{H}}$. Thus $\bar{\mathcal{H}}'$ will normally include, for example, second, third, ..., ℓ-neighbor interactions, and three-spin couplings, $s_i s_j s_k$ of all sorts, four-spin couplings $s_i s_j s_k s_\ell$, etc., etc. In short, one must introduce a large space of Hamiltonians, $\mathbb{H}$, parametrized by, in most cases, at least a denumerably infinite number of "coupling constants" or "thermodynamic fields" [5]. A renormalization group transformation, as written formally in (6), is then a mapping in this space of Hamiltonians. In a discrete renormalization group, the rescaling factor b is fixed larger than unity; in a continuous renormalization group one considers $b \to 1$ and, under iteration, then generates families of continuous trajectories or flows in $\mathbb{H}$ [5].

One may write a rather general class of renormalization groups by introducing a renormalization group kernel, $\mathcal{R}_{N'N}(\{s'\};\{s\})$, in which the $s_j'(j = 1,2,\ldots N')$ are new or renormalized spins. Indeed, to obtain nontrivial behavior at interesting fixed points, rescaling or renormalization of the spin variables, in addition to the spatial rescaling (6), is normally essential [5,6]. Given a kernel, the renormalization transformation (7) may be defined more explicitly by

$$e^{\bar{\mathcal{H}}'(\{s'\})} = \mathrm{Tr}_N^s\left\{\mathcal{R}_{N'N}(\{s'\};\{s\})e^{\bar{\mathcal{H}}(\{s\})}\right\}. \tag{10}$$

The unitarity relation (9), and hence the basic flow equation (8), are ensured if the kernel satisfies

$$\mathrm{Tr}_{N'}^{s'}\left\{\mathcal{R}_{N'N}(\{s'\};\{s\})\right\} = 1 \qquad \text{all} \quad \{s\}\,. \tag{11}$$

Amongst important renormalization groups which can be cast in this form are the original momentum shell integration renormalization group of Wilson for models of type (c) with continuous spins [6], and various real space groups using lattice cells introduced by Kadanoff and Niemeijer and van Leeuwen which can be defined also for fixed length spin models of classes (a) and (b) (see [7]).

To employ a renormalization group one looks for nontrivial fixed point Hamiltonians, $\bar{\mathcal{H}}^*$. Because the correlation length, ξ, for the decay of correlations and hence the corresponding susceptibility χ, are infinite at a fixed point [by virtue of (6) which yields $\xi' = \xi/b$] a nontrivial fixed point Hamiltonian, $\mathcal{H}^*$, is necessarily critical. (At a trivial fixed point $\xi \cong 0$ and there are no couplings.) One then anticipates the existence of a critical manifold in $\mathbb{H}$, typically of co-dimension 2, divided up into domains of attraction for particular fixed points. If the initial critical point Hamiltonian under study flows into a fixed point $\mathcal{H}^*$, under the action of $\mathbb{R}$, then, granted sufficient smoothness in the flow equation $(8)_*$, the original critical exponents will be the same as those of $\mathcal{H}^*$. Thus domains of attraction on the critical manifold correspond to universality classes of Hamiltonians all exhibiting the same values of α, β, γ, etc. [4,5,6].

Linearization of $\mathbb{R}$ about a fixed point $\mathcal{H}^*$ will, in general, yield an infinite spectrum of critical operators Q_j and associated eigenvalues $\Lambda_j = b^{\lambda_j}$. Those operators for which λ_j is positive are called relevant; they correspond, typically, to the energy E, and magnetization M. The relevant eigenvalues determine the basic critical exponents via

$$2 - \alpha = d/\lambda_1 \qquad \text{and} \qquad \beta + \gamma = \lambda_2/\lambda_1 . \tag{12}$$

Furthermore, the scaling law (2), and its extensions, follows by iterating the linearized renormalization group and utilizing the flow equation (8) [5]. Operators for which λ_j is negative are irrelevant: they describe the differences between Hamiltonians within the same universality class but influence only the corrections to the asymptotic scaling laws. On the borderline, operators with $\lambda_j = 0$ are called marginal; when they occur, they can lead to logarithmic factors, like $(\ln|t|)^{\mu}$ in the critical behavior (1), etc. to breakdown of the simple scaling laws, and to loss of discrete universality classes.

As may be seen easily by looking at examples, the explicit construction of a renormalization group and the location of its fixed points is normally a task of major difficulty. One may introduce readily various approximate schemes, in which the dimensionality of $\mathbb{H}$ is artificially truncated. This approach using real space or cell renormalization groups is popular and has had some successes. However, it has proved rather hard to make systematic improvements or to gain any control of the accuracy or even of the qualitative reliability of such schemes. [Recently, Griffiths and Pearce [9] have also identified some profound mathematical problems with some of the most practiced approaches.]

On the other hand, it has been discovered that the momentum shell integration renormalization groups can be constructed in a systematic way by introducing a dimensionality expansion. Explicitly, one first regards the dimensionality of space, d, as a continuous variable. Next, in the most important cases, one introduces the expansion parameter [8]

$$\epsilon = 4 - d . \tag{13}$$

A perturbation calculation, following a field-theoretic mold [5, 6,8], then reveals that the renormalization group recursion relations can be constructed explicitly to a given order ϵ^m (m=1,2,...). Furthermore, at fixed order, the recursion relations close, so that the dimensionality of $\mathbb{H}$ becomes, effectively, finite. The leading order calculation suffices to show that for $d > 4$ ($\epsilon < 0$) there is only one nontrivial fixed point; this simply generates the classical phenomenological critical behavior with $\beta = \frac{1}{2}$, $\gamma = 1$, etc. Thus, the naive theory turns out to be right in a space of more than four dimensions! However, at $d = 4$ or $\epsilon = 0$, a one-sided bifurcation occurs and a new, nontrivial fixed point splits off, which on the critical manifold, is now the stable, attractive fixed point. This new fixed point yields eigenvalues λ_j, and hence critical exponents, which can be calculated as power series in $\epsilon > 0$ [5-8]. Thus, for the susceptibility exponent, for example, one finds for models of class (c) with n-component continuous spins the result

$$\gamma = 1 + \frac{(n+2)}{2(n+8)}\epsilon + \frac{(n+2)(n^2+22n+52)}{4(n+8)^3}\epsilon^2$$
$$+ \frac{(n+2)}{8(n+8)^3}\left[(n+2)^2 + 4\frac{(55n^2+268n+424)}{(n+8)^2}\right.$$
$$\left. + 24\frac{(n+2)(n+3) - (10n+44)\zeta(3)}{(n+8)}\right]\epsilon^3$$
$$+ O(\epsilon^4), \qquad (14)$$

valid for $\epsilon > 0$. [As indicated, one has $\gamma \equiv 1$ for $\epsilon < 0$.] In the limit $n \to \infty$ (which corresponds to the exactly soluble spherical model [5]) the ϵ-expansion series converge for $\epsilon < 2$ (i.e., $d > 2$) and yield the known exact answers. For general n it appears that the expansion is only asymptotic, but rather good numerical results can be obtained even at $\epsilon = 1$, corresponding to $d = 3$, by truncating at order ϵ^2 or by using Borel or Padé summation methods. The ϵ expansion has also proved very useful in elucidating the role of long-range interactions and of a wide range of symmetry breaking terms, in the original, physical Hamiltonian. Dimensionality expansions for $\epsilon > 0$ in the forms $2+\epsilon$, $6-\epsilon$, $3-\epsilon$, $4\frac{1}{2}-\epsilon$ have also been developed for various classes of systems [5,7] and provide much further insight [7].

The renormalization group approach is by now an important weapon in the armoury of the practicing mathematical physicist [4]. Nevertheless, much remains to be done to place the method on a fully satisfactory and complete, even if not fully rigorous mathematical basis and to delineate its limitations and subtleties more clearly [9].

References

[1] Fisher, M. E.: 1967, Rept. Prog. Phys. 30, pp. 615-731.

[2] Stanley, H. E.: 1971, "Introduction to Phase Transitions and Critical Phenomena," Oxford Univ. Press,

[3] Nelson, D. R., and Fisher, M. E.: 1975, Ann. Phys. (N.Y.), pp. 226-274.

[4] Pfeuty, P., and Toulouse, G.: 1977, "Introduction to the Renormalization Group and to Critical Phenomena" (Wiley & Sons).

[5] Fisher, M.E.: 1974, Revs. Mod. Phys. 46, pp. 597-616.

[6] Wilson, K. G., and Kogut, J.: 1974, Phys. Repts. 12 C, pp. 75-199.

[7] Domb, C., and Green, M.S. (Eds.): 1976, "Phase Transitions and Critical Phemomena", vol. 6 (Academic Press).

[8] Wilson, K. G., and Fisher, M. E.: 1972, Phys. Rev. Lett. 28, pp. 240-243.

[9] Griffiths, R. B., and Pearce, P. A.: 1979, J. Statist. Phys. 20, pp. 499-545.

LATTICE RENORMALIZATION OF NON-PERTURBATIVE QUANTUM FIELD THEORY

Barry M. Mc Coy

Institute for Theoretical Physics, State University of New York, Stony Brook, New York 11794, U.S.A.*)

and

Tai Tsun Wu
Max-Planck-Institut für Physik und Astrophysik, München, Fed. Rep. of Germany ; Gordon McKay Laboratory, Harvard University, Cambridge, Massachusetts 02138, U.S.A.**)

ABSTRACT : We discuss lattice renormalization as a basis for understanding non-perturbative effects in quantum field theory. We first study two solvable examples of quantum field theory in two dimensions where lattice renormalization may be explicitly carried out, namely the Ising field theory and the Thirring model. The Ising theory demonstrates that, with the same Lagrangian, two different relativistic quantum field theories can be constructed by using different lattice renormalizations. The Thirring model without a bare mass exhibits the remarkable phenomenon of dynamic generation of mass accompanied by dynamic determination of the coupling constant. We then turn to four dimensions and, after contrasting Euclidean with Minkowski lattice renormalization, discuss their application to non-perturbative phenomena in the Wess-Zumino model, scalar electrodynamics, and extended non-Abelian gauge theory.

* Work supported in part by the United States National Science Foundation under grant No. DMR-7707863A01.

** Permanent address.
Work supported in part by the United States Department of Energy under contract No.EY-76-S-02-3227.

C. Bardos and D. Bessis (eds.), Bifurcation Phenomena in Mathematical Physics and Related Topics, 69–90.

1 - INTRODUCTION

Renormalization is an essential ingredient in the construction of a quantum field theory. It is primarily renormalization which makes quantum field theory different from classical field theory. The purpose of this lecture is to illustrate and discuss the role of renormalization in non-perturbative phenomena such as the construction of different phases of a quantum field and the dynamic generation of mass.

Renormalization was first invented to give meaning to each term in a perturbation expansion[1]. The original idea is to introduce an ultraviolet momentum cutoff Λ, to carry out the perturbation expansion, and then to reinterpret all essential dependences on Λ as a redefinition of masses and coupling constants [1-3]. Since such renormalization of the perturbation series is quite well understood, it is not difficult, by using for example diffentiation with respect to momenta [2], to hide this cutoff Λ completely such that it formally does not appear at all [2,4]. Nevertheless, whenever a new situation is encountered, it is almost always desirable to study the Λ dependence. There are several different schemes of introducing Λ ; the most popular ones are perhaps

i) putting the quantum field theory on a lattice (lattice renormalization [5]) ;
ii) Pauli-Villars method [6] of multiplying propagators by $\Lambda^2/(p^2 + \Lambda^2)$; and
iii) replacing the dimensionality d of the quantum field by d-ε (dimensional regularization [7]).

In the context of perturbation theory, each of these is useful. However, as emphasized by Symanzik [8], they are quite different once we go beyond perturbation theory. In particular, no one has yet succeeded in making sense of dimensional regularization outside of perturbation theory. Furthermore, although the Pauli-Villars method has the apparent advantage of preserving manifest Lorentz invariance, the introduction of higher derivatives into the kinematic part of the Lagrangian is awkward to deal with and theoretically quite artificial. In addition, the only tool developed to implement the Pauli-Villars scheme is the formal summation of perturbation theory [8]. Therefore, neither of these two renormalization schemes is particularly useful for the study of non-perturbative effects.

Lattice renormalization, however, is clearly independent of perturbation expansion. Furthermore, it has the following two great virtues.

(1) It is natural. In classical field theories, an elementary method to treat the field equations (partial differential equations) is to replace them by partial difference equations. The reason is

that the process of shrinking the lattice distance to zero is surely the most obvious way of giving meaning to the concept of a continuum. This reason holds for quantum field theory as well as classical field theory.

(2) At least two distinct, non-trivial quantum field theories have been constructed exactly by lattice renormalization. They are the two-dimensional Ising field theory [9,10] and the Thirring model [11,12]. It is virtually certain that the Schwinger model [13], the vector meson model [14], and some of their combinations can also be treated by lattice renormalization, but to our knowledge this has not been carried out completely systematically.

Both the Ising field theory and the Thirring model exhibit phenomena which are intrinsically non-perturbative and non-classical in nature.

(A) The Ising field theory gives the first explicit illustration of the fact that one Lagrangian may give rise to different quantum field theories depending on how the renormalization is carried out [9]. This is the phenomenon of phases of a quantum field, and will be discussed in Section 2. This model also demonstrates that quantum fluctuations can completely destroy the relevance of classical solutions of the equations of motion.

(B) The massless Thirring model demonstrates explicitly the role renormalization plays in the dynamic generation of mass [15]. By the massless Thirring model we mean the one where the mass term in the Lagrangian density is absent. The existence of this remarkable phenomenon of dynamic mass generation has only been realized recently and will be discussed in Section 3.

It has become clear that the concepts of dynamic mass generation and of phases of a quantum field are necessary for understanding nature. In our two examples, these features are present and are non-perturbative and non-classical in character. Moreover, in the Thirring model, it is most striking that the requirement that dynamic mass generation occurs determines the value of the coupling constant. Such dynamic determination of coupling constants has long been a dream of physicists. In the remainder of this lecture, we shall attempt to apply the intuition that we have gained from these two-dimensional models to more realistic four-dimensional field theories. In particular, we discuss the Yukawa interaction, the ϕ^4 coupling, the Wess-Zumino supersymmetric model, scalar electrodynamics, and non-Abelian gauge theories. In particular, we suggest that the concept of a gauge theory may be usefully extended to take advantage of the full range of options provided by lattice renormalization.

Part I : QUANTUM FIELD THEORY IN TWO DIMENSIONS

2 - ISING FIELD THEORY

The first example we consider is the Ising field theory [9,10]. The non-perturbative renormalization effects to be exhibited are the existence of multiple phases coming from the same Lagrangian and the corresponding lack of relevance of classical paths.

The action for the two-dimensional Ising model is [16]

$$S = \frac{1}{2} K \sum_{\vec{x}} \sum_{\{\vec{\delta}\}} \sigma(\vec{x})\sigma(\vec{x} + \vec{\delta}), \tag{2.1}$$

where $\vec{x}$ takes values on a square lattice of spacing a, the sum $\{\vec{\delta}\}$ is over the nearest neighbors, and $\sigma(\vec{x})$ is constrained to take only the values +1 and -1. On the lattice, multi-spin correlation functions are calculated from the usual formula [17]

$$< \sigma(\vec{x}_1)\ldots\sigma(\vec{x}_n)> = Z^{-1} \sum_{\{\sigma\}} \sigma(\vec{x}_1)\ldots\sigma(\vec{x}_n)\, e^S, \tag{2.2}$$

where

$$Z = \sum_{\{\sigma\}} e^S. \tag{2.3}$$

Strictly speaking (2.3) has meaning only for a finite lattice ; but for the correlation functions (2.2) we will pass immediately to the thermodynamic limit of an infinite lattice.

In order to construct a relativistic quantum field theory from the action (2.1) we define a set of renormalized Green's function by [1]

$$< \sigma(\vec{x}_1)\ldots\sigma(\vec{x}_n) >_{ren} = \lim_{a\to 0} Z_3(a)^{-n/2} < \sigma(\vec{x}_1)\ldots\sigma(\vec{x}_n) >, \tag{2.4}$$

where on the right hand side the parameter K in (2.1) is allowed to depend on the lattice spacing a. If the functions K(a) and $Z_3(a)$ can be chosen such that the renormalized Green's functions (2.4) exist, are non-zero, and are rotationally invariant, then these Green's functions are those of a quantum field theory.

The action (2.1) may be rewritten in the form

$$S = -\frac{1}{4} K a^2 \sum_{\vec{x}} \sum_{\{\vec{\delta}\}} \left[\frac{\sigma(\vec{x}) - \sigma(\vec{x} + \vec{\delta})}{a}\right]^2 + \text{constant} \tag{2.5}$$

which, as $a \to 0$, formally approaches $-\frac{1}{2}K\int(\partial_\mu\sigma)^2 d^2x$. Accordingly, the Ising field theory may be formally regarded as either the $N = 1$ non-linear σ model or the Euclidean field theory with the Lagrangian density

$$-\frac{1}{2} K (\partial_\mu \sigma)^2 - \lambda \sigma^4 + m_o \sigma^2 \tag{2.6}$$

in the limit where $m_o = 2\lambda$ and $\lambda \to \infty$.

The existence and explicit calculation of the renormalized Green's functions (2.4) are first shown in Ref. [9].

It has been known for a long time that, as a function of K, a second-order phase transition [18,19] occurs in the two-dimensional Ising model when

$$K = K_c = \frac{1}{2} \ln (1 + \sqrt{2}). \tag{2.7}$$

For values of K larger than K_c there is a long-range order in the sense that, for a fixed and $\vec{x} \to \infty$ (or for $\vec{x}$ fixed and $a \to 0$), [20]

$$M^2(K) = \lim \langle\sigma(0)\, \sigma(\vec{x})\rangle = \left|1 - (\sinh 2K)^{-4}\right|^{1/4}. \tag{2.8}$$

More precisely for $\vec{x}$ fixed and $a \to 0$, we have [21] when $K_c < K$

$$\langle \sigma(0)\, \sigma(\vec{x})\rangle \sim M^2(K) \{1 + C_{1<} (\theta,K) [\kappa_< (\theta,K) |\vec{x}|/a]^{-2} e^{-|\vec{x}|\kappa_<(\theta,K)/a}\}, \tag{2.9}$$

while for $K_c > K$

$$\langle\sigma(0)\, \sigma(\vec{x})\rangle \sim M^2(K)\, C_{1>}(\theta,K) \;[\kappa_>(\theta,K) |\vec{x}|/a]^{-1/2} e^{-|\vec{x}|\, \kappa_>(\theta,K)/a}. \tag{2.10}$$

Here θ is the angle $\vec{x}$ makes with the lattice axis, and $C_{1>}$, $\kappa_>$, $C_{1<}$, and $\kappa_<$ are given in Appendix A.

In (2.9) and (2.10), the dependence of C_1 and κ on θ implies that for fixed $K \neq K_c$ it is impossible to construct a rotationally invariant two-point Green's function. However, when $K \to K_c$ we explicitly find that

$$M(K) \sim 2^{7/16}|K - K_c|^{1/8}, \tag{2.11}$$

and

$$\kappa_{>,<} (\theta,K) \sim C_{2>,<}|K - K_c|, \tag{2.12}$$

where $C_{2>}$ and $C_{2<}$ are constants independent of θ. Furthermore,

$C_{1>}$ (θ,K) and $C_{1<}$ (θ,K) both approach constants independent of θ. Therefore, it is necessary to choose the a dependence of K(a) according to

$$C_{2>,<}|K(a) - K_c|/a = m, \tag{2.13}$$

where m is a constant independent of a and let

$$Z_3(a) = 2^{7/16}|K(a) - K_c|^{1/4} \tag{2.14}$$

in order that, when the leading-order approximations to (2.9) and (2.10) are used in (2.4), the limits exist and are rotationally invariant. This necessary condition was proven to be sufficient for the two-point function in 1973 [22,23] and the extension to the general n-point functions was carried out in 1977 [9].

The most important feature of this example is that we have constructed not one but two different Lorentz-invariant quantum field theories [9] : one by taking the limit $K(a) \to K_c+$ and a second one by $K(a) \to K_c-$. Since in both cases K(a) in the limit takes on the value K_c, these two field theories have the same Lagrangian, including the same bare coupling constants. This phenomenon that two different lattice renormalizations of the same Lagrangian can give rise to different field theories is referred to as the existence of different phases of the same quantum field.

Since different renormalizations lead to different field theories from the same Lagrangian, it is clear that any semi-classical approximation scheme based on the minimum-action solution of the classical field equations can at best be relevant to only one of these quantum theories. Moreover, in the Ising field theory the field variables can take on only the values +1 and -1, and hence the only possible "classical" configurations are $\sigma(\vec{x}) = +1$ for all $\vec{x}$ and $\sigma(\vec{x}) = -1$ for all $\vec{x}$. Clearly neither of these "classical" configurations has any relevance to either phase of the Ising field theory. Note that this discussion is completely at odds with the usual method of applying Higgs' mechanism [24] for mass generation.

A vivid difference between the two phases of the Ising field theory may be seen from the leading approximations to the two-point functions constructed from (2.9) and (2.10) respectively. If Fourier transform is applied, the connected two-point Green's function $G_{2>}{}^c(p)$ for the phase $K_c > K$ has a pole as the singularity nearest to the real p-axis. This is the signal for the existence of a particle in this phase and is intimately related to the prefactor $|\vec{x}|^{-1/2}$ in (2.10). On the other hand, the leading singularity in $G_{2<}{}^c(p)$ for the phase $K_c < K$ is not a pole but a branch point, as reflected in the prefactor $|\vec{x}|^{-2}$ of (2.9). Therefore, in this phase the usual particle interpretation breaks down and novel physical phenomena appear. These phenomena are describable in terms of a new object, the indeterminate mass particle, and have been discussed phenomenologically elsewhere [25].

3. THIRRING MODEL

The second example we consider is the Thirring model. The non-perturbative renormalization effects to be exhibited are the dynamic generation of mass and the corresponding dynamic determination of the coupling constant.

The continuum Lagrangian density for the massive Thirring model is formally given by [11,12]

$$L_{MT}(x) = i\,\bar{\Psi}\gamma_\mu\partial^\mu\Psi - g(\bar{\Psi}\gamma_\mu\Psi)(\bar{\Psi}\gamma^\mu\Psi) - m_o\bar{\Psi}\,\Psi, \tag{3.1}$$

where we use the metric $g^{00} = -g^{11} = 1$.

This Lagrangian density is only a formal object because the interaction term is singular and ambiguous without a renormalization procedure. We will use the renormalization scheme of Luther [26] by retaining a continuous time variable and putting the spacial variable on a lattice. Correspondingly, we will work with the Hamiltonian rather than the Lagrangian. If the Sommerfield [12] definition of the current and the coupling constant is used, then the lattice Hamiltonian for the massive Thirring model, as given by Luther [26] and also by Lüscher [27], is

$$\begin{aligned} H_{MT} = \sum_{n=-N+1}^{N} \Big[& \frac{i}{2a} V\,(\phi_n^\dagger\phi_{n+1} - \phi_{n+1}^\dagger\phi_n) \\ & + (-1)^n \frac{m_o}{2}(\phi_n^\dagger\phi_{n+1} + \phi_{n+1}^\dagger\phi_n) \\ & - \frac{G}{2a}(\phi_n^\dagger\phi_n - \tfrac{1}{2})(\phi_{n+1}^\dagger\phi_{n+1} - \tfrac{1}{2})\Big] \end{aligned} \tag{3.2}$$

where

$$\{\phi_n, \phi_\ell\} = \{\phi_n^\dagger, \phi_\ell^\dagger\} = 0,$$

$$\{\phi_n, \phi_\ell^\dagger\} = \delta_{n\ell}\,, \tag{3.3}$$

and, for $g \geq -\pi/2$,

$$G = -\frac{4\varepsilon}{\pi}\cot\varepsilon, \tag{3.4}$$

$$V = \frac{2\varepsilon}{\pi\sin\varepsilon} \tag{3.5}$$

with

$$\epsilon = \frac{\pi\left(\frac{1}{2} + \frac{g}{\pi}\right)}{1 + \frac{g}{\pi}} \tag{3.6}$$

This lattice Hamiltonian (3.2) may be reduced to that of a well-known many-body problem by using the Jordan-Wigner transformation [28]

$$\phi_n^{\dagger} = e^{i\frac{\pi}{4}(2N+1)} \sigma_n^{+} \prod_{j=-N+1}^{n-1} (i\,\sigma_j^{z}), \tag{3.7}$$

where σ^i are the Pauli spin matrices and $\sigma^{\pm} = \sigma^x \pm i\sigma^y$. If periodic boundary conditions are used in (3.2), then up to a boundary term the Hamiltonian becomes [29-31]

$$H_{MT} = H_{xyz} = -\frac{1}{2} \sum_{n=-N+1}^{N} (J_x\sigma_n^{x}\sigma_{n+1}^{x} + J_y\sigma_n^{y}\sigma_{n+1}^{y} + J_z\sigma_n^{z}\sigma_{n+1}^{z}), \tag{3.8}$$

where

$$J_x = \frac{V}{2a} + \frac{m_o}{2},\ J_y = \frac{V}{2a} - \frac{m_o}{2},\ \text{and}\ J_z = \frac{G}{4a}. \tag{3.9}$$

Lattice renormalization for the Thirring model may now be discussed in detail because a great deal is known about the spectra of the low-lying excited states of the XYZ Hamiltonian H_{xyz} [32]. In the limit $N \to \infty$ for an infinite chain, the energy gap between the first excited state and the ground state corresponds to the lightest particle or excitation present in the system. Since lengths (energies) are measured on the scale of the lattice spacing a, Lorentz invariance cannot be recovered in the limit $a \to 0$ unless either the gap is identically zero or the gap multiplied by a approaches zero as $a \to 0$.

It is known that if [32]

$$J_x = J_y \geq |J_z|, \tag{3.10}$$

there is no gap in the spectrum of H_{xyz}. A gap opens up if either

$$J_x \neq J_y \tag{3.11}$$

or

$$0 < J_x = J_y < |J_z|. \tag{3.12}$$

The construction based on (3.11) was used by Luther [26] to renormalize the massive Thirring model, and in this way a spectrum of bound states was found. We concentrate here on the construction based on (3.12). A comparison of (3.9) with (3.12) shows that this is the case of $m_o = 0$, the massless Thirring model.

If $0 < J_x = J_y < J_z$ then the XYZ model is frozen in the ferromagnetic ground state. Thus we will consider only the case

$$0 < J_x = J_y < -J_z . \tag{3.12a}$$

In this case the spectrum of the lowest lying excited states is given by [32]

$$E(q) - E_o = \frac{V}{\pi a} K(k_1)(1 - k_1^2 \cos^2 q)^{1/2} \sinh \lambda , \tag{3.13}$$

where

$$\Delta = J_z/J_x = G/(2V) ,$$
$$\lambda = \cosh^{-1} (-\Delta) , \tag{3.14}$$

$K(k_1)$ is the complete elliptic integral of the second kind of modulus k_1, and this modulus is calculated from

$$\pi K[(1 - k_1^2)^{1/2}] = \lambda K(k_1) . \tag{3.15}$$

If $-J_z \to J_x$ then $\lambda \to 0$. In this limit $k_1 \to 1$, and we may approximate (3.15) by

$$\pi^2 [\ln 16(1 - k_1^2)^{-1/2}]^{-1} = \lambda . \tag{3.16}$$

Thus for small λ

$$1 - k_1^2 \sim 2^{-8} e^{-\pi^2/\lambda} \tag{3.17}$$

and, denoting the gap $E(0) - E_o$ by M, we find from (3.14)

$$M = \frac{V}{\pi a} K(k_1) (1 - k_1^2)^{1/2} \sinh \lambda$$
$$\sim \frac{V}{32a} \lambda e^{-\pi^2/\lambda} . \tag{3.18}$$

Therefore, if we choose λ as a function of V/a such that, when $V/a \to \infty$, $\lambda \to 0$ with M fixed and if we define the momentum p by scaling from q

$$k_1^2(1 - k_1^2)^{-1} q^2 = p^2/M^2 , \tag{3.19}$$

we find that

$$E(p) - E_o = (M^2 + p^2)^{1/2} , \tag{3.20}$$

which is the proper relativistic dispersion relation. Therefore, there is dynamic mass generation [15] in the massless Thirring model when $g = -\pi/2$. This mechanism of dynamic mass generation is qualitatively different from both the perturbative mass generation [13,33] of the Schwinger model and the classical Higgs mechanism [24].

4 - COMPARISONS AND DISCUSSIONS

The massive Thirring model, as given by the Lagrangian density (3.1) with a non-zero value of m_o, fails to make sense [34] when $g < -\pi/2$. On the other hand, previous work on the massless Thirring model seemed to indicate that there was no problem [11,12] so long as $g > -\pi$. This latter result was obtained by calculating certain Green's functions for $g > -\pi/2$, and then observing that these Green's functions are perfectly sensible as long as g is larger than $-\pi$. Since the ground-state energy per spin is not analytic [31] at $g = -\pi/2$, analytic continuation beyond this point cannot be correct. We therefore conclude that neither the massive nor the massless version of the Thirring model is a relativistic field theory when $g < -\pi/2$.

In order to have a better understanding of the special role played by this value $-\pi/2$, let us contrast the two cases $\Delta > -1$ and $\Delta < -1$ for the H_{xyz} of (3.8) with $J_x = J_y$. For the infinite chain $N \to \infty$, there is no gap in the former case but there is a gap in the latter case between the ground state and the excited states. *This gap is of paramount importance.* If Δ is fixed at a value less than -1, the presence of this finite gap means that Green's functions probe distances of the order of the lattice spacing, and hence Lorentz invariance cannot be recovered by taking the limit $a \to 0$.

Let D be the domain in the space of parameters where there is no gap. In the case of the massless Thirring model, D is given by $\Delta \geq -1$; in the case of the massive Thirring model, D is given by $\Delta \geq -1$ and $m_o a = 0$. From the discussion of the preceding section, we see that there are two distinct ways of constructing a relativistic quantum field theory :

(A) Choose parameters in the domain D ; keep them fixed while taking the limit $a \to 0$.

(B) Choose parameters *outside* of the domain D ; let them approach the boundary of D while $a \to 0$.

In procedure (A), the resulting quantum field theory is massless, at least so far as the particular particle under consideration is concerned. This is the way of constructing the massless phase of the massless Thirring model. On the other hand, in procedure (B), the resulting quantum field theory is, in general, massive. For the massive Thirring model, the limiting point on the boundary of D is $m_o a = 0$ and $\Delta > -1$; for the massive phase of the massless Thirring model discussed in Section 3, the limiting point is $m_o a = 0$ and $\Delta = -1$.

We reemphasize that the important point is the appearance of the gap. In some sense, this appearance of a gap may perhaps be thought of as an instability. If so, there may be a link between this instability and the result of Coleman [34] that the non-existence of the massive Thirring model when $g < -\pi/2$ can be attributed to ground-state collapse.

Finally, it must be remarked that at the point $\Delta = -1$ there is an infinitude of renormalizations possible corresponding to letting $J_x \to J_y$ and $J_z \to J_y$ in a joint fashion. Thus, even in the massive Thirring model at $g = -\pi/2$ *some* of the physical mass may be dynamically generated. Whether or not this modifies the Green's functions remains an interesting open question. If the Green's functions do depend on the method of joint approach, then there is an infinite number of field theories that can be constructed at $g = -\pi/2$. On the other hand, if the Green's functions do not depend on the method of approach, then at $g = -\pi/2$ the massive Thirring model and the massless Thirring model describe the same theory.

Part II : QUANTUM FIELD THEORY IN FOUR DIMENSIONS

5 - LATTICE RENORMALIZATION

We now turn to the application of lattice renormalization to four-dimensional field theories.

In the preceding sections lattice renormalization is carried out in two distinct ways. Either the time is continued to purely imaginary values [35] and the resulting four-dimensional Euclidean space is put on a lattice (Ising field theory) or the metric is left Minkowski and only the coordinate space components are put on a lattice (Thirring model). We discuss these two procedures separately.

A. Euclidean Renormalization

(1) Take the classical continuum action S and put it on a four-dimensional Euclidean lattice. In general the action depends on several numerical parameters : bare masses, bare coupling constants, and an overall scale.

(2) Calculate Green's functions by means of the general formula [36,37]

$$\langle\phi(x_1)\ldots\phi(x_n)\rangle = Z^{-1}\int\{\Pi d\phi_i\}\phi(x_1)\ldots\phi(x_n)\, e^S, \qquad (5.1)$$

with

$$Z = \int \{\Pi d\phi_i\}\, e^S. \qquad (5.2)$$

In (5.2) there is an independent field ϕ_i for each site of the lattice, and we integrate these ϕ_i over all allowed values. The ϕ_i commute (anticommute) for bosons (fermions). At this stage we take the limit of infinite lattice ; thus all infrared cutoffs are removed.

(3) Consider the two-point function on the lattice. In general, as the distance between the two points $|x_1 - x_2|$ approaches infinity, the two-point function approaches its limiting value as $e^{-\kappa|x_1 - x_2|}$. The inverse correlation length κ is a function of the parameters of the Lagrangian. To construct a field theory we must find those values of the parameters where $\kappa = 0$ on the lattice.

(4) Renormalization is carried out by letting the parameters approach (or equal) those special values where $\kappa = 0$ as the lattice spacing $a \to 0$. Renormalized Green's functions are defined by [1]

$$\langle\phi(x_1)\ldots\phi(x_n)\rangle_{ren} = \lim_{a\to 0} Z_3(a)^{-n/2}\langle\phi(x_1)\ldots\phi(x_n)\rangle , \qquad (5.3)$$

where $Z_3(a)$, called the wave-function renormalization constant, is chosen so that the limit in (5.3) exists and is not zero. If the renormalized Green's functions so defined are rotationally invariant, then after a Wick rotation a relativistic quantum field theory has been constructed from the classical action.

B. Minkowski Renormalization

(1') Calculate the Hamiltonian which corresponds to the Lagrangian and put the three space coordinates on a lattice.

(2') The many-body wave function for the ground state is calculated by solving the Schrödinger equation, and Green's functions are obtained by carrying out integrations over this ground-state wave function.

(3') The excitation spectrum of the Hamiltonian is calculated and the values of the parameters are found which make the mass gap vanish.

(4') Renormalization is now carried out as in (5.3) directly in Minkowski space.

There are at least four ways in which the above procedure, either version A.) or version B.), can fail to give a non-trivial field theory.

(a) There may be no values of the Lagrangian parameters which make the inverse correlation length (or mass gap) vanish. Recent Monte Carlo calculations [38] indicate that this may indeed be the case for the Z_2 lattice gauge theory.

(b) There may not exist a $Z_3(a)$ which renormalizes all the n-point functions.

(c) The resulting Green's function may not be Lorentz invariant.

(d) Even if the Green's functions are Lorentz invariant, they still may not be different from those of a free field.

We say that the quantum field theory is trivial if the difficulty (d) occurs. Several decades ago, Landau [39] conjectured that quantum electrodynamics is trivial, because summation of leading logarithms gives an unacceptable singularity, the Landau pole. Since then, there has been a great deal of work done on possible triviality of field theories [40], especially in connection with the ϕ^4 theory in four dimensions. Very recently, Baker and Kincaid [41] gave a most convincing numerical demonstration that this ϕ^4 theory in four dimensions is indeed trivial. This will be discussed by Baker in this School.

6. MASS GENERATION AND INSTABILITY

We now wish to combine the general procedure of lattice renormalization with the intuition obtained from the models solved in two dimensions in order to gain insight into non-perturbative phenomena in four dimensions. Because there are no solved four-dimensional models, our discussion must perforce be more speculative.

From the Thirring model we abstract the intuition that lattice renormalization can convert an instability into mass generation. The term instability in connection with field theory is somewhat vague. We shall here take inspiration from the far better understood situation of instability [42] in N-body non-relativistic quantum mechanics where a system is said to be unstable if the ground-state energy is not bounded below as

$$E_0(N) > -AN, \tag{6.1}$$

where A is independent of N. Systems which do not obey (6.1) are said to undergo catastrophic collapse, and they have no statistical mechanics. Their analogues in quantum field theory are prime candidates for instabilities to be converted into mass generation.

Consider first the famous interaction of Yukawa

$$g\, \bar{\Psi}\, \Psi\, \phi \tag{6.2}$$

where Ψ is a Fermi field and ϕ is a Bose field. The lowest-order non-relativistic potential for such an interaction is a purely attractive one between the fermions. In three or four dimensions, it is known that the repulsion due to the exclusion principle is not sufficient to overcome the attractive interaction [43], and thus the ground state fails to satisfy (6.1).

This argument for collapse is, of course, only heuristic but is very compelling. The catastrophe can never be seen in any finite order of perturbation theory, because it is a property of an N-particle system as $N \to \infty$.

We thus argue that the pure Yukawa field theory does not exist.

In order to stabilize this attractive collapse some short-distance repulsion is needed. The simplest candidate is a $\lambda\phi^4$ self-interaction of the bosons. Therefore we consider the following Lagrangian density

$$L = i\, \bar{\Psi}\gamma_\mu \partial^\mu \Psi - (\partial_\mu \phi)^2 - m_F \bar{\Psi}\Psi - m_B\, \phi^2 - g\, \bar{\Psi}\Psi\phi - \lambda\phi^4 . \tag{6.3}$$

We conjecture that the collapse at $\lambda = 0$ persists for some non-zero range of λ. Thus there is a domain D' in the (g,λ) plane, including the g axis, such that the system collapses for every point in D'. If D is the complement of D', then the system is stable for every point of D. If the analogue with the Thirring model holds, there are at least two types of theories possible provided that they are not trivial. One is obtained by keeping the parameters g and λ on the boundary of D while the limit $a \to 0$ is taken, while in the other one the parameters remain in D' but approach the boundary of D as $a \to 0$. In the second case, the Thirring example indicates that dynamic mass generation can occur so that, even if $m_F = 0$ in the Lagrangian (6.3), the physical excitations may all be massive.

Similarly, since the theory is trivial for $g = 0$, we speculate that there is a domain $\tilde{D}$ which includes the λ-axis such that the theory is trivial for every point of $\tilde{D}$.

A more elaborate version of this model is that of Wess and Zumino [44]. In this model, there is a scalar boson A, a pseudoscalar boson B, and a Majorana fermion Ψ, and the Lagrangian is

$$L_{WZ} = -\frac{1}{2}(\partial_\mu A)^2 - \frac{1}{2}(\partial_\mu B)^2 - \frac{1}{2}i\,\bar{\Psi}\gamma_\mu\partial^\mu\Psi - \frac{1}{2}m^2A^2 - \frac{1}{2}m^2B^2$$

$$-\frac{1}{2}im\,\bar{\Psi}\Psi - g_B mA(A^2+B^2) - \frac{1}{2}g_B{}^2(A^2+B^2)^2$$

$$-\,ig_F\,\bar{\Psi}(A-\gamma_5 B)\Psi\ . \qquad (6.4)$$

By the previous argument catastrophic collapse happens at $g_B = 0$. This is a very special model, and Wess and Zumino have shown that if $g_F = g_B$ there is a symmetry between the bosons and the fermions. This is known as supersymmetry. Furthermore, Zumino [45] has shown that, at this supersymmetric point, the energy operator is positive definite and hence the model is surely stable. It is therefore natural to believe that the boundary of D is in this case given by $g_F = g_B$. We therefore speculate that, if

$$g_B \to g_F^- \qquad (6.5)$$

at the same time as $a \to 0$, the mechanism for dynamic mass generation operates. Furthermore, it is also attractive to speculate that in this case $\tilde{D}$ is the interior of D so that the line $g_B = g_F$ is not only the boundary of stability but also the boundary of triviality and hence the supersymmetric theory is the only one of physical interest.

Stability questions for the Wess-Zumino model have been studied by Curtright and Ghandour [46], using the method of effective potential evaluated in the two-loop approximation. They argue that the conjecture of $g_B = g_F$ being the boundary of stability is consistent with their calculation. However, since true catastrophic collapse cannot be seen in any finite order of perturbation their calculation is not definitive for the purpose of discussing dynamic mass generation.

There is a second model where non-relativistic N-body calculations indicate that a collapse takes place ; it is scalar electrodynamics whose Lagrangian density is

$$L = -\frac{1}{4}(F_{\mu\nu})^2 - \frac{1}{2}\,|\partial_\mu\phi - ieA_\mu\phi|^2 - \mu^2|\phi|^2 - \lambda|\phi|^4 \qquad (6.6)$$

where ϕ is a complex scalar field and

$$F_{\mu\nu} = \partial_\mu A_\nu - \partial_\nu A_\mu\ . \qquad (6.7)$$

First let $\lambda = 0$, and consider the corresponding theory of N charged bosons of mass μ interacting with a Coulomb potential only. This problem was beautifully treated by Dyson [47] in 1967 who showed that the ground-state energy is bounded as

$$E_o(N) < -A\ N^{7/5}\ , \tag{6.8}$$

where A is a positive constant. Therefore he argued that this field theory undergoes catastrophic collapse and hence does not exist.

With g replaced by e, the considerations after (6.3) apply equally well to scalar electrodynamics. In particular, in the second type of theories, if all the physical excitations are massive, the photon is turned into a vector meson.

The classical way to turn the photon into a vector meson is to let $\mu^2 < 0$; this is the Higgs mechanism [24]. The term $\mu^2|\phi|^2 + \lambda|\phi|^4$ in the Hamiltonian coming from (6.6) is treated as a one-body potential and, if $\mu^2 < 0$, the minimum is not at $\phi = 0$, but at $|\phi|^2 = -\mu^2/(2\lambda)$. This calculation is completely classical, and it is by no means clear whether quantum fluctuation destroys the argument. On a completely phenomenological level, it is fair to say that there is so far absolutely no evidence for any Higgs particle ; on the purely theoretical level, no one knows how to deal with the vast arbitrariness in the coupling of the Higgs system. By contrast, the mechanism of dynamic mass generation based on the Thirring model introduces no new particle.

As our final discussion of the situations where lattice renormalization can be applied, we turn to non-Abelian gauge theories [48,49]. In these theories, there are gauge fields (gluons) which are invariant under an internal non-Abelian symmetry group such as SU(N), and there may be fermion fields (quarks) and perhaps scalar fields (Higgs particles) which couple to the gauge fields with the internal symmetry maintained.

Perhaps the first question to be asked is whether the pure gauge theory with neither fermions nor scalars is an acceptable quantum field theory. In perturbation expansion, this system is plagued by severe infrared divergences as well as having to be ultraviolet renormalized. It is sometimes hoped that due to the infrared behavior a mass is somehow generated ; but it seems to us an open question whether or not a non-trivial theory exists at all for this pure gauge case. As already mentioned, recent Monte Carlo calculations on the Z_2 gauge theory in four dimensions indicate that this pure gauge theory does not exist [38]. We speculate similarly that pure non-Abelian quantum gauge theories, including in particular the special case of SU(N), do not exist except for the trivial cases. If so, there are two possibilities : either scalars or fermions (or both) must be added to the pure gauge theory. The most attractive idea is that fermions are necessary for non-trivial gauge theories to exist, as has been argued by Oehme and Zimmermann [50].

For definiteness we consider the original case studied by Yang and Mills [48], namely an SU(2) gauge field interacting with fermions

$$L = -\frac{1}{4}\left|\partial_\nu \vec{B}_\mu - \partial_\mu \vec{B}_\nu - g_B \vec{B}_\mu \times \vec{B}_\nu\right|^2 - \bar{\Psi}\,\partial_\mu(\gamma^\mu - ig_F\,\vec{\tau}.\vec{B}_\mu)\,\Psi - m_o\,\bar{\Psi}\,\Psi\,, \tag{6.9}$$

where we have written g_B and g_F separately for the boson and fermion couplings respectively. The only two cases considered in the literature so far are $g_F = 0$ (pure gauge theory) and $g_F = g_B$ (where gauge invariance holds for the complete Lagrangian). However, on the basis of lattice renormalization and dynamic mass generation as discussed here, we suggest that there is a third case which may be physically important, namely

$$g_F \to g_B \tag{6.10}$$

as the lattice spacing $a \to 0$. We call this case extended gauge theory. For this case, all manipulations at the level of the classical Lagrangian continue to hold without modification. Nevertheless, on the lattice, conservation of non-Abelian charge (for example color in the case of quantum chromodynamics) will fail by a small amount. We speculate that this infinitesimal violation of color conservation may be responsible for collapse, and that dynamic mass generation is possible. With the similarity to scalar electrodynamics in mind, it is most intriguing to ask whether this dynamic mass generation gives mass to the colorless states only and is accompanied by a complete suppression of the colored states.

ACKNOWLEDGMENT

We wish to thank Professor Daniel Bessis and Professor George Baker for helpful discussions and for hospitality extended at the Cargèse Summer Institute. We are also most grateful to Professor Wolfhart Zimmermann for fruitful conversations and for hospitality at the Max-Planck-Institut für Physik und Astrophysik where this paper was completed.

APPENDIX

We list here the functions $\kappa_>(\theta,K)$, $\kappa_<(\theta,K)$, $C_{1>}(\theta,K)$ and $C_{1<}(\theta,K)$ used in Section 2. Note that the angle θ is between 0 and $\pi/2$. These functions are obtained by specializing the results of Refs. [21] and [51].

Let $$z = \tanh K, \tag{A.1}$$

and

$$b = \left|\frac{1 - 6z^2 + z^4}{(1 + z^2)^2}\right| . \tag{A.2}$$

Then

$$\kappa_>(\theta,K) = A(\theta) \sin\theta + A(\tfrac{\pi}{2} - \theta) \cos\theta, \tag{A.3}$$

and

$$\kappa_<(\theta,K) = 2\kappa_>(\theta,K) , \tag{A.4}$$

with

$$\sinh A(\theta) = 2\left(\frac{2}{1-b^2}\right)^{1/2} \frac{b \sin\theta}{(1-b\cos 2\theta)^{1/2} + (1+b\cos 2\theta)^{1/2}} \tag{A.5}$$

Furthermore,

$$8\pi C_{1<}(\theta,K)\, \kappa_<(\theta,K)^{-2} = (2\pi)^{1/2}\, C_{1>}(\theta,K)\, \kappa_>(\theta,K)^{-1/2}$$

$$= \frac{1 - b^2}{\sqrt{2}\, b} [(1 - b\cos 2\theta)^{-1/2} + (1 + b\cos 2\theta)^{-1/2}] . \tag{A.6}$$

As $K \to K_c$ we find that $b \to 0$, and (2.12) follows immediately.

REFERENCES

[1] - S. Tomonaga : Prog. Theor. Phys. 1, 27 (1946) ; Phys. Rev. 74, 224 (1948) ; J. Schwinger : Phys. Rev. 73, 416 (1948) ; 74, 1439 (1948) ; R.P. Feynman : Rev. Mod. Phys. 20, 367 (1948) ; Phys. Rev. 74, 939 and 1430 (1948) ; F.J. Dyson : Phys. Rev. 75, 486 (1949) ;

[2] - J.C. Ward : Proc. Phys. Soc. (London) A64, 54 (1951) ;

[3] - N.N. Bogoliubov and O.S. Parasiuk : Acta Math. 97, 227 (1957) ; K. Hepp : Comm. Math. Phys. 2, 301 (1966) ;

[4] - Robert L. Mills and Chen Ning Yang : Prog. Theor. Phys. Suppl. 37, 507 (1966) ; T.T. Wu : Phys. Rev. 125, 1436 (1962) ;

[5] - K.G. Wilson : Phys. Rev. D10, 2445 (1974) ;

[6] - W. Pauli and F. Villars : Rev. Mod. Phys. 31, 434 (1949) ;

[7] - C.G. Bollini and J.J. Giambiagi : Nuovo Cim. 12B, 20 (1972); G.'t Hooft and M. Veltman : Nucl. Phys. B44, 189 (1972) ; See also C.G. Bollini, J.J. Giambiagi, and A.G. Dominguez : Nuovo Cim. 31, 550 (1964) ;

[8] - K. Symanzik : in "New Developments in Quantum Field Theory and Statistical Mechanics" (ed.M. Levy and P. Mitter, Cargèse , 1976 p. 265 ;

[9] - B.M. McCoy, C.A. Tracy, and T.T. Wu : Phys Rev. Lett. 38, 793 (1977) ; M. Sato, T. Miwa and M. Jimbo : Proc. Japan Acad. 53A, 6 (1977) ;

[10] - B. Schroer and T.T. Truong : Phys. Lett. 73B, 149 (1978) ;

[11] - W.E. Thirring : Ann. Phys. 3, 91 (1958) ; V. Glaser : Nuovo Cim. 9, 990 (1958) ; K. Johnson : Nuovo Cim. 20, 773 (1961); B. Klaiber : Helv. Phys. Acta 37, 554 (1964) ; and in "Lectures in Theoretical Physics" (ed. A. Barut and W. Britten, Gordon and Breach, 1968 Vol. XA, p. 141 ;

[12] - C.M. Sommerfield : Ann. Phys. 26, 1 (1964) ;

[13] - I. Bialynicki-Birula : Nuovo Cim. 10, 1150 (1958) ; J. Schwinger : Phys. Rev. 128, 2425 (1962) ;

[14] - W.E. Thirring and J.E. Wess : Ann. Phys. 27, 331 (1964) ;

[15] - B.M. McCoy and T.T. Wu : Scientia Sinica (to be published) ;

[16] - E. Ising : Z. Physik 31, 253 (1925) ;

[17] - J. Willard Gibbs : "Elementary Principles in Statistical Mechanics developed with especial reference to the Rational Foundation of Thermodynamics" (Charles Scribner's Sons, New York, 1902) ;

[18] - H.A. Kramers and G.H. Wannier : Phys. Rev. 60, 252 and 263 (1941) ; E.W. Montroll : J. Chem. Phys. 9, 707 (1941) ;

[19] - L. Onsager : Phys. Rev. 65, 117 (1944) ;

[20] - Chen Ning Yang : Phys. Rev. 85, 808 (1952) ;

[21] - H. Cheng and T.T. Wu : Phys. Rev. 164, 719 (1967) ;

[22] - E. Barouch, B.M. McCoy, and T.T. Wu : Phys. Rev. Lett. 31, 1409 (1973) ; C.A. Tracy and B.M. McCoy : Phys. Rev. Lett.

31, 1500 (1973) ;

[23] - T.T. Wu, B.M. McCoy, C.A. Tracy, and E. Barouch : Phys. Rev. B13, 316 (1976) ;

[24] - P.W. Higgs : Phys. Lett. 12, 132 (1964) ; Phys. Rev. Lett. 13, 508 (1964) ; Phys. Rev. 145, 1156 (1966) ; F. Englert and R. Brout : Phys. Rev. Lett. 13, 32 (1964) ; G.S. Guralnik, C.R. Hagen, and T.W. Kibble : Phys.Rev.Lett.13, 585 (1964) ;

[25] - B.M. McCoy and T.T. Wu : Phys. Lett. 72B, 219 (1977) ; Phys. Reports 49C, 193 (1979) ;

[26] - A. Luther : Phys. Rev. B14, 2153 (1976) ;

[27] - M. Lüscher : Nucl. Phys. B117, 475 (1976) ;

[28] - P. Jordan and E. Wigner : Z. Physik 47, 631 (1928) ;

[29] - R. Baxter : Phys. Rev. Lett. 26, 834 (1971) ; Ann. Phys. 70, 323 (1972) ;

[30] - H.A. Bethe : Z. Physik 71, 205 (1931) ; L. Hulthén : Arkiv Mat. Astron. Fysik 26A, No. 11 (1938) ; R. Orbach : Phys. Rev. 112, 309 (1958) ; L.R. Walker : Phys. Rev. 116, 1089 (1959) ; E. Lieb, T. Schultz and D. Mattis, Ann. Phys. 16, 407 (1961) ; S. Katsura : Phys. Rev. 127, 1508 (1962) ; R.B. Griffiths : Phys. Rev. 133, A768 (1964) ;

[31] - Chen Ning Yang and Chen Ping Yang : Phys. Lett. 20, 9 (1966) ; 21, 719 (1966) ; Phys. Rev. 150, 321 and 327 (1966) ; 151, 258 (1966) ;

[32] - J.D. Johnson, S. Krinsky and B.M. McCoy : Phys. Rev. A8, 2526 (1973). The special case of $J_x = J_y$ was studied earlier by J. des Cloizeaux and M. Gaudin : J. Math. Phys. 7, 1384 (1966) ;

[33] - H. Georgi and J. Rawls : Phys. Rev. D3, 874 (1971) ; I.O. Stamatescu and T.T. Wu : Nucl. Phys. B143, 503 (1978) ;

[34] - S. Coleman : Phys. Rev. D11, 2088 (1975) ;

[35] - J. Schwinger : Proc. Nat. Acad. Sci. U.S.A. 44, 956 (1958); Phys. Rev. 115, 721 (1959) ; K. Symanzik : J. Math. Phys. 7, 510 (1966) ; in : "Proceedings of the International School of Physics Enrico Fermi", Varenna Course XLV (ed. R. Jost, Academic Press, New York, 1969) ; E. Nelson : J. Funct. Anal. 12, 97 (1973) ; K. Osterwalder and R. Schrader:

Comm. Math. Phys. 31, 83 (1973) ; 42, 281 (1975) ;

[36] - R.P. Feynman : Doctoral Dissertation, Princeton University (1942) ; Phys. Rev. 80, 440 (1950) ;

[37] - I.M. Gel'fand and A.M. Yaglom : Uspekhi Math. Nauk 11, 77 (1956) ; English translation : J. Math. Phys. 1, 48 (1960); L.D. Faddeev : in "Methods in Field Theory" (ed. R. Balian and J. Zinn-Justin, North-Holland Publishing Co., 1976) p. 1 ; V.N. Popov : "Functional Integrals in Quantum Field Theory and Statistical Physics" (Atomicdat, Moscow, 1976) ;

[38] - M. Creutz, L. Jacobs, and C. Rebbi : Phys. Rev. Lett. 42, 1390 (1979) ;

[39] - L.D. Landau : in "Niels Bohr and the Development of Physics" (ed. W. Pauli, Pergamon Press, London, 1955) p. 52 ;

[40] - H. Lehmann, K. Symanzik, and W. Zimmermann, Nuovo Cim. 2, 425 (1955) ; K.G. Wilson and J. Kogut : Phys. Reports 12C, 75 (1974) ;

[41] - G.A. Baker and J.M. Kincaid : Phys. Rev. Lett. 42, 1431 (1974) ;

[42] - D. Ruelle, "Statistical Mechanics : Rigorous Results" (Benjamin, New York, 1969) ;

[43] - M.E. Fisher and D. Ruelle : J. Math. Phys. 7, 260 (1966) ;

[44] - J. Wess and B. Zumino : Phys. Lett. 49B, 52 (1974) ;

[45] - B. Zumino : in "Proceedings of the 17th International Conference on High Energy Physics", London (1974) p. I-254 ;

[46] - T. Curtright and G. Ghandour : Phys. Lett. 59B, 387 (1975); Ann. Phys. 106, 209 (1977) ;

[47] - F.J. Dyson : J. Math. Phys. 8, 1538 (1967) ;

[48] - Chen Ning Yang and Robert L. Mills : Phys. Rev. 96, 191 (1954) ;

[49] - L.D. Faddeev and V.N. Popov : Phys. Lett. 25B, 29 (1967) ;

[50] - R. Oehme and W. Zimmermann : Phys. Lett. 79B, 93 (1978) ; Preprint No. EFI 79/28, University of Chicago (1979) ;

[51] - B.M. McCoy and T.T. Wu : "The Two-Dimensional Ising Model" (Harvard University Press, Cambridge, Mass., 1973) pp. 299, 303 and 305.

THE CONTINUOUS-SPIN, ISING MODEL OF FIELD THEORY AND THE RENORMALIZATION GROUP

George A. Baker, Jr.

Theoretical Division, Los Alamos Scientific Laboratory
University of California, Los Alamos, N.M. 87545, U.S.A.

This paper sketches briefly the ideas of the ferromagnetic critical point, or bifurcation point, as exemplified in the Ising model. Historically, the only reliable, general methods available to attack this problem have been series expansion methods. The development of scaling ideas and the realization that Euclidean, Boson, quantum field-theory is the same problem as the scaling limit of critical phenomena has led to the development of the renormalization group approach to critical phenomena. For clarity, attention is focused on that continuous-spin Ising model which is equivalent to a $g_0{:}\phi^4{:}_d$ field theory. A review of the ideas of the renormalization group approach in statistical mechanical language shows that the key assumption is that the limits as the bare coupling-constant goes to infinity and the ultra-violet cut-off is removed are independent of order. Calculations show that this assumption is satisfied in one and two dimensions, but fails in three and four dimensions where the renormalized coupling-constant is not a monotonic function of the bare coupling constant. The renormalization group approach to critical phenomena is seen as a theory of the maximum of the renormalized coupling-constant which may, or may not, be a theory of the Ising-model critical-point.

The context of this paper is that of the statistical mechanics of the bifurcation point, or critical point as it is customarily called. I will begin with a brief review of the quest for a general theory of the critical or bifurcation point. I will initially focus on the spin-½ Ising-model, as there has been the most work done here. In this model, there is a set of

C. Bardos and D. Bessis (eds.), Bifurcation Phenomena in Mathematical Physics and Related Topics, 91–114.

spin variables $\sigma_{\vec{i}}$ which take on the values plus or minus unity. The index (subscript) of the spin variables runs over either a finite portion, or the whole of a regular, space lattice in d-dimensions. The model is defined by the Hamiltonian

$$\mathcal{H} = - J \sum_{\substack{\text{nearest} \\ \text{neighbors}}} \sigma_{\vec{i}} \sigma_{\vec{j}} - m H \sum \sigma_{\vec{i}} , \tag{1}$$

and the partition function

$$Z = \sum_{\{\sigma_{\vec{i}}=\pm 1\}} \exp(-\beta\mathcal{H}) . \tag{2}$$

Some typical thermodynamic quantities for this model are the free energy per spin,

$$f = - (\beta N)^{-1} \ln Z , \tag{3}$$

where β is the reciprocal of Boltzmann's constant, k, times the absolute temperature, and N is the number of spins in the model. In addition there are the magnetization per site,

$$M = - \left.\frac{\partial f}{\partial H}\right|_{\beta} , \tag{4}$$

the susceptibility per site χ, and the specific heat per site C_H,

$$\chi = - \left.\frac{\partial^2 f}{\partial H^2}\right|_{\beta} , \quad C_H = - k\beta^2 \left.\frac{\partial^2(\beta f)}{\partial \beta^2}\right|_{H} . \tag{5}$$

In Fig. 1 we have sketched typical behavior for these quantities. It will be seen that for $T > T_c$ there is only one value of M(T). For any $T < T_c$, an infinite number of values are possible. The limiting curve of these possibilities is called the spontaneous magnetization. In this model the result is symmetric with respect to ±M. At the critical, or bifurcation point, the various thermodynamic quantities display non-analytic behavior. In particular

$$M \propto (T_c-T)^{\beta'} , \quad \chi \propto (T_c-T)^{-\gamma'} , \quad C_H \propto (T_c-T)^{-\alpha'} \tag{6}$$

where α', β' and γ' are the critical indices.

For d = 2, we know exactly,

$$\gamma' = 1.75, \quad \beta' = 1/8, \quad C_H \propto - \ln T_c-T . \tag{7}$$

We remark that under the usual definition of a critical index

$$\phi = \lim_{T \to T_c} \frac{\ln \psi(T)}{\ln |T_c - T|} , \tag{8}$$

we find $\alpha' = 0$ for this ($d = 2$) case. There is no exact theory in three and four dimensions, or for $d = 2$ in a magnetic field, although great progress is being made here in terms of correlation functions (1). For a long time the only reliable method

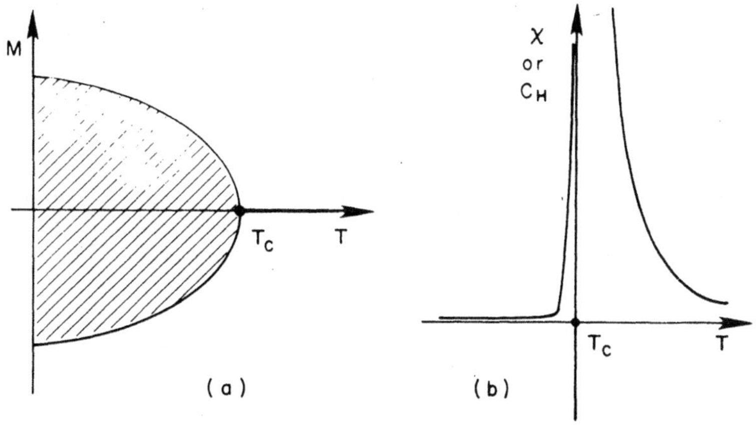

Fig. 1. In part (a) we show the typical behavior of the magnetization as a function of temperature for zero magnetic field. The point T_c is the critical or bifurcation point. The shaded region is the two phase region. In part (b), the typical behavior of χ or C_H is shown.

of obtaining results was from exact series expansions, when there were no closed form solutions. For example (2) for the three-dimensional Ising-model on the face-centered, cubic lattice one can compute

$$\chi/(\beta m^2) = 1 + 12w + 132w^2 + 1404w^3 + 14652w^4 + \dots + 1634743178420w^2 + \dots , \tag{9}$$

where

$$\omega = \tanh(\beta J) \ . \tag{10}$$

Powerful methods of approximate analytic continuation (3) were used to compute the various thermodynamic properties from series such as these. It is by these methods that the general picture of the critical point emerged in the first half of the 1960's and to this day these methods have provided the backbone of information on the critical indices. Still, by these methods, every model, every lattice, every interaction was a new case. There was no general theory to relate the different models, different critical indices, etc.

An important step forward was taken by Essam and Fisher (4) when they showed that the droplet model could be used to yield a two parameter family of critical indices which could be fit to the known critical values in two and three dimensions. They introduced the famous relation,

$$\alpha' + 2\beta' + \gamma' = 2, \tag{11}$$

in its modern form. Rushbrooke (5) soon showed that $\geq$ instead of = was the rigorous version of eq. (11). This formula was foreshadowed by Scott [implicitly in (6)] and Widom (7) in the forms

$$"\alpha' + 2\beta + 1 = 2" \quad \text{and} \quad "0 + 2\beta + \gamma' = 2" \ .$$

Shortly thereafter, in a further search for a unifying physical principle to explain these relations and to produce more, as well as a deeper understanding, Widom (8) and Kadanoff (9) looked at the scaling properties of the correlation functions. The idea, somewhat simplified, was that for T near to T_c, blocks of spins act like big spins. We have illustrated this concept in Fig. 2. Here, two square blocks of size D and separated by a distance, R, are nearly aligned near $T = T_c$. Each block is thought of as a big spin. The thought then is that a description of the effective size of the block and the large-distance, correlation behavior should suffice to describe the critical behavior and should be a simple scaling function of R and D. These strong ideas imply that the details of the microscopic behavior are unimportant compared to the behavior on the scale of the correlation length.

Another, confluent line of development starting with Schwinger (10) and Symanzik (11) relating field theory to statistical mechanics was given a rigorous basis by Nelson (12) who showed that Boson field-theory could be completely studied in Euclidean space, and that the corresponding Minkowski space theory could always be constructed from the Euclidean one.

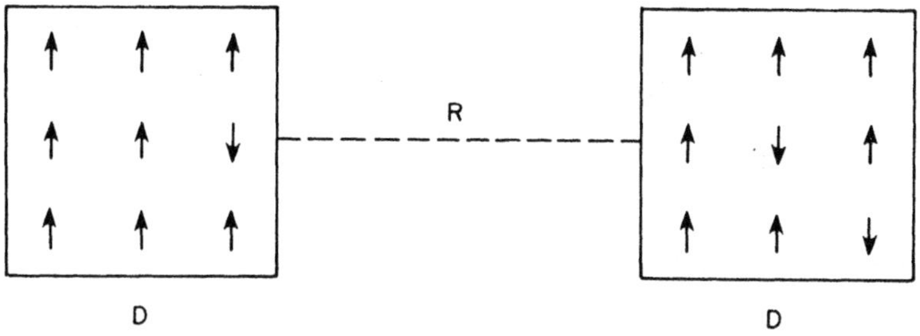

Fig. 2. The scaling picture of critical behavior near the critical point.

The background of these [and other (13)] ideas produced the atmosphere in which Migdal (14), Polyakov (15), and particularly K. Wilson (16) used field theoretic methods to try to form a general theory of the critical point. This theory goes under the name of the renormalization group theory. Since this theory is usually presented in such a way as to make its relationship to the usual statistical mechanical approach rather obscure, I will briefly review this theory from the statistical mechanical point of view in the context of the continuous-spin, Ising model. The standard presentation is covered in the reviews by Fisher (17) and others (18).

First, however, I will describe an alternate approach (19) which, while not so useful computationally, reveals certain aspects of the structure more clearly. Instead of describing the system by a Hamiltonian, as we did in eq. (1), one may follow Ruelle (20) and define the system in terms of the correlation functions.

$$u_0 = \langle 1\rangle,\ u_1(j) = \langle\sigma_j\rangle,$$

$$u_2(j,k) = \langle\sigma_j\sigma_k\rangle - \langle\sigma_j\rangle\langle\sigma_k\rangle\ , \tag{12}$$

$$u_3(j,k,\ell) = \langle\sigma_j\sigma_k\sigma_\ell\rangle - \langle\sigma_j\rangle\ \langle\sigma_k\sigma_\ell\rangle$$

$$- \langle\sigma_k\rangle\langle\sigma_j\sigma_\ell\rangle - \langle\sigma_\ell\rangle\langle\sigma_j\sigma_k\rangle + 2\ \langle\sigma_j\rangle\langle\sigma_k\rangle\langle\sigma_\ell\rangle,$$

where < > means the expectation value with respect to the summand in the partition function [eq. (2), for example] treated as proportional to a probability measure. The idea is similar to that of scaling. It is to study the behavior under a change of spin variables. For example, if one replaces all the spin-variables of a block by their scaled average, one has a new set of variables on a different scale than before. More generally, if one makes the linear transformation

$$s_j = \sum_k A(j, k)\sigma_k , \tag{13}$$

then it is easy to show that, if we treat the set of all the u's of eq. (12) as an element of an appropriate Banach space, the new set of u's for the variables s_j of eq. (13) are related to the old ones by

$$U(s) = R_A \, U(\sigma) , \tag{14}$$

where R_A is a linear mapping generated by eq. (13). If one looks at the U's which belong to a family of separable, reflexive Banach spaces (19), B_θ, then we find the following results. Let R_N be the mapping generated when the A of eq. (13) corresponds to the S_j being the average of N-σ-spin blocks. Let ω^* be the anomalous dimension of the vacuum. [See Baker and Krinsky (19) for a more detailed definition.] It is zero in usual examples like the two-dimensional, Ising model and always satisfies $\omega^* \geq 0$. Then for

$$T_N = N^{-\omega^*} R_N \tag{15}$$

one finds it natural to consider the fixed points

$$T_N \psi^* = \psi^* , \tag{16}$$

which in the usual examples with $\omega^* = 0$ are the fixed points of the renormalization group operator R_N. One can rigorously show that for any ψ in B_θ that

$$||N^{-\theta} T_N \psi||_\theta \leq ||\psi||_\theta , \tag{17}$$

i.e. that $N^{-\theta}T_N$ is a contraction operator. Unfortunately, the non-zero fixed points of T_N do not belong to B_o, and counter-examples show that B_o cannot be extended to both contain these non-zero fixed points and keep T_N a contraction. It has been shown (19, 21) however, that for ferromagnetic Ising models, the block-spin distributions approach independent Gaussian distributions, so long as we are not on the line, $T \leq T_c$, H = 0). One supposes that this result, appropriately stated, holds everywhere except at the critical point itself.

We now turn to the consideration of the ideas of the renormalization group from the statistical mechanical point of view in the context of the continuous-spin, Ising model (22). First, let us consider a $\lambda{:}\phi^4{:}_d$ Boson, Euclidean, quantum field theory. The usual action is

$$\int d\vec{x}\, \mathcal{L}(\phi(x)) = \tfrac{1}{2}\int_{-\infty}^{+\infty}\cdots\int d\vec{x}\left[(\nabla\phi(x))^2 + m_o{}^2{:}\phi^2{:} + \frac{2}{4!}\, g_o{:}\phi^4{:}\right], \tag{18}$$

where m_o is the bare mass and g_o is the bare coupling constant. The field theory is defined by the correlation (Schwinger) functions

$$\langle\phi_a\phi_b\cdots\phi_v\rangle = \frac{\partial^n\, \ell n\, Z}{\partial H_a\cdots\partial H_v}, \tag{19}$$

which are given in the usual way from the partition function. We directly put the theory on a lattice. Thus we have

$$Z(H) = M^{-1}\int_{-\infty}^{+\infty}\cdots\int\prod_{\vec{i}=1}^{N} d\phi_{\vec{i}}\ \exp\left[-\tfrac{1}{2}\, v\sum_{\vec{i}=1}^{N}\left\{\frac{2d}{q}\sum_{\{\vec{\delta}\}}\frac{(\phi_{\vec{i}}-\phi_{\vec{i}+\vec{\delta}})^2}{a^2} + m_o^2\ {:}\phi_{\vec{i}}^2{:} + 2g_o{:}\ \phi_{\vec{i}}^4\ {:}\right\} + \sum_{\vec{i}} H_{\vec{i}}\,\phi_{\vec{i}}\right], \tag{20}$$

where $v \propto a^d$ is the specific volume per lattice site, d is the spatial dimension, a is the lattice spacing, q is the coordination number, $\{\vec{\delta}\}$ is the sum over one-half the nearest neighbor sites, and $H_{\vec{i}}$ is the magnetic field at site $\vec{i}$. The normal ordered products are given by

$$:(\phi_{\vec{j}})^p: = \sum_{n=0}^{[p/2]} (-1)^n \frac{p!}{(p-2n)!\ n!}\ 2^{-n}\ C^n(\phi_{\vec{j}})^{p-2n}, \tag{21}$$

where $[p/2]$ is the greatest integer in $p/2$. The C in eq. (21) is from the field theory point of view a commutator, but from the statistical mechanical viewpoint it is just a sum over the usual lattice Green's function,

$$C = \frac{1}{V}\sum_{\vec{k}}\left[m^2 + \frac{8d}{qa^2}\sum_{\{\vec{\delta}\}} \sin^2(\pi\vec{k}\cdot\vec{\delta}a)\right]^{-1} , \tag{22}$$

where V is the spatial volume. It is to be noted that in the limit as $a \to 0$

$$\begin{aligned} &C \propto a^{2-d} , \quad d > 2 , \\ &C \propto - \ln(am) \quad d = 2 , \\ &C \text{ is finite} \quad d < 2 . \end{aligned} \tag{23}$$

The field theory motivation is to adjust the problem in such a way as to make perturbation expansions in g_o (or a very closely related variable) have all finite terms in the continuum limit, $a \to 0$. To this end, certain "renormalizations" are carried out. The first is "amplitude renormalization" where

$$H_{\vec{i}} \to H_{\vec{i}} Z_3^{-\frac{1}{2}} , \quad \psi_{\vec{i}} = Z_3^{-\frac{1}{2}}\phi_{\vec{i}} . \tag{24}$$

Thus the partition function becomes

$$\begin{aligned} Z(H) = M^{-1}\int_{-\infty}^{+\infty}\!\!\cdots\!\int \Pi\, d\psi_i \exp\Bigg[- \tfrac{1}{2} v\sum_{\vec{i}=1} & \left\{\frac{2dZ_3}{q}\sum_{\{\vec{\delta}\}}\frac{\left(\psi_{\vec{i}} - \psi_{\vec{i}+\vec{\delta}}\right)^2}{a^2} + m^2 Z_3(\psi_{\vec{i}}^2 - C Z_3^{-1})\right. \\ & + \frac{2}{4!} g_o Z_3^2\left[\psi_{\vec{i}}^{\,4} - 6CZ_3^{-1}\psi_{\vec{i}}^{\,2} + 3C^2Z_3^{-2}\right] \\ & \left. + \delta m^2 Z_3\left[\psi_{\vec{i}}^{\,2} - CZ_3^{-1}\right]\right\} + \sum H_{\vec{i}}\psi_{\vec{i}} \Bigg] , \end{aligned} \tag{25}$$

where M is a normalizing factor and is not necessarily the same as the same symbol in eq. (20). We have also introduced the mass renormalization by splitting $m_o^{\,2} = m^2 + \delta m^2$, where m^2 is the renormalized mass. A careful look at the structure of eq. (25) shows that it is of the same type as a continuous-spin Ising model. We will make this point clearer later.

For d < 4, we don't need to consider coupling-constant renormalization. A study of the perturbation series expansion in g_o leads to the following requirements in order that all the coefficients be finite. Let us define the two-point, vertex function

$$\Gamma_R^{(2)}(p,-p) = \left\{ v \sum_{\vec{j}=0}^{N-1} \left. \frac{\partial^2 \ln Z(H)}{\partial H_o \ \partial H_j} \right|_{H=0} \exp\left[-2\pi i \vec{p}\cdot\vec{j} a\right] \right\}^{-1}, \qquad (26)$$

and require

$$\Gamma^{(2)}(p,-p) = m^2 + 4\pi^2 p^2, \text{ as } p \to 0, \qquad (27)$$

which will pick the free parameters like Z_3 and δm^2.

To rewrite this condition in more familiar terms, we make the change of variables

$$[2d\ Z_3 v/(qKa^2)]^{\frac{1}{2}}\ \psi_{\vec{i}} = \sigma_{\vec{i}}\ , \qquad (28)$$

then

$$Z(\tilde{H}) = M^{-1} \int_{-\infty}^{+\infty}\!\!\cdots\!\int \prod_{\vec{i}} d\sigma_{\vec{i}}\ \exp\left[\sum_{\vec{i}} \left\{ K \sum_{\{\vec{\delta}\}} \sigma_{\vec{i}}\ \sigma_{\vec{i}+\vec{\delta}} - \tilde{g}_o\ \sigma_{\vec{i}}^{\ 4} - \tilde{A}\ \sigma_{\vec{i}}^{\ 2} + \tilde{H}_{\vec{i}} \sigma_{\vec{i}} \right\}\right], \qquad (29)$$

where we have introduced a new free parameter K in eq. (28) and we have relabeled

$$\tilde{g}_o = g_o\ K^2 q^2 a^4/(4d^2 v) \propto g_o\ a^{4-d}\ ,$$
$$\tilde{A} = qK(2d + m^2a^2 + \delta m^2 a^2 - 12Ca^2 g_o)/(4d)\ , \qquad (30)$$
$$\tilde{H}_i = H_i(2dZ_3 v/(qKa^2))^{-\frac{1}{2}}\ .$$

As the parameter K has been introduced for convenience, we can also introduce a supplementary condition for convenience. We impose

$$1 = \frac{\int_{-\infty}^{+\infty} dx\ x^2 \exp(-\ \tilde{g}_o\ x^4 - \tilde{A}x^2)}{\int_{-\infty}^{+\infty} dx\ \exp(-\ \tilde{g}_o\ x^4 - \tilde{A}x^2)} \qquad (31)$$

which determines $\tilde{A}(\tilde{g}_o)$. We remark that if $\tilde{g}_o = 0$, then $\tilde{A} = \frac{1}{2}$, the Gaussian model. Also if $\tilde{g}_o \to \infty$, then $\tilde{A} \simeq -\ 2\tilde{g}_o$

which leads to the Ising model. As $\tilde{g}_o$ varies the distribution function [the exponential in the denominator of eq. (31)] passes smoothly from the single-humped gaussian distribution to the two, delta-function distribution of the Ising model, and $\tilde{A}$ changes sign.

In terms of the σ's one may rewrite the two-point function as

$$\Gamma_R^{(2)}(p,-p) = \left\{ \frac{q\ Ka^2}{2d\ Z_3} \sum_{\vec{j}=0}^{N-1} \langle\sigma_o\sigma_{\vec{j}}\rangle \exp\left[-\ 2\pi i\vec{j}\cdot\vec{p}a\right]\right\}^{-1}$$
$$\simeq \frac{2d\ Z_3}{qK\ a^2}\ \chi^{-1}(1 + (2\pi)^2\ \xi^2\ a^2p^2 + \cdots , \tag{32}$$

in terms of the susceptibility and the correlation length

$$\chi = \sum_{j=0}^{N-1}\left[\langle\sigma_o\sigma_j\rangle - \langle\sigma_o\rangle^2\right],\ \xi^2 = \sum j^2(\langle\sigma_o\sigma_j\rangle - \langle\sigma_o\rangle^2)/2d\chi\ . \tag{33}$$

Thus by comparison between eq. (27) and (32) we get

$$m^2 = \frac{2dZ_3}{qKa^2}\ \chi^{-1},\ m^2\xi^2a^2 = 1\ . \tag{34}$$

If we fix $m^2 = 1$ to set our scale of distances, we have, in terms of K, $\tilde{g}_o$ as independent variables

$$a^2 = \frac{1}{\xi^2(K,\tilde{g}_o)}\ ,\ Z_3 = a^2(K,\tilde{g}_o)\tilde{\chi}(K,\tilde{g}_o)\ \frac{qK}{2d}\ . \tag{35}$$

We see that $a \to 0$ as $K \to K_c$, and the renormalization constants Z_3 and δm^2 are determined by eq. (30), (31), (33), and (35).

To review what we have found so far, we can, by re-expressing the lattice-cut-off field-theory in terms of a change of variables, [It is just this family of variable changes which is called the renormalization (semi-) group.] give direct expressions for the field theory constants in terms of the usual statistical mechanical ones. We have found that the continuum limit $a \to 0$ is the limit $K \to K_c$ where $K = \beta J$ in usual parlance [eq. (1) and (2)] and K_c is the critical or bifurcation point. However, field theory is made to give a well controlled continuum limit, so it should tell us about the critical point (a general theory?!).

Now field theory says that things are functions (in the continuum limit) of g_o, where by eq. (30) $\tilde{g}_o \propto a^{4-d}g_o$. Let us examine the renormalized coupling-constant.

$$g_R = \Gamma_R^{(4)}(0,0,0,0) = \frac{-v^3 \sum_{\vec{j},\vec{k},\vec{\ell}=0}^{N-1} \left.\frac{\partial^4 \ln Z(H)}{\partial H_o \, \partial H_{\vec{j}} \, \partial H_{\vec{k}} \, \partial H_{\vec{\ell}}}\right|_{H=0}}{\left(v \sum_{\vec{j}=0}^{N-1} \left.\frac{\partial^2 \ln Z(H)}{\partial H_o \, \partial H_j}\right|_{H=0} \right)^4} \tag{36}$$

which can be re-exposed using the σ variables as

$$g = g_R m^{d-4} = - \frac{v}{a^d} \frac{\frac{\partial^2 \chi}{\partial H^2}}{\chi^2 \xi^d} \tag{37}$$

where g is the dimensionless, renormalized coupling-constant. The resulting field-theory is trivial (23), that is to say no scattering, if and only if $g = 0$! The quantity g is basically the zero energy scattering amplitude. It can be rigorously shown (24) that g is bounded as $\xi \to \infty$, and also that g is never negative.

We can write g in statistical mechanical notation [following eq. (6), but here $T > T_c$]

$$\begin{aligned} &\chi \simeq A_+ (1 - K/K_c)^{-\gamma}, \quad \xi = D_+(1 - K/K_c)^{-\nu} \\ &\frac{\partial^2 \chi}{\partial H^2} = - B_+(1 - K/K_c)^{-\gamma-2\Delta}, \quad \gamma = (2 - \eta)\nu \end{aligned} \tag{38}$$

so that

$$g \approx \frac{v \, B_+}{A_+^2 D^d} (1 - K/K_c)^{\gamma+d\nu-2\Delta} \tag{39}$$

Now as we remarked above, g is finite as $K \to K_c$ so that

$$\gamma + d\nu \geq 2\Delta \tag{40}$$

If eq. (40) holds as an equality, then we call it a hyper-scaling relation. Plainly, if it is a strict inequality, then $g \to 0$ as $K \to K_c$ so that the field theory is trivial. In this case hyper-scaling is said to fail. It is necessary for hyper-scaling to hold in order to have a non-trivial field theory. We can re-express some of the other quantities near $K = K_c$ as

$$a \approx \frac{1}{mD_+}(K_c - K)^\nu$$

$$Z_3 \approx \frac{A_+ \, qK}{2dD_+{}^{2-\eta}} (ma)^\eta$$

$$g \approx \frac{v}{(a)^d} \frac{B_+}{A_+{}^2 D_+{}^d} (mD_+ a)^{\omega^*} \tag{41}$$

$$\omega^* = \frac{d\nu + \gamma - 2\Delta}{\nu}$$

In particular, the conventional wisdom of field theorists says that all renormalized quantities (e.g., g) should be nice (continuously differentiable enough times) in the neighborhood of $a = 0$, $0 \leqslant g_o \leqslant \infty$. Except for $d = 2, 3$ with g_o sufficiently small where it has been proven, this statement constitutes the key assumption of the renormalization group method. Let us examine what this assumption means.

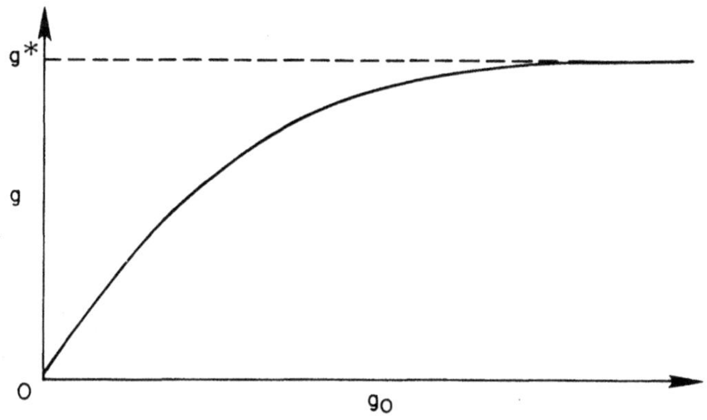

Fig. 3. The conventional wisdom picture of $g(g_o)$. The value g^* is the asymptotic value of g as $g_o \to \infty$.

We have sketched in Fig. 3 what the expected behavior is. This behavior is known rigorously for g_o small enough for $d = 2, 3$. The monotonicity is Schrader's monotonicity conjecture. It is assumed that this curve is smoothly approached as $a \to 0$, and explicitly assumed that there exists a unique, non-zero limit at $a = 0$, $g_o = \infty$, independently of the way in which $a \to 0$ and $g_o \to \infty$.

With this background, we can explain renormalization group theory's main point very simply. Remember that in the partition function, eq. (29), for statistical mechanical problems of the usual type

$$\tilde{g}_o = \text{finite, fixed,} \tag{42}$$

but by eq. (30),

$$g_o \propto \tilde{g}_o \, a^{d-4} \, . \tag{43}$$

Therefore, as $a \to 0$, $g_o \to \infty$ for any $d < 4$. Therefore by the renormalization group hypothesis,

$$g \to g^*, \tag{44}$$

and all statistical mechanical models correspond to the same g^* and so are the same. As the approach is smooth, by hypothesis, we get certain scaling behavior, exponent relations, etc.

These results can be made quantitative by means of the Callan-Symanzik approach (18, 25). One defines

$$\beta(g) = -\,(4 - d)\, g_o \left.\frac{\partial g}{\partial g_o}\right|_{a=0} \tag{45}$$

Plainly, if $g(g_o) \to g^*$ as $g_o \to \infty$ then it must be that $\beta(g^*) = 0$. As one can write β from its formal definition as a power series in g_o, and as g_o can be written as a function of g by reverting the formal power series giving g of g_o, it follows that β can be written as a formal power series in powers of g. Thus since g^* is the limit desired and it is finite, it is appealing to look for it by using $\beta(g^*) = 0$. To get the other critical indices such as η, we remember by eq. (41) tht $Z_3 \propto (ma)^{\eta}$ so that

$$\left.\frac{\partial \ln Z_3}{\partial \ln a}\right|_{\tilde{g}_o} = \eta, \text{ as } a \to 0. \tag{46}$$

But, as everything is a function of g_o [see eq. (43)], by hypothesis, it is easy to show that

$$\eta = \lim_{g\to g^*} \beta(g) \left.\frac{\partial \ln Z_3}{\partial g}\right|_{a=0} (g) \quad . \tag{47}$$

This formula is "turning the derivative" and follows by the assumed smoothness near $g_o = \infty$, $a = 0$.

When looked at in this way, the following questions beg to be asked, "How do we know that the limits $g_o \to \infty$, $a \to 0$ don't depend on the order in which they are taken?" and "How can one find out if he should believe this assertion?" Our answer (22) is to use the high-temperature series-method to calculate directly $g(K, \tilde{g}_o)$ and to see. The procedure is to expand χ, ξ^2, m, $(\partial^2\chi/\partial H^2)$, etc. in powers of K [eq. (29)]. These series are produced by standard methods [the Wortis method (26)]. For input to these methods, one needs a list of all star-graphs with less than or equal to L lines, where L is the order in K desired. A star graph is a linear graph which is multiply connected and has at most one line (or edge) connecting any pair of vertices. We have taken the necessary list from Baker, et al. (27) except for the 10 line, 8 and 9 vertex lists which were kindly supplied to us by M.F. Sykes (28).

The standard methods we use are based on Taylor's theorem

$$f(x) = \exp(x \frac{\partial}{\partial \xi}) f(\xi) \Big|_{\xi=0} = \sum_{n=0}^{\infty} \frac{x^n}{n!} f^{(n)}(0) \quad . \tag{48}$$

The idea is to re-express the partition function [eq.(29)]

$$Z = \int_{-\infty}^{+\infty}\cdots\int \prod_{\vec{i}} d\sigma_{\vec{i}} \exp\left[\sum_{\vec{i}\vec{j}} K_{\vec{i}\vec{j}} \sigma_{\vec{i}}\sigma_{\vec{j}} - \tilde{g}_o \sigma_{\vec{i}}^4 - \tilde{A}\sigma_{\vec{i}}^2 - H_{\vec{i}}\sigma_{\vec{i}}\right]$$
$$= \exp[W(\vec{K},\vec{H})] \ , \tag{49}$$

as

$$W(\vec{K},\vec{H}) = \left[\exp\left(\sum_{\vec{i}<\vec{j}} K_{\vec{i}\vec{j}} \frac{\partial}{\partial \mathcal{K}_{\vec{i}\vec{j}}}\right)\right] W(\vec{\mathcal{K}},\vec{H}) \Big|_{\vec{\mathcal{K}}=0} \quad . \tag{50}$$

A key step is to convert the K derivatives to H derivatives. The simplest conversion formula is

$$\frac{\partial W}{\partial K_{ij}} = \frac{\partial^2 W}{\partial H_i \partial H_j} + \frac{\partial W}{\partial H_i}\frac{\partial W}{\partial H_j} . \tag{51}$$

The formula for W can be given in terms of the unrenormalized semi-invariants,

$$\dot{M}_n^o(h) = \frac{d^n}{dh^n}\left[\ell n \int_{-\infty}^{+\infty} dx \exp(-\tilde{g}_o x^4 - \tilde{A} x^2 + hx)\right] . \tag{52}$$

If we set all the non-zero K_{ij} to K, then

$$W(K,\vec{H}) = N \sum_{\tau} \left\{ \frac{M_f(\tau)}{S(\tau)} \left[\prod_{v_i \epsilon \tau} M^o_{m(i)}(H_i) \right] K^{\ell(\tau)} \right\} \tag{53}$$

where the sum over τ is the sum over topologically distinct, unrooted, multi-line, connected graphs; $M_f(\tau)$ is the free multiplicity of graph τ; $S(\tau)$ is the symmetry number of graph τ; the product over v_i is the product over the vertex set of graph τ; $m(i)$ is the multiplicity of the vertex v_i; and $\ell(\tau)$ is the number of edges in graph τ. If we re-express eq. (53) in terms of the renormalized semi-invariants, M_n, then we gain a great topological simplification in that the sum over τ is replaced by the much more restrictive sum over only multiply-connected graphs. The cost of this simplification is an increase in the algebraic complexity of the calculation. An example of the type of replacement that occurs in going from the unrenormalized to the renormalized semi-invariants is shown in figure 4. The renormalized

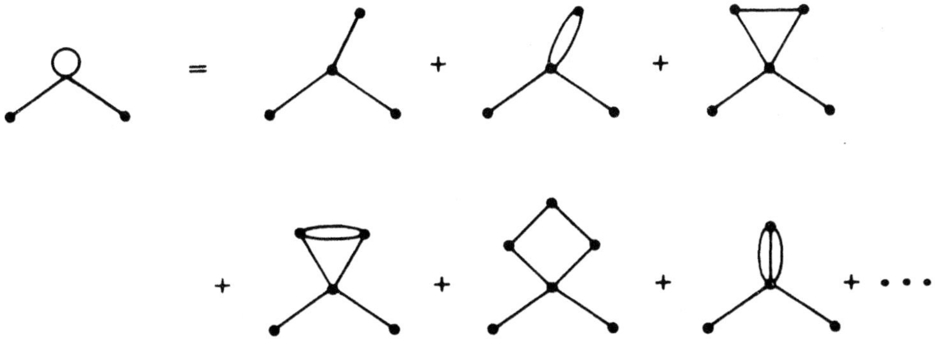

Fig. 4. The open dot (a renormalized semi-invariant) replaces the infinite sum of graphs in terms closed dots (unrenormalized semi-invariants).

semi-invariant at vertex i is determined by

$$M_n(i) = M_n^o(i) + \sum_{\ell=1}^{\infty} G_\ell(i) M_{n+\ell}^o(i) + \frac{1}{2!} \sum_{\ell,m=1}^{\infty} G_\ell(i) G_m(i) M_{n+\ell+m}^o(i)$$

$$+ \frac{1}{3!} \sum_{k,\ell,m=1}^{\infty} G_k(i) G_\ell(i) G_m(i) M_{n+k+\ell+m}^o(i) + \cdots \tag{54}$$

and the equations for the ℓ-valent, self-fields

$$G_\ell(i) = \sum_\tau \left\{ \frac{M_f(\tau)}{S(\tau)} \left[\prod_{v_j \varepsilon \tau \backslash i} M_{m(j)}(H_j) \right] K^{\ell(\tau)} \right\} \tag{55}$$

where now the sum over τ is over all ℓ-valent, singly rooted, multi-line stars, and the product is over all the vertices of τ except the root point i. The other notation is as in eq. (53). The magnetization is given by M_1 and various other thermodynamic quantities can be obtained by derivatives of it.

The necessary combinatorial data revolves around a list of the singly-rooted, multi-line stars, and their lattice embeddings. The number of single-lined stars is given in Table 1. Here the cyclotromic index $c = \ell+1-v$.

Table 1. Number of single line star graphs.

c \ ℓ	1	2	3	4	5	6	7	8	9	10	Total
0	1	0	0	0	0	0	0	0	0	0	1
1			1	1	1	1	1	1	1	1	8
2					1	2	3	4	6	7	23
3						1	3	9	20	40	73
4								2	14	50	66
5									1	12	13
6										1	1
Total	1	0	1	1	2	4	7	16	42	111	185

is used to break down the tabulation. This table agrees with that of Domb and Heap, (29). After adding the multi-lines and a root point we get the results listed in Table 2. The number of these stars has been broken down by the valence of the root point, i.e. root points with one line attached, with two lines attached, etc.

Table 2. Number of singly-rooted, multi-lined star graphs.

ℓ	1	2	3	4	5	6	7	8	9	10	Total
G1	1	0	0	0	0	0	0	0	0	0	1
G2		1	1	2	4	11	31	104	369	1439	1962
G3			1	1	3	9	28	97	371	1468	1978
G4				1	2	6	19	68	252	1020	1368
G5					1	2	8	30	123	514	678
G6						1	3	12	50	217	283
G7							1	3	15	70	89
G8								1	4	20	25
G9									1	4	5
G10										1	1
Total	1	1	2	4	10	29	90	315	1185	4753	6390

To compute the square of the correlation length [eq. (33)], we also need the labeled, two-rooted, multi-line stars. However, if we restrict ourselves to zero magnetic-field, then we need only those stars which have odd-valent roots and all other vertices even-valent. For technical reasons it is convenient to separate these graphs into two classes, class A in which the roots are nearest neighbors, and class B in which they are not. We list the number of these graphs in Table 3.

Table 3. Number of magnetic, multi-line star graphs.

ℓ	1	2	3	4	5	6	7	8	9	10	Total
A	1	0	1	1	4	6	19	45	142	411	630
B						2	5	24	88	350	469
Total	1	0	1	1	4	8	24	69	230	761	1099

In addition to the lattice embeddings (free multiplicities) we also need for these graphs the free, total, squared, root-to-root distance, i.e.

$$\mu_2(\tau) = \sum_{\substack{\text{free} \\ \text{embeddings}}} (\vec{j})^2 \ . \tag{56}$$

The results that we get when we combine all this information come out expressed as polynomials (for H = 0) in terms of the

$$I_{2n} = \langle\sigma^{2n}\rangle = \frac{\int_{-\infty}^{+\infty} x^{2n} \exp(-\tilde{g}_o x^4 - \tilde{A} x^2)dx}{\int_{-\infty}^{+\infty} \exp(-\tilde{g}_o x^4 - \tilde{A} x^2)dx} \tag{57}$$

Note that by eq. (31) $I_o = I_2 = 1.0$. Baker et al. (22) have produced among other results, the series through the tenth order in K for χ, ξ^2, and $(\partial^2\chi/\partial H^2)$ on the linear chain, plane square, triangular, simple cubic, body-centered cubic, face-centered cubic, hyper, simple cubic, and the hyper, body-centered, cubic lattices. As an illustration, we quote the first few terms of the susceptibility for the hyper, body-centered, cubic lattice

$$\begin{aligned}\chi = {} & I_2 + 16\, I_2^2\, K + (464\, I_2^3 + 16\, I_2\, I_4)K^2/2! \\ & + (20112\, I_2^4 + 1440\, I_2^2\, I_4 + 16\, I_4^2)K^3/3! \\ & + (1137696\, I_2^5 + 131184\, I_2^3\, I_4 + 720\, I_2^2\, I_6 \\ & + 3344\, I_2\, I_4^2 + 16\, I_4\, I_6)K^4/4! + (80417280\, I_2^6 \\ & + 12782400\, I_2^4\, I_4 + 100800\, I_2^3\, I_6 + 580560\, I_2^2\, I_4^2 \\ & + 7200\, I_2\, I_4\, I_6 + 16\, I_6^2)K^5/5! + \cdots\end{aligned} \tag{58}$$

Some previous results of this type have been given by Camp and Van Dyke (30).

We now turn to the analysis of these series. First it is convenient to re-express our series in terms of ξ^2 instead of K so that we can easily consider things at fixed lattice spacing, a, or which is the same, at fixed ultra-violet cut-off. This change is easily accomplished, since

$$\xi^2 = qK/2d + O(K^2)\ . \tag{59}$$

By the usual process of series reversion,

$$K = 2d\xi^2/q + O(\xi^4), \tag{60}$$

and by back substitution into eq. (37) we can accomplish this task for the renormalized coupling constant, g. We now seek to study the behavior of g vs. g_o at constant ξ^2 as $\xi^2 \to \infty$. In order to do this computation, we have computed, for a large number of $\tilde{g}_o$'s, the values of g for a set table of values of ξ^2. Then by means of eq. (30) and a proper analysis of $K(\xi^2)$, we are able to parameterize each of these points as a function of ξ^2, and g_o or $g_o a^{4-d}$.

As an analysis technique, we have relied mainly on the method of integral approximants (31), although these results have been supplemented and cross-checked by other methods. Briefly, the procedure is as follows. Suppose we know the first several terms of the power series expansion of a function f(x)

$$f(x) = \sum_{j=0}^{\infty} f_j x^j . \tag{61}$$

We can then (there are some exceptional cases) define the polynomials Q_M, P_L and R_N by the equations

$$Q_M(x)f'(x) + P_L(x)f(x) + R_N(x) = 0(x^{L+M+N+2}), \tag{62}$$

where there is an overall undefined scale-factor common to the polynomials. The maximum degree of each polynomial is indicated by its subscript. We then construct an approximation, denoted by [N/L; M], to f(x) by the solution to

$$\begin{aligned} Q_M(x)y' + P_L(x)y + R_N(x) &= 0 \\ y(0) &= f(0) \end{aligned} \tag{63}$$

These approximants have singular points at the M roots of the equation

$$Q_M(x_i) = 0 \tag{64}$$

and the structure of the approximants at these singularities is (there are, however, some exceptional cases)

$$y = \phi_i(x)(x - x_i)^{-\gamma_i} + \psi_i(x) \tag{65}$$

where ϕ_i and χ_i are regular functions in the neighborhood of x_i. This structure is rather well suited for the anticipated behavior of g vs ξ^2 at fixed $\tilde{g}_o$. Baker and Kincaid (22) have done a confluent singularity analysis and believe that, at least for $\tilde{g}_o$ not too small, these possible effects are not too serious.

When we make plots of g vs g_o and vs $g_o a^{4-d}$, we find for the one- and two-dimensional cases pictures very similar to Fig. 3, with smooth convergence as $a \to 0$. Thus in these low dimensions, the general renormalization group picture seems to hold. The situation is different in three dimensions. The results here are more complex and more interesting. For small values of the correlation length, the curve of g vs. g_o is again the same as in Fig. 3. When, however, the correlation length gets larger, of the order of 10 lattice spacings, then a definite peak appears. This peak violates the Schrader monotonicity conjecture mentioned previously and its failure has serious consequences for the renormalization group picture of critical phenomena as currently conceived. We tabulate in Table 4 a number of results which bear on this question. The values of $\tilde{g}_o$ used are approximately the peak values for each lattice. In the best cases reported, the failure of monotonicity is something like ten times the apparent error quoted. An important point to notice, besides the failure of monotonicity is that the peak height appears, within error, to take on a universal value near the renormalization group result (25) of 23.73 ± 0.03.

Table 4. The Renormalized Coupling Constant, $g(\xi^2,\tilde{g}_o)$.

SC	ξ^2	$\tilde{g}_o = 0.35$	$\tilde{g}_o = \infty$
	256	23.5 ± 1.0	21.1 ± 1.3
	2048	23.3 ± 1.7	17.8 ± 2.2
	10^6	23.0 ± 3.4	11.7 ± 2.2
BCC	ξ^2	$\tilde{g}_o = 0.75$	$\tilde{g}_o = \infty$
	512	23.6 ± 0.2	20.8 ± 0.3
	2048	23.6 ± 0.2	19.4 ± 0.5
	10^6	23.3 ± 0.2	14.4 ± 0.7
FCC	ξ^2	$\tilde{g}_o = 0.8$	$\tilde{g}_o = \infty$
	512	23.5 ± 0.2	21.1 ± 0.3
	4096	23.5 ± 0.2	19.3 ± 0.8
	10^6	23.5 ± 0.3	15.2 ± 1.2

From our numerical results then, it appears that since

$$\lim_{g_o\to\infty} \lim_{a\to 0} g \simeq 23.73 \qquad (66)$$

and that if we reverse the order of these limits a smaller [probably zero (32)] result is obtained that the idea of a unique, strong-coupling limit, independent of the order in which the bare coupling-constant goes to infinity and the ultraviolet cut-off is removed, is not well supported by the numerical evidence. Once the idea of a unique limit fails, the observation that for a range of $\tilde{g}_o$ near $\tilde{g}_o = \infty$, that $\omega^* = 0.044$ in agreement with the spin-½, Ising series result is much less surprising.

In Figure 5 we have plotted the curve obtained for the BCC lattice. The left-face and initial flat top are in agreement with the expectations of field theory. The decrease in value at the right-side is an unexpected effect from that point of

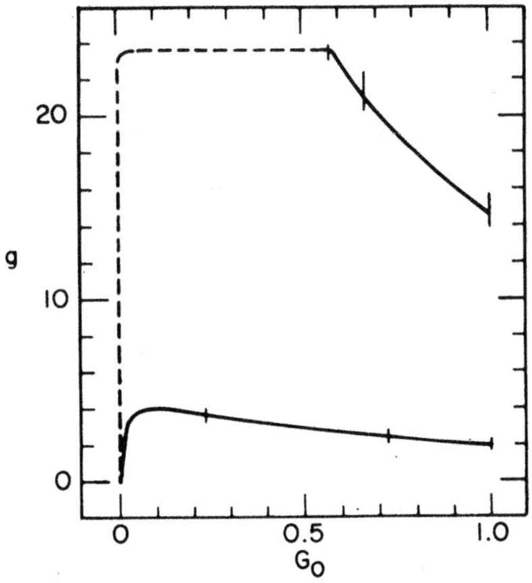

Fig. 5. The dimensionless, renormalized, coupling-constant, g, on the BCC_{4-d} (d=3) and $HBCC_{4-d}$ (d-4) lattice as a function of $G_o = g_o a^{4-d}/(10 + g_o a^{4-d})$ for a large but fixed cutoff, $\xi^2 = 10^6$. The vertical bars indicate the apparent error in that region. The curve for d = 3 in the dashed region is basically beyond the range of good convergence of Baker and Kincaid's (22) approximants with only ten terms. This curve has been supplied from the convergent behavior of g vs. g_o plots for smaller values of ξ^2.

view and shows that although these results support, and indeed compute $g_o:\phi^4:_3$ field theory over the whole range of g_o (after $a \to 0$), there is a different behavior in the statistical mechanical, or hyper-strong, coupling-constant region.

In Figure 5 there are also plotted results for four dimensions. The general conclusion is in line with that of Wilson (18), i.e. that $g_o:\phi^4:_4$ field theory is trivial, e.g. no scattering. By going beyond his 6 term work to 10 terms, a more detailed description of the behavior has been possible. Again, as in three dimensions for small correlation lengths, a monotonic g vs. g_o curve is found. For larger correlation lengths, a peaked curve, such as shown in Fig. 5, is found. The rate of decrease as a function of ξ^2 of the portion of the curve corresponding to the neighborhood of the Ising model is consistent with the spin-½ Ising-model result (32) of $\omega^* = 0.54 \pm 0.07$. The peak of the curve decays much more slowly, and it seems to be going to zero like $(1/\ell n\xi^2)$. The location is also approaching the origin in a manner consistent with the same rate.

The general picture that emerges is that the renormalization group theory is a theory of the (first) maximum of $g(g_o)$. This maximum may (d = 1,2) or may not (d = 3,4) coincide with the Ising model critical point. Because ω^* is rather small in three dimensions, the renormalization group approximation is not too bad numerically there.

Because the renormalization group is a theory of the maximum, the ϵ-expansion method (33) is in fact geared to follow the behavior of this maximum from dimension 4 where the maximum shrinks to $g_o = 0$ (which is equivalent to $g = 0$ there and corresponds to the Gaussian model with logarithmic corrections) when $a = 0$. As the dimension decreases, the value of g at the maximum increases and we track motion of the maximum, g^*, and the other quantities dependent on it as smooth functions of the dimension.

REFERENCES

1. McCoy, B. M., Tracy, C. A., and Wu, T. T.: 1977 Phys. Rev. Lett. 38, pg. 793; Abraham, D. B.: 1977 Phys. Lett. 61A, pg. 271; Sato, M., Miwa, T., and Jimbo, M.: 1977 Proc. Jpn. Acad. 53A, pg. 6; McCoy, B. M., and Wu, T. T.: 1978 Phys. Rev. D 18, pgs. 1243, 1253, 1259, and 4886.
2. Domb, C., in "Phase Transitions and Critical Phenomena", edited by C. Domb and M. S. Green (Academic, New York, 1974) Vol. 3, pg. 357.
3. See, for example, Baker, G. A., Jr.: "Essentials of Padé Approximants" (Academic, New York, 1975).
4. Essam, J. W., and Fisher, M. E.: 1963 J. Chem. Phys. 38, pg. 802.
5. Rushbrooke, G. S.: 1963 J. Chem. Phys. 39, pg. 842.
6. Scott, R. L.: 1953 J. Chem. Phys. 21, pg. 209.
7. Widom, B.: 1962 J. Chem. Phys. 37, pg. 2703.
8. Widom, B.: 1965 J. Chem. Phys. 43, pg. 3898.
9. Kadanoff, L. P.: 1966 Physics 2, pg. 263.
10. Schwinger, J.: 1958 Proc. Natl. Acad. Sci. U.S. 44, pg. 956; 1959 Phys. Rev. 115, pg. 721.
11. Symanzik, K.: 1966 J. Math. Phys. 7, pg. 510.
12. Nelson, E.: 1973 J. Funct. Anal. 12, pg. 97.
13. Gell-Mann, M., and Low, F. E.: 1954 Phys. Rev. 95, pg. 1300; Stueckelburg, E. C. G., and Petermann, A.: 1953 Helv. Phys. Acta 26, pg. 499; Bogoliubov, N. N., and Shirkov, D. V.: "Introduction to the Theory of Quantized Fields" (Interscience, New York, 1959) Ch. VIII.
14. Migdal, A. A.: 1971 Phys. Lett. 37B, pg. 386.
15. Polyakov, A. M.: 1970 Zhetp. Pis. Red. 12, pg. 538; [translation: 1970 JEPT Lett. 12, pg. 381].
16. Wilson, K. G.: 1971 Phys. Rev. B 4, pgs. 3174, and 3184.
17. Fisher, M. E.: 1974 Rev. Mod. Phys. 46, pg. 597.
18. Ma, S. K.: 1973 Rev. Mod. Phys. 45, pg. 589; Wilson, K. G. and Kogut, J.: 1974 Phys. Reports 12C, pg. 75; Brézin, E., LeGuillou, J. C., and Zinn-Justin, J., in "Phase Transitions and Critical Phenomena" edited by C. Domb and M. S. Green (Academic, New York, 1976) Vol. 6, pg. 127.
19. Baker, G. A., Jr., and Krinsky, S.: 1977 J. Math. Phys. 18, pg. 590.
20. Ruelle, D.: "Statistical Mechanics, Rigorous Results" (Benjamin, New York, 1969).
21. Iagolnitzer, D., and Souillard, B.: 1979 Phys. Rev. B 19, pg. 1515.
22. Baker, G. A., Jr., and Kincaid, J. M.: 1979 Phys. Rev. Lett. 42, pg. 1431; and in preparation; Kincaid, J. M., Baker, G. A., Jr., and Fullerton, L. W.,: Los Alamos Scientific Laboratory Report, LA-UR-79-1575.

23. Glimm, J., and Jaffe, A.: 1979 preprint, "The Coupling Constant in a ϕ^4 Field Theory".
24. Schrader, R.: 1976 Phys. Rev. B 14, pg. 172.
25. Baker, G. A., Jr., Nickel, B. G., Green, M. S., and Meiron, D. I.: 1976 Phys. Rev. Lett. 36, pg. 1351; Baker, G. A., Jr., Nickel, B. G., and Meiron, D. I.: 1978 Phys. Rev. B 17, pg. 1365; LeGuillou, J. C., and Zinn-Justin, J.: 1977 Phys. Rev. Lett. 39, pg. 95.
26. Wortis, M., in "Phase Transitions and Critical Phenomena", edited by C. Domb and M. S. Green (Academic, New York, 1974) Vol. 3, pg. 113.
27. Baker, G. A., Jr., Gilbert, H. E., Eve, J., and Rushbrooke, G. S.: 1967 Brookhaven National Laboratory Report No. BNL 50053.
28. Sykes, M. F.: 1977 private communication.
29. Domb, C., and Heap, B. R.: 1967 Proc. Phys. Soc. 90, pg. 985.
30. Van Dyke, J. P., and Camp, W. J.: 1975 Phys. Rev. Lett. 35, pg. 323.
31. Hunter, D. L., and Baker, G. A., Jr.: 1979 Phys. Rev. B 19, pg. 3808.
32. Baker, G. A., Jr.: 1977 Phys. Rev. B 15, pg. 1552.
33. Wilson, K. G., and Fisher, M. E.: 1972 Phys. Rev. Lett. 28, pg. 248.

BIFURCATIONS FOR MAPS

J.P. Eckmann

Département de Physique Théorique
Université de Genève
1211 Genève 4, Suisse.

BIFURCATIONS FOR MAPS

The study of complicated dynamical systems is, of course, intimately related to the differential equation governing the evolution of such a system, which we suppose of the form

$$\frac{dx}{dt} = F(x), \; x \in \mathbb{R}^{\nu} . \tag{1}$$

This system is too complicated for our purposes, and we consider a quite strong simplification

$$x_{n+1} = f(x_n) \qquad x_n \in \mathbb{R} \;, \; n = 1,2, \ldots \tag{2}$$

The equation (2) can be obtained from (1) in several more or less realistic ways which I shall briefly sketch.

1) Let $x(t)$ be a solution of (1). Then one defines $x_{n+1} = x(n+1)$, that is f is the time one map associated to the flow defined by F (for $\nu=1$).

2) One approximates (for $\nu=1$), $\frac{dx}{dt} \sim x(t+1) - x(t)$, so that $x_{n+1} = F(x_n) + x_n = f(x_n)$ is the difference approximation of (1).

3) Even for $\nu>1$, f may be viewed as the Poincaré map of the flow $t \to x(t)$ with respect to some special surface.{This method is in fact often quite successful in concrete examples and was a motivation for Hénon to find his map on $\mathbb{R}^2$[H], or for Landford, who showed that the Lorenz model [Lo] can be studied as a map on $\mathbb{R}^1$[La]}. This approach is the most tedious, but also the most interesting, because it shows that the study of maps on $\mathbb{R}$ may be realistically related to flows in $\mathbb{R}^{\nu}$.

We shall now concentrate on maps on $\mathbb{R}$, which have the additional property that they map $[-1,1]$ into itself . Since

C. Bardos and D. Bessis (eds.), Bifurcation Phenomena in Mathematical Physics and Related Topics, 115–135.

we are interested in the long time behaviour of the dynamical systems whose evolution this map is supposed to mimic, we shall study the iterates of maps from [-1,1] to itself. Such iterates can be conveniently described graphically. The following figure shows the graph of f and the iterates x_1, x_2, x_3 of a point x_o.

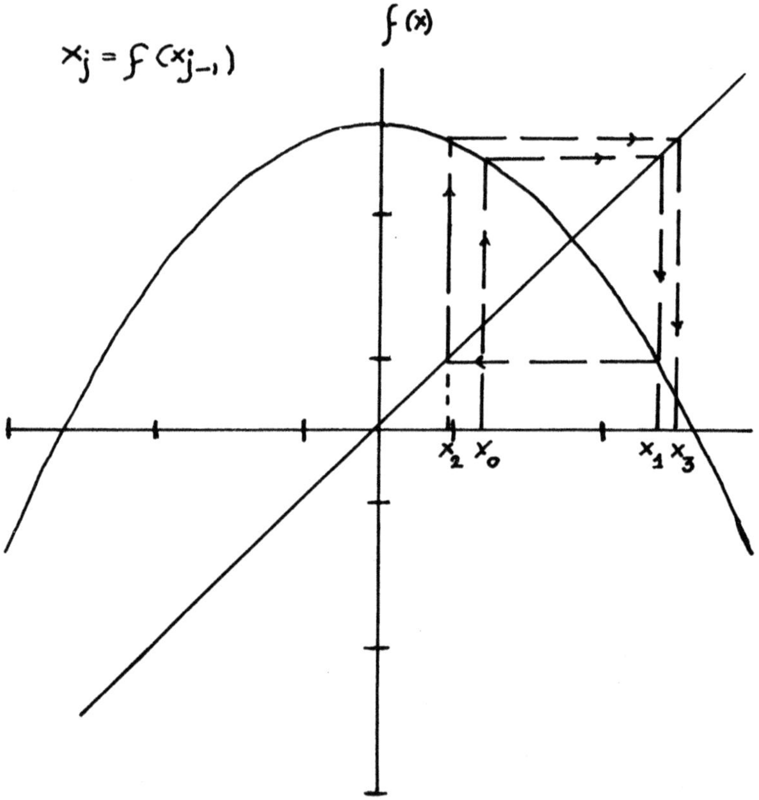

Let us perform, as a first exercise, an analysis of this map near x^1, defined by $x^1 = f(x^1)$. It is straight-forward to see, that if $|f'(x^1)| < 1$ [' denotes derivative] then points near x^1 will eventually evolve towards x^1 while if $|f (x^1)| > 1$ they will tend to evolve away from x^1.

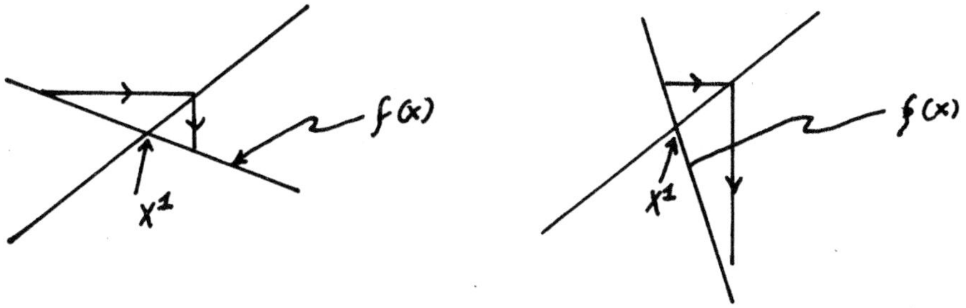

We shall call x^1 a stable, respectively unstable fixed point. Globally, the situation may be a lot more complex. There may occur stable and unstable fixed points or periodic orbits of any order (= number of points on the orbit), but it is also possible that for almost all initial points x_o, the orbit never settles down on an orbit, but wanders truly ergodically through the interval. In addition two nearby points may evolue into totally different histories, giving rise to mixing [Ru].

We shall now concentrate on maps from $[-1,1]$ to itself, <u>depending on a parameter</u> μ. One can think of this parameter as coming from a parameter in F (Eq. 1) where it has maybe the role of a Reynolds number of a temperature difference (in hydrodynamics) or a (re)production rate (in chemistry or biology). We may now ask how the number and position of periodic orbits evolves as one looks at $x_n, n \to \infty$ for different μ. There is the following experimental evidence.

1°) For some "decent" functions f_μ there is a range $\mu\epsilon(\mu_n, \mu_{n+1})$ of parameters where f_μ has exactly a stable period 2^n, and no other stable periods. In particular, as suggested by the notation, these ranges are contiguous.

2°) In each interval (μ_n, μ_{n+1}), there is one point $\bar{\mu}_n$ for which $f_{\bar{\mu}_n}$ has the property

$$\frac{d}{dx} f_{\bar{\mu}_n}^{2^n}(x) = 0 \tag{3}$$

for any point x on the period 2^n. By the chain rule

$$\frac{d}{dx} f^k(x) = f'(f^{k-1}(x))\, f'(f^{k-2}(x)) \ldots f'(x), \tag{4}$$

so that (3) says that one of the points of the period is the x for which of takes its maximum, cf Fig. 1. Such periods are called superstable.

3°) Feigenbaum has discovered the following regularity

1. $\lim_{n \to \infty} \bar{\mu}_n = \bar{\mu}_\infty$ exists.

2. For "all decent" families f_μ,

$$\lim_{n \to \infty} \frac{\log|\bar{\mu}_n - \bar{\mu}_\infty|}{n} = \log \delta^{-1}$$

where δ in independent of the family f_μ, $\delta = 4.66920160903\ldots$.

3. There is for each period 2^n a point x_n^1 nearest to the $\bar{x}$ for which $f_{\bar{\mu}_n}$ takes its maximum. In fact $x_n^1 = f_{\bar{\mu}_n}^{2n-1}(\bar{x})$. For "all decent" families f_μ

$$\lim_{n \to \infty} \frac{\log|x_n^1 - \bar{x}|}{n} = \log \lambda$$

where λ is independent of the family f_μ, $\lambda = 1/2.5029078750957\ldots$.

In addition, $(-1)^n[\hat{x}_n - \bar{x}]$ has constant sign as a function of n. I would like to illustrate this universality for the following examples :

- $f_\mu(x) = \mu x(1-x^2)$ Feigenbaum [F]; shown is μ_n
- $f_\mu(x) = (1-2x^2)$ Feigenbaum [F]; shown is $\underline{\mu_n}$

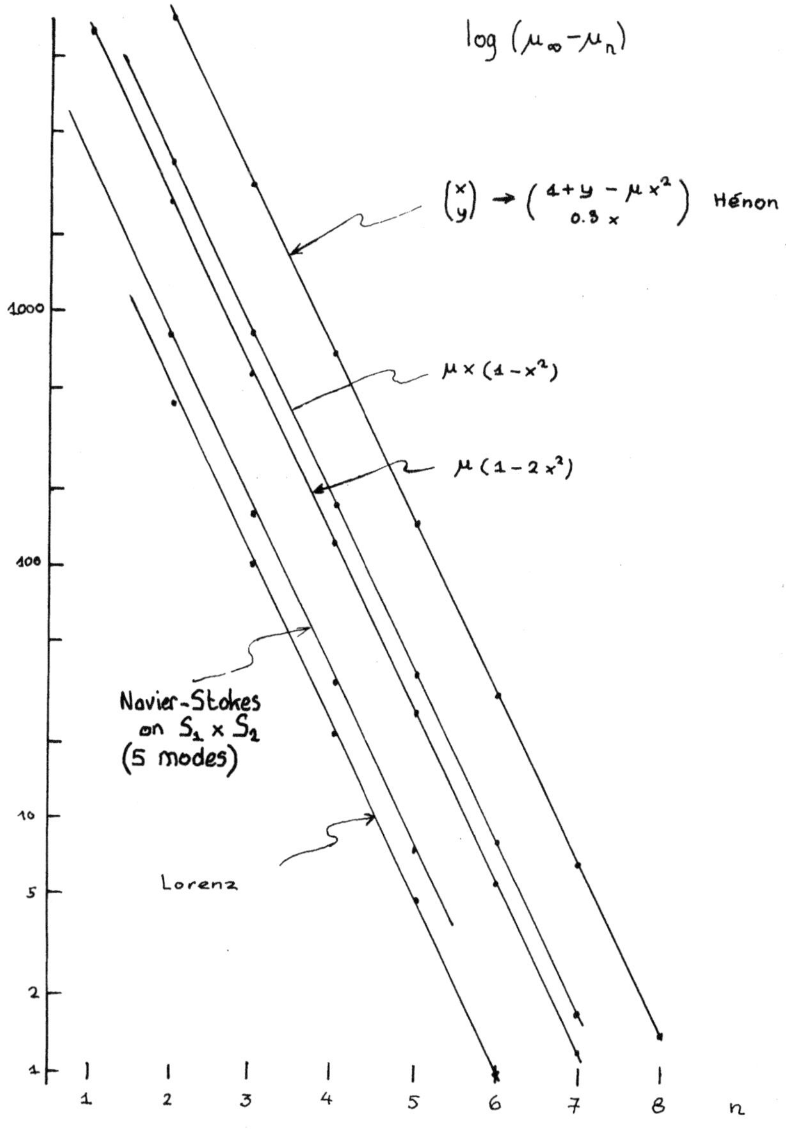
log (μ∞ − μn)
(x y) → (1 + y − μx² ; 0.3 x) Hénon
μ x (1 − x²)
μ (1 − 2 x²)
Navier-Stokes on S1 × S2 (5 modes)
Lorenz
1000
100
10
5
2
1
1
2
3
4
5
6
7
8
n

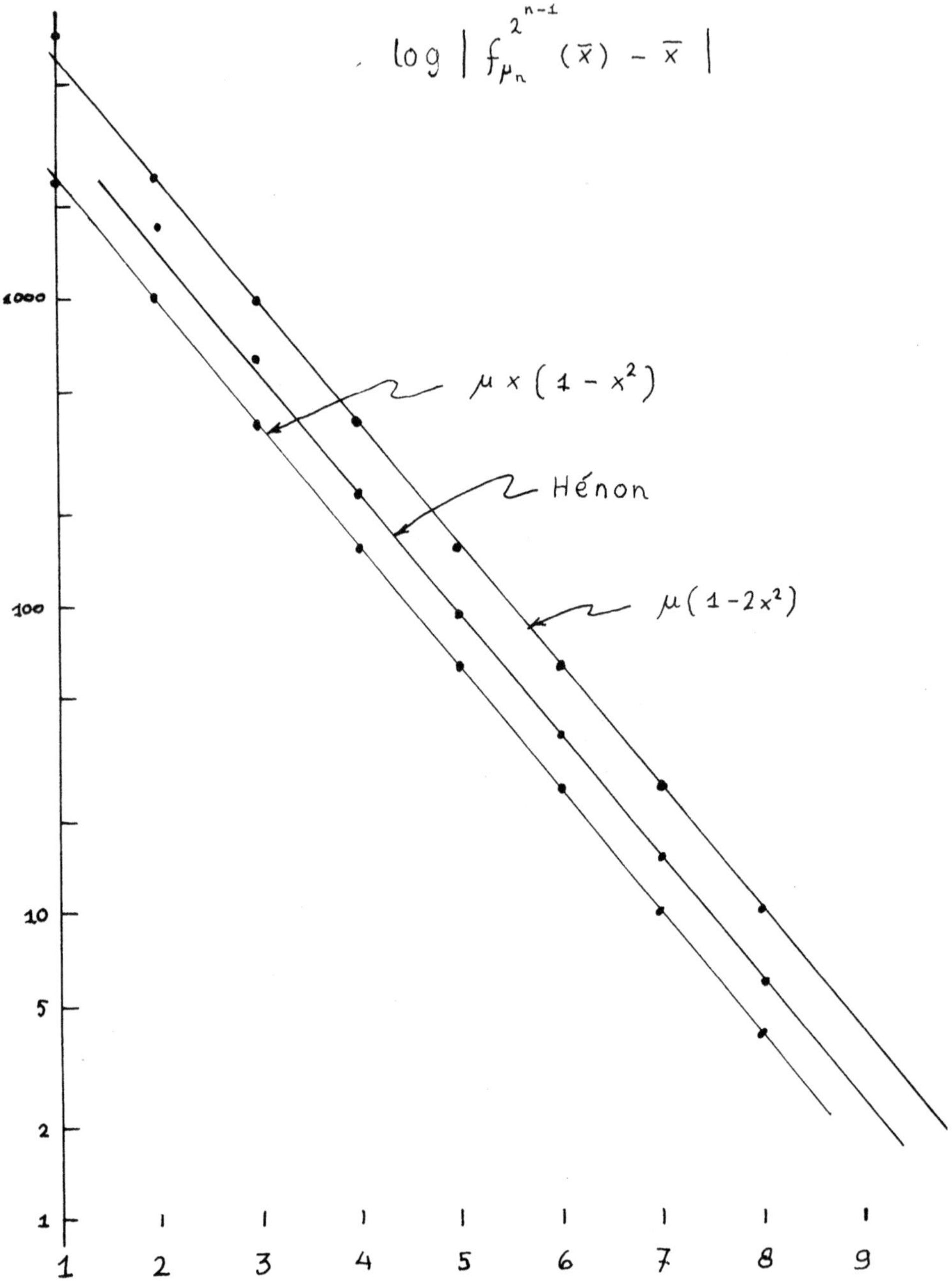

$\log | f_{\mu_n}^{2^{n-1}}(\bar{x}) - \bar{x} |$
1000
100
10
5
2
1
$\mu x(1-x^2)$
Hénon
$\mu(1-2x^2)$
1
2
3
4
5
6
7
8
9

- $f_\mu\binom{x}{y} = \binom{1 + y - x^2}{0.3x}$ Derrida et al [D] shown is μ_n
- Navier Stokes on torus $S_1 \times S_1$ (5 Fourier components)[BF] shown is $\bar{\mu}_n$
- Lorenz System [L]. Calculation done by Franceschini [Fr]. Shown is $\bar{\mu}_n$

Note that the 3 latter examples are not maps on $\mathbb{R}$!

I shall now try to argue why the non-linear map

$f \to Nf$

$$(Nf)(x) = f(1)^{-1} f{\circ}f(f(1)x) \qquad (5)$$

is important in studying the above facts. We view the facts discovered by Feigenbaum as expressing that there is an intimate connection between "any" map $x \to f^{2^n}(x)$ and $x \to f^{2^{n+1}}(x)$ provided n is very large [on purpose, we have omitted the μ-dependence]. In particular the most intimate connection possible would be $f^{2^n}(x) = f^{2^{n+1}}(x)$. This is however not possible except in the trivial case $f(x) = x$ but we could ask that the two functions are at least similar, i.e.

$$f^{2^n}(x) = -\lambda^{-1} f^{2^{n+1}}(\lambda x)$$

for some λ. Choosing $\lambda = f(1)$ amounts to break the dilatation symmetry of the problem, and we may fix $f(0) = 1$. Our "most intimate" connection has thus become the question of looking for a fixed point of the map N on a space of functions which take the value 1 for $x = 0$. (N preserves this property !)

The possibility that such a function may indeed exist is indicated through Fig. 3 below, where we plot $1 - x^2$ and $(1 - 1.411x^2)^{\circ 2}$.

Note the similarity of the two square pieces

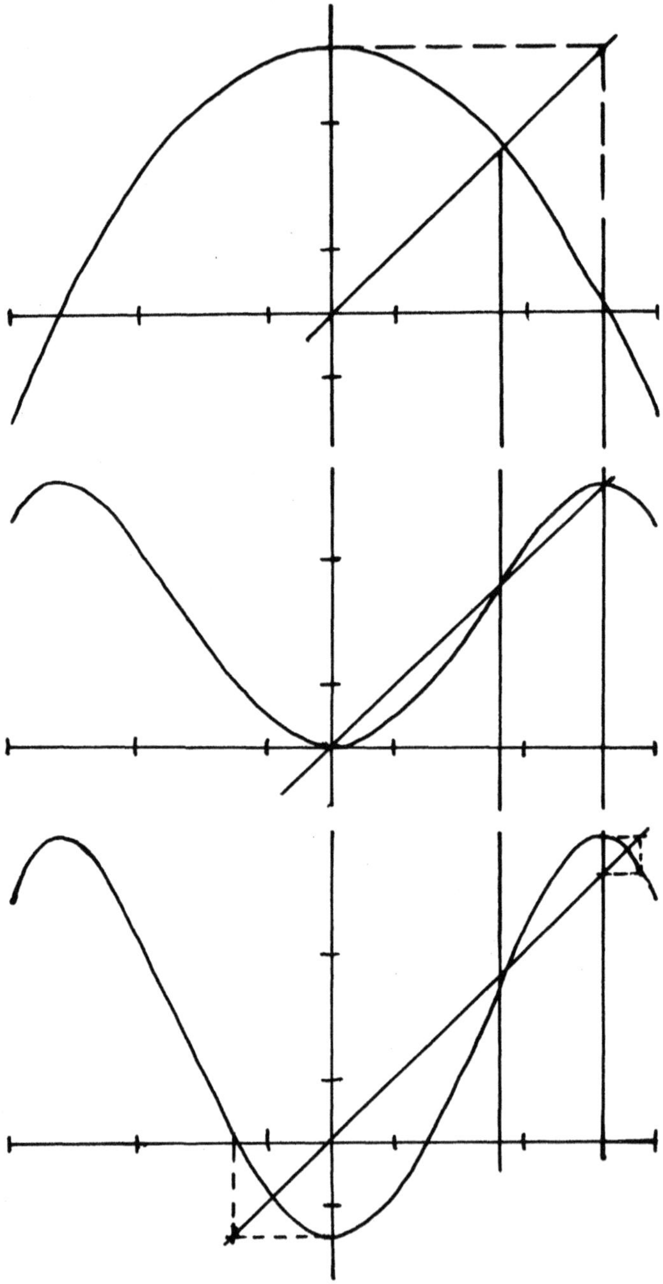

Unfortunately, to date, existence of a solution to $Nf = f$ is not known in C^2*. We now generalize the problem and explain the mechanism in special cases. We look for solutions ϕ to $N\phi = \phi$ which are of the form

$$\phi_\varepsilon(x) = f_\varepsilon(|x|^{1+\varepsilon}) \tag{6}$$

with $t \to f(t)$ being C^2. We claim that there is some sort of bifurcation associated to the problem

$$\phi(1)\,\phi(x) = \phi o \phi(\phi(1)x). \tag{7}$$

Indeed, (7) has for all ε the trivial solution $\phi(x) = 1 - |x|^{1+\varepsilon}$, corresponding to $\phi(1) = 0$. At $\varepsilon = 0$ another, non-trivial solution bifurcates, as a formal power series in ε, $\log\varepsilon$ and t which is of the form

$$\begin{aligned} f_\varepsilon(t) &= 1 - (1 + \lambda)t + \lambda^2 g(t) \\ \lambda &= \lambda(\varepsilon) = \varepsilon\,(-1 + \log 1/\varepsilon)\,(1 + O(\varepsilon)). \end{aligned} \tag{8}$$

This branch <u>does not</u> come from the sort of singular eigenvalue described in the lectures of Crandall and Rabinowitz, as you may easily check.

Our first aim will then be to show that to (8) there exists a solution $\phi_\varepsilon(x) = f_\varepsilon(|x|^{1+\varepsilon})$ for small $\varepsilon > 0$, with $\lambda > 0$. From this we shall then derive the following theorem. Let B be the Banach space of functions analytic in $|t| < 2$, equipped with the sup norm.

---

* There is evidence of existence, O. Lanford, privat communication.

<u>THEOREM</u> : <u>Fix</u> $\varepsilon > 0$ <u>sufficiently small, let</u> $\psi(x) = h(|x|^{1+\varepsilon})$, <u>with</u> $h \in \mathcal{B}$ <u>and</u> $\|h - f_\varepsilon\|$ <u>small. Let</u> $\psi_\mu(x) = \psi(\mu x)$. <u>For any such family</u> $\mu \to \psi_\mu$ <u>of functions, the</u> μ_n (or $\bar{\mu}_n$) <u>defined as before satisfy</u>

i) $\lim_{n\to\infty} \mu_n = \mu_\infty$ exists

ii) $$\lim_{n\to\infty} \frac{\log|\mu_n - \mu_\infty|}{n} = \log \delta(\varepsilon)^{-1}$$

<u>where</u> $\delta(\varepsilon)$ <u>depends on</u> ε <u>but not on</u> ψ

iii) $$\lim_{n\to\infty} \frac{\log(-1)^{n+1}\psi_{\mu_n}^{2^{n+1}}(0)}{n} = \log \lambda(\varepsilon)$$

<u>where</u> $\lambda(\varepsilon)$ <u>depends on</u> ε <u>but not on</u> ψ

Note : the condition $h \in \mathcal{B}$ is probably too strong. On the other hand $h \in C^1$ is definitely too weak.

We now come to the existence proof for f_ε. For technical reasons, it is easier to consider f_ε as a function of λ and to write an equation for $\varepsilon(\lambda)$. Since $\varepsilon(\lambda)$ will turn out to be monotone, this is certainly a legitimate procedure. The result is the following

THEOREM : There is a constant K such that for sufficiently small $\lambda > 0$, the Equation (7) has a unique solution $\phi = \phi_\lambda$ of the form $\phi_\lambda(x) = f_\lambda(|x|^{1 + \varepsilon(\lambda)})$, satisfying

i) $t \to f_\lambda(t)$ is C^2 in $0 < t < 1$,

ii) $f_\lambda(t) = 1 - (1 + \lambda)t + \lambda^2 g_\lambda(t)$, $\|g_\lambda\|_{C^2} < K$,

iii) $\varepsilon(\lambda) = -2/(1 + \log\lambda) + \lambda^2 E(\lambda)$, $|E(\lambda)| < K$.

For sufficiently small $\lambda > 0$, f_λ is analytic in $|t| < 2$, and both f_λ and $\varepsilon(\lambda)$ are analytic in $\{\lambda | \mathrm{Re}|\lambda| > 0, |\mathrm{Arg}\lambda| < \Theta_o\}$ for some small $\lambda_o, \Theta_o > 0$.

Once we have found out how to set up the problem, the proof is really not that hard. I will only sketch it. Because λ is small, $f_\lambda(\lambda^{1 + \varepsilon}t)$ will be positive, and it suffices to show existence of a solution of

$$\lambda f(t) + f(f(\lambda^{1 + \varepsilon}t)^{1 + \varepsilon}) = 0, \qquad (9)$$

where $f(t) = 1 - (1 + \lambda)t + \lambda^2 g(\lambda,t)$ and ε, g are unknowns, $g(\lambda,0) = g(\lambda,1) = 0$.

Differentiating Eq. (9) twice with respect to t and solving for ε and g one gets equations of the following form :

$$E = N_\lambda(g,\varepsilon)$$

and (10)

$$g = L\,(1 - \lambda K_\lambda(g,\varepsilon))^{-1} M_\lambda(g,\varepsilon).$$

The quantities K, L, M, N are defined as follows :
Let $\mathcal{D} = \{(g,\varepsilon),\ g \in C^2[0,1],\ g(0) = g(1) = 0,\ \varepsilon \in C\}$ equipped with norm

$$\|(g,\varepsilon)\| = \sup_{x \varepsilon\ 0,1} |g''(x)| + |\varepsilon|\ [\log 1/\lambda - 1]/2\lambda,$$

let $\mathcal{D}_1$ be the unit ball in $\mathcal{D}$.

$$N_\lambda(g,\varepsilon) = \lambda/(-1 + \log^1/\lambda) + \text{remainder}$$

and in fact $N_\lambda(g,\varepsilon)$ is a bounded, contracting map from $\mathcal{D}_1$ to C.

$$[M_\lambda(g,\varepsilon)](t) = -\lambda^{2\varepsilon}(\varepsilon/\lambda)\, f'(f(\lambda^{1+\varepsilon}t))^{\varepsilon-1}(1+\varepsilon)\, f'(\lambda^{1+\varepsilon}t)^2$$

is a bounded, contracting map from $\mathcal{D}_1$ to $C^1[0,1]$.
$K_\lambda(g,\varepsilon)$ is a bounded linear operator from $C^0[0,1]$ to $C^0[0,1]$ and from $C^1[0,1]$ to $C^1[0,1]$ and

$$\|[K_\lambda(g_1,\varepsilon_1)-K_\lambda(g_2,\varepsilon_2)]h\|_{C^0} < O(1)\,\|(g_1,\varepsilon_1)-(g_2,\varepsilon_2)\|_{\mathcal{D}}|h|_{C^1}$$

provided $g_i,\varepsilon_i \in \mathcal{D}_1$, $i = 1,2$.
The explicit definition is

$$(K_\lambda(g,\varepsilon)h)(t) = -\lambda^{2\varepsilon}\{h(f(\lambda^{1+\varepsilon}t)^{1+\varepsilon})f(\lambda^{1+\varepsilon}t)^{2\varepsilon}(1+\varepsilon)^2 f'(\lambda^{1+\varepsilon}t)$$
$$+\, h(\lambda^{1+\varepsilon}t)f'(f(\lambda^{1+\varepsilon}t)^{1+\varepsilon})f(\lambda^{1+\varepsilon}t)^{\varepsilon}(1+\varepsilon)\}$$

Finally L is bounded linear from $C^0[0,1]$ to $C^2_0[0,1]$;

$$(Lh)(t) = \int_0^t (t-\tau)h(\tau)d\tau - t\int_0^1 (1-\tau)h(\tau)d\tau.$$

The above properties are easily verified, and by using the identity

$$[1 - \lambda K(X_1)]^{-1} - [1 - \lambda K(X_2)]^{-1}$$
$$= [1 - \lambda K(X_1)]^{-1}[K(X_1) - K(X_2)][1 - \lambda K(X_2)]^{-1}$$

where $X_i = (g_i,\varepsilon_i)$ $i = 1,2$,
one verifies that the equations (10) have a unique solution, and that this solution solves (9), and hence it solves (7).

Existence (and, of cours uniqueness) of an analytic solution can be shown in a similar way. I refer to the original paper.

Let us now come back to the map

$$f \to Nf$$
$$(Nf)(x) = f(1)^{-1}\, f\circ f(f(1)x).$$

It should be obvious that a good understanding of N will bring us nearer to an understanding of f^{2^n} since, up to a change of scale, f^{2^n} equals $N^n f$. The importance of having found a fixed point of the map $f \to Nf$ lies then in the fact that in a neighborhood of the fixed point f_λ, the nonlinear map N can be essentially understood in terms of the (linear) tangent map DN at f_λ.

Note that if $f(0) = 1$ then $(Nf)(0) = 1$. Therefore, since $f_\lambda(0) = 1$, DN at f_λ will have the property that if $g(0) = 0$ then $(DNg)(0) = 0$.

A natural space of functions to work on is therefore the space E of functions (of t) analytic in $|t| < 2$ and vanishing at $t = 0$.

We now <u>fix</u> $\varepsilon > 0$ sufficiently small and we study DN. Viewed as a map on B, the Banach space of functions analytic in $|t| < 2$, (recall $t = |x|^{1+\varepsilon}$) the formula for $A = DN$ at $f = f_\lambda$ is

$$(Ah)(t) = -h(y)/\lambda - f'(y)(1+\varepsilon)f^{\varepsilon}(z)h(z)/\lambda$$
$$+ h(1)(f(t) - tf'(t)(1+\varepsilon)/\lambda,$$

where $\lambda = -f(1)$, $z = \lambda^{1+\varepsilon}t$, $y = f(z)^{1+\varepsilon}$.

In order to study A we shall first analyze its first order approximation (in λ), A_o, which is given by

$$(A_o h)(t) = -h(1-\lambda t)/\lambda + (1+\varepsilon)(1+\lambda)h(\lambda t)/\lambda$$
$$+ h(1)[1 + (1+\lambda)\varepsilon t]/\lambda.$$

<u>LEMMA</u> : <u>For small</u> $\lambda > 0$, <u>the operator</u> A_o <u>is compact on the space</u> E. <u>Its spectrum is the closure of the set</u> $\{\delta_i, i = 1,2,..\}$, <u>where</u> $\delta_1 = 1 + (1+\lambda)(1+\varepsilon/\lambda+\varepsilon)$, $\delta_n = \lambda^{n-1}((-1)^{n-1} + (1+\lambda)(1+\varepsilon))$, $n = 2,3\ldots$. <u>The corresponding eigenfunctions are polynomials of increasing degree. In particular the eigenfunction corresponding to</u> δ_1 <u>is the function</u> t.

Since, in these lectures, we insist more about the kind of theorems which it is necessary to prove in a "renormalization group" situation, rather than to elaborate on the technical details, we again only give some clues concerning the proof of the lemma. The details can be found again in [CEL].

The operator A_o is a sum of three pieces, of which the third is manifestly a rank one operator. On the other hand, the maps $h(.) \to h(\lambda .)$ and $h(.) \to h(1 - \lambda .)$ are both compact, as can be seen by using Montel's theorem. A combinatorial argument shows that for every n, there is (up to a factor) exactly one polynomial of degree n in E which is an eigenvector of A_o.

The following calculation shows that A_o is uniformly bounded for small λ, i.e. the factor $^1/\lambda$ does not matter :

$$- h(1 - \lambda t)/\lambda + h(1)/\lambda + (1 + \lambda + \varepsilon)h(\lambda t)/\lambda$$

$$= -\frac{1}{\lambda}\int_{1-\lambda t}^{1} d\tau h'(\tau) + \frac{1 + \lambda + \varepsilon}{\lambda}\int_{0}^{\lambda t} d\tau h'(\tau) \quad (\text{since } h(0) = 0)$$

$$= (2\pi i\lambda)^{-1} \int_{1-\lambda t}^{1} d\tau \int_{\Gamma} dz_o\, h(z_o)\, (\tau - z_o)^{-2}$$

$$- (1 + \lambda + \varepsilon)\int_{0}^{\lambda t} d\tau \int_{\Gamma} dz_o\, h(z_o)(\tau - z_o)^{-2}$$

$$= (2\pi i\lambda)^{-1}\int_{\Gamma} dz_o h(z_o)$$

$$\left[\frac{1}{1 - \lambda t - z_o} - \frac{1}{1 - z_o} + (1 + \lambda + \varepsilon)\left[\frac{1}{\lambda t - z_o} - \frac{1}{z_o}\right]\right]$$

(using Fubini's theorem)

$$= (2\pi i)^{-1}\int_{\Gamma} dz_o h(z_o)\left[\frac{t}{(1 - z_o)(1 - \lambda t - z_o)} + \frac{t(1 + \varepsilon + \lambda)}{z_o(\lambda t - z_o)}\right] .$$

A similar calculation shows $||A - A_o|| = O(\lambda)$ and from this we conclude that A is bounded, has one eigenvalue of modulus greater than 1 and the remainder of the spectrum in $|z| < O(\lambda)$. Let us call $\delta = \delta(\varepsilon)$ the largest eigenvalue. We have

$$\delta = 2 + (\varepsilon/\lambda) + O(\lambda),$$

by standard methods [K] and the corresponding eigenfunction e_1 satisfies

$$e_1(t) = t + O(\lambda), \qquad \text{in } E.$$

We now have, so to speak, an infinitesimal picture of what happens in a neighborhood of the fixed point. What we want to know next, is, whether to that infinitesimal picture there corresponds a global picture in a neighborhood of the fixed point. We use here the machinery from [*HPS*].

The following estimate is crucial in establishing the connection between the infinitesimal and the local picture. Note that we only get a local picture and not a global one at this stage, since N is not defined on functions f with $f(1) = 0$.

<u>LEMMA</u> : <u>Let</u> $\sigma < \lambda$. <u>For</u> $||h_1||_E, ||h_2||_E, ||h_1 - h_2||_E < \sigma$, <u>one has</u> $||(N - A)(h_1) - (N - A)(h_2)|| = O(\sigma/\lambda)\, ||h_1 - h_2||_E$.

The proof of this lemma is just done by brute force by writing the difference as an integral of the second derivative.

Let $(Th) = N(\phi + h) - \phi$, let E_s and E_u be the stable and unstable spectral subspaces of $DT = A$.

Define now $E(\sigma)$ to be the ball of radius σ in E, centered at zero, and let $E_u(\sigma) = E(\sigma)\, E_u$ and $E_s(\sigma) = E(\sigma)\, E_s$. By multiplying T by a cutoff function χ which is equal to one to $E(\sigma)$ we have thus

LEMMA : For a fixed ratio $\sigma/\lambda < 1$ the map $\chi T - \mathcal{D}\chi T_{g=0}$ is for sufficiently small $\lambda > 0$ a Lipshitz continuous map from E to itself with Lipshitz constant $\mathcal{O}(\sigma/\lambda)$. In addition, $\chi T - \mathcal{D}\chi T_{g=0} = \mathcal{O}(\sigma)$.

The preceding lemma allows us to apply Corollary 5.4 of [HPS] and we get the

THEOREM : The stable and unstable manifolds W_s, W_u through 0, tangent to E_s, E_u exist, are unique and are locally the graphs of two C^∞ functions $f_s : E_u(\sigma) \to E_s(\sigma)$, $f_u : E_s(\sigma) \to E_u(\sigma)$.

The next operation is to find a C^1 map U of E to E which diagonalizes $T : E \to E$, in the following sense. Write an arbitrary vector in E as $x \oplus y$ with x in the subspace of e_1, y in its spectral complement. Then we want to find U such that

$$UTU^{-1} \binom{x}{y} = \begin{bmatrix} \delta x \\ Ay + M(x,y) \end{bmatrix}$$

with $M(x,0) = 0$. Note that in general it is not possible to find a U which is C^1 and which would completely linearize T, since there are an infinity of eigenvalues of $\mathcal{D}T = A$.

However, for the purpose of renormalization group theory, it is totally sufficient to diagonalize T in the unstable direction only. On the other hand it is crucial to have this diagonalization at least once differentiable, because we want to take the distance on the unstable manifold to be used as a coordinate, after linearization.

The proof of existence of U is through a sequence of transformations using the implicit function theorem; in fact U is near to the identity [CE pp 131 - 134].

We can now harvest the fruits of our efforts. The geometical situation is as depicted in the figure below. Note that we draw not E but $f_\varepsilon + E$, so that the relevant map is N. The manifold $\mathcal{W}_o$ is defined to be those functions $f \in E + f_\varepsilon$ for which $f(1) = 0$ (and any way, $f(0) = 1$). That its position is as indicated in the figure follows at once from our knowledge of the approximate direction of $\mathcal{W}_u$.

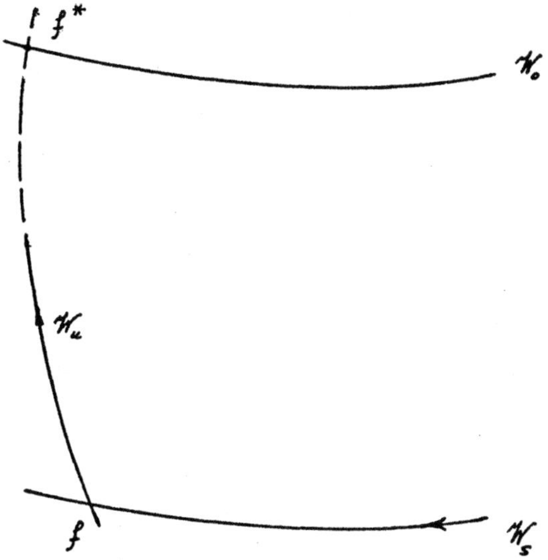

To this picture, we want to add one more curve, namely the objects which have motivated this study. This curve is obtained as follows.

1) Choose any point in $E + f_\varepsilon$ in a neighborhood (sufficiently small) of f_ε call it Ψ.

2) Define $\psi(x) = \Psi(|x|^{1+\varepsilon})$.

3) Define $\psi_\mu(x) = \psi(\mu x)$. It is for the family of functions $\mu \to \psi_\mu$ that we shall prove the universality properties.

4) We next analyze the behaviour of the curve $\mu \to \psi_\mu$ in a neighborhood of f_ε and in particular we claim it is transversal to $\mathcal{W}_u$. To see this we just have to check its derivative with respect to μ. Since Ψ_μ is very near to $f_{\varepsilon,\mu}$ it obviously suffices to verify this property for the

curve $\mu \to f_{E,\mu}$, where

$$f_{\varepsilon,\mu}(t) = f_{\varepsilon}(\mu^{1+\varepsilon}t).$$

Now $d_{\mu} f_{\varepsilon}(\lambda^{1+\varepsilon}t)\Big|_{\mu=1} = f'_{\varepsilon}(t)(1+\varepsilon)t$

$$= -(1+\lambda)(1+\varepsilon)t + \mathcal{O}(\lambda^2)$$

and this is certainly not in the complement of E_u for sufficiently small ε. Hence we have transversality in this case and so the same by continuity if Ψ is near f_{ε}.

5) For simplicity of the argument, we assume now that the curve $\mu \to \Psi_{\mu}$ extends as much as shown in the next figure.

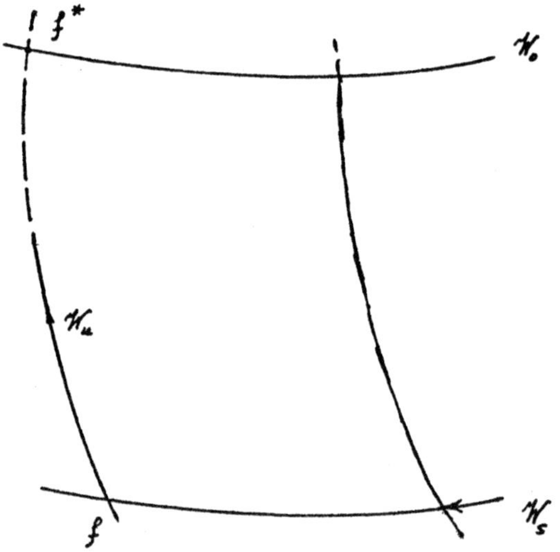

At the point μ_0, Ψ_{μ_0} lies in W_0 and thus has the property $\Psi_{\mu_0}(0) = 1, \Psi_{\mu_0}(1) = 0$, i.e. Ψ_{μ_0}, and hence ψ_{μ_0} has a superstable period of length two.

6) By using the implicit function theorem, we can show that for every $n > n_0$, n_0 sufficiently large, such that there is a unique μ_n such that

$$N^n(\psi_{\mu_n}) \in W_0 \,.$$

This means

$$[\mathcal{N}^n(\psi_{\mu_n})]^2(0) = 0,$$

i.e.

$$\prod_{j=1}^{n-1} \psi_{\mu_n}^{2^j}(1)^{-1} \psi_{\mu_n}^{[2^n]}\left(x \prod_{j=1}^{n-1} \psi_{\mu_n}^{2^j}(1)\right)\Big|_{x=0} = 0$$

so that

$$\psi_{\mu_n}^{2^n}(0) = 0;$$

ψ_{μ_n} is superstable of period 2^n !

Let $h_{n,\mu} = \mathcal{N}^n(\psi_\mu) - f$

We define an approximate μ_n as follows. According to the transversality statement, and by virtue of the fact that $\mathcal{W}_u$ is tangent to $e_1(t) = t + \mathcal{O}(\lambda)$, we can write, for some constant $a > 0$:

$$h_{o,\mu}(t) = (1 - \mu)\, a\, e_1(t) + r_\mu$$

where $r_\mu = \mathcal{O}(1 - \mu)$ is in $\mathcal{W}_s$. Having linearized $\mathcal{T}$, we find

$$h_{n,\mu}(t) = \delta^n(1 - \mu)\, a\, e_1(t) + r_{n,\mu},$$

where $r_{n,\mu} = \mathcal{O}((2\lambda)^n(1 - \mu))$. We now define $\tilde{\mu}_{n+1} = -\lambda\delta^{-n}(1/a) + 1$, so that $h_{n,\mu}(1) + f_\lambda(1)$ is approximately zero. The value of μ_n can now be found in principle by solving a fixed point problem for a_n in $\mu_{n+1} = 1 - \lambda\delta^{-n}(1/a_n)$, where one will find $a_n \to a$ as $n \to \infty$, by continuity.

To analyze the second problem, which is the scaling of the solutions, it is more convenient to consider the functions

$$\psi_{m,\mu}(x) = f_\lambda(|x|^{1 + \varepsilon(\lambda)}) + h_{m,\mu}(|x|^{1 + \varepsilon(\lambda)}).$$

We intend to compare $\psi_{n,\mu_n}(x)$ to $\psi_{n-1,\mu_{n-1}}(x)$. By the definition of T, we find

$$\psi_{m,\mu}(x) = \frac{1}{\psi_{m-1,\mu}(1)} \psi^{(2)}_{m-1,\mu}(\psi_{m-1,\mu}(1)x),$$

and interating this,

$$\psi_{m,\mu}(x) = \prod_{j=0}^{m-1} \psi_{j,\mu}(1)^{-1} \psi^{(2m)}_{o,\mu}(x. \prod_{j=0}^{m-1} \psi_{j,\mu}(1)).$$

By construction, the sequence $\psi_{n,\mu_n}(x)$ converges to $f^*(|x|^{1+(\lambda)})$, and we may therefore compare the relative total scales of ψ_{n,μ_n} and $\psi_{n-1,\mu_{n-1}}$. The scale

$$\rho_m = \prod_{j=0}^{m-1} \psi_{j,\mu_m}(1) \quad \text{satisfies, due to the linearization,}$$

$$\rho_m = (-\lambda)^m \prod_{j=0}^{m-1} (1 - \delta^{j-m} \frac{a + o(j^{-1})}{a + o(n^{-1})} + (Q\lambda)^j \delta^{-m} O(1)),$$

where we have used

$$\psi_{j,\mu_m}(1) = -\lambda(1 - \delta^{j-m} a_j/a_m + (Q\lambda)^{j-1}(1-\mu_m)O(1)).$$

Therefore, we find $\log|\rho_m|/m \to \log\lambda$ as $m \to \infty$.
We consider now,

$$\begin{aligned} x^1_n &= \psi^{(2^n-1)}(\mu_n 0) \\ &= \psi^{(2^n-1)}_{o,\mu_n}(0) \\ &= [\prod_{j=0}^{n-2} \mu_{j,\mu_n}(0)]\, \mu_{n-1,\mu_n}(0) \\ &= \rho_n, \end{aligned}$$

so for the problem scaled by $\psi(\mu_J)$, the assertion is proved.

REFERENCES

[H] M. Hénon, Commun. Math. Phys. 50, 69 (1976).

[Lo] E.N. Lorenz, J. Atmos. Sci 20, 130-141 (1963).

[F] M. Feigenbaum, J. Stat. Physics 19, 25 (1978).

[Ru] D. Ruelle, In Mathematical Problems in Theoretical Physics. Lecture Notes in Physics, Vol 80. Springer 1978.

[La] O. Lanford, Bressanone 1977, Proceedings.

[D] B. Derrida, A. Gervois, Y. Pomeau, Preprint, Saclay.

[BF] V. Franceschini, C. Tebaldi. Sequences of Infinite Bifurcations and Turbulence in a 5-Modes Truncation of the Navier-Stokes Equations (preprint).

[Fr] V. Franceschini. A Feigenbaum Sequence of Bifurcations in the Lorenz Model (preprint).

[CEL] P. Collet, J-P. Eckmann, O. Lanford (preprint).

CRITICAL PROPERTIES OF ONE DIMENSIONAL MAPPINGS

Bernard Derrida

Laboratoire de Physique de l'Ecole Normale Supérieure
24 rue Lhomond, 75231 Paris Cedex 05 (France)

INTRODUCTION

Bifurcation theory deals with the asymptotic (long time) behaviour of systems of differential equations (continuous time) or dynamical systems (discrete time). Generally these evolution laws are non-linear and involve a few adjustable parameters (for instance temperature or boundary conditions in hydrodynamics) (1,2). For certain critical values of these parameters, the asymptotic behaviour of the solution changes suddenly and qualitatively : this phenomenum is a bifurcation. This can be the transition from a stationary state to a limit cycle or from a periodic behaviour to a more erratic behaviour.

In many fields, the time dependence of phenomena is studied by writing differential equations connecting the relevant quantities of the problem. These evolution laws, even if they are deterministic, may present very complicated solutions (3) : in some cases, two initial conditions, even very near, may lead to completely different long time behaviours. This means that the knowledge of the initial conditions should be infinitely precise to choose between all the possible asymptotic behaviours of the system. The Lorenz model (1,4,5) which is an example of this kind has been extensively studied. This model introduced in the Benard problem is a set of three non linear differential equations. It shows that an erratic behaviour may exist for systems with a small number of degrees of freedom, and that a turbulent behaviour can occur after a few bifurcations (6,7). Usually these kinds of models are solved numerically and so it is difficult to distinguish between periodic with a long period, quasiperiodic and turbulent behaviours.

C. Bardos and D. Bessis (eds.), Bifurcation Phenomena in Mathematical Physics and Related Topics, 137–154.

The study of non linear differential systems is related to the study of mappings : on the one hand, it is necessary to use a discrete time to solve these systems numerically ; on the other hand, the behaviour in the neighbourhood of closed trajectories is often analysed by looking at the Poincaré mapping (8,9).

The study of mappings for themselves is also interesting : in many problems, iterative processes appear (10). For example, the Newton method gives the roots of a function f by iterating $x' = x - f(x)/f'(x)$. In ecology (11), the dynamics of some populations can be described by a mapping which relates the magnitude of a population in one generation to this magnitude in the preceding generation. For theoretical physicists, mappings are a very useful tool which enables one to understand phase diagrams and to calculate critical behaviours from the knowledge of the neighbourhood of the fixed points.

In mathematics, mappings play a prominent part in ergodic theory (3). It is possible to construct mappings presenting all the kinds of chaotic behaviour (ergodicity, mixing ...). For instance, in some cases, it is possible to generate sequences of random numbers using perfectly deterministic mappings.

The study of the fixed points and the periods of rational functions is a rather old problem (12,13). In the last few years, some mappings have attracted attention because they present strange attractors (14) comparable to the attractors of the Lorenz model or related models. In some cases, it is necessary to look at the structural stability (2,3) of these attractors, namely if they persist for a small deformation of the mapping.

One interest of non linear mappings is that it is sometimes possible to follow all the transitions from a stationary state to a turbulent behaviour. This is an advantage of mappings over differential systems for which, when a limit cycle becomes unstable, it is only possible to describe analytically the neighbourhood of the bifurcation using a perturbation theory (8).

In this work, we are going to focus our attention on one dimensional mappings with a single maximum (15,16,17,18). Even in one dimension, mappings may give very complicated behaviours like cycles, strange attractors, ergodic regions (this is also an advantage over differential systems). After recalling the notions of fixed point, periodic and non periodic attractors, we shall especially discuss the global properties of the bifurcation diagram (19). We shall use the internal similarity (20,21) of this diagram to understand the critical properties around the accumulation points of cascades (22) of bifurcations. We shall see that a "real space" like renormalisation theory allows one to describe the critical behaviours qualitatively and quantitatively. The situation is more complicated than in phase transition problems because the bifurcation diagram of these mappings has an infinite number of critical points and critical exponents. It is possible to construct series of renormalisation methods(23) which give more and more precise results in the case of quadratic mapping. These results will be compared with the numbers obtained from numerical study (24) with a very high precision.

We shall conclude by the study of piecewise linear transforms for which one can find explicitly the stability zones of all periods and one can write exact renormalisation formulas.

1. BIFURCATION DIAGRAM

1.1. Fixed points and periods

Let T be a mapping of the real axis.
A set of p distinct points $y_1, y_2 --- y_p$ form a cycle or a period of length p if the mapping T acts on these points like a circular permutation :

$$y_{i+1} = T\,(y_i) \qquad 1 \leqslant i \leqslant p-1$$

$$y_1 = T\,(y_p)$$

This cycle is stable (or attractive) if this set of points verify:

$$-1 < \prod_{i=1}^{p} \left.\frac{dT}{dx}\right|_{y_i} < 1$$

Then one can show that all the points of an open neighbourhood of the set $y_1, y_2 --- y_p$ ultimately fall on this cycle.

A fixed point is a period of length 1.

Generally a given mapping has many cycles : most of them are unstable (or repulsive) and only a few are stable. In some cases, for example when T is a polynomial, there exists only a finite number of stable cycles (12) (when T is a polynomial, this number is bounded by the number of roots of $T'(x)$).

1.2. Bifurcation diagram

For a given mapping T, we are going to study the behaviour of $T^n(x)$ when $n \to \infty$:

$$T^n(x) = \underbrace{ToTo \; --- \; oT}_{n \text{ times}}(x)$$

To describe what happens when one deforms a mapping, we take the example of the quadratic mapping which has been extensively studied :

$$T_a(x) = 1 - a\, x^2 \qquad (1)$$

One can show that when $-\frac{1}{4} < a \leqslant 2$, there exists an interval $I = [1-a, 1]$ stable under the action of T_a. (For other values of a : $a < -1/4$ or $a > 2$, the mapping drives almost all the points to infinity).

• If $-\frac{1}{4} < a < \frac{3}{4}$, T_a has a stable fixed point $x = \frac{-1+\sqrt{1+4a}}{2a}$

and the iterates of any point of the interval I converge on this fixed point.

• If a = 2, we have a turbulent behaviour : for almost all the points x of I, the iterates of x are dense in I and fill this interval with a density $\rho(x) = \frac{1}{\sqrt{1-x^2}}$. Moreover, the action of T always separates two different points even if they are very near initially.

The <u>bifurcation diagram</u> describes what happens between $a = \frac{3}{4}$ (stationary state) and a = 2 (turbulent behaviour).

• When a crosses $\frac{3}{4}$, the fixed point becomes unstable and a stable period 2 appears. Then the iterates of a point x (when $n \to \infty$) waver between the two points of the cycle ($x = \frac{\sqrt{4a-3}\pm 1}{2a}$) This cycle 2 remains stable as far as $\frac{3}{4} < a < \frac{5}{4}$ and becomes unstable when $a = \frac{5}{4}$.

• At this value $a = \frac{5}{4}$, a stable period 4 appears. We have therefore a normal bifurcation from a period 2 to a period 4. When a increases, this period 4 remains stable until a new bifurcation where a stable period 8 appears. One can see in the same manner a succession of bifurcations : a period 2^n becomes unstable and a period 2^{n+1} appears. These bifurcations happen quicker and quicker until an accumulation point (22) $a_c=1.4011552$ where all the periods 2^n have appeared and have become unstable.

• Above a_c, one can see stability zones of other periods (3, 5 etc...).

In Figure 1, we have drawn the stability zones of the periods $\leqslant 5$ in the bifurcation diagram.

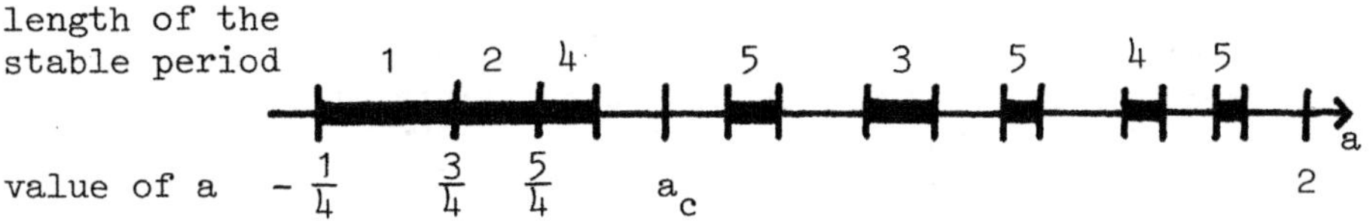

Figure 1

We can note that :

i) A given period may have many stability zones. It is possible to prove that the number of stability zones of a period n increases like $2^{n-1}/n$ when n is large. So these zones are very numerous and very narrow.

ii) Two stability zones of a same period can be distinguished by the order of the points of the cycle on the real axis. For example, for the 2 stability zones of period 4, the ordering of the points is :

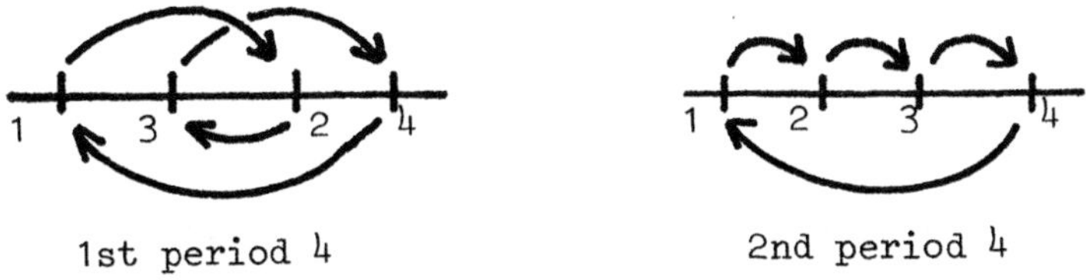

Figure 2

iii) The stability zones of periods are not contiguous in the bifurcation diagram, except when a period p becomes unstable (with a increasing); in this case a period 2p appears when the period p becomes unstable, and then the whole cascade of periods p 2^n and the stability zones of all these periods are contiguous.

Metropolis Stein and Stein (19) have given a complete description of the bifurcation diagram for mappings of the interval which depend on a parameter λ :

$$T_\lambda(x) = \lambda\ f(x)$$

They claim that the bifurcation diagram is the same for a large class of functions f (with a single maximum).

1.3. Values of a without stable cycle

The stability zones do not completely fill the bifurcation diagram. One can separate the values of a without limit cycle in two classes :

i) the points like $a_c = 1.4011552$ (or $a = 1.786440$) where there exists an attractor which is a Cantor set (strange attractor). These values of a are accumulation points of cascades of bifurcations. For example, the value $a_c = 1.4011552$ is the accumulation point of the bifurcations of periods 2^n when $n \to \infty$. The attractor is infinite but its Lebesgue measure is zero. It is the study of the neighbourhood of these critical values of a that we are going to look at in the remainder of this work. For the scaling properties of these strange attractors see reference (25).

ii) The points like $a = 2$ (or $a = 1.54368901$). The iterates of almost all the points x are dense in one or many intervals (26, 27). In some cases, one can find the density of these iterates (for example, when $a = 2$, this density is $d\mu(x) = (1-x^2)^{-1/2}\,dx$). These particular values of a occur when the iterates of the maximum of T fall on an unstable period after a few iterations.

Remark : an interesting question is to know what part of the measure of the bifurcation diagram is occupied by the stability zones of all the periods. We shall see in Section 4 that, in the case of piecewise linear mappings, the stability zones fill the whole measure of the bifurcation diagram.

2. CRITICAL BEHAVIOUR AND INTERNAL SIMILARITY

2.1. Critical behaviour

We have seen that when the fixed point becomes unstable, a stable period 2 appears, and so on : a period 2^n becomes unstable when a period 2^{n+1} appears.

This cascade of bifurcations has an accumulation point a_c. Call $n(a)$ the length of the stable cycle of the mapping T_a. If one draws this function of a to represent the sequence of bifurcations, one obtains the following picture :

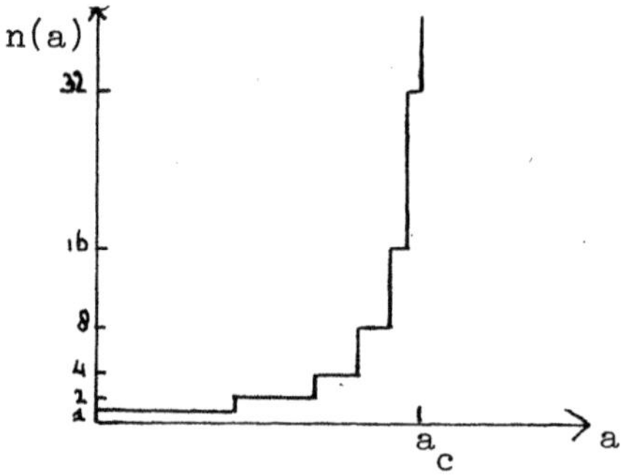

Figure 3

It turns out that the function n(a) diverges like a power law :

$$n(a) \sim |a_c - a|^{\nu_2} \quad \text{when } a \to a_c$$

As a matter of fact, the function n(a) has an infinite number of jumps near a_c and the correct way of writing the critical behaviour of n(a) is :

$$n(a) \sim |a_c - a|^{\nu_2} g_2 \left(\nu_2 \frac{\ln(a_c-a)}{\ln 2}\right) \tag{2}$$

where the function g_2 must verify

$$g_2(z) = g_2(z + 1)$$

One can measure the critical exponent ν_2 very precisely by calculating the successive values of bifurcations (24). In the case of the quadratic mapping (1), one finds :

$$\nu_2 = - .449807 \tag{3}$$

We have seen that when a period p becomes unstable, a period 2p appears, then a period 4p and so on. The cascade of these periods $2^n p$ has an accumulation point $a_c(p)$. (For example, $a_c(3) = 1.777982$). One can verify numerically that near these critical points $a_c(p)$, the function n(a) is given by a power law as in formula (2) with the same exponent ν_2. One finds again the same exponent ν_2 for cascades of bifurcations of other mappings (for example, $T_a(x) = 1 - a \sinh^2 x$) with a quadratic maximum.

2.2 Internal similarity (20,21)

The ordering of stable periods in the bifurcation diagram looks rather complicated. Nevertheless, it has a simple property of internal similarity : the whole bifurcation diagram is similar to some of its parts.

As shown in Figure 4, between two appropriate values of a (a_2, a_2') in the bifurcation diagram (Fig.4a), all the stable periods are even and the stability zones (Fig.4b) follow one another in the same order as in (Fig.4a), provided that the periods are multiplied by 2.

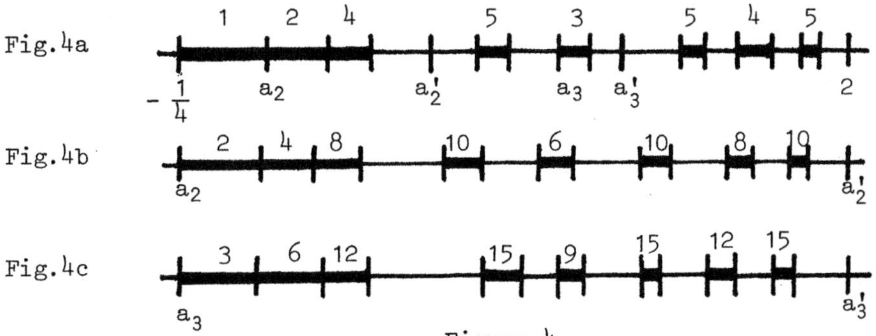

Figure 4

So there exists a monotonic function R_2 which assures the correspondence between diagram (4a) and diagram (4b) :

$$\text{If} \quad A = R_2(a) \qquad \text{then} \quad n(A) = 2n(a) \tag{4}$$

In the same way, one can select two points a_3 and a_3' between which the bifurcation diagram is reproduced as a whole, provided that the length of periods is multiplied by 3.

So there exists another monotonic function R_3 such that

$$n(R_3(a)) = 3n(a) \tag{5}$$

More generally, one can prove the internal similarity around the stability zone of any period p and there exists a family of functions R_p mapping the bifurcation diagram on its parts

$$n(R_p(a)) = p\, n(a) \tag{6}$$

Using these functions R_p, it is easy to obtain the critical points and the exponents.

Take the cascade $1 \to 2 \to 2^2 \to --- \to 2^n \to ---$. Call a_n the value of a, where period 2^n bifurcates to period 2^{n+1}.

One has $\quad a_{n+1} = R_2(a_n)$

Then $a_c = 1.4011552$ and ν_2 are given by :

$$a_c = R_2(a_c)$$

$$\nu_2 = \frac{\ln 2}{\ln\left(\left.\frac{dR_2}{da}\right|_{a_c} \right)} \tag{7}$$

Take now the cascade of bifurcations $p \to 2p \to --- \to 2^n p \to ---$. If b_n is the bifurcation value from period $2^n p$ to period $2^{n+1} p$ one has :

$$b_{n+1} = R_p(R_2(R_p^{-1}(b_n))) \tag{8}$$

If we call b_c the accumulation point of this cascade, b_c is given by :

$$b_c = R_p(R_2(R_p^{-1}(b_c))) = R_p(a_c)$$

Using (8), one can prove that the critical exponent of this cascade of bifurcations is still given by (7).

In the same way that the fixed point of R_2 is a critical value of a where the function n(a) diverges, the fixed points of renormalisations R_p are also critical points.

Using (6), one can show that in the neighbourhood of the fixed point α_c of R_p, the function n(a) behaves like :

$$n(a) \sim (\alpha_c - a)^{\nu_p} g_p\left(\nu_p \frac{\ln(\alpha_c - a)}{\ln p}\right) \tag{9}$$

where the exponent ν_p is given by :

$$\nu_p = \frac{\ln p}{\ln\left(\left.\frac{dR_p}{da}\right|_{\alpha_c}\right)}$$

and where g_p is a periodic function

$$g_p(z+1) = g_p(z)$$

When p = 3, we have measured ν_3 and found :

$$\nu_3 = -\ .273\ 84 \tag{10}$$

It is possible to measure ν_p for all p but it seems that there are no simple relations between these exponents.
We shall see in Section 4 that the renormalisations R_p can be obtained exactly in the case of the piecewise linear mappings. Nevertheless, generally, it is only possible to find approximate forms of these renormalisations : this will be explained in Section 3.

2.3. Comparison with phase transition problems

The method of finding exponents and critical points from renormalisation formulas is usual in phase transition theory. Our problem here has two important peculiarities :

i) In this problem there exist many distinct renormalisations R_p. These renormalisations do not commute. This means that we have an infinite number of critical points where the function n(a) diverges. These critical points are the fixed points of all these renormalisations, but also the iterates of one of these fixed points by the other renormalisations.
The simple fact that there exist 2 distinct renormalisations (R_2 and R_3) is enough to prove that the set of critical points is non countable : if α is a critical point, then all the points $R_{i_1}(R_{i_2} --- R_{i_n}(\alpha) ---)$ are critical points for any choice of i_1,

i_2 --- i_n. This shows that the set of critical points is a Cantor set.

ii) In phase transition theory, the renormalisation formulas are sometimes discrete but it is only an artefact of the method and in principle the renormalisation can be infinitesimal. Here the discrete aspect of the renormalisation is fundamental. R_2 multiplies periods by 2, R_3 by 3 but it is impossible to make them infinitesimal. This is the reason for the presence of function g which modulates the critical behaviour in formula (9).

3. APPROXIMATE RENORMALISATIONS

It seems very hard to find exactly the functions R_p. But it is possible to find approximations which give good values for the critical points and the exponents in comparison with measured values (obtained from computer analysis).

3.1. Calculation of ν_2 and a_c

We have seen that the interval $I = [1-a, 1]$ is invariant under the action of $T_a(x) = 1-a\,x^2$ if $-\frac{1}{4} < a < 2$. On can show that if $a_2 < a < a_2'$ ($a_2 = \frac{3}{4}$ and $a_2' = 1.543689$ are the values of figure 4), the interval I can be cut in two sub intervals $I_1 = [1-a, a-1]$ and $I_2 = [1-a + 2a^2 - a^3, 1]$ such that

$$T_a(I_1) \subset I_2 \qquad (11)$$
$$T_a(I_2) \subset I_1$$

Remark : It is obvious that when $a_2 < a < a_2'$, any stable period is even.

Using (11), it is easy to see that I_1 and I_2 are both stable under the action of T_a^2

$$T_a^2\,(I_1) \subset I_1$$

$$T_a^2\,(I_2) \subset I_2$$

Searching the renormalisation R_2 is equivalent to finding the mapping T_A which is homeomorphic to the restriction of T_a^2 on I_1 or I_2. This can be done only approximately.
We have : $T_a^2(x) = 1-a + 2a^2x^2 - a^3x^4$.

On the interval I_1, T_a^2 has a single maximum for $x = 0$. By neglecting the terms of order x^4, one can show that T_a^2 and T_A are homeomorphic provided that :

$$A = 2a^2(a - 1) \qquad (12)$$

This means that $n(a) = 2n(A)$.
These simple relations give

$$a_c = \frac{1+\sqrt{3}}{2} \quad \text{and} \quad \nu_2 = \frac{\ell n\, 2}{\ell n\left(\frac{4-\sqrt{3}}{13}\right)}$$

(the numerical values are in Table 1).

This method can be improved by considering a mapping $T_{a,b,c}$.. depending on several parameters. In the case of 3 parameters, this can be done by neglecting terms of order x^8 in T^2 :

If $T_{a,b,c} = 1-a\ x^2 + b\ x^4 + c\ x^6$

and $T^2_{a,b,c,} = \alpha-\beta\ x^2 + \gamma\ x^4 + \delta\ x^6 + O(x^8)$

Then $A = \alpha\ \beta$
$B = \alpha^3\gamma$
$C = \alpha^5\delta$

These approximate renormalisations are possible for any number of parameters, but the calculations become quickly tedious.

The results obtained by these different methods are given in Table I.

Method	a_c	ν_2
renormalisation with		
1 parameter	1.366 025	- . 3970
2 parameters	1.400 030	- . 4393
3 parameters	1.401 393	- . 4597
measured values	1.401 155	- . 4498

TABLE I

These results get better when the number of parameters increases but not in a monotonic way.

It is possible to find other methods of renormalisation : the best value of ν_2 we haye found is $\nu_2 = -\ .4486$.

3.2 Calculation of ν_3 and of the fixed point of R_3

When $a_3 < a < a_3'$ ($a_3 = \frac{7}{4}$ and $a_3' = 1.7903275$ are the values of Figure 4c), it is possible to find 3 intervals I_1, I_2, I_3 such that

$$T(I_1) \subset I_2 \quad ; \quad T(I_2) \subset I_3 \quad ; \quad T(I_3) \subset I_1$$

so the same calculations can be done to approximate the renormalisation R_3 :

$$T^3(x) = 1-a + 2a^2 - a^3 + 4a^3(a-1)x^2 + O(x^4)$$

This gives, by neglecting terms of order x^4 :

$$A = 4a^3(a-1)(a^3 - 2a^2 + a-1)$$

with $n(a) = 3n(A)$.

This is the formula for the 1 parameter renormalisation.

In Table II, we give the numerical results of this renormalisation with 1 or 2 parameters :

Method	fixed point of R_3	ν_3
renormalisation		
1 parameter	1.784 998	- . 26191
2 parameters	1.786 466	- . 27391
measured values	1.786 440	- . 27384

TABLE II

3.3 Mapping $T_a(x) = 1 - a|x|^{1+\varepsilon}$

We have seen that the exponents ν are the same for all the mappings with a quadratic maximum.
These exponents depend continuously on the shape of the mapping around its maximum.
For mappings like $T_a(x) = 1-a|x|^{1+\varepsilon}$, one can make an ε expansion when ε is small and positive.

One obtains :

$$\nu_2 \simeq -1 + \frac{1}{2\,\ell n\, 2}\,\frac{\varepsilon}{\phi}$$

$$a_{\bar{c}} = 1 + \phi$$

where ϕ is given by $\phi + \varepsilon + \varepsilon\,\ell n\,\phi = 0$.

These results were found by calculating the bifurcation values of this mapping. It is possible to recover them by a renormalisation as in formula (12) :

$$A = (1+\varepsilon)|a-1|^{\varepsilon} a^2$$

4. PIECEWISE LINEAR MAPPINGS

In this Section, we consider a piecewise linear mapping $T_{b,\lambda}$ which depends on 2 parameters :

$$\begin{aligned} T_{b,\lambda}(x) &= 1 - \lambda b \qquad \text{if} \quad |x| < b \\ &= 1 - \lambda|x| \qquad \text{if} \quad |x| > b \end{aligned} \tag{13}$$

(The condition $\frac{\lambda-2}{\lambda(\lambda-1)} < b < \frac{1}{\lambda-1}$ defines the bifurcation diagram, because this is the condition that an interval $I = [1-\lambda+\lambda^2 b,\ 1-\lambda b]$ is stable under the action of T).

We shall give, for $\lambda = 2$, the stability zones of all the periods and prove that they fill all the Lebesgue measure of the bifurcation diagram.
Then we shall discuss the case of any value of λ to write the exact renormalisation formulas and the critical curves.

4.1 Stability zones of all the periods for $\lambda = 2$

$$\begin{aligned} T_b(x) &= 1 - 2b \qquad \text{if} \quad |x| < b \\ &= 1 - 2|x| \qquad \text{if} \quad |x| > b \end{aligned} \tag{14}$$

Though $\lambda > 1$, the stability of periods comes from the flat maximum of the mapping T. So if a period n is stable, one of the points of the period is $1 - 2b$.
A period n is stable if and only if :

$$|T^i\ (1-2b)| > b \qquad 0 \leqslant i \leqslant n-2$$

$$|T^{n-1}(1-2b)| < b$$

Using (14), this proves that there exists an integer p such that:

$$\frac{2p+1}{2^n+1} < b < \frac{2p+1}{2^n-1}$$

Reciprocally, one can show that if $\frac{2p+1}{2^n+1} < b < \frac{2p+1}{2^n-1}$ where n and p are integers, then the mapping T has <u>a stable period of length $\leqslant$ n</u>.

Now we can find the stability zones of all the periods.

$$\text{Call } A_n = \bigcup_{p=0}^{(2^{n-1}-1)} I_{n\,p} \quad \text{where } I_{np} = \left] \frac{2p+1}{2^n+1} , \frac{2p+1}{2^n-1} \right[$$

$$\text{and } B_n = \bigcup_{q=1}^{n} A_q \qquad (15)$$

According to what we have said, B_n is the set of the stability zones of all the periods of length $\leqslant$ n.
So the set C_n of values of b for which a period n is stable is :

$$C_n = B_n - B_{n-1} \qquad (16)$$

Notice that generally $C_n \neq A_n$: for example

$$C_2 = \left] \frac{1}{5} , \frac{1}{3} \right[\quad \text{whereas} \quad A_2 = \left]\frac{1}{5} , \frac{1}{3}\right[\cup \left] \frac{3}{5} , 1 \right[$$

4.2. Measure of these stability zones

All the bifurcations of the mapping T (14) occur for $0 < b < 1$.

Some values of b ($b = \frac{1}{6} , \frac{3}{28} , \frac{7}{120}$ etc...) do not belong to any stability zone (it is impossible to find 2 integers p and n such that $\frac{2p+1}{2^n+1} < \frac{1}{6} < \frac{2p+1}{2^n-1}$). Nevertheless we are going to show that the stability zones fill the whole Lebesgue measure of the bifurcation diagram.
Using (15) and (16), we have :

$$\mu\Big(\bigcup_{n=1}^{\infty} C_n \Big) = \mu\Big(\bigcup_{n=1}^{\infty} B_n \Big) = \mu\Big(\bigcup_{n=1}^{\infty} A_n \Big)$$

This implies :

$$\mu(\bigcup_{n=1}^{\infty} C_n) \geqslant \mu(A_{n_1})+\mu(A_{n_2}-A_{n_1})+\mu(A_{n_3}-(A_{n_2} \cup A_{n_1}))+ \dots$$

$$\dots +\mu(A_{n_p}-(A_{n_{p-1}} \cup \dots \cup A_{n_1}))$$

for any choice of the sequence of integers $n_1, n_2 \dots n_p$.

Now, we can deduce that $\mu(\bigcup_{n=1}^{\infty} C_n) = 1$ by showing that $\mu(A_{n_1})$, $\mu(A_{n_2}-A_{n_1}) \dots \mu(A_{n_p}-(A_{n_{p-1}} \cup \dots \cup A_{n_1}))$ are as near as one wants from $\frac{1}{2}$, $\frac{1}{2\times 3}$ .... $\frac{1}{p(p+1)}$.

First, if n_1 is large enough, A_{n_1} consists of a large number of narrow intervals $I_{n_1,p}$. Take one part of the bifurcation diagram $]b, b+db[$: this part contains many intervals $\left]\frac{2p+1}{2^{n+1}}, \frac{2p+3}{2^{n+1}}\right[$ and only the length $\frac{4p+2}{2^{2n}-1}$ of these intervals is occupied by the set A_{n_1}. This means that A_{n_1} fills in $]b, b+db[$ a part of measure $b\,d\,b$ (if n_1 is large enough).

So $\mu(A_{n_1}) \to \int_0^1 b\,d\,b = \frac{1}{2}$ when $n_1 \to \infty$.

One can say that the filling rate by A_{n_1} of $]b,b+db[$ is equal to b.

There remain many holes and these holes are very narrow. Now by choosing n_2 much larger than n_1, it is easy to see that each hole contains many intervals I_{n_2p}. So A_{n_2} will fill these holes with the filling rate b.

$$\Rightarrow \quad \mu(A_{n_2}-A_{n_1}) \to \int_0^1 (1-b)b\,db = \frac{1}{2\times 3}$$

It is possible to go on and prove that

$$\mu(A_{n_p}-(A_{n_{p-1}} \cup \dots \cup A_{n_1})) \to \int_0^1 (1-b)^{p-1} b\,d\,b = \frac{1}{p(p+1)}$$

This means that $\mu(\bigcup_{n=1}^{p} A_{n_i})$ can be as near as required from

$$\sum_{i=1}^{p} \frac{1}{i(i+1)} = 1 - \frac{1}{p+1}$$

This proves that $\mu(\bigcup_{n=1}^{\infty} C_n) = 1.$

4.3 Renormalisations for a general value of λ

Some results obtained when $\lambda = 2$ remain true for any value of λ : we have seen that we can find the stability zones from the intervals $I_{n,p} = \left]\frac{2p+1}{2^n+1}, \frac{2p+1}{2^n-1}\right[$. This can be generalized for $\lambda > 2$ by replacing these intervals by intervals of the form : $\left]\frac{P(\lambda)}{\lambda^n+1}, \frac{P(\lambda)}{\lambda^n-1}\right[$ where $P(\lambda)$ is the polynomial of degree n-1 whose coefficients are + 1 or − 1 such that $P(2) = 2p+1$.

We can again take the arguments of Section 3 to construct the renormalisations R_p.
One can show that when

$$\frac{\lambda^2-2}{\lambda^2}\,\frac{\lambda-1}{\lambda^2-1} < b < \frac{\lambda-1}{\lambda^2-1}$$

it is possible to find two intervals I_1 and I_2.
The restriction of $T^2_{\lambda,b}$ on one of these two intervals is exactly homeomorphic to the mapping $T_{\Lambda,B}$ with

$$\Lambda = \lambda^2$$
$$B = \frac{b}{\lambda-1}$$

So we can write the renormalisation R_2 :

$$R_2(\Lambda,B) = (\sqrt{\Lambda},(\sqrt{\Lambda}-1)B) \tag{17}$$

This renormalisation maps the whole bifurcation diagram into one of its parts multiplying the period by 2

$$n(\sqrt{\Lambda},\ (\sqrt{\Lambda}-1)B) = 2n\ (\Lambda,B) \tag{18}$$

Using this renormalisation, we can find the equation of critical line corresponding to the accumulation of the bifurcations $1 \to 2 \to 2^2 \to \ldots \to 2^n \to \ldots$

The equation of this line is given by :

$$b_c = \frac{1}{\lambda} \prod_{i=0}^{\infty} (1 - \lambda^{-2^i})$$

One can verify that this line is stable under the action of R_2. If $n(\lambda,b)$ is the length of the stable limit cycle of $T_{\lambda,b}$, it is possible to prove that :

$$n(\lambda,b) \sim - \log(b-b_c) \quad \text{when} \quad b \to b_c \tag{19}$$

So we have $\nu_2 = 0$
One can again take these calculations for all the other renormalisations.
For example, for the renormalisation R_3, we can find 3 intervals I_1, I_2, I_3 as in Section 3,2 provided that :

$$\frac{\lambda^3-2}{\lambda^3} \times \frac{\lambda^2-\lambda-1}{\lambda^3-1} < b < \frac{\lambda^2-\lambda-1}{\lambda^3-1}$$

It is then possible to show that the restriction of $T^3_{\lambda,b}$ on one of these intervals is homeomorphic to $T_{\Lambda,B}$ with

$$\Lambda = \lambda^3$$
$$b = (\lambda^2 - \lambda - 1)B$$

This implies :

$$R_3(\Lambda,B) = (\Lambda^{1/3},(\Lambda^{2/3}-\Lambda^{1/3}-1)B)$$

The corresponding critical line (accumulation of the cascade $3 \to 3^2 \to \ldots. \to 3^n \to \ldots$) is then :

$$b_c = \frac{1}{\lambda} \prod_{i=0}^{\infty} (1 - \lambda^{-3^i} - \lambda^{-2\times 3^i})$$

The function $n(\lambda,b)$ has here also a logarithmic divergence and so $\nu_3 = 0$.

References

1. Martin, P.C., 1976, J. de Physique, C1, 37, p.57.
2. Chenciner, A., 1978, J. de Physique, C5, 39, p.69.
3. Arnold, V., Avez, A., "Problèmes ergodiques de la mécanique classique", Gauthier-Villars, 1967.
4. Lorenz, E.N., 1963, J. Atmos. Sci., 20, p.130.
5. Ibañez, J.L., Pomeau, Y., 1978, J. Non-Equil. Therm., 3, p.135.
6. Ruelle, D., Takens, F., 1971, Comm. Math. Phys., 20, p.167.
7. Smale, S., 1967, Bull. Amer. Math. Soc., 73, p.747.
8. Iooss, G., 1978, J. de Physique, C5, 39, p.99.

9. Henon, M., 1970, Quat. Appl. Math. 27, p.291.
10. Colloque du C.N.R.S. : "Transformations Ponctuelles et Leurs Applications", 1976, Editions du C.N.R.S., Paris.
11. Meyer, J.A., 1978, J. de Physique, C5,39, p. 29.
12. Fatou, P., Bull. Soc. Math. de France, 1919, 47, p.161 ; 1920, 48, p.33 ; 1920, 48, p. 208.
13. Julia, G., "Mémoire sur les itérations des fonctions rationnelles", 1918, J. Math. Pures Appl. 4,p.47.
14. Henon, M., 1976, Comm. Math. Phys. 50, p.69.
15. May, R.M., 1976, Nature, 261, p.459.
16. May, R.M., Oster, G.F., 1976, Am. Nat. 110, p.573.
17. Guckenheimer, J., 1977, "Inventiones Mathematicae", 39,p.165.
18. Li, T.Y., Yorke, J.A., 1975, Am. Math. Monthly,82, p. 985.
19. Metropolis, N., Stein, M.L., Stein, P.R., 1973, J. Comb.Theory A 15, p.25.
20. Mira, C., VII International Conference on Non Linear Oscillations, 1975, Berlin ; R.A.I.O.R. Automatique/Systems Analysis and Control, 1978, 12,p.63.
21. Derrida, B., Gervois, A., Pomeau, Y., 1978, Ann. Inst. Henri Poincaré, 29, p. 305.
22. Myrberg, P.J., 1962, J. Math. Pures et Appl. (9), 41, p.339.
23. Derrida, B., Gervois, A., Pomeau, Y., 1979, J. Phys. A, 12, p. 269.
24. Feigenbaum, M.J., 1978, J. Stat. Phys. 19, p.25.
25. Coullet, P., Tresser, C., 1978, J. de Physique, C5, 39, p.25.
26. Lasota, A., Yorke, J.A., 1973, Trans. Amer. Math. Soc. 186, p.481.
27. Ruelle, D., 1977, Comm. Math. Phys. 55,p.47.

INTERMITTENCY : A SIMPLE MECHANISM OF CONTINUOUS TRANSITION FROM ORDER TO CHAOS.

Y. Pomeau

DPh.T. CEA-CEN Saclay, 91190 Gif-sur-Yvette, France

Prompted by experiments on Rayleigh-Bénard thermoconvection in normal liquid Helium made by A. Libchaber and J. Maurer at the Ecole Normale Supérieure Lab. (Paris), we have undertook a study of a particular mechanism of transition to turbulence : the intermittent transition.

A well known mathematical model for the transition to turbulence in thermoconvective flows is the Lorenz system [1]

$$\begin{aligned} \frac{dx}{dt} &= \sigma(y-x) \\ \frac{dy}{dt} &= -xz + rx - y \\ \frac{dz}{dt} &= xy - bz \end{aligned} \qquad (1)$$

These three ordinary differential equations have been thorougly studied [1,2] for parameters (r, σ and b) close to the one given by Lorenz. However other mechanisms of generation of intrinsic stochasticity have been found [3] for larger values of the external constraint r. In particular the transition to the turbulence via cascade of frequency halving has been shown to exist, leading to a strange attractor with a Poincaré section close to the picture obtained by Hénon [4] for the Cremona mapping.

In systematic studies of the Lorenz system on an analog computer, I had noticed a "bursting" transition to turbulence which I did not properly understood at that time. It is relevant however to quote the corresponding observations to make clear this mechanism. I followed on a T.V. screen the point of coordinates, say x(t),

C. Bardos and D. Bessis (eds.), Bifurcation Phenomena in Mathematical Physics and Related Topics, 155–162.

z(t) where the time evolution of x and y is governed by eq. (1). When a stable limit cycle was reached, the (analog) computation was so fast that it was impossible to follow visually the point on the trajectory. The impression was the one of a steady luminous closed curve. In a stochastic regime, the visual impression was that a full surface was covered by the motion (actually this surface was a plane projection of the strange attractor). On the contrary, around some well definite values of the control parameter, r, an intermediate picture was observed : an apparently closed curve was destroyed by intermittent bursts : during these bursts, a full surface was covered by the representation point, then a closed curve was recovered and so on.

This is only when we began to think about the experimental results of Libchaber and Maurer, that we realized that this mechanism of destruction of order had a quite general character and could be explained in simple terms.

When it is stable, the closed limit cycle in a d-dimensional flows has (d-1) stable manifolds (mfolds) : these mfolds are the surfaces along which the motion tends to the limit cycle, each surface, which is globally invariant by the motion, is characterized by a rate of convergence of the motion toward the limit cycle. If one looks at the Poincaré map, the limit cycle is a stable fixed point of this map. When considering the linear stability of this fixed point, one has to study the eigenvectors of the tangent first return map. These eigenvectors, when the corresponding eigenvalues are real and non degenerate (as we shall assume it) define the tangent direction to the stable mfolds near the fixed point (see Fig. 1). By a method due to Hadamard [5], it is possible to construct the stable (and unstable) mfolds at finite distances of the fixed point. This is (one of) the purpose of the stable mfold theorem.

At the bifurcation to turbulence in which we are interested, one of the stable mfolds becomes unstable. That is the modulus of the corresponding eigenvalue for the tangent map becomes larger than 1. This may occur in three different ways : if this eigenvalue is real, it crosses the unit circle at (+1) or (-1) or it may cross the unit circle at $e^{\pm i\varphi}$; $\varphi \neq 0 \pmod{\pi}$. When the real eigenvalue crosses at (-1), a pair of stable fixed points appears in the Poincaré map when the bifurcation is supercritical [6]. For a differential equation, this gives the well known phenomena of frequency halving. In a subcritical bifurcation, on the contrary, that is if an unstable period two collapse with the fixed point, there is a possibility of intermittent transition to turbulence [7]. A more detailed analysis of this sort of transition poses the same difficult problems than the subcritical transition when a complex eigenvalue crosses the unit circle [8]. In that follows I shall only consider the intermittency generated when the eigenvalue crosses the unit circle at (+1).

Generically, when one eigenvalue of the tangent map crosses the unit circle, the modulus of the other ones remains strictly less than one, if the limit circle was stable. If, during the motion, a point falls in the vicinity of the previous stable closed trajectory, it is first attracted to the unstable mfold, since the attraction toward the unstable mfold dominates the weak instability along the unstable mfold. This remark allows one to undertand the above picture for the intermittent behavior : during the bursts the point of coordinates $x(t)$, $y(t)$, $z(t)$ explores randomly the phase space. This bursting phase ends up when the point falls in the vicinity of a stable mfold of the ghost of the limit cycle. As the instability is very weak, the point is attracted very near of this ghost before to feel the weak repulsion along the unstable mfold. At the end of this attraction process, the motion follows practically a slowly diverging almost closed curve. This explains the observation of apparently closed trajectories on the TV screen of the analog computer. But this is *not* a stable trajectory and, as time goes on, the divergence becomes so large that a burst begins, where no periodic motion can be recognized. Then a new cycle begins, and so on.

All this qualitative picture is in close agreement with the behavior of the Lorenz model (1) in the parameter range $r \simeq 166$, $b = 8/3$, $\sigma = 10$. We leave b and σ fixed and make r vary. For r a little smaller than a critical value, say r_T ($\simeq 166.07$) a pair of stable limit cycle exist. They are image of each other in the transform $(x,y,z) \to (-x,-y,z)$ that leaves the Lorenz equations invariant.

In this subcritical subrange, the time evolution is perfectly periodic, after some transients, for almost all initial conditions (see Fig. 2a). As r becomes a little larger than r_T, this perfect periodicity (or laminar phase) is destroyed by bursting periods (Fig. 2b). The mean time lag between two bursts is infinite at r_T and depends on r near $r_T|_+$ as $[r-r_T]^{-1/2}$. Similarly, in the same intermittent phase, the time correlation functions decrease with oscillations over time interval of order $(r-r_T)^{-1/2}$. The Lyapunov number, which is a measure of the mean rate of instability of a trajectory grows from zero at r_T to values of order $(r - r_T)^{1/2}$.

The existence and value of these exponents may be explained very simply if one limits oneself to the Poincaré map near $r_T|_+$. As explained before, in the intermittent phase, the motion takes place during most of time on a surface that is the "remnant" of the stable mfold of the limit cycle. Thus, the Poincaré map is well represented by a one dimensional map along this mfold. This is what is pictured (from numerical experiment) in Fig. 3. The representative curve for the function $y_{n+1}(y_n)$ is very near to be tangent to the first bisectrix. By letting r vary around r_T, this function is deformed around the contact with the bisectrix

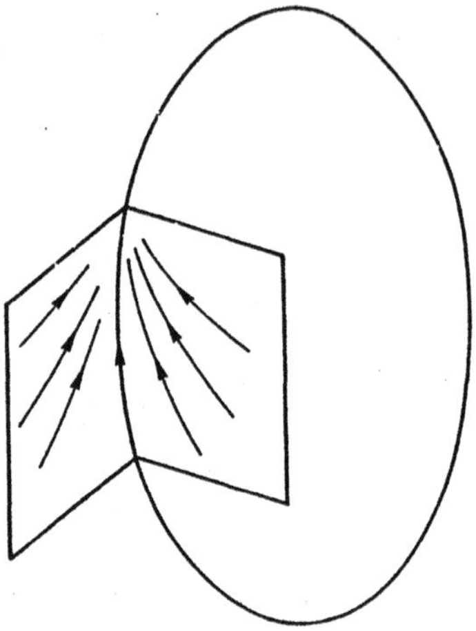

Fig.1. Invariant manifolds of a stable closed limit cycle for a flow in a 3d space.
The invariant manifolds are two susfaces which intersect along the limit cycle. Any trajectory starting in one of those surfaces remains on it and tends (exponentially if the limit cycle is stable) to the limit cycle.

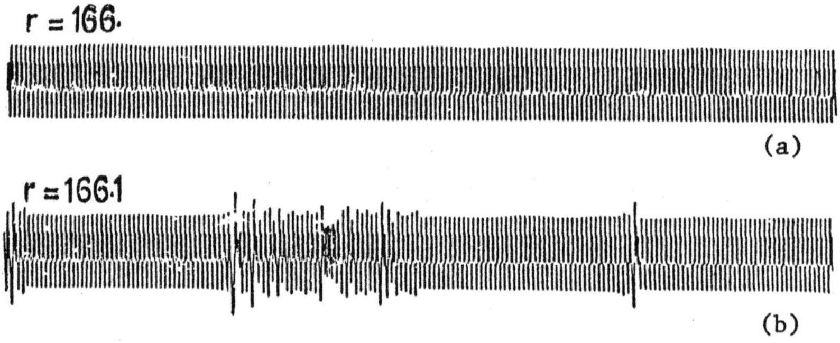

Fig.2. (a) Here are the time records of the stable oscillation of the Lorenz model at $r = 166$, $\sigma = 10$, $b = 8/3$.
(b) At $r = 166.1$ and from time to time, these oscillations are interrupted randomly by bursts.

as pictured in Fig. 4. For r less than r_T the curve intersects twice the bisectrix. This corresponds to two fixed points (= two closed trajectories for the Lorenz system) of the map $y_n \to y_{n+1}(y_n)$. One of these fixed points is stable, the other one being unstable. At $r = r_T$ these two fixed points collapse into a single one that is the point of tangency between the first bisectrix and the curve of equation $y_{n+1}(y_n)$. In the range $r = r_T|_+$ (i.e. in the intermittent phase) this unique fixed point disappears and a very narrow "channel" is open between the first bisectrix and the curve $y_{n+1}(y_n)$. During the periodic (or laminar) phase of the motion, the successive iterations allow the point to tunnel through this channel. This tunneling period corresponds, for the continuous time representation, to a nearly periodic motion.

When the point runs far from the channel region, the approximation of the Poincaré map (that is actually a mapping of a 2d surface on itself) by a 1d transform is no more valid. But during this bursting period, the points explores a domain that is quite large so that it falls again close to the channelling region and a laminar phase starts again.

To explain the 1/2 exponents given above, it is enough to get an estimate of the mean duration of a tunnelling (or laminar) period, as the coherence is completely destroyed during the bursts.

In the vicinity of the channel, the function $y_{n+1}(y_n)$ can be approximated as

$$y_{n+1} = y_n + \varepsilon + y_n^2, \tag{2}$$

where the irrelevant constants have been dropped and ε is proportional to the variation of the control parameter around its critical value ($\varepsilon \sim r - r_T$). In the channel the difference between y_{n+1} and y_n is so small that we may consider n as continuous and y as a smooth function of this index n. Thus the difference equation (2) becomes an ordinary differential equation

$$\frac{dy}{dn} = \varepsilon + y^2(n). \tag{3}$$

This is integrated at once, giving $n\sqrt{\varepsilon} = \text{Arctg}\,(y(n)/\sqrt{\varepsilon})$. Accordingly y(n) diverges (formally) at $n\sqrt{\varepsilon} = \pi/2$. This means that the iterates run through the channel after a number of iterations of order $\varepsilon^{-1/2}$. This corresponds to the exponent given above for the various quantities which are critical at the onset of intermittency.

As briefly explained in my lecture at the 1979 Cargèse meeting on intrinsic stochasticity in plasmas [8], it is not possible to get so simply the "critical exponents" for the growth from zero of

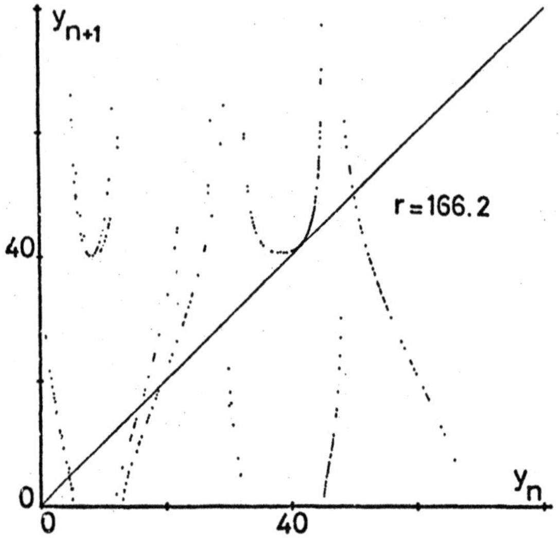

Fig.3. Here is the Poincaré map along the y-coordinates for $r = 166.2$ (a slightly beyond the onset of stochasticity). Notice the close contact between the curve $y_{n+1}(y_n)$ and the first bisectrix around $y_n \simeq 40$.

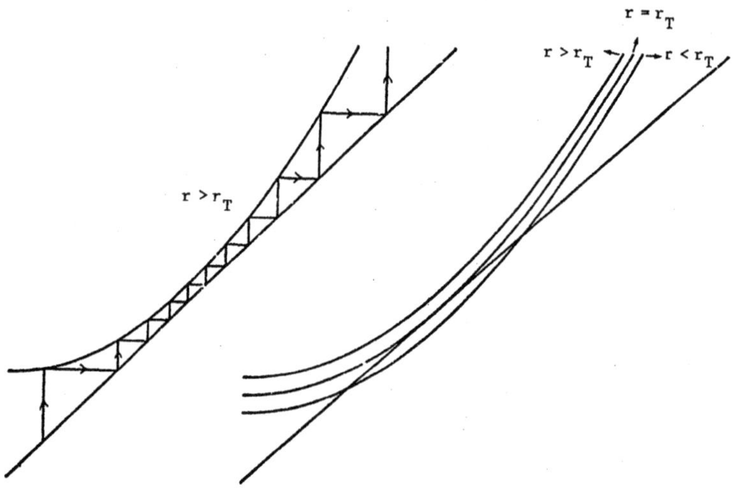

Fig.4. This is an idealized picture of the deformation of the curve $y_{n+1}(y_n)$ in the vicinity of the contact (that is around $y \sim 40$, as seen in fig.3) r crosses the onset of stochasticity, i.e. around $r_T \simeq 166.07$.

the Lyapunov number at the intermittency threshold when the turbulence comes from a subcritical bifurcation wherein the eigenvalue of the tangent map crosses the unit circle at (-1) or with an imaginary part.

In these cases, at the onset of intermittency, the time records of the fluctuations are intermittent in the probabilistic sense [9] : During a time lag T, as large as it is, there is a (laminar) period of apparently regular oscillations of the order of duration T, and one has to describe the situation as $T \to \infty$, by means of the fractal dimension of the set of turbulent bursts on the time scale. This is the subject of forthcoming publications.

An interesting suggestion was made recently [10] concerning the possibility of intermittency in *space dependent* structures. Consider the melting of a large 2d dissipative structure generated by a convective instability in nematics [11]. This structure remains well ordered (as, say, a crystal at 0°K) up to a threshold for the external constraint, for which it melts in a well definite way : grain boundaries appear spontaneously separating large crystallites which are randomly packed pieces of the subthreshold crystal structure. One might interpret these grain boundaries as sort of bursts in the intermittency phenomena, the crystallites being the "laminar phases". It is striking to notice, in particular, that at the onset of melting the average size of the crystallites diverges, as does the mean duration of the laminar phase in the intermittent time dependent phenomena. Perhaps this is one more manifestation of the subtle connections between time dependent phenomena and non trivial ground states, as studied in depth by S. Aubry.

As a conclusion, I would like to show the connection between the intermittent transition and general problems of bifurcation theory, which are the subject of this conference. This connection can be made by considering invariant measures of the flow. As shown by Ruelle [12], for a number of dissipative systems, there is an unique "relevant" invariant measure. This measure describes the average properties of the motion for almost all initial conditions. Furthermore this measure is selected when one adds to the deterministic equations of motion a small fluctuating external force, the amplitude of which tends to zero.

When a stable limit cycle exists, the (relevant) invariant measure for the Poincaré map, is a Dirac δ-function, with the fixed point as a support.When the limit cycle looses its stability by intermittent bursts, the invariant measures take a support that is continuous, this is the plane section of a "strange attractor". But this invariant measure keeps the memory of its δ-behavior before the onset of turbulence : it is heavily concentrated on the "ghost" of the fixed point. In this sense there is a smooth bifurcation from a δ-function invariant measure to a continuous invariant measure.

ACKNOWLEDGEMENTS

I am deeply indebted to Paul Manneville, with which I have done almost all this work.

REFERENCES

[1] E.N. Lorenz, J. of Atm. Sciences 20, 130 (1963).
[2] D. Ruelle, Lectures at the Bielefeld conf. (1978) Lecture Notes in Physics 80 (1978) 341, and Lecture Notes in Math. 565, 146 (1976).
[3] J.L. Ibanez, Y. Pomeau, J. of non eq. Therm. 3, 135 (1978).
[4] M. Hénon, Comm. Math. Phys. 50, 69 (1976).
[5] M.W. Hirsch, C.C. Pugh, M. Shub "*Invariant manifolds*" Lecture Notes in Math. 583, Springer-Verlag, Berlin, New-York (1977).
[6] G. Iooss "*Bifurcations of maps and applications*" North-Holland Math. Studies 36, N.H. Pub. Company, Amsterdam and New-York (1979).
[7] A. Arneodo, P. Coullet, C. Tresser, private communication.
[8] Y. Pomeau, Workshop on "*Intrinsic stochasticity in plasmas*", Cargèse (1979) to be published by Ed. de Physique (Paris).
[9] B. Mandelbrot "*Fractals form, chance and dimensions*" Freeman, New-York (1978).
[10] M. Dubois, E. Guyon, Private communication.
[11] J.M. Dreyfus, E. Guyon, Private Communication.
[12] D. Ruelle, Phys. Lett. 72A, 81 (1979), "On the measures which describe turbulence", IHES preprint (1978).

METAL-INSULATOR TRANSITION IN ONE-DIMENSIONAL DEFORMABLE LATTICES

Serge AUBRY

Laboratoire Léon Brillouin, CEA-CEN-Saclay, 91190 Gif-sur-Yvette, and D.R.P. Univ. P. et M. Curie, 75 Paris, France.

ABSTRACT

Metal-insulator transition at 0 K in a one-dimensional classical deformable lattice (Peierls transition) is viewed through two complementary models which consider properly the discreteness of the lattice in the case of a Fermi wave-vector k_F incommensurate with the reciprocal lattice vector. The electron-phonon interaction is represented in both models by the same parameter λ while the electron-electron interaction and the spin effects are neglected. The first model assumes the knowledge at each atom of the potential $\lambda V(x)$ with amplitude λ. This potential is created by the charge density wave (CDW) of the electronic distribution and produces the periodic lattice distortion (PLD) at the same wave-vector $2k_F$ as the CDW. (The existence of the CDW and of the PLD was proved by Peierls). When $2k_F$ is incommensurate with the reciprocal lattice wave-vector and when V(x) is an analytic function, a phase transition versus λ, occurs at $\lambda = \lambda_c$ in the PLD, which transforms from a quasi-sinusoidal modulation for $\lambda < \lambda_c$ into an array of equidistant and locked phase defects for $\lambda > \lambda_c$. This transition is represented by the breaking of analyticity of the PLD hull function. The zero frequency mode corresponding to the phase rotation of the PLD-CDW (phason or Frölich mode), which carries a current, locks at $\lambda = \lambda_c$ and transforms the system from a metal into an insulator. If V is not differentiable enough, this transition does not exist and the system always remains insulating. The second model assumes that the PLD is known and submits the electrons to a quasi-periodic potential with Fermi and lattice periods. This potential is represented in a tight-binding model by a potential defined at site i by the periodic hull function $\lambda V(i)$ with the Fermi period $2\pi/2k_F$. For a particular analytic potential V(i) we

C. Bardos and D. Bessis (eds.), Bifurcation Phenomena in Mathematical Physics and Related Topics, 163–184.

show that a transition occurs versus λ : for $\lambda < \lambda_c$ all the electrons are described by modulated plane waves with an analytic hull function, while for $\lambda > \lambda_c$ all the electrons are exponentially localized. This transition induces also a metal-insulator transition. When the potential V is not differentiable enough, the electrons are expected to be always localized. On this basis we explain that the CDW and PLD hull functions must be both analytic or both discontinuous. A well-defined metal-insulator transition versus the electron-phonon coupling λ is expected. The conducting phase would then exhibit a Frölich mode with extended electronic wave functions and the insulating phase would consist in locked phase defects plus localized electrons. This last phase can be physically reinterpreted as a crystal of localized polarons superimposed onto the lattice.

1. INTRODUCTION

As is well known the fundamental difference between a metal and an insulator is that a weak DC electric field can move the electrons in the first system, while it cannot in the second system. (The differentiation is ideal only at 0 K since at finite temperature, there exists also some conductivity in insulators by thermal activation of the electrons).

Before the discovery of quantum mechanics, theories[1] which considered the electrons as classical particles were almost unsuccessful in describing the properties of metals. However, the qualitative description of insulators was not fundamentally bad. The advent of quantum mechanics allowed to make a simple and natural explanation of both metallic and insulating behavior for non-interacting electrons in perfect crystals. As a consequence of the Bloch theorem, the electronic wave functions in a periodic lattice are modulated plane waves which extend over the whole crystal. The Pauli principle used for non-interacting electrons (fermions) allows to build the band theory which requires the lattice periodicity. A metal is then characterized by an unfilled band and a free Fermi surface. However, such a model (which we call "integrable" cf. Ref.[2]) must be improved to take into account the electron-electron interaction, the electron-phonon interaction, the spin effects, etc... These effects may induce metal-insulator transitions such as for example the Wigner or the Mott transitions[1].

When the translational invariance of the crystal is broken, for example in amorphous systems or in incommensurate crystals, (crystals with a superimposed static and periodic modulation, the period of which is incommensurate with the period of the crystal) the band concept cannot be used in a strict sense. Moreover, P.W. Anderson[3] has shown that the wave functions of non-interacting electrons in a random potential could become localized or in other words square summable. This theory has been improved by S.N. Mott[4] who predicted the existence of mobility edges which separate between regions of different energies, those for which the electrons are localized from others in which the electrons propagate as extended waves. If at the Fermi energy the electrons are localized, the system is an insulator, while if they propagate the system is conducting. In the insulating region, the system is better described by localized classical electrons than by a band theory (however, one must keep the Pauli principle). At finite temperature, the small conductivity is then produced by hopping of electrons from localized states to neighbouring empty states.

In this paper, we restrict ourselves to the study of conductivity at 0 K in one-dimensional deformable lattices which break spontaneously their translational invariance.

2. ONE-DIMENSIONAL CONDUCTIVITY : PEIERLS INSTABILITY AND FRÖLICH CONDUCTIVITY

Let us consider a conducting one-dimensional and deformable chain of atoms (or molecules) at 0 K. The atomic displacements are considered classically while the electrons are assumed non interacting and quantized. If a periodic distortion with wave-vector q is imposed to the chain, the perturbed Schrödinger equation of a single electron in the distorted chain becomes then

$$-\frac{\hbar^2}{2m}\frac{\partial^2\psi}{\partial x^2} + \left(V_o(x) + \delta V(x)\right)\psi(x) = E\,\psi(x) \qquad (1)$$

$\hbar$ and m are the Planck constant and the electronic mass, E is the eigen energy, $V_o(x)$ is the periodic potential with period ℓ at the abscissa x (the unit cell spacing of the chain) created by the undistorted chain. $\delta V(x)$ is the perturbation of this potential produced by the distortion; it is a quasi-periodic function of x with the two fundamental periods ℓ and $2\pi/q$. At first order, its amplitude is proportional to the amplitude of the distortion. For $\delta V(x) = 0$, the electronic wave functions satisfy the Bloch theorem and can be written as

$$\psi_k(x) = e^{ikx}\,\chi_k(x) = |k\rangle \qquad (2)$$

where k is a wave-vector and $\chi_k(x)$ has the period ℓ of $V_o(x)$. The set of eigen-energies has a usual band structure.

Standard perturbation theory at first order exhibits non-zero off-diagonal terms. The largest one corresponds generally to the first harmonic of δV with wave-vector q :

$$\Delta = \langle -\tfrac{q}{2} \mid \delta V \mid \tfrac{q}{2} \rangle \qquad (3)$$

If $q = 2k_F$ (k_F is the Fermi wave-vector of the unperturbed chain) a gap of

$$\Delta = \langle -k_F \mid \delta V \mid k_F \rangle$$

is open at the Fermi energy. The total energy of the electronic gas is reduced. More precise calculations and exact models (see for example Ref.[6]) show that at 0 K the electronic energy gain is always larger than the elastic energy lost for small deformations in one-dimensional chains (but not necessarily in two or three dimensions). This is the Peierls instability [5]. The perturbed eigenfunction with the lowest energy is now a combination of the unperturbed wave functions $|k_F\rangle$ and $|-k_F\rangle$. It generates a superimposed periodic modulation of the electronic density with wave-vector $2k_F$. Thus, at 0 K, a one-dimensional conductor is expected to exhibit a periodic lattice distortion (PLD) with wave vector $2k_F$ associated with the periodic modulation of the electronic density which is called charge density wave (CDW).

At first sight, perturbation methods in usual band theory would suggest that the opening of a gap Δ at the Fermi level makes the system an insulator at 0 K. However, when the period $2\pi/2k_F$ of the CDW is incommensurate with the lattice spacing ℓ, Frölich has noted that the phase of the PLD-CDW can be chosen arbitrarily. This is physically obvious and he argued that the translation of the PLD-CDW (which corresponds to the phase rotation) can be done without any draw-back force because the system energy remains unchanged. This translation corresponds to a zero frequency mode (it is the Frölich mode, the phason or a Goldstone mode following various terminologies). It carries both an electronic charge and a lattice distortion (A high renormalized mass results for the electrons since the atoms are also involved in their motions.) The system becomes in fact conducting since a weak DC electric field produces a current by translating the PLD-CDW. Such a conductivity mechanism was initially proposed by Frölich to explain superconductivity before the discovery of the BCS theory [8] in which the Cooper pairing seems essential. However, perturbation theory does not really prove the existence of the Frölich mode and we will see that in fact it does not exist necessarily.

In the recent years, many quasi-one-dimensional metals [9] (which are often made of parallel and stacked organic chains) have been discovered. In agreement with the Peierls theory, they have been found to exhibit a PLD. It was also expected that the Frölich superconductivity model would explain the very high conductivity of these compounds at low temperature [10]. But in most cases the conductivity drops down unexpectedly at lower temperature, sometimes after a very sharp metal-insulator transition (e.g. TTF-TCNQ, see Ref.[9]).

Thus the pinning of the Frölich mode is necessary to explain the insulating phase. Various explanations of this pinning have been proposed but none of them appeared to be convincing. The main purpose of this paper is to show that when the electron-phonon coupling is strong enough, a one-dimensional chain is necessarily insulating at 0 K. Indeed most one-dimensional metals are specially known for a large electron-phonon coupling (from which perturbative calculations allowed to hope superconductivity at high temperature [9]). Thus, our results are in contradistinction with superconductivity and do not require extra-mechanisms to explain the insulating phase.

Our study has been prompted by classical mechanics results showing paradoxically that dynamical systems the motion of which can be explicitly calculated (or which can be expressed formally by quadratures), are exceptional (they are called "integrable systems"). In general, the calculations are impossible to be done explicitly. We understand this fact by the breaking of analyticity of certain functions describing the system (due to the occurrence of stochasticity) which of course cannot be written down with the help of the familiar analytic functions. The same situation turns to appear in the problem that we study here for which the analyticity breaking is responsible of the metal-insulator transitions.

Thus well-chosen models which explicit analytic solutions miss this important aspect of the physics. For example, we believe that one must be careful in generalizing the results of models using the Sine-Gordon equation in the classical or quantum version just because of the highly exceptional properties of this equation, and that the domain of validity of such theories, which are in fact physically approximate, must be carefully limited. In the present study, we have chosen simple models which do not neglect the discreteness of the lattice but we do not force their "integrability". In counter part, we are limited to qualitative results which are only exact at 0 K for a classical lattice and quantized electrons. The quantitative aspects must be studied either by approximate methods or numerical calculations. However, this approach brings new physical features which are not obtained by the other treatments.

We study the analyticity of functions (without knowing them explicitly) which describe the configurations of two crude models which treats separately the PLD and the CDW at 0 K. Both models yield a metal-insulator transition produced by the breaking of analyticity of these functions. We conjecture that the same kind of transition should occur in models which treat self-consistently the lattice and the electrons.

3. BREAKING OF ANALYTICITY OF THE (CLASSICAL) PERIODIC LATTICE DISTORTION AT 0 K AND EXTINCTION OF THE FRÖLICH CONDUCTIVITY

We assume in this section that the Peierls instability appears at the wave vector $2k_F$ and that we know the CDW or more precisely the superimposed periodic potential $\lambda V(u_i)$ which it produces on the one-dimensional chain. We write the energy of the chain as

$$\phi(\{u_i\}) = \sum_i \lambda V(u_i) + W(u_{i+1} - u_i) \qquad (4)$$

in which u_i is the abscissa of the ith atom (or molecule) along the chain (we do not consider any eventual transverse component of the lattice distortion). W is a smooth potential between neighbouring atoms, minimized for $u_{i+1} - u_i = \ell$ which is the unit cell spacing in the absence of CDW. We assume that W is a convex function in the domain of variation of $(u_{i+1} - u_i)$; this is not physically an important restriction because of the necessary stability of the free chain ($\lambda = 0$) under deformations. $\lambda V(x)$ is assumed to have the period predicted by Peierls (cf. the previous section) : $2\pi/2k_F = 2a$ which is determined, according to the Pauli principle, by the number c of conduction electrons per atom

$$2a = \frac{2}{c}\,\ell \qquad (5)$$

Besides, V is assumed to be an even function of x . (This last condition is useful to explicit exact results by our methods but does not seem physically essential). λ is a positive adjustable parameter representing the strength of the electron-phonon coupling.

This potential V can either be an analytic function or have a discontinuous derivative according as the electrons are extended or localized. This will be explained in the last section.

3.1. Ground state structure

At 0 K, the system is in its ground state. The potential V creates a lattice distortion on the chain which is shown to be periodic (PLD). The ground state $\{u_i^o\}$ of (4) satisfies the stationary equations

$$\frac{\partial\phi}{\partial u_i^o} = -W'(u_{i+1}^o - u_i^o) + W'(u_i^o - u_{i-1}^o) + \lambda V'(u_i^o) = 0 \qquad (6)$$

with the boundary condition

$$\ell = ac = \lim_{N-N'\to\infty} \frac{1}{N-N'}(u_N^o - u_{N'}^o) \qquad (7)$$

In the absence of potential V (λ=0) the atoms are equidistant and the solution of (6) and (7) is

$$u_i^o = i\ell + \alpha \tag{8}$$

where α is an arbitrary phase (and also the position of the zero-th atom). For any finite λ, it is proved that the ground state is obtained from (8) by a finite displacement $g_i(\alpha)$:

$$u_i^o = i\ell + \alpha + g_i(\alpha) \tag{9}$$

and that $g_i(\alpha)$ can be determined by a periodic hull function $g(x)$, which has the period 2a of V, defined as

$$g_i(\alpha) = g(i\ell+\alpha) \quad . \tag{10}$$

Moreover the function $f(x) = x + g(x)$ is shown to be monotonous increasing and may have two determinations which are either left or right continuous. It depends on ℓ and on the potentials. (When $\ell/2a$ is rational, the hull function g can be easily defined from (12) and is a finite sum of step functions (with a period 2a)). Thus, g describes the PLD associated to the CDW which produces the potential V, and is its hull function.

3.2. Phason and conductivity

Assuming now, a mass m for each atom (or molecule), the linearized equation of the small motions at frequency ω : $\varepsilon_i(\omega) = \int e^{i\omega t}\left(u_i(t) - u_i^o\right) dt$ in the vicinity of this ground state is

$$m\omega^2\varepsilon_i = -W''(u_{i+1}^o - u_i^o)\,(\varepsilon_{i+1} - \varepsilon_i) + W''(u_i^o - u_{i-1}^o)\,(\varepsilon_i - \varepsilon_{i-1}) + \lambda V''(u_i^o)\,\varepsilon_i \tag{11}$$

Because of the stability of the ground state, all the eigenvalues ω^2 of these equations are positive. Conductivity at 0 K is possible only if there exists a solution $\{\varepsilon_i\}$ at $\omega^2 = 0$. (This one does not depend on the atomic mass m). Then, the sequence $\{\varepsilon_i\}$ at this minimum eigenfrequency is shown to have no nodes [13] (i.e. no change of sign). (This result is the same as for the ground state of a one-dimensional Schrödinger equation). Thus it corresponds to a translation of the whole chain with respect to the potential V (or the CDW) or equivalently to a translation of the CDW-PLD with respect to the chain, and carries electric charges. It is this mode, called phason or Frölich mode, which produces conductivity.

The existence of this zero frequency mode is a consequence of the analyticity of function g in (10). It seems that these two properties are equivalent. We mention first that the commensurate state obtained, when $\ell/2a = c/2 = r/s$ is a rational number, is insulating. Then, the atomic positions u_i in the ground state of (4) satisfy

$$u_{i+s} = u_i + 2ra \qquad \text{for any } i \tag{12}$$

showing that the chain is periodic with a unit cell of s atoms with size 2ra and is a one-dimensional crystal. It is expected that there is no zero frequency mode in equation (11) except perhaps in very particular cases (which in mathematical terms are called non-generic) and that there is a finite gap in the excitation spectrum of (11). (We have not yet obtained a rigorous proof of this result which corresponds to a well-admitted physical intuition and is easily confirmed by any perturbative or variational calculation.) The CDW-PLD is then locked on the lattice and the system is an insulator, while the hull function g is discontinuous.

More interesting is the incommensurate case when the electronic concentration is an irrational number c (as well as $\ell/2a$). When V is an analytic potential, it has been shown rigorously by using the Kolmogorov-Arnold-Moser theorems that for almost any irrational electronic concentration c (except a zero measure set) there exists $\lambda_c(\ell) > 0$ such that for $\lambda < \lambda_c(\ell)$ the hull function g (or f) of the PLD is analytic. Conversely, we proved [12] that for any ℓ there exists $\lambda_1(\ell) > 0$ such that for $\lambda > \lambda_1(\ell)$ the function

$$f(x) = x + g(x) \tag{13}$$

is a discrete function (defined as a sum of step functions). Here

$$f(x) = \sum_i f_i \, Y(x-x_i) \tag{14}$$

with $f_i > 0$ and

$$Y(x) = 1 \quad \text{for} \quad x > 0 \quad , \quad Y(x) = 0 \quad \text{for} \quad x < 0 \quad .$$

We can choose $Y(0) = 0$ or $Y(0) = 1$ which yields the two possible determinations for the function f . Moreover, the x_i are shown to make a dense and countable set of the real axis. Numerical calculations [12] exhibit a well defined transition in λ at which function f transforms from a smooth-like function into a discrete-like one. Then we conjecture that $\lambda_1(\ell) = \lambda_c(\ell)$ and that there is a unique transition in λ at which f transforms from an analytic to a discrete function.

When f is analytic, the derivative of f provides a zero frequency mode $\varepsilon_i = f'(i\ell+\alpha)$ in equation (11) which is simply the translation mode of the whole chain with respect to the CDW. The system is then conducting by the Frölich mechanism. When f is discrete, its derivative which is defined almost everywhere is zero and thus does not provide any zero frequency mode. Numerical calculation confirms the presence of a finite gap in the excitation spectrum of (11). The system is then insulating at 0 K.

The case $\lambda = \lambda_c(\ell)$ is singular since function f is still continuous but neither analytic nor discrete. At the threshold the system is just between a conducting and an insulating state (as a zero-gap semi-conductor). As in the other phase transitions, there is probably at this transition a non-trivial critical behaviour but it has not been yet studied. We guess however that renormalization group technics could be an appropriate tool.

3.3. Analyticity of the conducting state

Whatever $\ell/2a$ is rational or not, we suggest now that the conductivity property is simply expressible in terms of analyticity. The chain is conducting at 0 K if and only if there exists an analytic solution $f(x) = x + g(x)$ with g periodic with period 2π, for the finite difference equation derived from (6) and (7) :

$$-\mathcal{W}'\Big(f(x+\ell) - f(x)\Big) + \mathcal{W}'\Big(f(x) - f(x-\ell)\Big) + \lambda\,\mathcal{V}'\Big(f(x)\Big) = 0 \quad (15)$$

Equation (15) has an analytic solution only if the following two conditions are satisfied :

1) $\ell/2a$ is irrational enough, i.e. satisfies diophantine conditions given in Refs.[12] or [13].

2) $\lambda < \lambda_c(\ell)$.

Clearly the second condition is not satisfied when the electron-phonon coupling is strong enough ($\lambda > \operatorname{Sup}_\ell \lambda_c(\ell)$). Numerical calculations [12][13] have shown that $\lambda_c(\ell)$ is rather small and moreover becomes much smaller close to low order rational numbers $\ell/2a$. Such insulating phases are then physically possible.

When $\mathcal{V}$ is not analytic and has strictly more than 4 derivatives, application of the Moser theorem (see Ref.[13]) states that there exists $\lambda_c(\ell)$ (different of zero when $\ell/2a$ is irrational enough) such that for $\lambda < \lambda_c(\ell)$, equation (15) has a solution f with a continuous derivative. The physical results are unchanged, despite the non-analyticity of $\mathcal{V}$.

On the other hand when $\mathcal{V}$ has a discontinuous derivative, equation (15) cannot have any continuous solutions f . An example in Ref.[11] shows that for any λ, f is discrete and that there is no transition in λ . We conjecture that this result remains true for any other potential $\mathcal{V}$ which has a discrete derivative.

4. BREAKING OF ANALYTICITY OF THE CDW AT 0 K AND ELECTRON LOCALIZATION

We are assuming now that we know the PLD of the system as described by (9) and (10), and we study the electron propagation in the chain. Consequently the resulting potential for an electron in the distorted chain is quasi-periodic with two periods 2a and ℓ and can be written formally as

$$V(x) = \sum_{n_1,n_2} c_{n_1 n_2} e^{i(K_1 n_1 x + K_2 n_2 (x-\alpha))} = \bar{V}(K_1 x , K_2 (x-\alpha)) \qquad (16)$$

with

$$K_1 = \frac{2\pi}{\ell} \quad \text{and} \quad K_2 = \frac{2\pi}{2a} ,$$

the phase α of the PLD can be chosen arbitrarily. $\bar{V}$ has the period 2π with respect to each of its two variables. This result can be obtained for example, by assuming that V(x) is the sum $\sum_i v(x-u_i)$ of analytic and fast decaying local potentials $v(x-u_i)$ created by the atoms i at u_i . Given the set $\{u_i\}$ defined in (9), the potential V is analytic with respect to the variable x, but $\bar{V}$ is not necessarily analytic with respect to each of its two variables. It is analytic only when the PLD hull function g is analytic, and it is discontinuous when g is discrete.

The motion of an electron is described by the usual Schrödinger equation

$$-\frac{\hbar^2}{2m}\frac{\partial^2\psi}{\partial x^2} + V(x)\,\psi = E\,\psi \qquad (17)$$

If $K_2/K_1 = r/s = (\ell/2a = c/2)$ is rational with r and s two irreducible integers, the potential V is in fact periodic with period $s\ell = 2ar$. The well known Bloch theorem predicts that all the electronic wave functions are extended plane waves and that the eigenenergies have a non-overlapping band structure. Thus the rationality of the electron number per site $c = 2r/s$ ensures that r bands are completely filled, so that the system is an insulator with a finite gap, since there exists no Frölich mode.

4.1. Dinaburg and Sinai theorem for analytic potentials V

The non-trivial case is obtained when K_2/K_1 is irrational. Equation (17) has been studied by Dinaburg and Sinai [14] for a quasi-periodic potential (16) V with a finite number of wave-vectors $K_1, K_2, \ldots, K_p$. Their result applies only when $\bar{V}$ is analytic with respect to each of its variables and when the p wave-vectors $K_1,\ldots,K_p$ involved in

$$V(x) = \bar{V}(K_1 x , K_2 x , \ldots , K_p x) \qquad (18)$$

satisfy incommensurability relations called diophantine inequalities, i.e. there exists some positive constant C such that

$$\left| \sum_{i=1}^{p} n_i K_i \right| > \frac{C}{(\Sigma |n_i|)^{p+1}} \tag{20}$$

holds for any set of integers n_i (i=1,p). This turns out to be true for most choices of $K_1 \dots K_p$. They found that :

1) There exist eigen wave functions for equation (17) which can be written with a wave vector k as

$$\psi_k(x) = e^{ikx} \chi_k(x) \tag{21}$$

where $\chi_k(x)$ is quasi-periodic with the same periods as V.

2) The corresponding eigenenergies E_k belong to a Cantor-like set which has a non-zero measure and determines an absolutely continuous component in the measure of state density. (For more details see the original paper [14]).

Thus, this result extends partially the Bloch theorem which predicts form (21) for all the eigenwaves when the potential V is simply periodic. However, the Bloch theorem is not extensible for any wave vectors k in (21). Roughly speaking k must not be too much commensurate with the wave-vectors $K_1, K_2, \dots, K_p$. This result can be easily understood by examining series, generated by the perturbation expansion at first order in V, of plane waves (see Ref.[2]). Moreover, the propagating eigen wave functions may be obtained at rather large energies in a region out of physical interest. We have been led, in coauthorship with G. André [12], to study the non-propagating eigenfunctions (which have not the form (21)) on a tight-binding model for electron propagation.

4.2. Tight-binding model

The tight-binding approximation is valid for a Hamiltonian which has the form

$$H = \frac{p^2}{2m} + \sum_i v(x-u_i) \tag{22}$$

where $v(x-u_i)$ represents the potential created by the atom or the molecule i at u_i. It is assumed moreover that the eigenfunction of the single site Hamiltonian

$$H_i = \frac{p^2}{2m} + v(x-u_i) \tag{23}$$

at site i almost not overlap over the neighbouring sites at i+1 or i-1 . This is equivalent to saying that the tunelling energy from one site to its neighbours is much smaller than the gap between the lowest on-site eigenenergies. The eigen wave function

$\psi(x)$ is expanded as

$$\psi(x) = \sum_i \psi_i \; \varphi(x-u_i) \tag{24}$$

where $\varphi(x-u_i)$ is the single-site eigensolution of (23) which has its eigenenergy in the physical range of the Fermi level, and has amplitude ψ_i at site i . The variational principle applied to $\langle\psi| H |\psi\rangle$ with ψ of the form (24) yields discrete eigenequations on the amplitudes ψ_i instead of the continuous one (17). This approximation is known to be well adapted to most one-dimensional conductors (which have a narrow band), and seems to play the same role as the exact Poincaré's map by the technics of surface of section in dynamical systems (see Ref.[2][12] and references therein). We expect that it does not change qualitatively the physical results if we preserve the analyticity concept on V (except that we get an upper bound for the eigenenergies).

To study the effect of an incommensurate modulation, we propose the simplest tight-binding model for non-interacting electrons

$$\psi_{n+1} + \psi_{n-1} + \lambda V(qn+h) \psi_n = E \psi_n \tag{25.a}$$

or

$$H_D \psi = E \psi \tag{25.b}$$

in which V is now a periodic function with period 2π which describes the effect of the PLD with phase h and with wave vector $q = 2k_F$ incommensurate with 2π . λ is a positive adjustable parameter which describes as in model (4), the strength of the electron-phonon coupling. The period of the lattice ℓ is here taken into account by the discreteness of ψ_n . It could be also useful to consider that the coefficients of ψ_{n+1} and ψ_{n-1} are also modulated by the PLD but for sake of simplicity, we assume that they are constant and unity. Some of the qualitative ideas of this paper extend also if we do not make this assumption and include exchange terms between remote sites, even in several dimensional models.

The hull potential V can be considered for example as proportional to the PLD described by function g in (10). As an equivalent of the analyticity property of $\overline{V}$ in (16), V is an analytic function as well as g in this model. The rotation of the phase h in V corresponds to the translation of the PLD. The Frölich conductivity will be possible only if the CDW translates continuously with the phase h and keeps the same energy. The existence of discontinuities corresponds physically to energy barriers which have to be passed through, and thus implies that the PLD is locked on the lattice by the CDW.

4.3. Continuity of the eigen-hull functions

Let us assume that a given eigenvector ψ_n corresponding to a constant eigenvalue is a continuous function $\psi_n(h)$ of the phase h. We get readily that

$$\psi_n(h) = \psi_o(qn+h) \tag{26}$$

Following Azbel [15], we note that the translation operator T

$$T\,\psi_o(x) = \psi_o(x + 2\pi/q) \tag{27}$$

commutes with the discrete Hamiltonian H_D

$$H_D\,\psi_o(x) = \psi_o(x+1) + \psi_o(x-1) + \lambda\,V(qx+h)\,\psi_o(x) \tag{28}$$

and thus that $\psi_o(x)$ can be written as

$$\psi_o(x) = e^{ikx}\,\chi_k(x) \tag{29}$$

for some wave vector k , and $\chi_k(x)$ periodic with period $2\pi/q$. $\psi_o(x)$ is the hull function of the discrete sequence ψ_i and is necessarily an extended propagating wave function (such that $\sum_i |\psi_i|^2 = \infty$) . Conversely, no localized eigenfunction (such that $\sum_i |\psi_i|^2 < \infty$) can exist at constant eigenenergy as a continuous function (26) of the phase h.

4.4. Role of the analyticity of V in perturbation expansion

We examine now the transition versus λ in equation (25) which corresponds to the localization of extended eigen wave functions.

For $\lambda = 0$, equation (25) yields a band of eigenenergies $E_k^o = 2\cos k$ corresponding to plane waves $\psi_n = \exp(ikn)$ with wave-vector k . Perturbation theory at first order in λ shows that in fact infinitely many gaps Δ_n are opened at the wave-vectors k_n determined by

$$k_n = n\,\frac{q}{2}\ (\text{mod } 2\pi) \tag{30}$$

with n an arbitrary integer and

$$\Delta_n = |\langle -k_n|\lambda V(x)\,|k_n\rangle| = \lambda\,V_n \tag{31}$$

V_n is defined as the nth Fourier coefficient of V

$$V(x) = \sum_{n=-\infty}^{+\infty} V_n\,e^{inx} \tag{32}$$

If V(x) is an analytic function, the sequence V_n is exponentially decaying for $|n|$ going to infinity. The sum $\sum_n |V_n|$ is convergent, so that, for small λ, the total gap opening $\lambda \sum_n |V_n|$ is also small. If V(x) is a discontinuous function, series (32) cannot be absolutely convergent and $\sum_n |V_n| = \infty$. Thus for any non-zero λ, perturbation theory yields an infinite total gap opening, showing then the inconsistency of the method. Simultaneously the perturbation expansion at first order of the plane waves exp(ikn)

$$\psi_n^{(k)} = e^{ikn} \left(1 + \lambda \sum_m \frac{V_m \; e^{im(qn+h)}}{2(\cos(mq+k) - \cos k)} \right) \tag{33}$$

generally diverges for any wave-vector k due to small denominators of (33) (see Ref.[12]).

4.5. Duality

Hamiltonian H_D lets invariant the subspace G_o of square summable (localized) $\{\psi_n\}$. When V is analytic, the subspace $F_k^{(a)}$ determined for a given k by the $\{\psi_n\}$ such that

$$\psi_n^{(k)} = e^{ikn} \chi(n) \tag{34}$$

for all the analytic periodic functions with period $2\pi/q$ $\chi(x)$, is also invariant by the operator H_D. Despite each subspace $F_k^{(a)}$ is invariant by H_D, H_D may have no eigenvectors in it. However, we can try to find the eiven wave vectors of (25) inside each subspace $F_k^{(a)}$ as

$$\psi_n^{(k)} = e^{ikn} \sum_{-\infty}^{+\infty} f_m \; e^{im(qn+h)} \tag{35}$$

where $f_m \; e^{imh}$ are the Fourier components of χ in (34). Sequence $|f_m|$ is such that $\sum_m |f_m|^2 < \infty$ and moreover decreases exponentially when $|m|$ goes to infinity since χ is analytic. Equation (25) yields a similar set of equations for f_m in the subspace G_o, which is called dual, in which f_n is a localized eigenfunction. (These dual equations can be generalized for any quasi-periodic potential V in any dimensional space).

4.6. A self-dual model

Particularly, the initial equations (25) can be chosen in such a way that the transformation keeps exactly the same form for the equations on f_m. This situation which is called self-dual occurs in the simplest case, in one-dimension and with first neighbour coupling, in model (25) with

$$V(x) = \cos x \quad . \tag{36}$$

Thus

$$\psi_{n+1} + \psi_{n-1} + \lambda \cos(q\, n + h)\, \psi_n = E\, \psi_n \tag{37}$$

is transformed by duality at the wave vector k, into

$$f_{m+1} + f_{m-1} + \frac{4}{\lambda} \cos\,(qm + k) f_m = \frac{2E}{\lambda} f_m \tag{38}$$

This particular property used together with the Thouless formula [12] allows to prove that for $\lambda < \lambda_c = 2$ all the eigen-functions of (38) are extended and twice degenerate. They are described alternatively by a complex analytic hull function (26), or its complex conjugate which corresponds to a reversed propagation, and form an uncountable set. This one is the uncountable union of countable subsets, each of which spanning a subspace $F_k^{(a)}$. We found numerically that the eigen-energies set have a finite measure which, in agreement with the perturbation approach (4.4), exhibits infinitely many gaps. With increasing λ, this measure decreases and just vanishes when the eigen wave functions begin to localize. These extended wave functions correspond to the extended wave functions described by Dinaburg and Sinai.

For $\lambda > \lambda_c = 2$ all the eigen-functions are exponentially localized over the range $\xi = 1/\log(\lambda/2)$ and form a countable set spanning G_o. They have no analytic hull function and are non degenerate. The set of eigen-energies depends on the phase h (cf. 2 while none of the eigen-functions can be determined as continuous function of the phase h, but the closure of the eigen-energies (called spectrum) remains independent of h.

The case $\lambda = \lambda_c = 2$ exhibits a singular spectrum : the eigen-wave functions are still non degenerate but they do not decrease exponentially at infinity. They decrease likely with some power law. A numerical study of the spectrum has been done in [16] but the eigen-functions have not yet been carefully studied.

Generally, models for electron propagation in incommensurate lattices are not self dual and do not yield the same results. However, the localization of the eigen-wave functions still shows up and remains a consequence of the breaking of analyticity of their hull functions. Numerical investigations have shown that when λ is not zero, there exist exponentially localized states which first replace the extended eigen-wave-functions, the wave-vector k of which are close to simple resonances [12] ($k \sim qn \pmod{2\pi}$). With increasing λ, all the extended eigen-wave functions successively disappear. When λ is larger than some critical λ_c, they are all replaced by localized eigen-wave-functions.

4.7. Non-analytic potentials.

If the hull potential V in (25) is non-analytic but is p-differentiable, operator H_D does not preserve subspace $F_k(a)$ defined in (34) but only subspaces F_k^p of the function defined in (34) with the new condition that χ is now p-differentiable. When p is sufficiently large, (p > 2 seems to be sufficient by duality argument) we conjecture that there exists extended states for λ small enough in the eigen-equation (25) and that their hull function (29) is differentiable. This conjecture would be in analogy. the Moser theorem applied to the epitaxial model (4). Thus for p sufficiently large there would be no qualitative change with respect to the case of an analytic function V.

Conversely, when produced by a PLD with a discrete hull function, becomes discontinuous. The dual transformation yields

$$\lambda \sum V_m f_{m+n} + 2 \cos (qn + k) f_n = E f_n \tag{39}$$

where V_n are the harmonics of V (as defined in (32)) which decay slowly as $1/|n|$ for n going to infinity. Interactions in model (39) are long range which suggests that most eigen-states cannot be localized and are modulated plane-waves. Consequently, in the initial model (25), a discontinuous hull potential would generally imply that all the eigen-states are exponentially localized.

A particular case is to choose V(x) in order that V(qn + h) is a random variable a_n. Thus V is a highly pathological function which has infinitely many discontinuities and has a non-bounded variation. The above conjecture is thus consistent with the known result that at one-dimension most electrons are localized in a random potential [17]. However, when V(x) has only a few discontinuities, numerical calculations suggest that the localization length is much larger than for purely random potential with the same mean amplitude.

4.8. Analyticity of the CDW

$c = q/\pi$ is the mean number of electron per site. At OK they occupy the lowest energy states according to the Pauli principle (two electrons per state) and the electron density at site n is

$$\rho_n(h) = 2 \lim_{N \to \infty} \sum_{\alpha=1}^{Nc/2} |\psi_n^\alpha(h)|^2 \tag{40}$$

where α indexes the eigen-states with increasing energies for a chain of N atoms (N going to infinity). Clearly $\langle\rho_n(h)\rangle = c$ and the CDW is the oscillating part of $\rho_n(h)$:

$$w_n(h) = \rho_n(h) - c \tag{41}$$

If all the ψ^α are extended wave-functions with analytic hull functions $\psi^\alpha(qn+h)$ we reasonably expect that the CDW also have an analytic, hull function $w(qn+h)$.

Conversely if some electronic states are localized, they vary discontinuously with the phase h but this condition is not sufficient to have a non-analytic hull function for the electronic density. (For example if all the states were occupied, we would obtain $\rho_n \equiv 1$ with the trivial hull function $\rho = 1$ whenever the electrons are localized or not). The effect of the electron localization can be examined in our model, in a simple case, for $\lambda = \infty$. Then the eigen-states of (25) are localized at each site and the electron density is

$$\rho_n(h) = 2Y\left[V(\pi - \tfrac{q}{2}) - V(qn+h)\right] \tag{42}$$

where $Y(x)$ is the Heaviside function. Clearly, it is discontinuous functions of the phase h. The discontinuities are interpreted physically by the fact that for a small change of the phase h, some eigen-energies which were just smaller than the Fermi energy, get an energy just larger while it is the converse for some other eigen-energies. Thus, to keep the total energy minimum, some electrons must hop discontinuously from the first states to the second ones, which produces the discontinuities of $\rho_n(h)$. The same picture can be reasonably expected for finite λ, if the electrons are localized at the Fermi level. Then we conjecture that when the electrons are localized at the Fermi level (for an unfilled band $c \neq 2$), the electron density (and the CDW) is a discontinuous function of the phase h and thus the system is an insulator. Conversely, if the electronic wave functions are modulated plane waves at the Fermi level and if the electronic concentration is irrational, the CDW is expected to be an analytic function of the phase h, and its energy does not depend on the phase h. Thus the system is conducting by the Frölich mechanism and the above study on electron localization suggests that the system exhibits a metal-insulator transition when the electron phonon coupling is increased.

5. CONCLUDING REMARKS : A NEW LOOK UPON THE PEIERLS INSTABILITY

The effects of the CDW on the PLD and the PLD on the CDW have been studied separately at OK via two crude models which assume that the lattice and the electronic distorsion are periodic with the wave-vector $2k_F = c$. Both models yield an insulating state at OK when the electron number per site c is rational. When the number c is irrational, both models exhibit a metal-insulator transition at OK when the electron-phonon coupling is increased. These two approaches of the same physical phenomenon must be consistent with each other, which implies that the breaking of analyticity which characterizes these metal-insulator transitions, must occur simultaneously, on the respective hull functions of the PLD and of the CDW.

Indeed, if the PLD has an analytic hull function (9), the electrons are submitted to an analytic hull potential in (25) and their propagation is possible. Then the CDW (41) must also have an analytic hull function because if the CDW has a discontinuous hull function (41), it would produce a force on the atoms which depends discontinuously on the phase h, so that the PLD would also have a discontinuous hull function. Conversely if the PLD has a discrete hull function, either the electrons localize in the incommensurate case, or fill up a finite number of bands with a finite gap, in the commensurate case. In both cases, the system is an insulator at OK.

Clearly the exact results we have used in this paper are interpreted and extended as a guide for physical intuition, but these models do not fit all the aspects of the physical problem. A better model should consider self-consistently the lattice and the electrons without assuming the initial existence of a Peierls instability at $2k_F$. An example of such a model could be described by the tight-binding Hamiltonian of non-interacting electrons in a deformable lattice :

$$H = \sum_i [c^+_{i+1} c_i + c^+_i c_{i+1} + \lambda \varepsilon_i c^+_i c_i - \mu c^+_i c_i] + \phi_{ela}(\{\varepsilon_j\}) \qquad (43)$$

where c^+_i and c_i are the usual creation and annihilation operators of an electron at site i, ε_i is the phonon variable at site i ; λ is the strength of the electron-phonon coupling and μ is the chemical potential of the electrons ; $\phi_{ela}(\{\varepsilon_j\})$ is the elastic energy of the lattice after deformation $\{\varepsilon_j\}$ and can be chosen as a quadratic form of ε_j. The atoms are assumed to be classical particles. The ground state of this model is obtained for the configuration $\{\varepsilon_j\}$ which minimizes the energy of the electrons $E_o(\{\varepsilon_i\})$ in their ground state. Following Peierls the lattice distortion and the electronic density are expected to be periodic with wave vector $2k_F = \pi c$.

The intuitive ideas developed in this paper, suggest that this model with a fixed electron number per site c, has the same qualitative properties as those of the epitaxial model (4) : there exists a critical $\lambda_c(c)$ such that for $\lambda < \lambda_c(c)$ the system exhibits a PLD-CDW with analytic hull functions (for ε_i and $n_i = n_i = \langle c_i^+ c_i \rangle$) and thus is conducting. For $\lambda > \lambda_c(c)$ the hull function becomes discontinuous and the system is an insulator at OK. $\lambda_c(c)$ is strictly positive for most irrational c, but vanishes for rational c (at which it is, presumably and paradoxically continuous !) and this is a highly pathological function. One could believe that such discontinuities are physically unacceptable. In fact the physical continuity is maintained. When c is a rational r/s but with a large s there is no real transition but a sharp cross-over at which the system transforms from an insulating state with a very small gap in the phonon spectrum (The Frolich mode almost exists !) into another insulating state with a large gap in the phonon spectrum. When the Frolich mode almost exists, the conductivity will remain practically important at low (but non-zero) temperature because the easy hopping of the localized electrons produced by low frequency phonons (phonon drag). When s becomes larger and larger, the cross over becomes sharper and sharper and transforms into a true transition only at $s = \infty$, i.e. for irrational electron number per site c.

When the electron number per site c is not fixed but is a function of the chemical potential μ , we also expect the same qualitative properties as those of the epitaxial model (4) with an external pressure (the reader is refered to ref. 11 and 17). The variation of the electron number per site c versus the chemical potential μ should vary as a devil's stair case. This one would be complete (i.e. entirely composed of steps) if all the phases versus c are insulating ($\lambda > \sup_c \lambda_c(c)$) and then c is almost always rational. The devil's stair case $c(\mu)$ should become incomplete for small enough λ , while at the irrational $c(\mu)$ which are obtained for a finite measure of μ the system would be conducting.

Comparison with the epitaxial model (4) can be further developed. The concept of epitaxy dislocation is replaced by the concept of polaron ; in the epitaxial model (4) the structure of the chain can be interpreted as equidistant defects (epitaxy dislocation) which can form either a "fluid" phase for $\lambda < \lambda_c(\ell)$ (then the hull function is analytic) or a locked phase of quasi independent defects for $\lambda > \lambda_c(\ell)$ ((then the hull function is discontinuous). A single defect is generally locked. As for the epitaxy dislocation, the one-electron ground-state of model (43) (choosing $\phi_{ela}(\{\varepsilon_j\}) = \sum_j \frac{1}{2} m\omega_o^2 \varepsilon_j^2$) can also be represented by the trajectory of a homoclinic point (see ref. 11 and 17) of an adequate canonical transformation. The resulting configuration is an electron localized on a local deformation of the lattice which is called a polaron [22]. This single polaron is locked on the lattice because it is assumed to be classical (but the electrons are quantized). Thus a many electron structure is also a many-polaron structure. It can be either "fluid" with delocalized electrons and a quasi-sinusoidal PLD in the case of analytic hull functions, or locked on the lattice with electrons either localized or forming a filled band structure with a PLD rather rich in harmonics in the case of discontinuous hull functions. The system is a Frölich superconductor in the first case and an insulator in the second case.

One can wonder for physical applications if the assumption that the lattice is classical is valid. We believe that this assumption can be valid if the one-dimensional chains are imbedded in a three dimensional system with weak interchain coupling. Then, it is well-known that most of the quantum and thermal fluctuations of the lattice can be frozen by a three dimensional ordering at sufficiently low temperature if the amplitude of the atomic zero point motion is not too large compared with the amplitude of the PLD. This assumption becomes true in the case of a large electron-phonon coupling. Thus the incommensurate insulating phase could be viewed as a three dimensional classical crystal of localized polarons superimposed to the lattice structure, while the Frölich conducting phase can be viewed as a quantum fluid of delocalized polarons. The locked polaron structure appears to be similar to a Wigner crystal [18] of electrons. But this one is produced by the repelling interaction between the electrons which is a fundamentally different mechanism of the electron-phonon coupling in a *discrete* lattice as in our theory (electron phonon interaction is known to result into an attractive electron-electron interaction after the (approximate) elimination of the phonon variables).

This study should help to understand the mysterious metal insulator transitions which occur at low temperature in many stoichiometric one-dimensional conductors [19]. Most theories invoke the impurities as the main cause of the transition; but in fact recent experiments seem to show that more perfect the sample is, sharper the metal-insulator transition. We believe that the explanation of the phenomenon lies mostly in the strong electron-phonon coupling and in the discrete lattice. A Frölich conducting chain at OK remains in principle possible in our theory but requires two conditions :

1) the electron-phonon coupling is not too strong,
2) the electron-number per site is sufficiently far from simple rational numbers.

The PLD is expected to have a rather small amplitude or not to exist if the quantum fluctuations on the atoms become important. (Let us mention that these compounds also involve distortions with wave-vector $4k_F$ which are likely produced by a strong electron-electron repelling interaction [20]. At low temperature this interaction is probably compensated by the equivalent of an attractive component created by the PLD, which particularly develops if the system becomes an insulator).

This paper has shown the importance of commensurability effects in a conductivity problem in one-dimension. We expect that the rationality of the electronic concentration could play also an essential role in other one-dimensional models with electron-electron interaction. Exact solutions are only available for well chosen models (for example the Hubbard model [21]). They can be considered in some sense as integrable and have a particular behavior. In fact, slight changes of these models could already exhibits non-integrability and commensurability effects.

Finally, from a more general point of view, we associated the intuitive idea of a coherent and smooth conducting state with analyticity and of a non-coherent and rough insulating state with discontinuity. This could help to understand better the non-periodic matter for which the usual but non-generic band theory do not extend.

REFERENCES

[1] W. Kohn in "Many Body Physics" Edited by C. de Witt and R. Balian, Gordon and Breach (1968)

[2] S. Aubry, Application of the Kolmogorov-Arnold-Moser theorems to structure problems in condensed matter. International Workshop on "Intrinsic stochasticity in Plasmas" June 1979 Cargèse, To be published.

[3] P.W. Anderson, Phys. Rev. 109, 1492 (1958).

[4] S.N. Mott, Proc. Phys. Soc. (London) 62 416 (1949).

[5] R.E. Peierls, Quantum theory of Solids, Clarendon, Oxford (1955) p.108.

[6] J.Y. Dubois and J.P. Carton, J. de Physique 35, 371 (1974).

[7] H. Frolich, Proc. Roy. Soc. A223, 296 (1955).

[8] J. Bardeen, L.N. Cooper and J.R. Schrieffer, Phys. Rev. 108 1175

[9] Quasi-one dimensional conductors, Edited by S. Barisic, A. Bjelis, J.R. Cooper and B. Leontic, Lecture Notes in Physics Springer Verlag, Vol. 95 and 96 (1979).

[10] J. Bardeen, Solid State Comm. 13, 357 (1973).
D. Allender, J.W. Bray and J. Bardeen, Phys. Rev. B 9, 119 (1974).
P.A. Lee, T.M. Rice and P.W. Anderson, Phys. Rev. Lett. 31 462 (1973).

[11] S. Aubry "The new concept of transition by breaking of analyticity" In "Soliton and Condensed matter physics", Edited A.R. Bishop and T. Schneider, Springer Verlag, Solid. State Sciences 8, 264 (1978).

[12] S. Aubry and G. André,"Analyticity breaking and Anderson localization in incommensurate lattices". Colloquium on Group theoretical methods in physics, Kiriat Anavim, Israel, March 1979, to be published by Springer-Verlag.
[13] G. André, Thesis (in preparation).
[14] E.I. Dinaburg and Y.G. Sinai, Fundamental Analysis, 9, 279 (1976).
[15] M.J. Azbel, Soviet Mathematics 5, 1549 (1964)
[16] D.R. Hofstadter, Phys. Rev. B14, 2239 (1976).
[17] S. Aubry,"Defectibility and Frustration in incommensurate structures. The devil's stair case transformation". Fourth European Meeting on Ferroelectricity, Portoroz, Yugoslavia, Sept. 1979.
[18] E. Wigner, Trans. Far. Soc. 34, 678 (1938).
[19] A.J. Epstein and J.S. Miller, Scientific American p. 48 (1979)
[20] J.P. Pouget, S. Kahna, F. Desnoyer, R. Comes, Phys. Rev. Lett. 37, 347 (1976).
V. Emery, Phys. Rev. Lett. 37, 107 (1976).
[21] J. Hubbard, Proc. Roy. Soc. A 276, 238 (1963) and 277, 237 (1964).
E. Lieb and F.Y. Wu, Phys. Rev. Lett. 20, 1445 (1968).
[22] J.R. Cooper, Phys. Rev. B 19, 2404 (1979).

SETS OF MINIMUM CAPACITY, PADÉ APPROXIMANTS AND THE BUBBLE PROBLEM

J. Nuttall

University of Western Ontario

We give a characterization of a set of minimum capacity whose components include certain prescribed points, but the proof of uniqueness of such a set remains to be completed. It is speculated that Padé approximants to certain functions with branch points converge away from a corresponding set of minimum capacity. In some cases this speculation has been proved. These ideas are used to study the expansion for a function describing the shape of a cylindrical bubble rising through a fluid.

1. INTRODUCTION

We report progress on three non-linear problems whose relation to each other should become apparent. The first concerns the determination of a set in the complex plane of minimum capacity whose connected components each include certain given points. The second problem is the convergence of sequences of Padé approximants to certain classes of function and the third is to determine how a bubble of gas rises through a fluid placed in a vertical gravitational field. Our work brings up interesting open questions in each area.

2. SETS OF MINIMUM CAPACITY

Suppose that we are given n distinct, finite points a_i, $i=1,\ldots n$, in the complex plane arranged into sets σ_k, $k=1,\ldots,\nu$, each containing at least two points, such that each point is in one and only one of the σ_k. We shall call such a choice of sets σ_k a partition P_α. Let M be a closed, bounded set in the complex

C. Bardos and D. Bessis (eds.), Bifurcation Phenomena in Mathematical Physics and Related Topics, 185–201.

plane containing no more than ν disconnected components, each of which are simply connected and therefore mutually exclusive, with the property that, for each $k=1, \ldots, \nu$, all the points a_i of σ_k are contained in the same component of M, which may depend on k. Define M_α as the set of all such M.

We have proved the following theorem [1].

THEOREM. There exists at least one set $S_\alpha \varepsilon M_\alpha$ such that $\mathrm{cap}(M) \geq \mathrm{cap}(S_\alpha)$ for all $M \varepsilon M_\alpha$. Any such set S_α may be described as $S_\alpha = \{z:\mathrm{Re}\, \phi(z) = 0\}$ where

$$\phi(z) = \int_{a_1}^{z} dt\; Q^{\frac{1}{2}}(t) X^{-\frac{1}{2}}(t) \tag{1}$$

where

$$X(t) = \prod_{i=1}^{n} (t-a_i), \quad Q(t) = \prod_{i=1}^{n-2} (t-c_i).$$

The c_i must satisfy

$$\mathrm{Re}\, \phi(a_i) = 0, \quad i = 2, \ldots, n-1. \tag{2}$$

$\mathrm{Re}\, \phi(c_i) = 0$ for all $i = 1, \ldots, n-2$ such that c_i is a zero of $Q(t)$ with odd multiplicity. (3)

The symbol cap(M) means the logarithmic capacity of M. It may be defined in terms of $g(z)$, the Green's function with pole at ∞ for the complement of M, which vanishes on (most of) M, is harmonic outside M and satisfies at $z = \infty$

$$g(z) = \ell n|z| + V + O(z^{-1}). \tag{4}$$

We have (see [2]) $\mathrm{cap}(M) = e^{-V}$.

The proof of the existence of a set of minimum capacity may be obtained by modifying standard arguments. The characterization of a minimum set S_α in terms of $\phi(z)$ follows from an application of the Schiffer variational technique [3], using an extension of Bergman's method [4]. A brief sketch follows and full details will appear elsewhere.

Suppose that $\phi(z)$ is the function analytic outside S_α of which the Green's function $g(z)$ for S_α is the real part. Suppose that z_0 is a finite point outside S_α and consider the transformation

$$z^* = z + \varepsilon\, X(z)(z-z_o)^{-1}. \tag{5}$$

For $|\varepsilon|$ sufficiently small, S_α will be mapped into another set $S_\alpha^* \in M_\alpha$, with corresponding function $\phi^*(z)$. We have for large $|z|$

$$\begin{aligned} \phi(z) &\sim \ell nz + W \\ \phi^*(z) &\sim \ell nz + W^* \end{aligned} \tag{6}$$

where $\mathrm{cap}(S_\alpha) = \exp(-\mathrm{Re}W)$, $\mathrm{cap}(S_\alpha^*) = \exp(-\mathrm{Re}W^*)$.

Let γ_R be a circle, centre $z = 0$, of radius $R = \varepsilon^{-\frac{1}{n}}$, which contains S_α. Let s be the boundary of S_α, γ_o a small circle, centre z_o, and let $\gamma_1, \ldots, \gamma_K$ (where S_α has $K + 1$ components) be (both sides of) a set of analytic cuts which transform that part of the complement of S_α inside γ_R into a simply connected domain. Then Cauchy's theorem gives

$$\mathrm{Re}\left\{ \frac{1}{2\pi i} \int_{\gamma_R+s+\gamma} [\phi^*(z^*) - \phi(z)]\frac{d\phi(z)}{dz} dz \right\}$$

$$= \mathrm{Re}\left\{ \frac{1}{2\pi i} \int_{\gamma_o} [\phi^*(z^*) - \phi(z)]\frac{d\phi(z)}{dz}\, dz \right\} \tag{7}$$

where $\gamma = \bigcup_{i=1}^{K} \gamma_i$. Bergman [4] shows that the contribution from $s + \gamma$ is zero and that the right hand side gives

$$\mathrm{Re}\{\varepsilon\phi'^2(z_o)X(z_o)\} + o(\varepsilon).$$

Further analysis shows that the term on the left of (7) coming from γ_R has the form

$$\mathrm{Re}(W^*-W) + \mathrm{Re}(\varepsilon\, Q(z_o)) + o(\varepsilon)$$

where $Q(z)$ is a polynomial of degree $n-2$ such that for large z

$$X(z)\phi'^2(z) = Q(z) + O(z^{-1}).$$

Thus we deduce

$$\mathrm{Re}(W^*-W) = \mathrm{Re}\{\varepsilon(\phi'^2(z_o)X(z_o) - Q(z_o))\} + o(\varepsilon). \tag{8}$$

Since $\mathrm{cap}(S_\alpha) \leqslant \mathrm{cap}(S_\alpha^*)$, it follows that $\mathrm{Re}(W^*-W) \geqslant 0$, and that, since ε may be any sufficiently small complex number,

$$\phi'^2(z_o)X(z_o) = Q(z_o), \tag{9}$$

from which we obtain (1). Because $a_i \in S_\alpha$, we have (2), and (3) follows since $\phi(z)$ is singular at $z=c_i$ if c_i is a zero of odd multiplicity, and $\phi(z)$ cannot be singular outside S_α.

We deduce that S_α must consist of a number of analytic arcs ending at the points a_i and those c_i of odd multiplicity. Some examples are shown in Figs. 1-3.

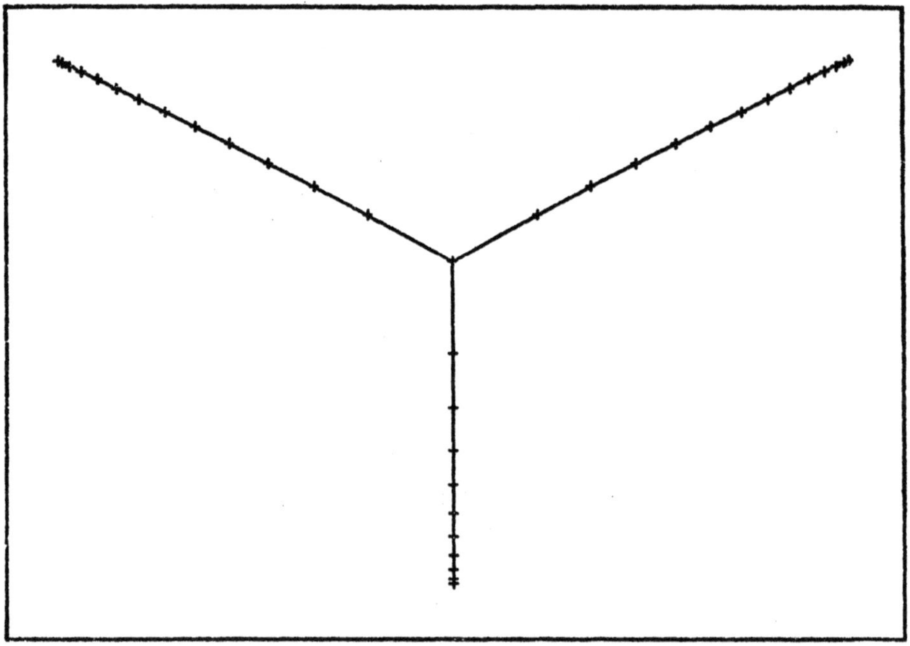

Fig. 1. The set of minimum capacity (solid line) for n=3, k=1, $a_1=-i0.2, a_2, a_3 = \pm 0.75 + i0.8125$. Also shown are the zeroes of $p_{37}(z)$ for $f_1(z)$ with this choice of a_i.

We conjecture that S_α is unique, and can prove this in the case n=3, ν=1. In this case we have

$$\phi(z) = \int_{a_1}^{z} dt\,(t-c)^{\frac{1}{2}}\, X^{-\frac{1}{2}}(t)$$

with

$$\text{Re}\ \phi(a_2) = \text{Re}\ \phi(c) = 0. \tag{10}$$

When a_3 lies on the line segment joining a_1 and a_2, S_α is obviously unique, and $c = a_3$. Now suppose a_3 moves away from this point along a path $a_3 = a_3(s)$, s real. From (10) we have

$$\begin{bmatrix} \frac{\partial\phi(a_2)}{\partial c} & \left(\frac{\partial\phi(a_2)}{\partial c}\right)^* \\ \frac{\partial\phi(c)}{\partial c} & \left(\frac{\partial\phi(c)}{\partial c}\right)^* \end{bmatrix} \begin{bmatrix} \frac{dc}{ds} \\ \frac{dc^*}{ds} \end{bmatrix} = - \begin{bmatrix} \frac{\partial\phi(a_2)}{da_3} \\ \frac{\partial\phi(c)}{\partial a_3} \end{bmatrix} \frac{da_3}{ds} - \begin{bmatrix} \left(\frac{\partial\phi(a_2)}{\partial a_3}\right)^* \\ \left(\frac{\partial\phi(c)}{\partial a_3}\right)^* \end{bmatrix} \frac{da_3}{ds} \quad (11)$$

The square matrix on the left turns out to be the Riemann period matrix [5] for the Riemann surface $y^2=X(z)(t-c)$ and is non-singular provided c, a_i are all distinct, leading to uniqueness with this proviso. The situation when c is near to a_3 can be dealt with directly, completing the proof.

The approach of (11) and the relation to the Riemann period matrix apply for general n, ν but the analysis when two or more points are coincident has not been performed.

A related problem where uniqueness has also been proved is to find the set of minimum capacity from $\bigcup_\alpha M_\alpha$, where α runs over all partitions with $\nu=m$, $n=2m$. Each σ_k contains two points. It is easy to show in this case that

$$Q(t) = R^2(t), \; R(t) = \sum_{i=1}^{m-1} (t-c_i),$$

so that ϕ is a hyperelliptic integral of the third kind with pure imaginary periods. Such an integral is unique [5].

3. CONVERGENCE OF PADÉ APPROXIMANTS

If we have a function of the form

$$f(z) = \int_L dt \; \omega(t)(t-z)^{-1} , \quad (12)$$

where L is a set of arcs, then the [N/N] Padé approximant to f expanded about infinity is

$$[N/N] = V(z^{-1})/W(z^{-1}), \quad (13)$$

where V(t),W(t) are polynomials in t of degree N which satisfy

$$f(z)\,W(z^{-1}) - V(z^{-1}) = O(z^{-(2N+1)}),\quad z \to \infty .$$

If $p_N(t)$ is a degree N orthogonal polynomial satisfying

$$\int_L dt\ \omega(t)\ t^k p_N(t) = 0,\ k = 0,\ \ldots,\ N-1, \tag{14}$$

then it may be shown that [6]

$$f(z) - [N/N] = p_N^{-1}(z) \int_L dt\ \omega(t)\ p_N(t)(t-z)^{-1} .$$

To understand the possible convergence of the sequence [N/N] to f(z), we need to study the form of $p_N(z)$ for large N. An explicit formula for $p_N(z)$ is [7]

$$p_N(z) = \int_L dt_1 \omega(t_1) \ldots . \int_L dt_N\ \omega(t_N) \prod_{i=1}^{N} (z-t_i)\Gamma \tag{15}$$

where

$$\Gamma = \prod_{i<j=1}^{N} (t_i - t_j)^2$$

Suppose now that L is the line segment [-1, 1]. We give a heuristic argument for obtaining the asymptotic form of $p_N(z)$ that agrees with results proved rigorously for certain classes of $\omega(t)$. Application of the argument in a more general case leads to a conjecture about the convergence of Padé approximants to functions with several branch points.

By symmetry we can assume in (15) that $-1<t_1<t_2<\ldots<t_N<1$. In this domain it is known [7] that Γ has a unique maximum at $t_i=t_i^o$, i=1, ..., N, where t_i^o,i=2,..., N-1 are the zeros of the Jacobi polynomial $p_{N-2}^{(1,1)}$. Having most factors, Γ is in some sense the dominant part of the integrand of (15), and we can approximate the integral for large N by evaluating the other factors at $t_i=t_i^o$, leading to

$$p_N(z) \propto \prod_{i=1}^{N} (z-t_i^o)$$

where the constant of proportionality is independent of z. It is known that asymptotically the t_i^o are distributed with density $N(1-t^2)^{-\frac{1}{2}}$ so that we obtain

$$p_N(z) \sim \exp\left(N\int_{-1}^{1} dt(1-t^2)^{-\frac{1}{2}} \ln(z-t)\right) \tag{16}$$

which agrees with the rigorous results for certain real, positive $\omega(t)$ [7].

Now suppose that L consists of several arcs ending at the points a_i, i=1,...,n, and that $\omega(t)$ is analytic in a suitable neighborhood of the arcs, except perhaps at the ends. It is clear that the definition of $p_N(z)$ and [N/N] is independent of the precise location of the arcs. Using the philosophy of the method of steepest descent to approximate (15), we choose a location for the arcs for which the maximum value of $|\Gamma|$ is minimum. But in the limit as $N \to \infty$ the N^{-2} root of this maximum value is just cap(L), so we are choosing L to minimize its capacity.

The points t_i^o that maximize $|\Gamma|$ are asymptotically distributed with density proportional to the charge density required to make L an equipotential, which means that

$$\prod_{i=1}^{N} (z-t_i^o) = \exp\{N \sum_{i=1}^{N} \ln(z-t_i^o)\}$$

$$\sim \exp\{N \times (\text{the potential due to unit charges at } t_i^o,\ i=1,\ldots,N)\}$$

This leads to

$$p_N(z) \sim \exp\{N\phi(z)\} \tag{17}$$

where $\phi(z)$ is the function of (1) for the correct set of minimum capacity. We are led to the following conjecture.

CONJECTURE. Suppose that a function f(z) may be written

$$f(z) = \int_S dt\ \omega(t)(t-z)^{-1},$$

where S is a set of minimum capacity as obtained in Sect. 2, and that, on S, $\omega(t)$ is sufficiently smooth and does not vanish at more than a finite number of points. Then the sequence [N/N] converges in capacity to f on any closed bounded set not intersecting S.

Let us illustrate by considering a few examples. First take the function

$$f_1(z) = \prod_{i=1}^{3} (z-a_i)^{-1/3} \tag{18}$$

with $a_1=-0.2$, a_2, $a_3=0.8125 \pm i0.75$

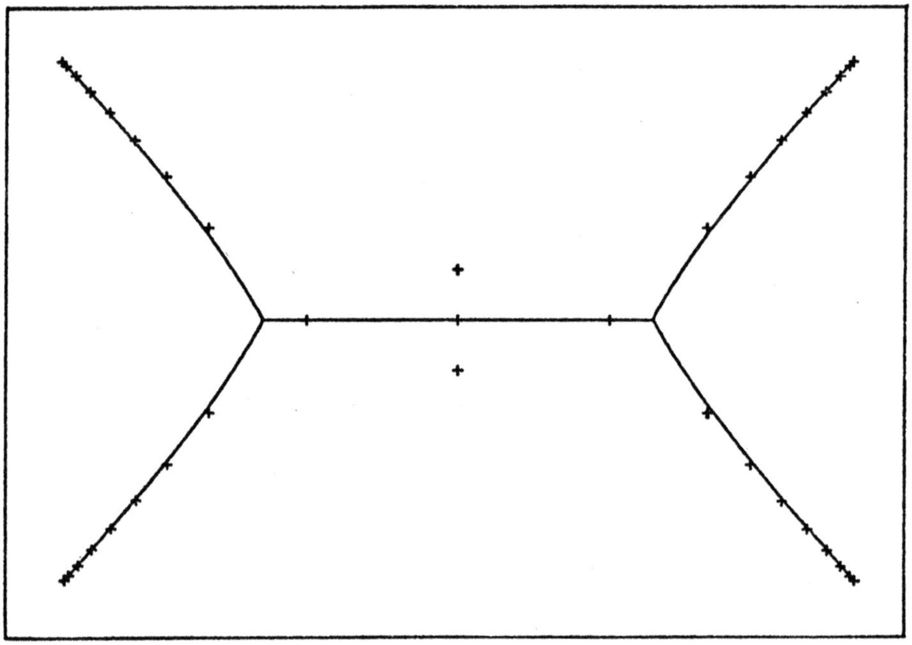

Fig. 2. The set of minimum capacity (solid line) for n=4, k=1, a_1, a_3 = 1.5±i0.5, a_2, a_4 = ± i0.5. Also shown are the zeroes of $p_{37}(z)$ for $f_2(z)$ with this choice of a_i.

For three points a_i, i=1,...,3, we must choose $\nu=1$, $\sigma_1=\{a_1,a_2.a_3\}$ and S is shown in Fig. 1. Clearly $f_1(z)$ may be written in the form (12) with $\omega(t)\neq 0$, $t \in S$. The zeroes of $p_{37}(z)$ are also plotted in Fig. 1. The numerical values of the Padé approximants do appear to converge at typical points not on S.

As another example consider

$$f_2(z)=[(1-a_1/z)(1-a_2/z)]^{\frac{1}{2}} + [(1-a_3/z)(1-a_4/z)]^{\frac{1}{2}}, \tag{19}$$

which has branch points at a_i, $i=1,\ldots,4$. There are two possible sets of minimum capacity. One set, corresponding to $\nu=1$, has one component and two different points $c_i, i=1, 2$ on S. If c_1 (and c_2) is joined by arcs of this S directly to any pair of a_i except the pairs (a_1,a_2), (a_3,a_4), then $f_2(z)$ may be written in the form (12). satisfying the non-vanishing condition of the conjecture. This is the case for $a_1=1.5+i0.5$, $a_2=i0.5$, $a_3=1.5-i0.5$, $a_4=-i0.5$ and the corresponding S is shown in Fig. 2, along with the zeroes of $p_{37}(z)$ for f_2. Again we have apparent convergence away from S.

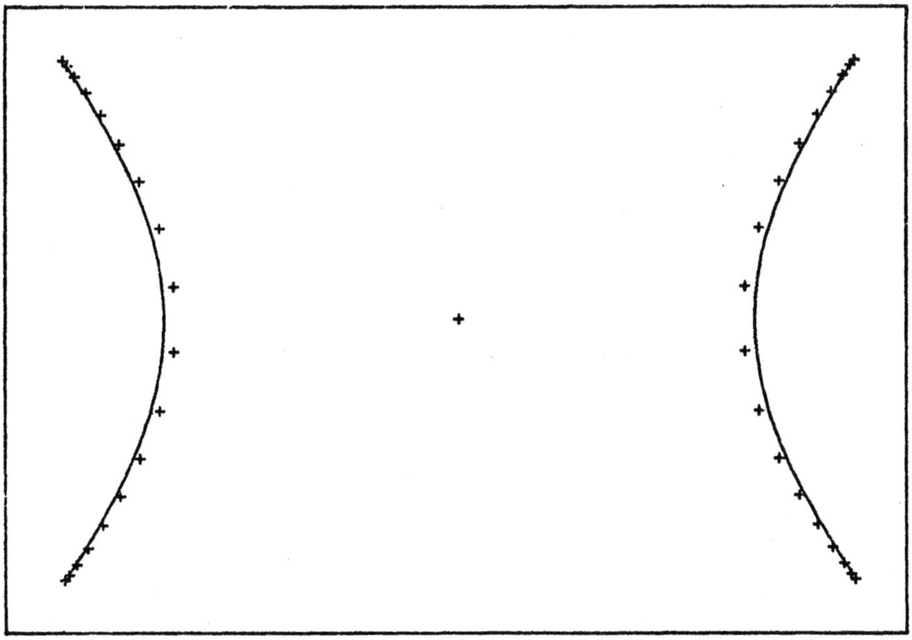

Fig. 3. The set of minimum capacity (solid line) for $n=4$, $k=2$, $\sigma_1=(a_1,a_2)$, $\sigma_2=(a_3,a_4)$ with $a_1,a_2=1.5\pm i0.5$, $a_3,a_4=\pm i0.5$. Also shown are the zeroes of $p_{37}(z)$ for $f_2(z)$ with this choice of a_i.

However, if it turns out that, in the S above, c_1(say) is joined directly to a_1, a_2, then $f_2(z)$ written in the form (12) would have $\omega(t)=0$ for t on the arc of S joining c_1,c_2. This is the signal that we should be using the S for a different minimum capacity problem, namely $\nu=2$ with $\sigma_1=\{a_1,a_2\}$, $\sigma_2=\{a_3,a_4\}$. For

$a_1, a_2 = 1.5 \pm i0.5$, $a_3, a_4 = \pm i0.5$ this is the situation and S is shown in Fig. 3 along with the zeroes of $p_{37}(z)$ for f_2. In this case the capacity of S with $\nu=2$ is less than that with $\nu=1$. It appears that the rule is to choose the set of minimum capacity of all partitions consistent with the way in which f(z) is single-valued on closed curves including some of the a_i.

When S is a set of the type referred to in the last paragraph of Sect. 2 the conjecture has been proved [6] for $\omega(t)$ which vary in proportion to $(t-a_i)^{-\frac{1}{2}}$ near each a_i.

A more detailed conjecture about the asymptotic form of $p_N(z)$ is obtained by analogy with the case n=2 and a study of special cases for higher n. Suppose S is such that

$$\phi(z) = \int_{a_1}^{z} dt\ Y^{-\frac{1}{2}}(t)\ R(t) \tag{20}$$

where Y is a monic polynomial whose zeroes are $\{a_i\}$ and those c_i of odd multiplicity (each taken once only). The degree of Y is even, say 2m, and the degree of the monic polynomial R is m-1.

Now suppose that a function f(z) has the form

$$f(z) = \int_S dt\ Y^{-\frac{1}{2}}(t)\ \sigma(t)(t-z)^{-1},$$

with suitable restrictions on $\sigma(t)$. Then a conjecture for the asymptotic form of the corresponding $p_N(z)$ for large N, $z \notin S$, is

$$p_N(z) \sim F_N(z), \tag{21}$$

where

i) $F_N(z)$ is analytic, single-valued outside S

ii) On S, $F_N^+ F_N^- \sigma = P_N$, where P_N is a polynomial of degree m-1 (+,- refer to opposite sides of S)

iii) As $z \to \infty$, $F_N(z) \sim z^N$.

iv) Outside S, $P_N(z)/F_N(z)$ is analytic.

This information is enough to specify $F_N(z)$, and an explicit form is given in [6] in the case where $\sigma(t)$ is the inverse of a polynomial. This form applies also in the general case. It is shown in [6] that (21) is consistent with (17), but we see that there is a possibility of zeroes in $p_N(z)$ not near to S leading to 'spurious' poles in the Padé approximant. In an example studied by Dumas [8], where f(z) involved the square root of a fourth degree polynomial, one such spurious zero is present for some N. We also see spurious zeroes in Figs. 2, 3.

4. THE BUBBLE PROBLEM

Suppose that we have an infinite, homogeneous, incompressible, inviscid fluid in a uniform (vertical) gravitational field contain-

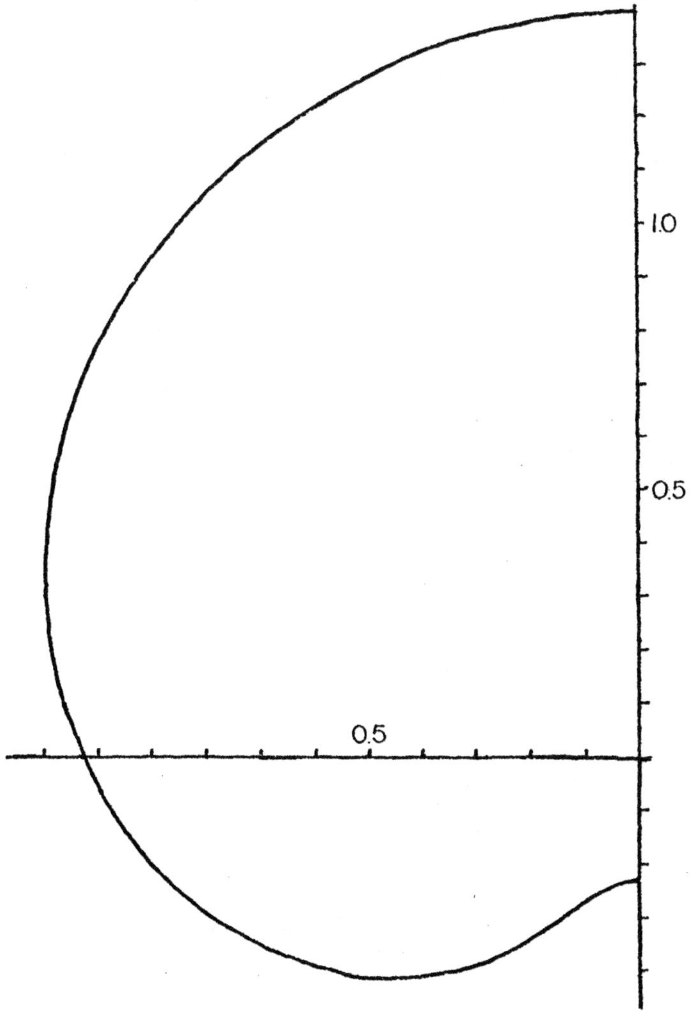

Fig. 4. The boundary of the bubble at t=1. The half of the boundary not shown is obtained by reflection in the x-axis, which is vertical.

ing at time t=0 a cavity in the form of a infinite, horizontal, circular cylinder filled with a homogeneous, incompressible massless gas (a bubble). We wish to describe the subsequent motion. (For an experimental investigation see [9]).

Take XY axes, vertical and horizontal respectively, in a plane perpendicular to the axis of the cylinder and set $Z=X+iY$. Then so long as the interior of the bubble remains simply connected the Riemann mapping theorem [10] shows that there exists a univalent function $F(z,t)$, analytic in $z=x+iy$ for $|z|>1$, so that $Z=F(z,t)$ maps the exterior of the unit circle into the exterior of the bubble in the Z-plane and such that $F(z,t)\sim c(t)z$ as $|z| \to \infty$, with $c(t)$ real.

N	[N/N] t=1	[N/N] t=1.15
1	-.250000000000	.182563338301
5	-.269671903517	-1.385989550195
9	-.205240065164	.650323943818
13	-.223750431783	.382811365837
17	-.224843129499	.094955689149
21	-.224849912942	1.164779239576
25	-.224637951899	.384563891665
29	-.224666965461	.295675727274
33	-.224669831775	.077346030034
37	-.224670169055	.507740740049
41	-.224669226592	.288476947662
45	-.224669290771	.238685951908
49	-.224669299965	-.066674193804
53	-.224669295847	.252484468731
57	-.224669296502	.231515531956
61	-.224669296653	.202755067671
65	-.224669296666	.156006709436
69	-.224669296667	.193864458499

Table 1. Some Padé approximants to the series in t^2 for the location of the bottom of the bubble.

There will also be a velocity potential function $\Phi(Z,t)$, analytic in Z outside the bubble. We write $\phi(z,t)=\Phi(F(z,t),t)$. If $V(z,t)=v_x - iv_y$ is the complex function such that v_x, v_y are the components of the velocity of the fluid at $Z=F(z,t)$ then we have

$$V(z,t) = \frac{\partial \Phi}{\partial Z} = \phi'/F'$$

where ' means $\frac{\partial}{\partial z}$. Thus

$$VF' = \phi' \tag{22}$$

Because the pressure is uniform in the interior of the bubble, Bernoulli's equation gives (after a suitable choice of units)

$$\mathrm{Re}(\dot{\phi}-V\dot{F}+F) + \tfrac{1}{2}VV^{*} = f(t), \quad |z| = 1 \tag{23}$$

where $\cdot$ means $\frac{\partial}{\partial t}$ and $f(t)$ is arbitrary. Because the velocity of the boundary in the normal direction must equal the same component of the velocity of the fluid, we find

$$\mathrm{Re}(\dot{F}^{*}F'z) = \mathrm{Re}(z\phi'), \quad |z| = 1 \tag{24}$$

The task is to solve (22) - (24). One approach is to solve the Riemann-Hilbert problems (23) and (24) to obtain $F(z,t)$, $\phi(z,t)$ in terms of functionals of F,ϕ at time t, and to integrate numerically over time. We [11] were unable to carry this approach beyond t=1 due to the rapid growth of instabilities in the numerical procedure. The shape of the bubble at t=1 is shown in Fig. 4.

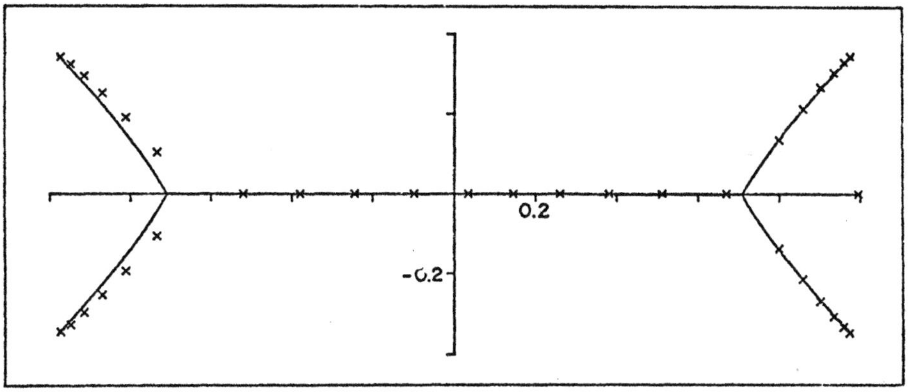

Fig. 5. The poles (X) of [36/36] to the series in t^2 for the bottom of the bubble (z=-1) plotted in the t^{-2} complex plane. One pole at $t^{-2}=-1.382$ is not included. Also shown is the set of minimum capacity (solid line) in the t^{-2} plane containing the four points $t^2= \pm 0.901 \pm i0.32$.

Here we give a progress report on an alternative method. A formal solution of the equations may be obtained in the form

$$F(z,t)=\sum_{m=0}^{\infty} t^{2m} \sum_{k=m(\mathrm{mod}\ 2)-1}^{m-1} z^{-k} a_{mk} \qquad (25)$$

with similar expansions for ϕ, V. The somewhat complicated recurrence relations that result from substituting (24) into (22)-(24) have been solved numerically up to m=152. (It is essential to use multiple precision arithmetic because of the high degree of cancellation occurring in the recurrence relations. The first few terms are

$$F(z,t)=z+\frac{1}{2}t^2 - \frac{1}{6}z^{-1}t^4 + (-\frac{1}{18} + \frac{2}{15} z^{-2})t^6$$

$$+ (\frac{1}{72} z + \frac{11}{140}z^{-1} - \frac{169}{1260} z^{-3})t^8 + \ldots$$

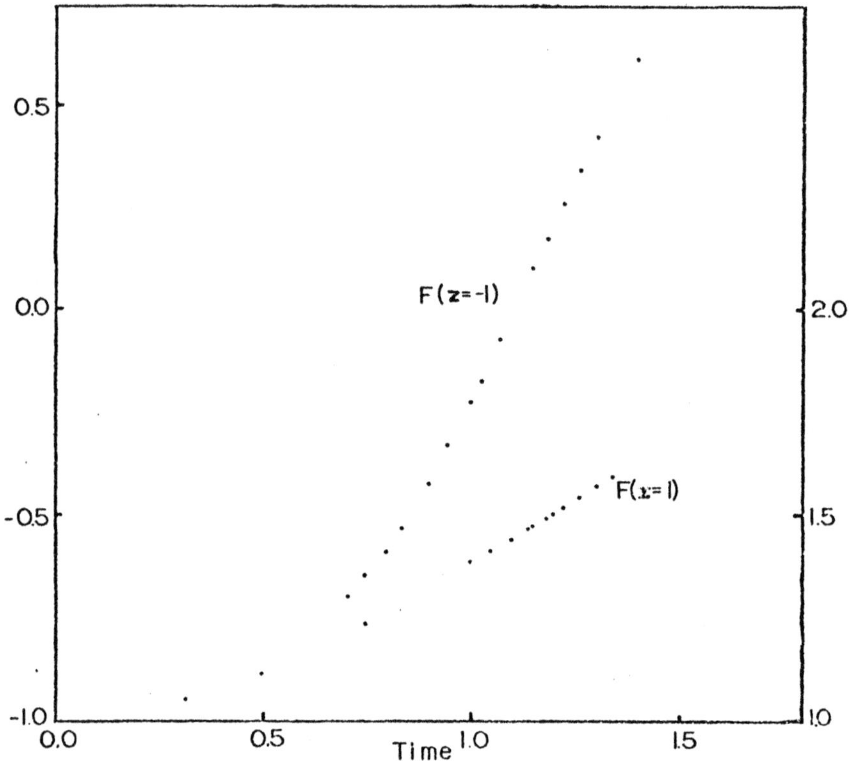

Fig. 6. The location of the top (F(z=1), right-hand scale) and bottom (F(z=-1), left-hand scale) of the bubble at various times.

For each value of z with $|z|=1$, (25) gives a power series in t^2 for the location of a point on the boundary of the bubble. We have studied mostly the top and the bottom, corresponding to $z = \pm 1$. One series is obtained from the other by changing $t^2 \to -t^2$ and multiplying by -1. We have 153 coefficients in each series. As m increases the coefficients oscillate with period about 19 under an envelope which, after decreasing until $m \simeq 45$, slowly increases. This suggests a radius of convergence for the series in t^2 of slightly less than 1. To use the series for larger times requires a method of summing a divergent series.

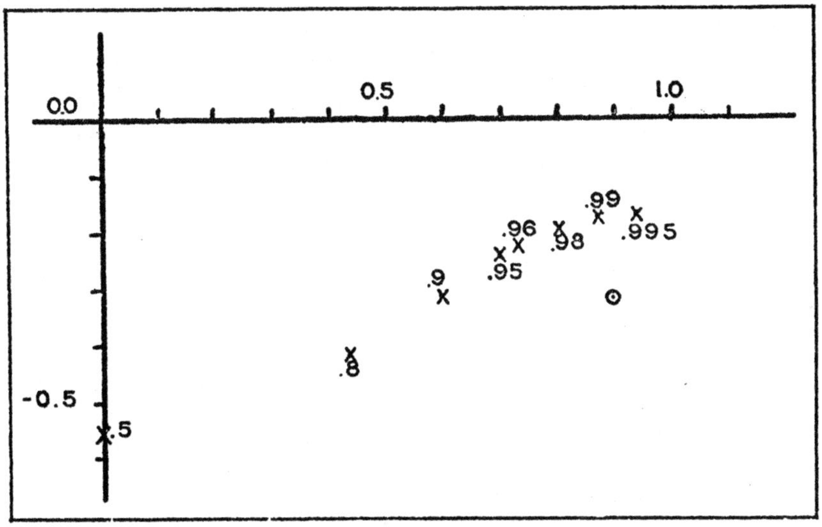

Fig. 7. The location of the square root singularity of $F(e^{i\theta},t^2)$ for $\theta=\lambda\pi$, plotted in the t^2-plane. The value of λ is shown near the corresponding point. The circled point is one of the singularities whose position appears to be independent of θ.

Consequently, we calculated the sequence of diagonal Padé approximants. Results for t=1, 1.15 are shown in Table 1. There is apparently excellent convergence at t=1 but virtually none at t=1.15. To investigate this behavior, we show (Fig. 5) the poles of [36/36] to the series in t^2 for the bottom of the bubble, plotted in terms of the variable t^{-2}. The poles are close to the set of minimum capacity associated with a function with branch points at $t^2 = \pm 0.901 \pm i0.32$. To explain the observed behavior on the basis of the conjecture of the previous section we need only to assume that $F(-1,t)$ has these four branch points. Unfortunately, the Padé approximants choose to place a cut for real $t^2 > \sim 1.4$, even though we have no reason to believe that $F(-1,t)$ is singular for such t.

A somewhat crude way of overcoming this difficulty is to use Padé approximants to evaluate the various series for $F(-1,t)$ and its derivatives at say $t^2=0.8$. We obtain meaningful results for about 15 derivatives. These quantities are then used to work out Padé approximants for an expansion about $t^2=0.8$. In this way we obtain reasonable results up to $t \sim 1.4$. The results for the top and bottom of the bubble are shown in Fig. 6.

The same procedure could be used for other values of z, but we feel it would be better to incorporate into the extrapolation procedure information about the nature of the singularities of $F(z,t)$. From a study of the location of the poles in t^2 of [N/N] to $F(e^{i\theta},t)$ it is plausible to assume that, for real θ other than $0,\pi$, $F(e^{i\theta},t)$ has singularities at the same four points as $F(\pm 1,t)$. In addition, there is another singularity in the t^2-plane whose position depends on θ. Since this singularity is nearest to the origin it is easy to analyze. We have calculated polynomials $P_i(x)$ of degree N so that $(x=t^2)$

$$\frac{\partial F}{\partial x}(e^{i\theta},t)\, P_1(x) + F(e^{i\theta},t)P_2(x) + P_3(x) = O(x^{3N+2}), \quad (26)$$

using a set of recurrence relations for vector orthogonal polynomials. An approximation to F is obtained by replacing the r.h.s. of (26) by zero and solving. (The differential equation Padé approximant [12], sometimes called the method of integral approximants [13]). Singularities of F may occur at zeros x_0 of $P_1(x)$, with exponent $-P_2(x_0)/P_1'(x_0)$. For $0.5\pi \leqslant \theta \leqslant 0.995\pi$ this method shows that the moving singularity is square root and gives its location with a precision of more than 20 figures when $\theta=.5\pi$ (4 figures at $\theta=0.995\pi$). The results are shown in Fig. 7.

Further progress would appear to depend on using the information about the singularities we have already gained and on learning more about their nature. The precise results we have already obtained should be useful in testing methods for solving moving boundary problems of more general applicability than the present approach.

5. ACKNOWLEDGEMENTS

The author is indebted to D. Freeman, S. Burley and I.Zmora for assistance with the calculations reported here. He is particularly grateful to J.L.Gammel, who introduced him to the subject of Padé approximants and who has provided help and encouragement for many years. Without the multiple precision arithmetic routine of K. Aashamar most of the calculations would have been impossible. The research was supported in part by the National Research Council of Canada.

REFERENCES

1. Nuttall, J.: "On sets of minimum capacity" (unpublished)
2. Hille, E.: "Analytic Function Theory", Vol. 2, Ginn and Co., Waltham, Mass. 1962, p 275
3. Schiffer, M.: appendix to Courant, R.: "Dirichlet's Principle, Conformal Mapping and Minimal Surfaces," Interscience Publishers Inc., New York, 1950.
4. Bergman, S.: "The Kernel Function and Conformal Mapping", American Mathematical Society, Providence, 1970.
5. Siegel, C.L.: "Topics in Complex Function Theory," Vol. 2, Interscience Publishers Inc., New York, 1970.
6. Nuttall, J., and Singh, S.R.: 1977, J. Approx. Theory 21, pp 1-42
7. Szego, G.: "Orthogonal Polynomials", American Mathematical Society, New York, 1959.
8. Dumas, S.: "Sur le Développement des Fonctions Elliptiques en Fractions Continues," Thesis, Zurich, 1908
9. Walters, J.K., and Davidson, J.F.: J. Fluid Mech. 12, pp 408-416
10. Pommerenke, C.: "Univalent Functions," Vandenhoeck and Ruprecht, Göttingen 1975
11. Freeman, D.F., and Nuttall, J.: "Time Development of a cylindrical fluid without viscosity", (unpublished)
12. Gammel, J.L.: in "Padé Approximants and their Applications" edited by P.R. Graves-Morris, Academic Press, London, 1973, pp 3-9
13. Hunter, D.L., and Baker, G.A.: "Methods of Series Analysis, III: Integral Approximant Methods", preprint, 1978

BIFURCATIONS IN THE DIOPHANT MOMENT PROBLEM

D. Bessis

Service de Physique Théorique, CEN-Saclay
91190 Gif-sur-Yvette, France.

ABSTRACT : We consider the diophant moment problem

$$n_k = \int_0^\Lambda \lambda^k d\mu(\lambda) \qquad k = 0,1,2,...$$

where $d\mu(\lambda)$ is a positive measure and the n_k's are positive integers. The fact that the n_k's are integers puts stringent constraints on the measure $d\mu(\lambda)$. One of the simplest results we obtain is that for $\Lambda < 4$, $d\mu(\lambda)$ is a finite sum of Dirac measures. An infinite number of bifurcation points in the parameter Λ occur in this problem. They are located at the points $4\cos^2 \pi/m$, $m = 2,3,4,...$ with an accumulation point at $\Lambda = 4$.

I. INTRODUCTION :

The diophant moment problem occurs in statistical mechanics, for the ferromagnetic Ising model[1].

The mathematical problem associated with it, is the following : Consider the function

$$G(z) = \int_0^\Lambda \frac{d\mu(\lambda)}{1-\lambda z} \qquad \Lambda > 0 \qquad (I.1)$$

where $d\mu(\lambda)$ is a positive measure. This defines an analytic function in the cut plane $C - \left[\frac{1}{\Lambda}, +\infty\right]$. On the other hand by expanding $G(z)$ around the origin we get :

$$G(z) = \sum_{k=0}^{\infty} n_k z^k \qquad (I.2)$$

C. Bardos and D. Bessis (eds.), Bifurcation Phenomena in Mathematical Physics and Related Topics, 203–215.

with :

$$n_k = \int_0^{\Lambda} \lambda^k d\mu(\lambda) \tag{I.3}$$

In general, one is interested in reconstructing $d\mu(\lambda)$ or some functional of it from the knowledge of the n_k's. This program has last been achieved by Krein[2], who has been able to give the best upper and lower bounds to $\mu(\lambda)$, when only a finite number of moments is known.

Here we shall be concerned with a different problem. What kind of constraints on the measure $\mu(\lambda)$ can be expected when the n_k's as a whole belong to a specific field ? For instance, if the n_k's are integers. Rather unexpectedly, in this last case, very stringent constraints come out, as we shall see in the sequel.

Before plunging into the problem, let us make a remark. Clearly from (I.1), the radius of convergence of (I.2) is $1/\Lambda$, therefore the n_k's will grow at most like Λ^k, and we expect the problem to become more and more difficult with larger values of Λ. In fact we shall see that this is only true in part.

II. THE CASE $\Lambda = 1$

As a first insight in this problem, let us consider the case $\Lambda = 1$. In this case we have :

$$n_k = \int_0^1 \lambda^k d\mu(\lambda) \tag{II.1}$$

Clearly

$$0 \leq \ldots \leq n_{k+1} \leq n_k \leq \ldots \leq n_1 \leq n_o \tag{II.2}$$

The sequence n_k is monotonic non-increasing, and therefore converges to a non negative value $\bar{n}$. However the sequence must have converged after a finite number of steps, the n_k's being *integers*. Therefore it exists a finite k_o such that :

$$n_{k_o} = n_{k_o+1} = \ldots \tag{II.3}$$

Or :

$$n_{k_o} - n_{k_o+1} = \int_0^1 \lambda^{k_o}[1-\lambda]d\mu(\lambda) = 0 \tag{II.4}$$

Equation (II.4) implies that

$$d\mu(\lambda) = \{\sigma_o\delta(\lambda) + \sigma_1\delta(1-\lambda)\}d\lambda \tag{II.5}$$

with

$$\left.\begin{aligned}\sigma_o &= n_o - n_1 \geq 0\\ \sigma_1 &= n_1 \geq 0\end{aligned}\right\} \tag{II.6}$$

and

$$G(z) = (n_o - n_1) + \frac{n_1}{1-z} \quad . \tag{II.7}$$

We have the proposition :

Proposition I

The Diophant Moment Problem (D.M.P) on [0,1] has a solution if $n_o \geq n_1 = n_2 = n_3 = \ldots \geq 0$ and in this case the generating functions of the n_k's is given by (II.7).

We remark that the singularity of G(z) is universal and does not depend on the n_k's. This fact is general. G(z) itself in this case depends only on the knowledge of n_o and n_1.

III. THE CASE $\Lambda = 2$

To have a slightly more general view of the problem, let us now consider the case $\Lambda = 2$.
In this case, we associate to :

$$n_k = \int_o^2 \lambda^k d\mu(\lambda) \tag{III.1}$$

the set

$$m_k = \int_o^2 \lambda^k (2-\lambda)^k d\mu(\lambda) \tag{III.2}$$

Clearly

$$m_k = \sum_{p=o}^{p=k} (-)^p \binom{k}{p} 2^{k-p}\, n_{p+k} \tag{III-3}$$

which shows that the set m_k is a set of positive integers. On the other hand the sup on [0,2] of the function $\lambda(2-\lambda)$ being 1, clearly we have again :

$$0 \leq \ldots \leq m_{k+1} \leq m_k \leq \ldots \leq m_1 \leq m_o \qquad \text{(III.4)}$$

And the sequence $\{m_k\}$ must converge after finitely many steps. It exists k_o such that :

$$m_{k_o} - m_{k_o+1} = \int_o^2 \lambda^{k_o}(2-\lambda)^{k_o}(1-\lambda)^2 d\mu(\lambda) = 0 \qquad \text{(III-5)}$$

Therefore

$$d\mu(\lambda) = \{\sigma_o\delta(\lambda) + \sigma_1\delta(1-\lambda) + \sigma_2\delta(2-\lambda)\}d\lambda \qquad \text{(III.6)}$$

Identification with the three first moments gives :

$$\sigma_o = n_o - \frac{3n_1}{2} + \frac{n_2}{2} \geq 0 \qquad \text{(III-7)}$$

$$\sigma_1 = 2n_1 - n_2 \geq 0 \qquad \text{(III-8)}$$

$$\sigma_2 = \frac{n_2 - n_1}{2} \geq 0 \qquad \text{(III-9)}$$

and

$$G(z) = \frac{n_o-(3n_o-n_1)z + (2n_o - 3n_1 + n_2)z^2}{1 - 3z + 2z^2} \qquad \text{(III-10)}$$

we have :

Proposition II

The D.M.P. on [0,2] has a solution if and only if $2n_1 \geq n_2 \geq \text{Max}\{n_1, 3n_1-2n_o\}$, $n_o \geq 0$ and $n_k = (2n_1-n_2)+(n_2-n_1)2^{k-1}$ for $k \geq 1$. In such a case the generating function of the n_k's is given by (III-10).

Again we verify that the singularities of G(z) are universal (they do not depend on the n_k's). Also the solution when it exists depends only on the knowledge of the first three moments.

Up to now we have encountered two bifurcation points in the parameter Λ (the length of the support of the measure $d\mu(\lambda)$).

When $0 < \Lambda < 1$, we have a generating function :

$$G_o(z) = n_o \tag{III.11}$$

For $1 \leq \Lambda < 2$, we have a new class which embeds the old one :

$$G_1(z) = n_o + \frac{zn_1}{1-z} \tag{III.12}$$

For $\Lambda = 2$ appears an even richer class :

$$G_2(z) = \frac{n_o - (3n_o - n_1)z + (2n_o - 3n_1 + n_2)z^2}{1 - 3z + 2z^2} \tag{III.13}$$

IV. THE PROBLEM FOR $\Lambda \leq 4$ REDUCED TO A TRIGONOMETRICAL MOMENT PROBLEM

We shall now consider the more general case $\Lambda \leq 4$. The value $\Lambda = 4$ will appear as a "critical" value, because it will be shown to be the limiting point of an infinite sequence of bifurcation pointsin the parameter Λ. We shall first transform the Diophant Moment Problem for $0 \leq \Lambda \leq 4$ into a trigonometrical moment problem.

We define θ_o by :

$$\Lambda = 4 \cos^2\theta_o/2 \qquad 0 \leq \theta_o \leq \pi \tag{IV.1}$$

and we make in

$$n_k = \int_o^{\Lambda} \lambda^k d\mu(\lambda) \tag{IV.2}$$

the change of variables

$$\lambda = 4 \cos^2\theta/2 \qquad 0 \leq \theta \leq \pi \tag{IV.3}$$

Then (IV.2) becomes :

$$\frac{2n_k}{4^k} = \int_{\theta_o}^{\pi} [\cos^2\theta/2]^k \, d\Sigma(\theta) \tag{IV.4}$$

where the measure $d\Sigma(\theta)$ is a positive measure, the measures are connected by :

$$\left.\begin{aligned} \Sigma(\theta) &= -2\mu(4\cos^2\theta/2) \\ \mu(\lambda) &= -\frac{1}{2}\,\Sigma(2\ \mathrm{Arc}\cos\frac{\sqrt{\lambda}}{2}) \end{aligned}\right\} \qquad \text{(IV.5)}$$

We now introduce the trigonometrical moments associated with $G(\theta)$:

$$c_k = \int_{\theta_o}^{\pi} \cos k\theta \, d\,\Sigma(\theta) \qquad \text{(IV.6)}$$

We have, introducing the (even) Tchebychef polynomial of order 2k:

$$\cos k\theta = T_{2k}(\cos\frac{\theta}{2}) = \bar{T}_k(\cos^2\theta/2) \qquad \text{(IV.7)}$$

where

$$\bar{T}_k(\cos^2\theta/2) = \sum_{p=0}^{k} \bar{T}_k^p(\cos^2\theta/2)^p \qquad \text{(IV.8)}$$

and

$$T_k^o = (-)^k \qquad \bar{T}_k^p = (-)^{k-p}\binom{k+p-1}{k-p}\frac{k}{p}\,\frac{4^p}{2} \qquad \text{(IV.9)}$$

Then, the formula connecting the c_k's to the u_k's reads :

$$c_k = \sum_{p=0}^{k} \bar{T}_k^p \frac{2n_p}{4^p} = (-)^k\, 2n_o + \sum_{p=1}^{k} (-)^{k-p}\frac{k}{p}\binom{k+p-1}{k-p} n_p \qquad \text{(IV.10)}$$

Using the identity :

$$\frac{k}{p}\binom{k+p-1}{k-p} = 2\binom{k+p}{k-p} - \binom{k+p-1}{k-p} \qquad \text{(IV.11)}$$

We find that the c_k's are integers when the n_k's are. We have therefore transformed the D.M.P on [0,4] into a T.M.P. on $[0,\pi]$.

The main interest of this transformation is that while the n_k's were allowed to grow as fast as 4^k, the c_k's are all bounded by c_o, in absolute value. Through the transformation (IV.5) there is a one-to-one correspondence between these two moment problems.

This is best understood, if we introduce the generating function of the n_k's and of the c_k's :

$$G(z) = \sum_{k=o}^{\infty} n_k z^k = \int_o^{\Lambda} \frac{d\mu(\lambda)}{1-\lambda z} \qquad \text{(IV.12)}$$

and

$$I(z) = \sum_{k=o}^{\infty} c_k z^k = \int_{\theta_o}^{\pi} \frac{(1 - z\cos\theta)}{1-2z\cos\theta + z^2}\, d\Sigma(\theta) \qquad \text{(IV.13)}$$

I(z) and G(z) are connected by the following relations :

$$I(z) = \frac{1}{2} G(0) + \frac{1-z}{1+z} G\left(\frac{z}{(1+z)^2}\right) \qquad \text{(IV.14)}$$

$$G(z) = \frac{1}{\sqrt{1-4z}} \left\{ I\left(\frac{1-\sqrt{1-4z}}{1+\sqrt{1-4z}}\right) - \frac{1}{2} I(0) \right\} \qquad \text{(IV.15)}$$

when $0 < \Lambda < 4$, I(z) is clearly analytic in the full z plane less in arc of circle of radius 1, and the values of I(z) inside and outside the circle of radius 1 are connected by :

$$I(z) + I(1/z) = I(0) \qquad \text{(IV.16)}$$

In particular, in this case we have $I(\infty) = 0$ as implied by (IV.16).

When Λ is exactly 4, $\theta_o = 0$ and (IV.13) defines now *two* analytic functions $I_+(z)$ inside the unit circle and $I_-(z)$ outside, these functions are still connected by :

$$I_+(z) + I_-\left(\frac{1}{z}\right) = I_+(0) \qquad \text{(IV.17)}$$

however this relation does not tell anything about $I_+(z)$ and its hypothetical analytic continuation outside $|z| = 1$.

V. SOLUTION OF THE TRIGONOMETRICAL MOMENT PROBLEM

We consider now the trigonometrical moment problem :

$$c_k = \int_{\theta_o}^{\pi} \cos k\theta \; d\Sigma(\theta) \qquad 0 \leq \theta_o \leq \pi \tag{V.1}$$

where $d\Sigma(\theta)$ is a positive measure.

To the c_k's we associate their generating function :

$$I(z) = \sum_{k=o}^{\infty} c_k z^k \tag{V.2}$$

or equivalently :

$$\tilde{I}(z) = \sum_{k=o}^{\infty} (c_k + c_o) z^k = I(z) + \frac{c_o}{1-z} \tag{V.3}$$

Setting $\tilde{c}_k = c_k + c_o$ (V.4), we see that the set $\tilde{c}_k$ is a *bounded* set of integers :

$$0 \leq \tilde{c}_k \leq 2c_o = \tilde{c}_o \tag{V.5}$$

It is convenient to associate to $\tilde{I}(z)$ the number :

$$\tilde{c} = \tilde{c}_o . \tilde{c}_1 \tilde{c}_2 \ldots \tilde{c}_k \ldots \tag{V.6}$$

written in the basis $\tilde{c}_o$. If this number is a rational number, the sequence $\{\tilde{c}_k\}$ must ultimately be periodic, so that there exist integers m and n with $\tilde{c}_k = \tilde{c}_{k+m}$ (V.7) for all k whenever $k > n$.

This implies

$$\tilde{I}(z) = c_o + c_1 z + \ldots + c_n z^n + z^{n+1} \frac{(c_{n+1} + c_{n+2} z + \ldots + c_{n+m} z^{m-1})}{1 - z^m} \tag{V.7}$$

and therefore $\tilde{I}(z)$ or $I(z)$ is a rational fraction having its poles on the unit circle as roots of unity. If the number $\tilde{c}$ is irrational, one can think of it as a limit of sequence of rational numbers and therefore one expects a natural boundary for $\tilde{I}(z)$ on $|z| = 1$. This last statement is made rigorous, by Szegö theorem[3] :

If among the coefficients $\tilde{c}_k$ of a Taylor series $\tilde{I}(z) = \sum_{k=o}^{\infty} \tilde{c}_k z_k$ there appear a finite number of different values, then either $\tilde{I}(z) = \frac{\tilde{P}(z)}{1-z^m}$ where $\tilde{P}(z)$ is a polynomial with integer coefficients and m a non negative integer, or else $\tilde{I}(z)$ cannot be analytically continued beyond the unit circle.

For $\Lambda < 4$, that is $\theta_o > 0$, we have seen that I(z) admits an analytic continuation through all the complex plane defined by (IV.13). Therefore

$$I(z) = \frac{P_{m-1}(z)}{1-z^m} \tag{V.8}$$

where $P_{m-1}(z)$ is a polynomial of degree m-1 at most, taking into account that $I(\infty)$ is zero. Therefore :

$$P_{m-1}(z) = c_o + c_1 z + \ldots + c_{m-1} z^{m-1} \tag{V.9}$$

On the other hand the identity (IV.16) implies

$$P_{m-1}(z) - z^m P_{m-1}\left(\frac{1}{z}\right) \equiv P_{m-1}(0)[1-z^m] \tag{V.10}$$

That is

$$\left.\begin{array}{l} c_1 = c_{m-1} \\ c_2 = c_{m-2} \\ \text{---------} \\ c_k = c_{m-k} \\ \text{---------} \end{array}\right\} \tag{V.11}$$

We must now introduce the fact that the measured $\Sigma(\theta)$ is positive.

This is most conveniently expressed in terms of the Toeplitz[4] determinants

$$T_k(c_o, c_1, \ldots c_k) = \begin{vmatrix} c_o c_1 c_2 \ldots c_k \\ c_1 c_o c_1 \quad c_{k-1} \\ c_2 c_1 c_o \ldots c_{k-2} \\ \vdots \\ c_k c_{k-1} \ldots c_o \end{vmatrix} \tag{V.12}$$

A necessary and sufficient condition that $d\Sigma(\theta)$ is non negative on $0 \leq \theta \leq \pi$ is that

$$T_k(c_o,c_1,\ldots c_k) \geq 0 \qquad \text{for all } k=0,1,2,\ldots \tag{V.13}$$

while $d\Sigma(\theta)$ is non negative for $\theta_o \leq \theta \leq \pi$ and vanishes on $0 \leq \theta \leq \theta_o$, if and only if we have both (V.13) and :

$$T_k(d_o,d_1,\ldots d_k) \geq 0 \qquad \text{for all } k=0,1,2,\ldots \tag{V.14}$$

where

$$d_k = 2c_k \cos\theta_o - c_{|k+1|} - c_{|k-1|} \tag{V.15}$$

Furthermore if $T_{k_o}(c_o,c_1,\ldots c_{k_o})=0$ then $T_k(c_o,c_1,\ldots k)=0$ for all $k \geq k_o$ and this is the case if and only if $I(z)$ takes the form (V.8) and (V.9) for some $m \leq k_o$. To end this paragraph, we notice that the singularities of $I(z)$ (for $\theta_o > 0$) are poles on the unit circle with arguments :

$$\theta_{p,m} = \frac{2\pi p}{m} \qquad p=0,1,2,\ldots m-1 \tag{V.16}$$

Necessarily (V.8) cannot be irreducible (for instance the value $p=0$ can never be a singularity). The true poles must necessarily fulfil :

$$\theta_{p,m} \geq \theta_o \tag{V.17}$$

Therefore we shall have bifurcations any time

$$\theta_o = \frac{2\pi}{m} \qquad m=2,3,4,5,\ldots \tag{V.18}$$

However, it may happen that there are more bifurcations than this minimal set, if we can find a polynomial $P_{m-1}(z)$ with integer coefficient fulfilling all requirements (V.11), (V.13), (V.14) which has a zero at $z=e^{\pm 2i\pi/m}$.

Finally, if we consider the case $\theta_o = 0$ $(\Lambda = 4)$, we cannot exclude a priori a natural boundary on the unit circle. However, we have in the case $c_o = 1$, the following result.

Proposition
In the case $c_o = 1$, the trigonometrical moment problem on $[0,\pi]$, has for generating function of the moments either :

$$I(z) = 1 \tag{V.19}$$

$$\text{or } I(z) = \frac{P_{m-1}(z)}{1-z^m}\text{, the previous set found for } \theta_o > 0.$$

Clearly either for some k_o

$$c_{k_o} = \pm 1 \tag{V.20}$$

Then

$$c_o \mp c_{k_o} = \int_o^{\pi} (1 \mp \cos k_o \theta) d\Sigma(\theta) = 0 \tag{V.21}$$

$$= 2 \int_o^{\pi} \begin{pmatrix} \sin^2 k_o \frac{\theta}{2} \\ \cos^2 k_o \frac{\theta}{2} \end{pmatrix} d\Sigma(\theta) = 0 \tag{V.22}$$

and $d\Sigma(\theta)$ is necessarily a finite sum of Dirac measures.
Or

$$c_k = 0 \qquad \text{for all } k \neq 0$$

in which case $I(z)$ reduces to 1.

VI. CONCLUSION

By transforming back the trigonometrical moment problem to the Diophant Moment Problem, we see, using formula (IV.5), that a finite sum of Dirac measures will transform into a finite sum of Dirac measures.

Therefore for $\Lambda < 4$, the generating function $G(z)$ reduces to a rational fraction, having its poles located at :

$$z_k^{(m)} = \frac{1}{4 \cos^2 \frac{k\pi}{m}} \qquad k=1,2,3,\ldots m \tag{VI.1}$$

the bifurcations will occur for

$$\Lambda_m = 4 \cos^2 \frac{\pi}{m} \qquad m=2,3,4,\ldots \qquad \text{(VI.2)}$$

The first values are :

$$\begin{cases} m = 2 & \Lambda_2 = 0 \\ m = 3 & \Lambda_3 = 1 \\ m = 4 & \Lambda_4 = 2 \\ m = 5 & \Lambda_5 = \dfrac{3+\sqrt{5}}{2} \\ m = 6 & \Lambda_6 = 3 \\ \ldots\ldots\ldots\ldots\ldots\ldots & \end{cases} \qquad \text{(VI.3)}$$

For $\Lambda = 4$, we have not been able to exclude a natural boundary, except for $n_o = 1/2$, all n_k integers. This corresponds to a physical situation in the Ising model.

In this case we get :

$$G_{\Lambda=4}(z) = \frac{1}{2} \frac{1}{\sqrt{1-4z}} \qquad \text{(VI.4)}$$

which corresponds to $I(z) = +1$, otherwise $G_{\Lambda=4}(z)$ is a rational fraction.

From $\Lambda > 4$, a great variety of functions belonging to various classes are known. All Diophant problem with $\Lambda > 6$, can be reduced to $\Lambda = 6$, by a finite algebraic transformation which transforms integers into integers and positive measures into positive measures[1] .

However a real understanding of the case $\Lambda > 4$, could be achieved, if it were possible to discover an isomorphism between the solutions for $\Lambda < 4$ and $\Lambda > 4$. This last approach is in progress.

References

[1] - Barnsley M., Bessis D, and Moussa P., 1979 J. Math. Phys. 20 (4)
[2] - Karlin S., and Studden W.J., Tchebycheff systems with applications in Analysis and Statistics. Interscience (1966)
[3] - Dienes P., The Taylor series (Dover, New-York 1957) p.324
[4] - Grenander U and Szegö G., Toeplitz forms and their applications, University of California Press,Berkeley and Los Angeles 1958

III

YANG–MILLS FIELD THEORY

YANG-MILLS FIELDS : SEMI CLASSICAL ASPECTS

R. Stora

Centre de Physique Théorique du CNRS Marseille
and
Theory Division CERN Geneva

INTRODUCTION

Gauge theories have been formulated in 1954 by Yang and Mills [1] as natural non abelian generalizations of Maxwell's theory of the electromagnetic field. Since the electromagnetic field mediates electromagnetic interactions between charged systems it was natural to seek for the non abelian analog susceptible to accomodate the non abelian internal symmetries discovered in the realm of elementary particle physics. The naturalness of this construction seems also to have drawn the attention of other physicists who were simultaneously aware of a number of difficulties which remain unsolved even now. As time went, more approximate internal symmetries [2] were discovered in particle physics. Also, a number of fundamental remarks of a technical nature [3] allowed to put gauge theories to practical use. There resulted a net progress in unifying "weak", "electromagnetic", [4][5], and "strong" interactions [5] between elementary particles, to such an extent that at present, it has become standard to think about particle physics in terms of gauge theories [6]. The naturalness of these theories has also appeared in mathematics where it constitutes a branch of differential geometry : the study of fiber bundles and connections on them [7], associated with a Lie group G-[4][5][8]. The existing mathematical knowledge turns out to be well adapted to the construction of the perturbative expansions of the quantized gauge theories analogous to those of quantum electrodynamics [3],[9]. Such expansions are formal power series in an argument $\hbar$ which has ultimately to be understood as Planck's quantum of action. In the present context, the properties of local quantum field theory combine with the non abelian internal local symmetry in such a way that new phenomena of a topological nature

C. Bardos and D. Bessis (eds.), Bifurcation Phenomena in Mathematical Physics and Related Topics, 219–229.

have to be taken into account, some of which will be described here.
These brief notes are divided into five sections :
Section II provides the definition of Yang Mills fields and of the Yang Mills action.

Section III motivates the study of some classes of solutions of the Yang Mills equations on the basis of unavoidable properties of local quantum theories.

Section IV is devoted to a description of some Yang Mills fields with self dual curvature.

Section V is devoted to a description of some analogies between four dimensional Yang Mills fields and two dimensional "chiral" fields.

II YANG MILLS FIELDS [1]

For technical reasons which have to do with the structure of local quantum field theory [10], we shall consider euclidean fields : E^4 is four dimensional euclidean space with the usual metric :

$$\underline{x} = (x_1, x_2, x_3, x_4) \in E^4 \qquad x^2 = (\underline{x}, \underline{x}) = \sum_1^4 (x_i)^2 \tag{1}$$

Let G be a compact Lie group with Lie algebra $\mathfrak{g}$, which we shall assume to be simple, and, unless otherwise stated, simply connected [11].

A Yang Mills field $\{a_\mu^\alpha\}$ on an open set Λ of E^4 is defined in terms of a $\mathfrak{g}$ valued one form

$$a = \sum_{\alpha,\mu} a_\mu^\alpha \, dx^\mu \, e_\alpha, \tag{2}$$

where e_α is a basis of $\mathfrak{g}$. The corresponding field strength $\{F_{\mu\nu}^\alpha\}$ is defined in terms of the corresponding curvature two form :

$$\begin{aligned} F &= da + \frac{1}{2}[a,a] = \frac{1}{2} \sum_{\mu,\nu,\alpha} F_{\mu\nu}^\alpha \, dx^\mu \wedge dx^\nu \, e_\alpha \\ F_{\mu\nu}^\alpha &= \partial_\mu a_\nu^\alpha - \partial_\nu a_\mu^\alpha + f_{\beta\gamma}^\alpha a_\mu^\beta a_\nu^\gamma \end{aligned} \tag{3}$$

where the $f_{\beta\gamma}^\alpha$'s are the structure constants of $\mathfrak{g}$.

Note the Bianchi identity

$$dF + [a,F] = 0 \tag{4}$$

The Yang Mills action is defined as the functional [1]

$$S_{YM}(a) = \frac{1}{4}\int F^{\alpha}_{\mu\nu}\, g_{\alpha\beta}\, F^{\alpha}_{\mu\nu}\, d^4x \tag{5}$$

where g is the Killing form of $\mathcal{G}$ (for a suitable choice of the basis $\{e_\alpha\}$, $g_{\alpha\beta} = \delta_{\alpha\beta}$).

In more intrinsic terms one can express the Yang Mills action as

$$S_{YM}(a) = \int F \wedge (* F) \tag{6}$$

where $*$ is the Hodge duality operation for the metric on E_4 :

$$(* F)_{\mu\nu} = \frac{1}{2}\epsilon_{\mu\nu\rho\sigma}\, F^{\rho\sigma} \tag{7}$$

ϵ being the Levi-Cevita totally antisymmetric symbol.

"Matter fields" can easily and have to be incorporated into the Yang Mills action but will be omitted here.

The Yang Mills action has a basic property : it is gauge invariant, i.e. invariant under the substitution

$$a \rightarrow {}^{g}a = g^{-1}ag + g^{-1}dg \tag{8}$$

where g is a map from Λ to G, the first term in Eq.(8) involves the pointwise adjoint action of G on $\mathcal{G}$ and the second term is the pull back of the Maurer-Cartan form of G under g.

III QUANTIZATION OF THE YANG MILLS ACTION

The basic object to be defined can be formally written as [10]

$$d\mu(a) = \text{"} e^{-\frac{S_{YM}}{g^2\hbar}} da \Big/ \int e^{-\frac{S_{YM}}{g^2\hbar}} da \text{"} \tag{9}$$

It is supposed to be a measure on some space of classical Yang Mills fields to be made precise. If this measure could be constructed, it would serve to calculate expectation values of observables constructed from a :

$$< \mathcal{O}(a) > = \frac{\text{"}\int e^{-\frac{S_{YM}(a)}{g^2\hbar}}\, \mathcal{O}(a)\, da\text{"}}{\int e^{-\frac{S_{YM}(a)}{g^2\hbar}}\, da} \tag{10}$$

$\mathcal{O}(a)$ is supposed to be gauge invariant. One may also want to define fields $\psi(a)$, not necessarily gauge invariant. Both observables, and, possibly fields are taken to be of the form

$$\mathcal{O}(a) = \prod_{i=1}^{n} \mathcal{O}_i(a) \tag{11}$$

where $\mathcal{O}_i(a)$ is some local polynomial in a and its derivative smeared with some test function

$$\mathcal{O}_i(a) = \int f_i(x) \, [\tilde{\mathcal{O}}_i(a)](x) \, d^4x \tag{12}$$

In formula (10), $\hbar$ is Planck's constant and g the Yang Mills coupling constant such that $g^2/\hbar c$ is dimensionless.

The usual methods used to attempt a definition of such objects as $d\mu(a)$ go through at least two necessary steps [10] :

Step 1) enclosing the system in a bounded region $\Lambda \subset E^4$ outside which boundary conditions are prescribed.

Step 2) Performing an ultraviolet regularization e.g. discretizing Λ into a lattice of mesh ε. The construction of $d\mu(a)$ should then go through the investigation of the limits $\varepsilon \to 0$ for a suitable dependence of g upon ε, and $\Lambda \to \infty$ in a suitable sense.

This program which has been successfully carried through for a number of field theory models [10] looks at present formidable in the case at hand.

The matter becomes a great deal simpler if one is willing to construct the corresponding formal perturbative expansions [3] in powers of $\hbar$, in which case step 2) is rather well understood, provided step 1) is sufficiently well understood. It is there, however that topological complications arise :

Let us try to impose the Dirichlet boundary conditions

$$\mathcal{O}(a) = 0 \text{ outside } \Lambda.$$

for local gauge invariant observables (13)

$$\mathcal{O}(a)$$

For smooth a, this implies

$$a = g^{-1}dg \text{ outside } \Lambda \tag{14}$$

Modulo smooth gauge transformations we see that these boundary conditions split into sectors labelled by the homotopy class of the map

$$S^3 \overset{g}{\to} G \tag{15}$$

obtained by restricting g to any 3-sphere which is the boundary of a ball containing Λ.

Although there is no solid justification at the moment that the construction of the perturbative expansions exclusively involves smooth configurations, this is quite likely to be true, and, at any rate, smooth configurations surely have to be considered.

Should we like to impose another type of boundary conditions, such as periodic boundary conditions on the boundary of a cubic box Λ, an analogous situation will prevail [11].

Taking advantage of the conformal invariance of the Yang Mills action, one can easily see that Dirichlet boundary conditions allow to interpret $\underline{a}$ in terms of a $\mathfrak{g}$-connection on a G-bundle ε over a compact Riemannian four-manifold which is conformally flat over a compact subset Λ, $\underline{a}$ being flat outside Λ. This bundle is topologically classified by the homotopy class of g(Eq. 15).

Thus, step 1) can be studied in terms of the connections on a G bundle ε over a compact Riemannian manifold M of volume $O(|\Lambda|)$ which is conformally flat over a region of size $O(|\Lambda|)$. Various choices correspond to various ways of smoothening out Dirichlet boundary conditions.

For practical purpose [7], the Yang Mills fields defined in section III turn into a collection of Yang Mills fields a_i defined on the open sets of a covering $\{\Lambda_i\}$ of M, with the fitting law

$$a_i = {}^{g_{ji}}a_j \quad \text{in} \quad \Lambda_i \cap \Lambda_j \tag{16}$$

where the transition functions g_{ij} of ε defined in $\Lambda_i \cap \Lambda_j$ fulfill the compatibility condition

$$g_{ij}\, g_{jk}\, g_{ki} = e \quad \text{in} \quad \Lambda_i \cap \Lambda_j \cap \Lambda_k \tag{17}$$

Then clearly, gauge invariant local objects define a corresponding global object on M. For instance the Yang Mills action density defines a four-form globally over M.

The essence of the construction of perturbative expansions consists in formally applying the saddle point method to "integrals" such as those occurring in Eq.(10).

One thus needs to find solutions of the classical Yang Mills equations (the Euler Lagrange equations derived from the Yang Mills action) :

$$D * F = 0 \tag{18}$$

$$D = d + \text{ad } a$$

Among those, for fixed integer

$$k = \frac{1}{8\pi^2} \int_M \sum_\alpha F^\alpha \wedge F^\alpha \tag{19}$$

the second Chern number of ε [7], which represents the homotopy class of g Eq.(15) are the solutions with self dual or antiself dual curvature

$$F = \pm * F \tag{20}$$

which make the Yang Mills action minimum :

$$\int F_1 * F > \left| \int F \wedge F \right| . \tag{21}$$

Indeed, from the Schwarz inequality

$$\int (F \pm * F) * F \pm * F) > 0 \tag{22}$$

with the equality holding iff Eq.(20) holds ($\pm$ = sgn k).

Such solutions are called instanton or anti-instanton solutions according to the sign of k.

IV SELF DUAL YANG MILLS FIELDS OVER S^4

As the manifold M, choose S^4, obtained from E^4 by stereographic projection. There the geometry is the simplest. The first of these solutions were discovered in 1975 by A. Belavin, A. Polyakov, A.S. Schwarz, Y. Tyupkin [12]. For G = SU2 = Sp1 = U(1,H) consider the Maurer Cartan form on Sp2 = U(2,H) which is a principle bundle over $S^4 \simeq P_1(H)$ with structure group Sp1 × Sp1, and restrict it to the Lie algebra of either Sp1 factor. This yields the U(2,H) invariant connection which under conformal transformations of $P_1(H)$ yields all solutions for $k = \pm 1$, parametrized by SL(2,H)/U(2,H).(Here H denotes the field of quaternions). A more general class of solutions was given by G.'t Hooft, R. Jackiw, C. Rebbi [12] under the form

$$a = * \omega \wedge d \ln \rho \tag{23}$$

where ω is a self dual or antiself dual two form whose space is isomorphic with su2, and

$$\rho(x) = \sum_{0}^{k} \frac{\lambda_i^2}{(x-x_i)^2} \tag{23}$$

The singularities at $x = x_i$ can be removed by gauge transformations, so that a defines a connection on a bundle with Chern number $\pm$ k.

It was then proved by A.S. Schwarz, R. Jackiw, C. Rebbi [13] that the manifold of solutions is 8k-3 dimensional (for G = SU2) so that some solutions were missing. This counting comes from the use of the index theorem, vanishing theorems due to positivity, and the implicit function theorem in infinite dimensional spaces, applied in the neighbourhood of a known solution.

The problem was then completely solved by streamlining and applying some of the mathematical ideas proposed by R. Penrose in the framework of the "twistor program" [14] which were put in shape by M.F. Atiyah, N.J. Hitchin, I.M. Singer [13] under the impulse of R.S. Ward [15] : the problem was transformed into that of classifying algebraic bundles over $P^3(\mathbb{C})$ with certain restrictive properties allowing to use some harmonic theory on S^4 to get vanishing theorems which simplify the algebra down to the point were the solutions can be written as follows (M.F. Atiyah, N.G. Drinfeld, N.J. Hitchin, Yu I Manin) [16] :

Let H^{k+1} be a k+1-dimensional quaternionic vector space.

$$W_i = A_i + B_i x \qquad i = 1\ldots k \qquad W_i \in H^{k+1} \tag{24}$$

x the quaternion representing a point in $P_1(H)$ in one chart. The k vectors A_i, B_i are subjected to the constraints

$$\begin{aligned} &A_i^+ A_j \text{ real symmetric} \\ &B_i^+ B_j \text{ real symmetric} \\ &A_i^+ B_j = A_j^+ B_i \end{aligned} \tag{25}$$

and $M = || W_i + W_j ||$ real symmetric positive for all x including ∞. Modulo the gauge transformations :

$$\begin{aligned} A_i &\to L_{ij} A_j \\ B_i &\to L_{ij} B_j \end{aligned} \qquad L \in GL(k,R) \tag{26}$$

and

$$\begin{aligned} A_i &\to UA_i \\ B_i &\to UB_i \end{aligned} \qquad U \in U(k+1,H) \tag{27}$$

Then construct $W \in H^{k+1}$ such that

$$W^+W_i = 0 \qquad W^+W = 0 \tag{28}$$

Then, up to gauge transformations all solutions of Eq.(20) are given by

$$a = W^+dW \tag{29}$$

The extension to G = Sp(n) is obvious. The further restrictions which reduce the connection to a U(n) or O(n) subgroup are not so easy to write down [16].

V. ANALOG MODELS IN TWO DIMENSIONS

The conspicuous fact that, although methods of complex algebraic geometry were used to carry out the relevant part of the twistor program, the answer can be formulated most conveniently in terms of quaternions, calls for analogies.

Replacing E^4,(S^4) by E^2,(S^2), consider harmonic maps [17] from S^2 into a Kähler manifold K, i.e. maps which minimize the "chiral" action :

$$S_{chiral} = \int d^2x \, \frac{\partial \bar{Z}^\lambda}{\partial x_i} h_{\bar{\lambda}\mu} \frac{\partial Z^\mu}{\partial x_i} \tag{30}$$

These maps are topologically characterized by the Chern number

$$k = \frac{i}{2\pi} \int d\bar{Z}^\lambda \wedge h_{\bar{\lambda}\mu} \, dZ^\mu \tag{31}$$

provided K is algebraic and

$$S_{chiral} > 2\pi |k|$$

with equality iff the mapping

$$S = x_1 \pm ix_2 \to \underline{Z} \in K \tag{32}$$

is holomorphic. Such mappings exist provided K is rational [17], e.g. $K = P_n(\mathbb{C})$ or complex Grassmannians. The Yang Mills instantons turn out to be restrictions to some quaternionic curves of $P_k(H)$ of the standard connexion on the Hopf bundle $H^{k+1}/\{0\} \to P_k(H)$.

The characterization of these curves has however not been completely understood yet in clear geometrical terms [18].

All these models share a number of properties both at the classical and at the quantum level, but also exhibit a number of puzzling differences. To quote only one : the $P_1(\mathbb{C})$ model admits no other continuous finite action solution over E^2 than those mentioned here (instantons, anti-instantons) (17). This is not the case for the $P_n(\mathbb{C})$ models $n > 1$ [22],[19]. One expects an analogous situation to hold for Yang Mills models. On the other hand one would expect configurations interpolating e.g. two well separated k instanton, k anti-instanton solution configurations to be important in the evaluation of the functional integrals Eq.(10). Some of the non instanton solutions in the $P_n(\mathbb{C})$ models resemble such configurations [19]. It would be desirable to find them all. There remains then a puzzle for the $P_1(\mathbb{C})$ model, which makes it worthwile to investigate also complex solutions [20]. This kind of questions gets amplified if instead of generalizing $P_1(\mathbb{C}) \simeq S^2$ into $P_n(\mathbb{C})$, one chooses to generalize to S^n : for $n>2$ there is no instanton. For even n all finite action solutions defined on $S^2 = E^2 \cup \infty$ are in principle known, but their role not yet sufficiently understood [21][19]. For odd n, no such solution exists that cannot be embedded in an S^p, p even.

CONCLUSION, OPEN PROBLEMS

We have sketched out some classical aspects of Yang Mills theories and their "chiral" two dimensional companions. In the Yang Mills case the matter field have been omitted and should of course be included.

Two classes of problems are essentially open :

1) relax the regularity assumptions keeping only the condition that the action be finite

2) compare the situations occurring for different manifolds M (boundary conditions) under the thermodynamic limit vol $M \to \infty$.

A third class is completely open : look for solutions in the complexified field manifold.

Given a "sufficiently complete" family of classical solutions, the problem of defining the corresponding perturbative series is far from completed. Some remarkable results have however been obtained for the $P_1(\mathbb{C})$ two dimensional model [22], whose extension definitely requires a thorough knowledge of the underlying classical theories.

It should be finally emphasized that this approach "from inside" can in no way be taken as a substitute for constructing a gauge theory and can at best provide some approximation scheme whose, soundness is at present subject to extremely strong criticisms.

REFERENCES* AND FOOTNOTES

* This bibliography is very incomplete, but hopefully allows to trace back the historical development.

[1] C.N. Yang, R.L. Mills, P.R. 96, 191(1954).

[2] Nuclear and subnuclear physics exhibit a hierarchy of approximate symmetries : isotopic spin SU(2), strangeness SU3, charm SU4, nowadays referred to as "flavour" groups. It has turned out however that these are not the groups relevant to modern gauge theories [4][5][8].

[3] P. Higgs, P.L. 12, 132 (1964) ; P.R. 145, 1156 (1966).
TWB Kibble, P.R. 155, 1554 (1967).
L.D. Faddeev, V.N. Popov, P.L. 25B, 29 (1967).
G.'t Hooft, P.R. D14, 3432 (1976).
G.'t Hooft, N.P. B33, 173 (1971) ; B35, 167 (1971).
G.'t Hooft, M. Veltman, N.P. B50, 318 (1972).
C.G. Callan, R.F. Dashen, D.J. Gross, P.R. D19, 1826 (1979).

[4] S. Weinberg, P.R.L. 19, 1264 (1967).

[5] H. Fritsch, M. Gell-Mann, H. Leutwyler, P.L. 47B, 365(1973).
S. Weinberg, P.R.L. 31, 494 (1973).

[6] Further efforts are currently being pursued to unify strong electromagnetic and weak interactions. One favoured approach consists in having colour SU3×flavour SU2×U(1) as a subgroup of a larger gauge group e.g. SU5 : H. Georgi, S.L. Glashow, P.R.L. 32, 438 (1974).

[7] S. Kobayashi, K. Nomizu, Foundations of Differential Geometry, Interscience, New York, 1963).

[8] In present day particle physics, G is "colour SU3" of strong interactions X "flavour" $SU2_L \times U(1)$ of weak and electromagnetic interactions.

[9] J.H. Lowenstein, B. Schroer, P.R. D6, 1556 (1972).

[10] B. Simon : The $P(\phi)_2$ Euclidean quantum field theory Princeton University Press, Princeton (1974).

[11] See however the qualitative discussion of static quark confinement in quantum chromo dynamics (QCD) : G.'t Hooft, N.P. B153, 141 (1979).

[12] A. Belavin, A. Polyakov, A.S. Schwarz, Y. Tyupkin, P.L. B59, 517 (1976).
G.'t Hooft, Unpublished
R. Jackiw, C. Rebbi, P.R. D14, 517 (1976).

[13] A.S. Schwarz, P.L. B67, 172 (1977).
R. Jackiw, C. Rebbi, P.R.L. B67, 189 (1977).
M.F. Atiyah, N.J. Hitchin, I.M. Singer, Proc. Nat. Ac. Sci. USA 74, 2662 (1977).
M.F. Atiyah, N.J. Hitchin, I.M. Singer, Proc. R. Soc. London, A362, 425 (1978).
A.S. Schwarz, C.M.P., 64, 233 (1979).

[14] R. Penrose, Rep. Math. Phys., 12, 65 (1977).

[15] R.S. Ward, P.L. 61A, 81 (1977).
M.F. Atiyah, R.S. Ward, C.M.P., 55, 117 (1977).

[16] M.F. Atiyah, V.G. Drinfeld, W.J. Hitchin, Yu I Manin, P.L. 65A, 185 (1978).
V.G. Drinfeld, Yu I Manin, Funct. An. Appl. (in russian), (1979) ; C.M.P., 63, 177 (1978).
M.F. Atiyah, Lectures given at the Scuola Normale Superiore, Pisa, 1978, to appear.
A. Dovady, J.L. Verdier, ed. E.N.S. Seminar (1978), to appear in Astérisque.

[17] J. Eells, L. Lemaire, Bull. London Math. Soc. 10, 1 (1978).
W.D. Garber, S.N.M. Ruijseaars, E. Seiler, D. Burns, Ann. Phys. 119, 305 (1979).
A. Perelomov, C.M.P., 63, 237 (1978).

[18] F. Gursey, H.C. Tzé, Yale preprint Oct. (1979).

[19] A.M. Din, W.J. Zakrzewski, CERN preprint TH 2721-2722.

[20] R. Balian, G. Parisi, A. Voros, P.R.L. 41, 1141 (1978).

[21] H.J. Borchers, W.D. Garber, C.M.P., to appear.

[22] V.A. Fateev, I.V. Frolov, A.S. Schwarz, N.P. B154, 1 (1979).
B. Berg, M. Lüscher, C.M.P., to appear.

PROPAGATION OF THE ENERGY OF YANG-MILLS FIELDS

R.T.Glassey and W.A.Strauss

Departement de Mathématiques
Université de Nice
Parc Valrose, NICE, France

I. THE EQUATIONS

Let us begin with the familiar Maxwell equations. We write a spatial point as $x = (x_1, x_2, x_3)$ and a space-time point as $(t,x) = (t, x_1, x_2, x_3)$; also $\partial_0 = \partial/\partial t$, $\partial_k = \partial/\partial x_k$ (k=1,2,3), and $\nabla\times$ = curl, $\nabla.$ = divergence. In terms of the electric field E and the magnetic field H, the Maxwell's equations are

$$\begin{cases} \partial_0 H = -\nabla\times E \\ \partial_0 E = \nabla\times H \\ \nabla.H = \nabla\cdot E = 0 \end{cases} \tag{1}$$

They may alternatively be written in a Lagrangian formulation with the Lagrangian

$$\mathcal{L} = 1/2\int(|E|^2 - |H|^2) \quad dxdt \tag{2}$$

subject to the constraints

$$\begin{cases} H = -\nabla\times A \\ E = \partial_0 A + \nabla A_0 \end{cases} \tag{3}$$

for some 3-vector $A = (A_1, A_2, A_3)$ and some scalar A_0. This 4-vector (A_0, A_1, A_2, A_3) is the <u>potential</u>. It is not unique: one can add to it a 4-dimensional gradient $(-\partial_0\chi, \partial_1\chi, \partial_2\chi, \partial_3\chi)$ without affecting E and H. The process of adding such a gradient is called a <u>gauge transformation</u>. (Equation (2) is to be understood formally ; the integral is not assumed to converge).

For convenience we write (1) and (3) in component form :

C. Bardos and D. Bessis (eds.), Bifurcation Phenomena in Mathematical Physics and Related Topics, 231–241.

$$\begin{cases} \partial_o H_1 = \partial_3 E_2 - \partial_2 E_3 & , \quad \text{etc} . \\ \partial_o E_1 = \partial_2 H_3 - \partial_3 H_2 & , \quad \text{etc} . \\ \partial_k H_k = \partial_k E_k = 0 \end{cases} \tag{1'}$$

$$\begin{cases} H_1 = \partial_3 A_2 - \partial_2 A_3 & , \quad \text{etc} . \\ E_1 = \partial_1 A_o + \partial_o A_1 & , \quad \text{etc} . \end{cases} \tag{3'}$$

We use the summation convention where repeated indices are to be summed from 1 to 3. The word "etc" indicates the equation for the second and third components (obtainable by cyclic permutation of the spatial indices). Equations (3') may be rewritten in operator notation as

$$\begin{cases} H_1 = [D_3, D_2] & , \quad \text{etc} . \\ E_1 = [D_o, D_1] & , \quad \text{etc} . \end{cases} \tag{3''}$$

where $D_k = \partial_k + A_k$ (k=1,2,3) and $D_0 = \partial_0 - A_0$ denote the covariant derivative operators. The bracket $[L,M]$ means LM-ML. (The notation could be made consistent with the metric, but we prefer to have ∂_0 denote a derivative and not the negative of a derivative).

After this long, elementary preamble, let's get on to the Yang-Mills equations. It involves 10 unknown functions

$$H_1, H_2, H_3, E_1, E_2, E_3, A_0, A_1, A_2 \text{ and } A_3.$$

Each of these functions is defined on Minkowski space (ordinary space-time with the Lorentz metric) with values in a certain Lie algebra $\mathfrak{g}$. We denote the multiplication in $\mathfrak{g}$ by $\times$ (rather than the usual [,]) and we assume that $\mathfrak{g}$ has an inner product, denoted by . , which satisfies the identity $(A\times B).C = A.(B\times C)$. We let $|E|^2 = E_1.E_1 + E_2.E_2 + E_3.E_3$, and so on. Then the Lagrangian

$$\mathcal{L} = \frac{1}{2} \int (|E|^2 - |H|^2) \, dxdt \quad , \tag{4}$$

is identical (!) with the Maxwell case (but of course the constraints have to be different). The covariant derivative operators are

$$D_k = \partial_k + A_k \times \quad (k=1,2,3) \text{ and } D_o = \partial_o - A_o \times \ . \tag{5}$$

Then the Yang-Mills equations (YM) are

$$D_0 H_1 = D_3 E_2 - D_2 E_3 \tag{6.1}$$

$$D_0 H_2 = D_1 E_3 - D_3 E_1 \tag{6.2}$$

$$D_0 H_3 = D_2 E_1 - D_1 E_2 \tag{6.3}$$

$$D_0 E_1 = D_2 H_3 - D_3 H_2 \tag{6.4}$$

$$D_0 E_2 = D_3 H_1 - D_1 H_3 \tag{6.5}$$

$$D_0E_3 = D_1H_2 - D_2H_1 \tag{6.6}$$
$$D_kH_k = 0 \tag{6.7}$$
$$D_kE_k = 0 \tag{6.8}$$

$$\begin{cases} H_1 \times = [D_3, D_2] \quad , \quad \text{etc.} \\ E_1 \times = [D_0, D_1] \quad , \quad \text{etc.} \end{cases} \tag{7}$$

That is, they are identical with the Maxwell equations if all derivatives are replaced by covariant derivatives ! Equations (7) may be written more explicitly as

$$\begin{cases} H_1 = \partial_3 A_2 - \partial_2 A_3 + A_3 \times A_2 \ , \ \text{etc.,} \\ E_1 = \partial_1 A_0 + \partial_0 A_1 + A_1 \times A_0 \ , \ \text{etc.} \end{cases} \tag{7'}$$

These are the equations. Now some comments.

(i) They are the original physical equations as proposed by Yang and Mills in 1954 [9]. Our underlying space-time manifold M possesses the Lorentz metric -+++. However 95 % of the recent literature is concerned with the Euclidean case ; that is, the Riemannian metric ++++. In the Euclidean case, time has become "imaginary-time" and hyperbolic partial differential equations have become elliptic ones. The wave equation has become the Laplace equation. So, on the classical (PDE) level there is barely any relation between the two theories. The Minkowski case, while manifestly more physical than the Euclidean one, has enjoyed less mathematical progress and so has been relatively neglected.

(ii) The purpose of Yang and Mills was to build into the Maxwell's equations certain observed symmetries of physics. These symmetries are given mathematically by a Lie group, the <u>gauge group</u> G. It is known that a Lie algebra $\mathfrak{g}$ possesses an invariant inner product (as above) if G is compact.

A function g from Minkowski space into the Lie group G induces the gauge transformation

$$E_k \to g^{-1} E_k g \ , \ H_k \to g^{-1} H_k g \ ,$$
$$A_k \to g^{-1} A_k g + g^{-1}(\partial_k g) \ .$$

This transformation carries a solution of YM into another solution of YM (gauge invariance of YM). The Maxwell equations are the special case where G is one-dimensional and $g = \exp\chi$.

(iii) The equations can be neatly expressed in differential-geometric language. Thus $-A_0 dt + \sum^3 A_k dx_k$ is a connection on the principle bundle M×G and $E_1^1, \ldots, H_3^1$ are components of the corresponding curvature form. We will not pursue this point of view here.

(iv) YM differs from Maxwell's equations in that the potential A is intrinsic to the problem, since it is explicitly contained in (5),(6). This is a major cause of difficulty in the mathematical analysis of the equations. On the other hand, the equations can be written entirely in terms of the potentials, if the expressions in (5) and (7) are substituted into (6). Then we have a system of second order equations with nonlinear terms of first order. This system is "almost" hyperbolic, a term like $\partial_o \partial_o A_o$ being missing. It is often assumed that a gauge transformation can be made which adjoins a further equation to the system. For instance, in the Lorentz gauge one assumes that $\partial_o A_o + \Sigma \partial_k A_k = 0$ and the resulting system is truly hyperbolic. A coupling constant appears in the usual form of YM, but we have set it equal to one with no loss of generality.

2. CONSERVATION LAWS

The Yang-Mills equations are devised so as to be gauge invariant. A different kind of invariance, conformal invariance, is inherited from the structure of the space-time manifold M. Consider a transformation T:M→M. It induces a transformation on functions defined on M. If T is a conformal transformation on M, it can be shown that the induced transformation leaves the Lagrangian $\mathcal{L}$ (see (4)) invariant. But Noether's theorem states that whenever a one-parameter group of transformations leaves a Lagrangian invariant, the equations of motion enjoy a conservation law.

What are the conformal transformations of M? They are the transformations which preserve Lorentz angles ; that is, $(x.y)^2(x.x)^{-1}(y.y)^{-1}$. They form a 15-dimensional group, consisting of the translations $T_a: x \to x+a$ (4 parameters), the Lorentz transformations $x \to Lx$ (6 parameters), the dilations $x \to x$ (one parameter), and the inversions VT_aV where $V: x \to x/x.x$ (4 parameters). (We have temporarily used x to denote a space-time point and a dot to denote the Lorentz inner product). Therefore we must have 15 conservation laws.

The easiest way to derive these laws is to calculate them directly from equations (5) and (6). To get the energy, multiply (take the dot product of) (6.1) by H_1, (6.2) by H_2, (6.3) by H_3, (6.4) by E_1, (6.5) by E_5 and (6.6) by E_6. Add these six equations. The 12 triple scalar products (involving the A's) all cancel and we are left with the identity

$$\partial_o e = \partial_k p_k \tag{8}$$

where $e = \frac{1}{2}(|E|^2 + (|H|^2)$ is the <u>energy</u> and

$$p_1 = H_2.E_3 - H_3.E_2 \quad ,$$

$$p_2 = H_3.E_1 - H_1.E_3 \quad ,$$

$$p_3 = H_1.E_2 - H_2.E_1 \quad ,$$ are the <u>momenta</u>.

Now calculate $\partial_o p_1$ as the sum of four terms. Use (6.2),(6.3),(6.5) and (6.6) to substitute for the time derivatives. Finally use (6.7) and (6.8). The equation greatly simplifies into the form (for j=1)

$$\partial_o p_j = \partial_j e - \partial_k(E_j.E_k + H_j.H_k) \quad . \tag{9}$$

The other eleven conservation laws follow immediately from (8) and (9). They express the conservation of the angular momenta $x_2p_1 - x_1p_2$, $x_3p_2 - x_2p_3$, $x_1p_3 - x_3p_1$, $x_je + tp_j$ (j=1,2,3), $te + x_kp_k$, $(t^2+r^2)e + 2tx_kp_k$ and $tx_je + (t^2-r^2)p_j/2 + x_jx_kp_k$ (j=1,2,3). Here $r=|x|$. For instance, if we multiply (8) by (t^2+r^2), multiply (9) by $2tx_j$ and sum on j, we obtain directly

$$\partial_o\,[(t^2+r^2)e + 2t\,x_kp_k] = \partial_k n_k \tag{10}$$

where

$$n_k = (t^2+r^2)p_k + 2tx_ke - 2tx_j(E_j.E_k + H_j.H_k) \quad . \tag{11}$$

For further details, see [3].

3. RADIATION OF THE ENERGY

Are there any static solutions of YM with finite energy ($\int e dx < \infty$)? This question must be answered if we are to understand how general solutions of YM behave. We will only consider solution with finite energy. The question was answered in the negative, heuristically and then rigorously, by Deser, Pagels, Coleman, Magg, and Weder. The result of their work is that for any solution (of finite energy), there is a sequence of times $t_n \to +\infty$ such that the local energy (the integral of e over a fixed bounded spatial region) tends to zero (see [8]). In fact a much stronger result follows trivially from identity (10) above. Integrate (10) and use the inequality

$$|\omega_k p_k| \leq e \tag{12}$$

where $\omega_k = x_k/r$, $r=|x|$. (With considerable abuse of notation (12) follows by writing $p = H\times E$, $|\omega.p| = |\omega.H\times E| \leq |H||E| \leq e$). We obtain from (10) the inequality

$$\int [(t^2+r^2)e - 2tr\,e]\; dx \leq \text{constant.}$$

The integrand is $(t-r)^2e$. Hence we have proved that :

$$\int_{r\leq(1-\varepsilon)t} e\,dx \leq \frac{c}{t^2} \tag{13}$$

for each $\varepsilon>0$. This is a cone of smaller aperture than the light cone. Precisely : if $\varepsilon>0$, there exists a constant c such that, for any solution for which $r^2 e$ is integrable over all space, (13) is valid. This means that all the energy radiates to infinity at the speed of light.

(13) immediately implies that there are no static solutions (independent of time) and no time-periodic solutions. Nor are there any solitary wave solutions $E=E(x-bt)$, $H=H(x-bt)$ travelling at less than the speed of light ($|b|<1$, a constant vector). The case $|b|=1$ can also be excluded ; we do this at the end of the next section.

No solution can travel at speed >1 because of the following argument. Integrate (8) over a piece of solid light cone, as in the sketch. Then

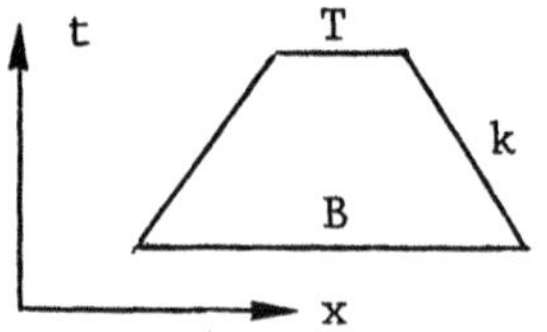

$$\int_T e - \int_B e + \frac{1}{\sqrt{2}} \int_K (\omega_k p_k + e) = 0.$$

The last integrand is non-negative. So

$$\int_T e\,dx \leq \int_B e\,dx \quad . \tag{14}$$

4. SPLITTING OF THE ENERGY

If we analyze the identity (10) more carefully, we can get the following stronger conclusion. There exists a decomposition

$$e = e_{for} + e_{back} + e_{ang} , \tag{15}$$

of the energy density into forward, backward and angular parts such that

$$\int (e_{back} + e_{ang})dx \to 0 \text{ as } t\to +\infty , \tag{16}$$

and

$$\int (e_{for} + e_{ang})dx \to 0 \text{ as } t\to -\infty . \tag{17}$$

Thus e_{for} carries all the energy forward in time and e_{back} carries it backward in time, asymptotically. The integrals in (16),(17) are taken over all of space.

Proof : As before, let $\omega = x/|x|$. Let α and β be a pair of vector fields in 3-space such that α, β, ω form an orthonormal basis with $\alpha\times\beta=\omega$. Denote $E\alpha = E_k\alpha_k$ (summed) and so on. Then an elementary calculation shows that $|E|^2 = |E\omega|^2 + |E\alpha|^2 + |E\beta|^2$,

$$|H|^2=|H\omega|^2+|H\alpha|^2+|H\beta|^2 \ , \quad \omega_k p_k = H\alpha . E\beta - H\beta . E\alpha \ .$$

Define

$$e_{ang} = \frac{1}{2}|E\omega|^2 + \frac{1}{2}|H\omega|^2 \ ,$$
$$e_{back} = \frac{1}{4}|E\alpha+H\beta|^2 + \frac{1}{4}|E\beta-H\alpha|^2 \ ,$$
$$e_{for} = \frac{1}{4}|E\alpha-H\beta|^2 + \frac{1}{4}|E\beta+H\alpha|^2 \ .$$

Obviously these three expressions add up to e. The conserved quantity in (10) is

$$(t^2+r^2)e + 2t\, x_k p_k$$
$$= (t^2+r^2)\, e_{ang} + (t+r)^2\, e_{back} + (t-r)^2\, e_{for} \ .$$

The integral of this quantity is conserved. As $t \to +\infty$ we use the first two terms to obtain (16). As $t \to -\infty$ we use the first and last terms to obtain (17).

We illustrate the significance of the energy splitting (15) in the case of Maxwell's equations, which is notationally simpler. (In this case, as well as for a class of linear hyperbolic problems, there is a completely different proof by transform methods [1]. We use the ordinary cross and dod products. Then

$$2e_{ang} + 8e_{back} = |E+\omega\times H|^2 + |H-\omega\times E|^2 \ .$$

Hence

$$\int \{|E+\omega\times H|^2 + |H-\omega\times E|^2\}\, dx \to 0 \quad \text{as} \quad t \to +\infty \ .$$

If we dot $E+\omega\times H$ with H, we find that $\int E.H\, dx \to 0$. Since $\int E.H\, dx$, $\int |E.\omega|^2 dx$ and $\int |H.\omega|^2 dx$ all tend to zero as $t \to +\infty$, we can say that <u>E, H and ω are asymptotically mutually orthogonal</u>. If we dot $E+\omega\times H$ with E, $H-\omega\times E$ with H, and subtract we see that

$$\int \{|E|^2 - |H|^2\}\, dx \to 0 \quad \text{as} \quad t \to +\infty \ .$$

This is known as <u>equipartition of the energy</u> : the electric and magnetic fields each possess half the energy asymptotically. Equipartition for Maxwell's equation was first proved in [2].

For another example, consider the ordinary wave equation $\varphi_{tt}=\nabla.\nabla\varphi$. Even in this case, energy splitting appears to be a new idea, although equipartition is not. In this case the principle of energy splitting states that

$$\int |\nabla\varphi - \omega\varphi_t|^2\, dx \to 0 \quad \text{as} \quad t \to +\infty \ .$$

This has two implications. One is equipartition : the energy is equally divided asymptotically between its kinetic and potential parts. The second is that $\nabla\varphi$ points approximately in a radial direction at large time. This is a precise mathematical formulation of Huygen's idea that a localized initial disturbance becomes, after the passage of a long time, an expanding spherical wave.

As a final example of energy splitting we show that the Yang–Mills equations do not possess solitary wave solutions which travel at characteristic speed, and are of finite energy.

Assume that

$$E=E(x-bt),\ H=H(x-bt),\ A=A(x-bt)$$

where b is a constant vector with $|b|=1$. Exactly as for Maxwell's equations, we have

$$\int \{|E+\omega\times H|^2 + |H-\omega\times E|^2\}\, dx \to 0 \quad \text{as} \quad t\to+\infty ,$$

where $\omega\times H$ is the 3×3 matrix whose first column is $\omega_2 H_3 - \omega_3 H_2$, etc... Thus we have

$$\int \{|E(y) + \frac{y+bt}{|y+bt|} \times H(y)|^2 + |H(y) - \frac{y+bt}{|y+bt|} \times E(y)|^2\}\, dy \to 0 ,$$

as $t\to\infty$.

It follows that

$$E + b\times H = 0$$

and

$$H - b\times E = 0$$

The above definition of $b\times H$ shows that the matrix products

$$(b\times H)b \quad \text{and} \quad (b\times E)b$$

vanish identically.

Hence $\quad Eb = Hb = 0$ and

$$E_1 = b_3H_2 - b_2H_3 \quad , \quad \text{etc..,}$$

$$H_1 = b_2E_3 - b_3E_2 \quad , \quad \text{etc.}$$

Next, we compute the first component p_1 of the momentum:

$$p_1 = H_2\cdot E_3 - E_2\cdot H_3$$

$$= H_2.(b_2H_1-b_1H_2)-H_3.(b_1H_3-b_3H_1)$$
$$= b_2H_1.H_2+b_3H_1.H_3-b_1(H_2.H_2+H_3.H_3)$$
$$= b_kH_1.H_k-b_1|H|^2$$

But $b_kH_k = Hb = 0$ and therefore

$$p_1 = -b_1|H|^2 \qquad \text{Thus in general}$$
$$p_j = -b_j|H|^2 \qquad (j=1,2,3) .$$

The energy identity (8) then gives

$$\partial_o e = -\partial_k(b_k|H|^2)$$

When we substitute $H=b\times E$ into the elementary identity

$$|E|^2 = |Eb|^2 + |b\times E|^2$$

we find that $|E|^2 = |H|^2$ from which we conclude that

$$e \equiv \frac{1}{2}(|E|^2 + |H|^2) = |H|^2 .$$

Now (8) takes the form

$$\partial_o e = -\partial_k(b_k e) = -b_k\partial_k e$$

which implies that e is a function of the single variable t-b.x. Thus we cannot have

$$\int_{\mathbb{R}^3} edx < \infty .$$

5. FURTHER REMARKS

It is of great physical interest to consider a Yang-Mills field coupled to a scalar field. Consider the Lagrangian

$$\mathcal{L} = \int \frac{1}{2}|E|^2 - \frac{1}{2}|H|^2 + \frac{1}{2}|D_o\varphi|^2 - \frac{1}{2}\sum_k |D_k\varphi|^2 - V(\varphi)$$

together with equations (5) and (7'), where $\varphi : M \to \mathcal{G}$ denotes the scalar field. If the self-coupling $V(\varphi)$ vanishes to sufficiently high order at $\varphi=0$, almost all the results given above extend to this system. More precisely, we have the following results.

If $V(\varphi)=c|\varphi|^4$, c=constant, the system is invariant under the conformal group and there are 15 conservation laws.

Assume $V(\varphi)=V_o(|\varphi|^2)$ for some real function V_o. The energy density is

$$e = \frac{1}{2}|E|^2 + \frac{1}{2}|H|^2 + \frac{1}{2}|D_o\varphi|^2 + \frac{1}{2}\sum_k |D_k\varphi|^2 + V(\varphi) \quad .$$

If $0 \leq 4V(\varphi) \leq \varphi.V'(\varphi)$, the energy radiation result (13) and the energy splitting result (16),(17) are valid.

If $0 \leq 2V(\varphi) \leq \varphi.V'(\varphi)$, then $V(\varphi)$ may contain a term $(m^2/2)|\varphi|^2$, the local energy tends to zero and there are no static solutions of finite energy. For the details, see[4].

Let us go back to pure YM. The Cauchy problem is to show the existence of a solution if initial data are given at a finite time, say t=0. Solving in a finite time interval $0 \leq t \leq T$, where T depends on the data, is called the local Cauchy problem and is not difficult (see [6]). The global problem is whether the solution can be extended to all times $0 \leq t \leq \infty$ or whether on the contrary it blows up in a finite time (in some norm). Conservation of energy implies bounds on the L^2 norms of E and H. A bound on the L^2 norm of A would follow from a gauge condition. A typical nonlinear term looks like $A \times E$ (or $A \times H$). We require bounds on a sequence A_n and E_n so that $A_n \times E_n$ converges weakly (in the sense of distributions). The L^2 bounds provide weak convergence (for a subsequence) of A_n and E_n. In order to obtain the weak convergence of $A_n \times E_n$, it would be enough to get A_n to converge strongly in L^2. This is true of ∇A_n are bounded in L^2.

If we now assume the Coulomb gauge $\Sigma\, \partial_k A_k = 0$, the conservation of energy gives.

$$\int |\nabla A|^2\, dx \leq c_1 + c_2 \int |A|^4\, dx \quad .$$

Thus it would be sufficient to get a bound for

$$\alpha(t) = (\int A^4 dx)^{1/4} \quad .$$

We can show that

$$\alpha(t) \leq c_3(1+t)^{-1/2} + c_4\int_0^t (t-\tau)^{-1/2}[\alpha(\tau) + (\alpha(\tau))^3]\, d\tau \; .$$

Unfortunately this estimate still permits $\alpha(t)$ to blow up in a finite time.

If we modify the system so that

$$\mathcal{L} = \frac{1}{2}\int \{|E|^2 - |H|^2 - \varepsilon^2|A|^2\}\, dx$$

for a given constant $\varepsilon > 0$, and assume the Coulomb gauge, then we can easily solve the global Cauchy problem. Unfortunately this system is no longer gauge—invariant.

Certain special types of initial data do lead to global solutions, but the general problem remains unsolved. See [5] for a detailed discussion.

REFERENCES

[1] Costa, D.G. and Strauss, W.A. : to appear.
[2] Dassios, G. : Equipartition of energy for Maxwell's equations, Quart. Appl. Math. : to appear.
[3] Glassey, R.T. and Strauss, W.A. : Decay of classical Yang—Mills fields, Comm. Math. Phys. 65 (1979), 1—13.
[4] ibid.; Decay of a Yang—Mills field coupled to a scalar field, Comm. Math. Phys. 67 (1979), 51—67.
[5] ibid. : to appear.
[6] Segal, I.E. : The Cauchy problem for the Yang—Mills equations, J. Funct. Anal. : to appear.
[7] ibid., Geeneral properties of the Yang—Mills equations in physical space.
[8] Weder, R. : Absence of classical lumps, Comm. Math. Phys. 57 (1977), 161—4.
[9] Yang, C.N. and Mills, R. : Conservation of isotropic spin and isotropic gauge invariance, Phys. Rev. 96 (1954), 191—5.

The authors were supported by NSF Grants MCS 77—01340 and MCS 79—01965.

EUCLIDEAN YANG-MILLS AND RELATED EQUATIONS

Basilis Gidas

Courant Institute of Mathematical Sciences
New York University

Lectures given at the Institut d'Études Scientifiques de Cargèse, on "Bifurcation Phenomena in Mathematical Physics and Related Topics" (June 24 - July 7, 1979).

I. INTRODUCTION

By a generally accepted philosophy, the basic constituents of hadrons (protons, neutrons, mesons, etc.) are quarks. But quarks have never been directly confirmed by experiment because they have never been isolated. The proposition has been that quarks are permanently confined inside hadrons and cannot be liberated in high energy particle collisions. It is expected that pairs (and some triplets) of quarks are attracted by forces which increase linearly with the distance between quarks. Thus when the quarks are near to each other they behave like free particles ("asymptotic freedom"), while when they are separated their attractive forces keep them together in the form of a bound state. According to this philosophy, we observe only these bound states (i.e. protons, neutrons, mesons, etc.).

The forces between quarks in a hadron are most likely described by a non-Abelian gauge theory. The best known example of a gauge theory (albeit Abelian) is Quantum Electrodynamics, QED, (classically described by Maxwell's equations). In electrodynamics, the forces between charged particles are "mediated" by

C. Bardos and D. Bessis (eds.), Bifurcation Phenomena in Mathematical Physics and Related Topics, 243–267.

photons (classically: electromagnetic waves). Similarly, in non-Abelian gauge theories, the quark forces are mediated by the so-called gluons (they are also supposed to be confined), and the non-Abelian gauge theory which underlies strong interactions is called Quantum Chromodynamics, QCD, and is classically described by the Yang-Mills (YM) equations. In these lectures we study the elliptic YM equations, and some related equations, from the point of view of non-linear PDE.

While the short distance behavior (i.e. quarks near to each other) of QCD is well understood and accurately described by conventional perturbation techniques, the long distance behavior (quark confinement) is a terra incognita, and it has been established that it cannot be controlled by any perturbative methods, i.e. quark confinement cannot be seen in any order of the perturbation series -- it is a non-perturbative effect. Many sophisticated and unconventional procedures have been invented to study the non-perturbative (non-analytic) effect and show that indeed the cumulative forces of the gluons keep quarks permanently together. It is generally believed that confinement is caused by the severe infrared singularities characteristic of non-Abelian gauge theories. Polyakov [1] proposed that the infrared behavior of such theories might be governed by solutions of the Euclidean YM equations. This led to the discovery of the instanton [2] and meron [3] (and multiinstantons [4] and mutimerons [5,6]). From the physical point of view, the instanton, although insufficient to control confinement, yielded a rich structure of the quantum vacuum [7,8] and provided an explanation of the famous U(1) problem [9]. From the mathematical point of view, it implied a deep connection between YM equations and Differential Geometry, in particular holomorphic vector bundles [10]. The relevance of merons to physics is not clear (see however [11,19]), but they seem to be interesting from the mathematical point of view. They are distribution solutions of the YM equations with isolated singularities. In Section III, we construct the instanton, the meron, and some singular solutions due to Loewner and Nirenberg [13]. We also summarize some recent uniqueness results on certain scalar equations related to YM equations.

The significance of studying the elliptic (Euclidean) YM equations, rather than the actual equations of motion, i.e. the hyperbolic (Minkowski) YM equations, is at least two-fold:
a) The elliptic problem provides a basis for understanding the quantization of the hyperbolic equations, by employing the powerful tools of functional integrals and the underlying stochastic processes. (In these lectures we shall not touch upon the quantization problem.) Furthermore, practical calculations (e.g. determining scattering amplitudes) are most frequently performed in Euclidean space (which is obtained by a Wick rotation $t \to it$), and then analytically continued to the Minkowski space.

b) Solutions of the elliptic equations have a direct physical interpretation at the quantum level (and only at the quantum level!): they signal the occurrence of tunneling, i.e. of a motion through a potential barrier, which is classically forbidden, but quantum-mechanically allowed because of quantum effects. This physical interpretation of instantons should be contrasted to the physical interpretation of solitons. Classically, the solitary waves are observed, but instantons have no meaning. Quantummechanically, the solitons are heavy particles (extended hadrons?), while the instantons reflect tunneling through classically forbidden barriers.

In Section II we set up the pure YM equations and summarize some results obtained via Differential Geometric techniques. In Section III we construct, via 't Hoost's ansatz, the instanton, the meron, and some singular solutions of Loewner and Nirenberg [13]. We also describe briefly some uniqueness theorems for certain equations related to YM equations. Finally, in Section IV, we briefly describe the beautiful work of K. Uhlenbeck [14] on the removability of isolated singularities in YM equations.

II. CLASSICAL YANG-MILLS EQUATIONS AND GENERAL RESULTS

A. Euclidean Pure Yang-Mills Equations

In these lectures we are primarily concerned with the pure Yang-Mills (YM) equations on $M = \mathbb{R}^4$, or $M = \mathbb{R}^4 - \{S\}$, where S is a submanifold of $\mathbb{R}^4$. (In general M could be any 4-dimensional Riemannian manifold, e.g. $M = S^4$.) The basic objects are the gluon (or YM) fields $A_\mu(x)$, $\mu = 0,1,2,3$ which take values in the Lie algebra $\mathcal{G}$ of a compact semisimple Lie group G -- the <u>gauge group</u>. For convenience we shall consider here the case G = SU(2) (the group relevant to strong interactions is SU(3); for QED, G = U(1)). Associated with $A_\mu(x)$ is the tensor

$$F_{\mu\nu}(x) = \partial_\mu A_\nu - \partial_\nu A_\mu + [A_\mu, A_\nu] \tag{2.1}$$

where $[\cdot,\cdot]$ denoted the commutator of the Lie algebra, and

$\partial_\mu = \frac{\partial}{\partial x_\mu}$, $\mu = 0,1,2,3$. The YM action is

$$S(A) = -\frac{1}{2}\int_M d^4x\, \mathrm{Tr}\{F_{\mu\nu}(x)F_{\mu\nu}(x)\} \equiv \int_M d^4x\, |F(x)|^2 \tag{2.2}$$

where Tr denotes the trace on the Lie algebra of SU(2); it defines a positive definite norm $|\cdot|$ on $\mathcal{G}$ (for a general G, $-\frac{1}{2}$ Tr denotes the Killing form on the Lie algebra). The YM equations are the Euler equations of S(A), and they read

$$\partial_\mu F_{\mu\nu}(x) + [A_\mu(x), F_{\mu\nu}(x)] = 0 \tag{2.3}$$

(we use summation convention).

Remark 1. Equations (2.3) are the "pure" YM equations. The full equations of QCD have a RHS which contains the matter fields (i.e. quark fields). The "pure" Maxwell equations (i.e. Maxwell's equations without electric currents) are trivial. In contradistinction, the pure YM equations are non-linear PDE with a rich structure.

Remark 2. In a Differential Geometric language, A_μ is globally defined as a connection which may be viewed in two ways: either as a 1-form on a principal G-bundle, or as a covariant derivative on the associated vector bundle (whose fibers are the Lie algebra of G). In this language, $F_{\mu\nu}$ are the components of the curvature of A, and (2.1) are the structure equations.

Next we define the self-dual equations. The components of the dual to the curvature 2-form are

$$^*F_{\mu\nu}(x) = \frac{1}{2} \varepsilon_{\mu\nu\kappa\lambda} F_{\mu\nu}(x) \tag{2.4}$$

where $\varepsilon_{\mu\nu\kappa\lambda}$ is the Levi-Cevita symbol with $\varepsilon_{0123} = 1$. The dual tensor $^*F$ is generally defined by the Hodge $*$-operation, and it satisfies the Bianchi identity

$$\partial_\mu {}^*F_{\mu\nu}(x) + [A_\mu(x), {}^*F_{\mu\nu}(x)] = 0 \tag{2.5}$$

The self-dual (s.d.) and anti-self-dual (a.s.d.) equations are

$$F_{\mu\nu}(x) = \pm {}^*F_{\mu\nu}(x) \tag{2.6}$$

Because of (2.5), any solution of (2.6) is also a solution of the YM equations (2.3). The converse is not true. However, all known solutions of the YM equations which are neither s.d. nor a.s.d., are distribution solutions with singularities (at isolated points, or on two- and three-dimensional submanifolds of $\mathbb{R}^4$; no solutions with 1-dimensional singularities are known). One of the attractive open problems is

Conjecture 1. All finite action (i.e. $S(A) < +\infty$) solutions of the YM equations are either s.d. or a.s.d.

Later we shall use a concrete basis of the Lie algebra $\mathcal{G}$ of

SU(2) given by the Pauli matrices σ_a, $a = 1,2,3$,

$$\sigma_1 = \begin{pmatrix} 0 & 1 \\ 1 & 0 \end{pmatrix}, \quad \sigma_2 = \begin{pmatrix} 0 & -i \\ i & 0 \end{pmatrix}, \quad \sigma_3 = \begin{pmatrix} 1 & 0 \\ 0 & -1 \end{pmatrix} \tag{2.7}$$

and the components (Christophel symbols) of the connection and its curvature defined by

$$A_\mu(x) = \frac{1}{2i} A^a_\mu(x)\sigma^a \ ; \qquad A^a_\mu = i \ \mathrm{Tr}\{A_\mu(x)\sigma^a\} \tag{2.8a}$$

$$F_{\mu\nu}(x) = \frac{1}{2i} F^a_{\mu\nu}(x)\sigma^a \ ; \ F^a_{\mu\nu}(x) = i \ \mathrm{Tr}\{F_{\mu\nu}(x)\sigma^a\} \tag{2.8b}$$

B. Special Properties

i) Conformal Invariance: The action (2.2) (and therefore the YM equations (2.3)) are invariant under conformal transformations. The conformal groups is a 15-parameter group and is isomorphic to SO(5,1). It is generated by

a) rotations (6-parameters)
b) translations (4-parameters)
c) dilatations: $x \to \lambda x$, $\lambda > 0$ (1-parameter)
d) special conformal transformations (4-parameters):

$$x \to \frac{x - a|x|^2}{1 - 2a\cdot x + |a|^2|x|^2}, \quad a \in \mathbb{R}^4$$

Conformal invariance of (2.3) means that if $A_\mu, F_{\mu\nu}(x)$ satisfy satisfy (2.3), and $x \to x'$ is a conformal map, then

$$A'_\mu(x) = \frac{\partial x'_\kappa}{\partial x_\mu} A_\kappa(x') \tag{2.9a}$$

$$F'_{\mu\nu}(x) = \frac{\partial x'_\kappa}{\partial x_\mu} \frac{\partial x'_\lambda}{\partial x_\nu} F_{\kappa\lambda}(x') \tag{2.9b}$$

is also a solution of (2.3), and $S(A) = S(A')$.

ii) Gauge Invariance: If $g(x) \in SU(2)$, then the transformation

$$A^g_\mu(x) = g^{-1}(x) A_\mu(x) g(x) + g^{-1}(x) \partial_\mu g(x) \tag{2.10a}$$

$$F^g_{\mu\nu}(x) = g^{-1}(x) F_{\mu\nu}(x) g(x) \tag{2.10b}$$

leaves (2.2) and (2.3) invariant. This invariance is known as (local) gauge invariance. In fact the YM action has been constructed to satisfy this invariance, and certain "minimal coupling" assumptions. Connections that are related by a gauge transformation (i.e. they belong to the same orbit under the action of the gauge group) are physically equivalent. In fact measurable quantities (e.g. energy) are gauge invariant. However, smoothness, asymptotic behavior and other properties of the connection and its curvature vary with gauge transformations -- hence the term choice of gauge, which refers to a particular element of the orbit (under the gauge group) of a connection.

Definition. We say that A is a weak YM field over $\Omega\subset M$, if there exists a gauge in which $A_\mu(x)$ is C^1, and

$$\int_\Omega d^4x \left(\partial_\mu\phi + [A_\mu,\phi]\right) F_{\mu\nu}(x) = 0 \qquad (2.11)$$

for any $C_0^\infty(\Omega)$ Lie algebra valued 1-form ϕ.

Theorem 2.1 Let A be a weak YM field over $\mathbb{R}^4$ with finite action. Then there exists a gauge in which A is C^∞, and a gauge transformation $g(x)\in SU(2)$ such that

$$A(x) \xrightarrow[x\to\infty]{} g^{-1}(x)dg(x) + O\left(\frac{1}{|x|^2}\right) \qquad (2.12)$$

where d is the exterior derivative on a large sphere.

Remark. Theorem (2.1) holds even if the solution $A_\mu(x)$ has isolated singularities which are integrable in the sense that the action is finite at the singular points. The theorem was proven by K. Uhlenbeck (see Section IV).

iii) Homotopic index: Let A,F be a smooth YM field over $\mathbb{R}^4$ with finite action. By theorem 2.1, there exists a gauge transformation $g(x)$ defined on a large sphere S^3_∞,

$$g(x):\ S^3_\infty \longrightarrow SU(2) \qquad (2.13)$$

Since $SU(2)$ is isomorphic to S^3, the maps (2.13) fall into homotopic classes indexed by $\pi_3(SU(2)) = \pi_3(S^3) \cong \mathbb{Z}$. The integer $k\in\pi_3(SU(2))$ will be referred to as Pontryagin index, or topological charge, or homotopic index. It can be expressed in terms of the connection

$$k = -\frac{1}{16\pi^2}\int_{\mathbb{R}^4} d^4x \, \mathrm{Tr}\{F_{\mu\nu}(x)^*F_{\mu\nu}(x)\} = \frac{1}{8\pi^2}\int_{\mathbb{R}^4} d^4\lambda\{|F^+|^2-|F^-|^2\} \quad (2.14)$$

where $F^\pm$ are the s.d. and a.s.d. parts of F, i.e.

$$F^\pm_{\mu\nu} = \frac{F_{\mu\nu} \pm {}^*F_{\mu\nu}}{2} \quad (2.15)$$

The integer k is the degree of the map (2.13), i.e. it expresses the number of times $SU(2) \simeq S^3$ is covered under the map (2.13).

Remark 3. k is invariant under gauge and continuous conformal transformations. It changes sign under an inversion (a discrete conformal transformation).

4) By Theorem 2.1, a smooth finite action solution of the YM equations on $\mathbb{R}^4$, extends to a smooth connection over S^4. In Differential Geometric terms, $k \in \pi_3(SU(2))$ classifies (smooth) principal $SU(2)$ - bundles over S^4, and it corresponds to the second Chern class of the associated vector bundle. The problem of finding all smooth solutions of the YM equations with a given k is equivalent to finding all smooth connections on a given bundle over S^4.

C. General Results

We start with a simple but important proposition.

Proposition 2.2 Absolute minima of the YM action (2.2) are necessarily either self-dual or anti-self-dual.

Proof: Consider

$$-\frac{1}{4}\int_{\mathbb{R}^4} d^4x \, \mathrm{Tr}\{(F_{\mu\nu} \pm {}^*F_{\mu\nu})(F_{\mu\nu} \pm {}^*F_{\mu\nu})\} \geq 0$$

by expanding and using $F_{\mu\nu}F_{\mu\nu} = {}^*F_{\mu\nu}{}^*F_{\mu\nu}$, we obtain

$$S(A) = \int_{\mathbb{R}^4} d^4x|F(x)|^2 \geq \left|\frac{1}{2}\int_{\mathbb{R}^4} d^4x \, \mathrm{Tr}(F_{\mu\nu}{}^*F_{\mu\nu})\right| = 8\pi^2|k| \quad (2.16)$$

with equality iff (2.6) is satisfied.

Remark 5. In relation to Conjecture 1, Bourgiugnon, Lawson, and Simon [15] have shown that any local minimum (i.e. weakly stable critical point) of S(A) is necessarily either s.d. or a.s.d.

Definition. The space of connections over $\mathbb{R}^4$ modulo gauge transformations is called moduli space. Let $\overline{M}(k)$ be the moduli space of s.d. (or a.s.d.) connections, and M(k) the moduli space of all smooth solutions of the full YM equations.

Theorem 2.3 [16,10] $\overline{M}(k)$ is a manifold of dimension $8|k| - 3$.

Theorem 2.4

1) $\overline{M}(1)$ is isomorphic to the open unit ball in $\mathbb{R}^5$. Hence it is a connected, contractible space.

2) $\overline{M}(2)$ is connected but not simply connected.

Remark 6. Part 1) of Theorem 2.4 was proven by Atiyah, Hitchin, and Singer [10] using Differential Geometric techniques; it was also obtained by Hartshorne [17] using Algebraic Geometry techniques. Part 2) was proven by Hartshorne [17]:

Open Problem: Is $\overline{M}(k)$ connected for $k \geq 3$?

III. SPECIAL SOLUTIONS AND RELATED EQUATIONS

In this section we construct 't Hooft's multi-instanton solutions as well as solutions of the YM equations with singularities on 0-, 2-, and 3-dimensional manifolds. Following 't Hooft [9] we define two tensors, $\eta_{a\mu\nu}$ and $\overline{\eta}_{a\mu\nu}$ by

$$\begin{aligned} \eta_{aij} &= \varepsilon_{aij} \quad (\varepsilon_{123} = 1) \\ \eta_{a\nu o} &= \delta_{a\nu} = -\eta_{ao\nu} \\ \eta_{aoo} &= 0 \\ \overline{\eta}_{a\mu\nu} &= (-1)^{\delta_{\mu o}+\delta_{\nu o}} \eta_{a\mu\nu} \end{aligned} \tag{3.1}$$

Throughout these notes we will use the notation: $\mu,\nu,\kappa,\lambda,\ldots = 0,1,2,3$, $i,j,k,\ldots = 1,2,3$. We will also use the matrices

$$\sigma_{\mu\nu} = \frac{1}{2}\eta_{a\mu\nu}\sigma^a \iff \eta_{a\mu\nu} = \mathrm{Tr}\left[\sigma_{\mu\nu}\sigma^a\right] \tag{3.2a}$$

$$\overline{\sigma}_{\mu\nu} = \frac{1}{2}\overline{\eta}_{a\mu\nu}\sigma^a \iff \overline{\eta}_{a\mu\nu} = \mathrm{Tr}\left[\overline{\sigma}_{\mu\nu}\sigma^a\right] \tag{3.2b}$$

The matrices $\sigma_{\mu\nu}$, $\bar{\sigma}_{\mu\nu}$ are the generators of the Lie algebra of SO(4), and satisfy

$$\sigma_{\mu\nu} = *\sigma_{\mu\nu} \tag{3.3a}$$

$$\bar{\sigma}_{\mu\nu} = -*\bar{\sigma}_{\mu\nu} \tag{3.3b}$$

We employee 't Hooft's ansatz

$$A_\mu^a = -\eta_{a\mu\nu}\partial_\nu \log u(x) \tag{3.4}$$

where u(x) is a scalar function. Inserting (3.4) into (2.3) we obtain the following equation for u(x)

$$\Delta u + \frac{1}{6}Ku^3 = 0 \tag{3.5}$$

where K is a constant. The homotopic index can also be expressed in terms of u(x).

$$k = -\frac{1}{16\pi^2}\int d^4x \operatorname{Tr}\{F_{\mu\nu}(x)F_{\mu\nu}(x)\} = -\frac{1}{16\pi^2}\int d^4x\ \Delta\Delta \log u(x) \tag{3.6}$$

Remark 1. The equality $-\frac{1}{2}\operatorname{Tr} F_{\mu\nu}{*F_{\mu\nu}} = -\frac{1}{2}\Delta\Delta \log u(x)$ for connections of the form (3.4) holds away from singularities of u(x). At a singularity of u(x), $\operatorname{Tr} F_{\mu\nu}{*F_{\mu\nu}}$ might be regular. We avoid this difficulty by excluding from the integration in (3.6) an infinitesimal neighborhood around the singularity.

Remark 2. Equation (3.5) rises in Differential Geometry as follows: Let $g_{\mu\nu} = u^2(x)\delta_{\mu\nu}$ be a metric on $\mathbb{R}^4$ conformally related to the standard metric $\delta_{\mu\nu}$. The new metric has constant scalar curvature K iff u(x) satisfies (3.5).

We are interested in positive solutions of equation (3.5). The equation is reduced to three distinct equations depending on whether $K = 0$, $K < 0$, or $K > 0$.

Case i: $$\Delta u = 0 \tag{3.7}$$

This equation is obtained if (3.4) is inserted into the self-dual equations (2.6), and it is this equation which was considered by 't Hooft (unpublished). He takes the solution

$$u(x) = 1 + \frac{\lambda_1^2}{|x-a_2|^2} + \frac{\lambda_2^2}{|x-a_2|^2} + \ldots + \frac{\lambda_n^2}{|x-a_n|^2} \qquad (3.8)$$

where $\lambda_1,\ldots,\lambda_n > 0$, and $a_1,\ldots,a_n \in \mathbb{R}^4$. Using (3.6) and Remark 1, one finds that $k=-n$. Although (3.8) has singularities at $a_1,\ldots,a_n$, the corresponding connection is gauge equivalent to a smooth connection. We demonstrate it for $n=1$. Set $\lambda_1 = \lambda$, and $a_1 = a$. Then from (3.8) and (3.4)

$$A'^a_\mu(x) = 2\eta_{a\mu\nu} \frac{(x-a)_\nu}{\lambda^2+|x-a|^2} \frac{\lambda^2}{|x-a|^2} \qquad (3.9)$$

Using the gauge transformation

$$g(x-a) = \frac{x_o I - i(x-a)_a \sigma_a}{|x-a|} \qquad (3.10)$$

we obtain

$$A'^a_\mu(x) \longrightarrow A^a_\mu(x) = 2\bar{\eta}_{a\mu\nu} \frac{(x-a)_\nu}{\lambda^2+|x-a|^2}$$

or

$$A_\mu = -i\bar{\sigma}_{\mu\nu} \frac{(x-a)_\nu}{\lambda^2+|x-a|^2} \qquad (3.11a)$$

and

$$F_{\mu\nu} = -i\bar{\sigma}_{\mu\nu} \frac{4\lambda^2}{(\lambda^2+|x-a|^2)^2} \qquad (3.11b)$$

Thus the transformed connection A has no singularities. By (3.3b), it is anti-self-dual. It is the 1-antiinstanton (if we use $\bar{\eta}_{a\mu\nu}$ in (3.4), instead of $\eta_{a\mu\nu}$, we obtain instantons). The general connection obtained from (3.8) is an a.s.d. connection with $5|k|$ $(k=-n)$ parameters. These are 't Hooft's multi(anti-)-instantons. For $k=-1$, we obtain the number of parameters expected from Theorem 2.3. For $k=-2$, the number of parameters is 10. But by using the conformal group, it can be shown [4] that three more effective (i.e. parameter which cannot be gauged away) parameters can be introduced. Thus the total number is 13, which again agrees with Theorem 2.3. For $|k| \geq 3$, the number of effective parameters is $5|k|+4$, which is smaller than $8|k|-3$. The general instanton solution with $8|k|-3$ parameter has been constructed in [18,19] by different methods.

Case ii

$$\Delta u - u^3 = 0 \tag{3.12}$$

This equation has been studied by Loewner and Nirenberg [13]. It has no regular solutions, but Loewner and Nirenberg [13] have shown the existence of positive singular solutions. One of their results is

Theorem 3.1 Let S be a smooth non-self-intersecting compact surface of dimension 2 or 3. Then there exists a positive solution u(x) of (3.12) such that $u(x) \to +\infty$, like $(\mathrm{dist}(x,s))^{-1}$, as $x \to S$.

Remark 3. Theorem 3.1 holds [13] for any compact surface S of Haussdorff dimension less than 1.

Let S be the hypersphere $S = \{x \in \mathbb{R}^4 : |x-a| = \lambda\}$. Then (3.12) has the explicit solution [13]

$$u(x) = \frac{2\sqrt{2}\,\lambda}{\left|\lambda^2 - |x-a|^2\right|} \tag{3.13}$$

which yields the connection

$$A^a_\mu = 2\eta_{a\mu\nu} \frac{(x-a)_\nu}{\lambda^2 - |x-a|^2} \tag{3.14}$$

and its curvature

$$F^a_{\mu\nu} = -\lambda_{a\mu\nu} \left(\frac{2\lambda}{\lambda^2 - |x-a|^2}\right)^2 \tag{3.14'}$$

From (3.14') we see that this connection is self-dual and its action density is

$$-\frac{1}{2}\,\mathrm{Tr}(F_{\mu\nu}F_{\mu\nu}) = \frac{1}{4} F^a_{\mu\nu}F^a_{\mu\nu} = 48\left(\frac{\lambda}{\lambda^2 - |x-a|^2}\right)^4 \tag{3.15}$$

Since the action density is a gauge invariant quantity, we

conclude that the singularity is real and cannot be gauged away. The action and the topological charge of (3.14) are

$$S(a) \cong 8\pi^2 + 4\pi^2 \left(\frac{\lambda}{\varepsilon}\right)^3 \tag{3.16}$$

$$k \cong 1 + \frac{1}{2}\left(\frac{\lambda}{\varepsilon}\right)^3 \tag{3.17}$$

where ε is a cut-off about the hypersphere. The geometry corresponding to $u(x)$ (recall Remark 2) is the hyperbolic square H^4 defined by

$$\xi_4^2 - \sum_{\mu=0}^{3} \xi_\mu \xi_\mu = 1 \qquad \xi_4 \geq 1 \tag{3.18}$$

Each $x \in R^4$ corresponds to a $\xi \in H^4$ by

$$\xi_4 = \frac{\lambda^2 + |x-a|^2}{\left|\lambda^2 - |x-a|^2\right|}, \quad \xi_\mu = \frac{2\lambda(x-a)_\mu}{\left|\lambda^2 - |x-a|^2\right|}, \quad = 0,1,2,3, \tag{3.19}$$

The connection (3.14) becomes the Riemannian SU(2)-connection over H^4. This Riemannian connection (and hence (3.14)) is invariant under the O(4,1) subgroup of the conformal group SO(5,1). Therefore (3.14) gives rise to a 15 - 10 = 5 parameter family of connections. The parameter space is the homogeneous space SO(5,1)/O(4,1).

Another explicit solution of (3.12) which yields an interesting solution of the YM equation is the following: Let S be a two-dimensional plane in $\mathbb{R}^4$. Without loss of generality we assume that S is the (x_α, x_β)-plane. Then

$$u(x) = \frac{1}{\rho}; \quad \rho = \left(x_\alpha^2 + x_\beta^2\right)^{-\frac{1}{2}} \tag{3.20}$$

is a singular solution of (3.12). The corresponding connection and its curvature are

$$A_\mu^a = \eta_{a\mu\theta} \frac{x_\theta}{\rho^2} \tag{3.21a}$$

$$F^a_{\mu\nu} = \eta_{a\mu\nu}\frac{1}{\rho^2} + \eta_{a\nu\theta}\frac{\delta_{\mu\theta}}{\rho^2} - \eta_{a\mu\theta}\frac{\delta_{\nu\theta}}{\rho^2} + \eta_{a\mu\theta}\frac{x_\theta x_\phi \delta_{\nu\phi}}{\rho^4} - \eta_{a\nu\theta}\frac{x_\theta x_\phi \delta_{\mu\phi}}{\rho^4} \qquad (3.21b)$$

where we use summation convention and $\theta,\phi = \alpha,\beta$. The connection (3.21) is neither s.d. nor a.s.d., and its action diverges quadratically at the (x_α,x_β)-plane (and at infinity).

Remark 4. Since the connection (3.21) has a two-dimensional singularity one expects, on physical grounds [20], to be significant for confinement (perhaps more significant than merons). Therefore, it would be interesting to study the spectrum of its stability operator (the solution is easily seen to be weakly stable), and compute its contribution to the functional integrals.

Case iii

$$\Delta u + u^3 = 0 \qquad (3.22)$$

The spherical symmetric solutions of (3.22) can be explicitly computed [21,22]: Assuming that u is spherically symmetric and setting

$$t = \log r \qquad (3.33a)$$

$$ru(r) = \psi(t) \qquad (3.33b)$$

we find that

$$u'' + \frac{3}{r}u' + u^3 = 0 \qquad (3.34a)$$

becomes

$$\ddot{\psi} - \psi + \psi^3 = 0 \qquad (3.34b)$$

where $\ddot{\psi} = \frac{d^2\psi}{dt^2}$. Multiplying (3.34b) by $\dot{\psi}$ and integrating we find

$$\dot{\psi}^2 + \frac{1}{2}(\psi^2-1)^2 = c^2 \qquad (3.35)$$

where c is a positive constant.

From (3.35) we construct the phase-space diagram:

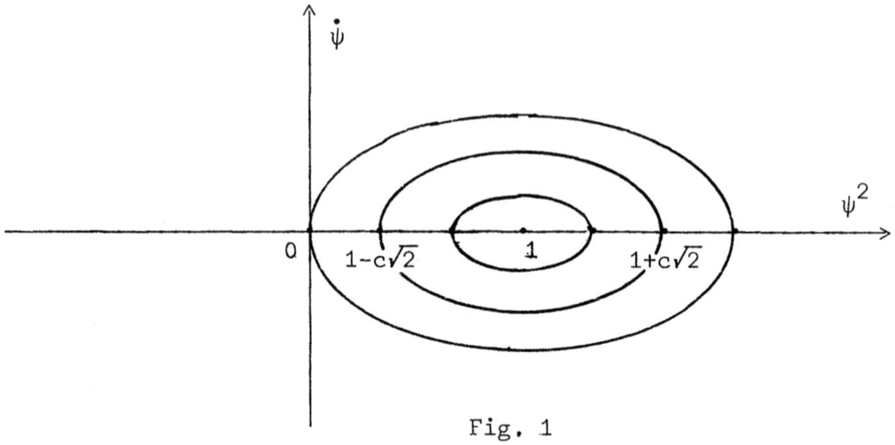

Fig. 1

From Fig. 1, we see that $0 \leq c \leq \frac{\sqrt{2}}{2}$. For $c = \frac{\sqrt{2}}{2}$

$$\psi(t) = \frac{2\sqrt{2}\,\lambda}{\lambda^2 e^{-t} + e^t} \tag{3.36}$$

for $c = 0$

$$\psi(t) = 1 \tag{3.37}$$

and for $0 < c < \frac{\sqrt{2}}{2}$, we obtain a two-parameter family of solutions satisfying

$$c_1 \leq \psi(t) \leq c_2 < \sqrt{2} \tag{3.38}$$

where c_1, c_2 are functionally related and $c_1 \to 0$ as $c_2 \to \sqrt{2}$, and $c_1 \to \frac{\sqrt{2}}{2}$, as $c_2 \to \frac{\sqrt{2}}{2}$. All solutions may be explicitly written down in terms of elliptic functions

$$\psi(t) = \sqrt{\frac{2}{2-m}}\, \mathrm{dn}\left(-\log\lambda + (2-m)^{-\frac{1}{2}}\, t \,\middle|\, m\right)$$

$$= \left(\frac{2}{2-m}\right)^{-\frac{1}{2}} \left\{1 - m + m\, \mathrm{cn}^2\left(-\log\lambda + (2-m)^{-\frac{1}{2}}\, t \,\middle|\, m\right)\right\}^{\frac{1}{2}} \tag{3.39}$$

where $m = \frac{2c\sqrt{2}}{1 + c\sqrt{2}}$. From (3.36), (3.37), and (3.38) we obtain, respectively,

$$u(r) = \frac{2\sqrt{2}\,\lambda}{\lambda^2 + r^2} \tag{3.36'}$$

$$u(r) = \frac{1}{r} \tag{3.37'}$$

$$\frac{c_1}{r} \le u(r) = \frac{\psi(\log r)}{r} \le \frac{c_2}{r} \tag{3.38'}$$

The curve $u(r) = \frac{\sqrt{2}}{r}$ is the envelope of (3.36')-(3.38'). We note that (3.36') is the only regular solution of (3.34a). Through each point (u_o, r_o) which satisfies $u_o < \frac{\sqrt{2}}{r_o}$, there pass four solutions of (3.34a): two regular solutions and two singular solutions. The slopes of the regular solutions enclose the slopes of the singular solutions. By performing a conformal transformation which takes $x = 0$ to $x = a$, and $x = \infty$ to $x = b$, we obtain solutions of (3.22):

$$u(x) = \frac{\psi\left(\log \frac{|x-a|}{|x-b|}\right)}{|x-a|\,|x-b|} \tag{3.39}$$

Corresponding to (3.36') and (3.37') we have

$$u(x) = \frac{2\sqrt{2}\,\lambda}{\lambda^2 + |x-a|^2} \tag{3.40}$$

$$u(x) = \frac{|a-b|}{|x-a|\,|x-b|} \tag{3.41}$$

Solutions (3.40) and (3.41) yield the 1-instanton (compare with (3.11)), and the 1-meron solutions of the YM equations. Properties of these solutions have been studied extensively.

We next consider the question: Are there positive solutions of (3.22) other than (3.39)-(3.41)?

Theorem 3.2 Let u(x) be a positive regular solution of (3.22) having the asymptotic behavior at ∞

$$u(x) = O\left(\frac{1}{|x|^{\varepsilon}}\right) \tag{3.42a}$$

$$u_{x_\mu} = O\left(\frac{1}{|x|^{1+\varepsilon}}\right) \tag{3.42b}$$

$$u_{x_\mu x_\nu} = O\left(\frac{1}{|x|^{2+\varepsilon}}\right) \tag{3.42c}$$

for some $\varepsilon > 0$. Then u(x) is precisely (3.40).

Proof: Set $u = w^{-1}$, and consider the vector field (summation convention)

$$v_\mu = w^{-3}\left(w_{\mu\nu}w_\nu - \frac{1}{4} w_\mu \Delta w\right) \tag{3.43}$$

then

$$\nabla\cdot\vec{v} = w^{-3}\{w_{\mu\nu}w_{\mu\nu} - \tfrac{1}{4}(\Delta w)^2\} \tag{3.45}$$

By Schwartz inequality

$$(\Delta w)^2 \leq 4 \sum_{\mu=0}^{3} w_{\mu\nu}w_{\mu\nu} \leq 4 \sum_{\mu,\nu=0}^{3} w_{\mu\nu}w_{\mu\nu} \tag{3.46}$$

Thus $\nabla\cdot\vec{v} \geq 0$. Applying the divergence theorem, and using (3.42), we find that

$$\nabla\cdot\vec{v} = w^{-3}\{w_{\mu\nu}w_{\mu\nu} - \tfrac{1}{4}(\Delta w)^2\} = 0 \tag{3.47}$$

Hence (3.46) should be an equality. This gives

$$\begin{aligned} w_{\mu\nu} &= 0 \quad \text{for} \quad \mu \neq \nu \\ w_{oo} &= w_{11} = w_{22} = w_{33} \end{aligned} \tag{3.48}$$

Equations (3.48) imply that $w(x) = c_1 + c_2|x-a|^2$ with $a \in \mathbb{R}^4$, $c_1, c_2 > 0$. This is (within a scaling of the coordinates) equivalent to (3.40).

Remark 5. The above proof is an adaptation of the following geometric result of Obata [23] : A Riemannian metric on S^n which is conformal to the usual one and having the same constant scalar curvature, is in fact the pullback of the usual one under a conformal map of S^n to itself.

Remark 6. Recently we have shown that the result of Theorem 3.2 holds without assuming (3.42), but only assuming that $u \in C^1(\mathbb{R}^n)$.

Concerning singular solutions of (3.22) we have

Theorem 3.3

1) Equation (3.22) has no positive solutions which are singular (i.e. $u \to +\infty$) at a single point (the point could be infinity).

2) If $u(x)$ is a positive solution of (3.22) in $\mathbb{R}^4 - \{a,b\}$, and $u(x) \to +\infty$ as $x \to a$ and $x \to b$, then $u(x)$ is necessarily given by (3.39) with ψ satisfying (3.38).

Remark 7. The proof of Theorem 3.3 will appear in [24].

The following theorem partially classifies the isolated singularities of (3.22).

Theorem 3.4 Let $u(x)$ be a positive solution of (3.22) in $0 < |x| < R$ such that the action

$$S_r(u) = \int_{r \leq |x| \leq R} d^4x \ \{\tfrac{1}{2} |\nabla u|^2 - \tfrac{1}{4} u^4\} \tag{3.49}$$

does not diverge faster than logarithmically as $r \to 0$. Then either $u(x)$ can be extended to a positive regular solution on $|x| \leq R$, or it has the asymptotic behavior

$$u(x) = 0\left(\frac{1}{|x|}\right) \quad \text{near} \quad x = 0 \tag{3.50}$$

Remark 8. We believe that the logarithmically divergent assumption can be disposed. The main tools in proving Theorem 3.4 are the de Giorgi-Nash-Moser iterative techniques [25], and the existence of a Harnack inequality for solutions satisfying the assumptions of the theorem

IV. REMOVABILITY OF ISOLATED SINGULARITIES

In this section we describe briefly the important work of K. Uhlenbeck [14]. Her main results are

Theorem 4.1 (Regularity): Let A,F be a weak YM solution on a domain $\Omega \subset \mathbb{R}^4$. Then around every point $x \in \Omega$, there exists a neighborhood U and a gauge over U in which A is $C^\infty(U)$, and estimates on the derivatives of A depend only on the distance to the boundary $\partial\Omega$ and $\|F\|_\infty$.

Theorem 4.2 (Removability of Isolated Singularities).
Let A,F be a smooth solution of the YM equations on $D_1 - \{0\} = \{x \in \mathbb{R}^4 : 0 < |x| < 1\}$ such that

$$\int_{D_1} d^4x \; |F(x)|^2 < +\infty \tag{4.1}$$

Then $|F(x)|$ is bounded for $|x| \leq \frac{1}{2}$, and there exists a gauge in which A can be extended to a smooth solution on the entire ball $|x| \leq 1$.

Corollary 4.3 Let A,F be a (smooth) solution of the YM equations in $\{x \in \mathbb{R}^4 : |x| \geq R$, such that

$$\int_{|x| \geq R} d^4x \; |F(x)|^2 < +\infty \tag{4.1'}$$

Then

$$|F(x)| = 0\left(\frac{1}{|x|^4}\right) \quad \text{at infinity.} \tag{4.2}$$

In particular, a smooth finite action solution of the YM equations on $\mathbb{R}^4$, extends to a smooth solution on $S^4 = \mathbb{R}^4 \cup \{\infty\}$.

Remark. Throughout this section, $|F(x)|, |A(x)|$ will denote the norm of the Lie algebra of SU(2) (see (2.2)), and $\|F\|_p, \|A\|_p$, $1 \leq p \leq +\infty$, will denote the usual L_p norms of $|F(x)|, |A(x)|$ respectively. Note that $|A(x)|$ is gauge dependent, but $|F(x)|$ is gauge independent.

Proof of Theorem 4.1 (An Outline):

Step 1. The main problem is that the YM equations (2.3), as they stand, are not elliptic. We will use the property of gauge invariance to achieve ellipticity. The non-ellipticity is easily seen by considering the linearization of (2.3) which reads

$$\partial_\mu \partial_\mu A_\nu - \partial_\mu \partial_\nu A_\mu = 0 \tag{4.2'}$$

This equation would be elliptic if

$$\partial_\mu A_\mu \;=\; 0 \tag{4.3}$$

Equation (4.3) may be achieved by making use of the gauge invariance. In physics literature, (4.3) is called the <u>Landau gauge</u>. Its existence is proven by the implicit function theorem: Let us denote by $\tilde{A}_\mu(x)$ the connection in a given gauge. Suppose that it satisfies

$$\|\tilde{A}\|_p \;=\; \left(\int_U d^4x \; |\tilde{A}(x)|^p \right)^{\frac{1}{p}} \;\leq\; \varepsilon, \quad \text{for some} \quad p \geq 8 \tag{4.4}$$

for sufficiently small ε. Then we claim that there exists a gauge transformation $g(x) \in SU(2)$ so that the transformed connection

$$A_\mu(x) \;=\; g^{-1}(x)\, \tilde{A}_\mu(x)\, g(x) + g^{-1}(x)\, \partial_\mu\, g(x) \tag{4.5}$$

satisfies (4.3). If $g(x) = e^{\Lambda(x)}$, $\Lambda(x) \in \mathfrak{g}$, then the linearization of (4.3) reads

$$\partial_\mu \partial_\mu \Lambda(x) + \partial_\mu [\tilde{A}_\mu(x), \Lambda(x)] + \partial_\mu \tilde{A}_\mu(x) \;=\; 0 \tag{4.6}$$

If $\tilde{A}$ satisfies (4.4), an application of the implicit function theorem gives the existence of g. Condition (4.4) is achieved in Steps 2 and 3 below.

<u>Step 2</u>. Let $\tilde{\tilde{A}}_\mu$ be the weak solution of Theorem 4.1, i.e. we assume $\tilde{\tilde{A}} \in c^1(\Omega)$. For the neighborhood U we will construct a gauge A' in which

$$|A'(x)| \;\leq\; c|x| \max_{|y| \leq |x|} |F(y)| \tag{4.7}$$

(recall that $|F(y)|$ is gauge independent). To this end, let

$r = |x|$, $s = \frac{x}{|x|} \in S^3$. For a fixed $s \in S^3$, we define a gauge transformation $g(x) = g(r,\sigma)$ by the ordinary differential equation

$$\frac{dg}{dr} \;=\; -\tilde{\tilde{A}}_r g \tag{4.6a}$$

$$g(0,s) \;=\; I \in SU(2) \tag{4.6b}$$

where

$$\tilde{\tilde{A}}_r(x) = \sum_{\mu=0}^{3} \frac{x_\mu}{|x|} \tilde{\tilde{A}}_\mu(x) \tag{4.7}$$

It is easily seen that

$$A'_\mu(x) = g^{-1}(x)\, \tilde{\tilde{A}}_\mu\, g(x) + g^{-1}(x)\, \partial_\mu g(x)$$

satisfies

$$A'_r(x) = 0 \tag{4.8}$$

In this gauge (called exponential gauge), one easily obtains from (2.1)

$$A'_\nu(x) = \sum_{\mu=0}^{3} \int_0^1 d\tau\, \tau x_\mu\, F'_{\mu\nu}(\tau x) \tag{4.9}$$

which quickly yields (4.7).

Step 3. Using (4.7), we will choose a gauge over the neighborhood U so that (4.4) holds. Suppose that U has radius ρ. By setting $y = \frac{x}{\rho}$, U becomes the unit ball, the connection is multiplied by ρ (see (2.9a)), and the curvature by ρ^2 (see 2.9b)). Thus we may assume that $\tilde{\tilde{A}}$ is a solution in $U = \{|x| < 1\}$, and that $\|\tilde{\tilde{F}}\|_\infty = \|F\|_\infty \leq \varepsilon$ for sufficiently small ε. Choosing an exponential gauge from the north pole of $\partial U = S^3$, and another one from the south pole, and matching the two gauge (by an appropriate gauge transformation -- see Lemma 2,2 of [14]), we obtain via (4.7) a gauge on $\partial U = S^3$ so that in this gauge $|\tilde{A}_s(x)| \leq \|F\|_\infty \leq \varepsilon$. Here $\tilde{A}_s$ denotes the tangential component of $\tilde{A}$. Then using an exponential gauge from the center of U, and another exponential gauge from $\partial U = S^3$ (see Lemma 4.4 of [14]) we obtain a gauge over U so that (recall (4.7)) $\|\tilde{A}\|_\infty \leq \|F\|_\infty \leq \varepsilon$. This implies (4.4).

Step 4. For a small neighborhood U of $x \in \Omega$, we use Steps 2 and 3 to go from the given gauge $\tilde{\tilde{A}}$ over U to the gauge $\tilde{A}$ which satisfies (4.4). Then (employing the implicit function theorem) we obtain the Landau gauge (4.3) over U. In this gauge and neighborhood the YM equations are elliptic. By standard ellipticity, solutions are C^∞ in U, and estimates depend on $\|A\|_\infty, \|F\|_\infty$, and the distance to the boundary

Proof of Theorem 4.2 (An Outline): The proof has two main components:

i) The starting point is the Weitzenböck formula which implies that $|F(x)|^2$ is a subsolution of a differential inequality. This allows the application of the de Giogi-Nash-Moser iteration arguments to derive the upper bound

$$|F(x)| \leq \frac{c}{|x|^2}\left(\int_{D(x;|x|)} d^4y\ |F(y)|^2\right)^{\frac{1}{2}}, \quad \text{for} \quad 0 < |x| < \frac{1}{2} \tag{4.10}$$

where $D(x_o,r)$ denotes the ball with center x_o and radius r. The proof of (4.10) makes no use of the YM equations themselves, nor involves gauges.

ii) The second component involves the construction of a "broken" Landau gauge about the singular point $x = 0$. This gauge is then used to integrate by parts $\int d^4y\ |F|^2$, and show that

$$\int_{D(x;|x|)} d^4y|F(y)|^2 = O(|x|^4) \quad \text{near} \quad x = 0 \tag{4.11}$$

A combination of (4.10) and (4.11) yields the boundedness of $|F(x)|$. The proof, then, of the theorem is completed by arguments similar to the ones involved in the proof of Theorem 4.1. We provide now some more details.

Step 1. If (4.1) holds, then we may assume

$$\int_{D_1} d^4x|F(x)|^2 \leq \varepsilon^2 \tag{4.12}$$

To see this we perform a scaling $y = \frac{x}{r}$. Since ((2.9b)) $F'_{\mu\nu}(y) = r^2 \cdot F_{\mu\nu}(x)$, we have

$$\int_{|y| \leq 1} d^4y\ |F'(y)|^2 = \int_{|x| \leq r} d^4x\ |F(x)|^2 \leq \varepsilon^2$$

for sufficiently small r. From now on we will work in the unit ball and assume that (4.12) holds.

Step 2. Here we prove (4.10). From (2.3) and (2.5), we may derive Wetzenböck's formula

$$D_\kappa D_\kappa F_{\mu\nu} + 2[F_{\mu\kappa}, F_{\kappa\nu}] = 0 \tag{4.13}$$

where D_κ is the covariant derivative $D_\kappa F = \partial_\kappa F + [A_\kappa, F]$. Equation (4.13) yields

$$0 \leq \frac{1}{2}\ \Delta|F|^2 - \Big|\nabla|F|\Big|^2 - 2\ \mathrm{Tr}\{F_{\mu\nu}[F_{\mu\kappa}, F_{\kappa\nu}]\}$$

or

$$\Delta|F(x)|^2 + 4|F(x)|^3 \geq 0 \tag{4.14}$$

Let $B \subset D_1 - \{0\} = \{x + \mathbb{R}^4:\ 0 < |x| < 1\}$, $\eta(x) \in C_0^\infty(B)$, and

$$\phi(x) = \eta^2(x)\ |F(x)|^{2(p-1)}$$

Then (4.14) yields

$$\int_B \{\nabla\phi\cdot\nabla(|F|^2) - 4\phi\ |F|^3\} \leq 0$$

From this we obtain (via Sobolev's inequality) for any sufficiently small $\delta > 0$

$$\left\{\frac{2(p-1-\delta)}{p^2}\ \frac{1}{C_4^2} - 4\left(\int_B |F(x)|^2\right)^{\frac{1}{2}}\right\}\left(\int_B \eta^4(x)\ |F(x)|^{4p}\right)^{\frac{1}{2}}$$

$$\leq \left[\frac{4(p-1-\delta)}{p^2} + \frac{1}{\delta}\right]\left(\int_B |\nabla\eta|^2|F|^{2p}\right)^{\frac{1}{2}} \tag{4.15}$$

where C_4 is the Sobolev's constant in $W^{1,2}(B) \subset L^4(B)$. From (4.15), by the de Giorgi-Nash-Moser [25] iteration procedure, we obtain (4.10), which also yields (by (4.12))

$$F(x) \leq \frac{c\varepsilon}{|x|^2}, \quad 0 < |x| < \frac{1}{2} \tag{4.16}$$

Step 3. We now construct a "broken" Landau gauge over $D_{1/2} - \{0\}$, $D_{1/2} = D(0;\frac{1}{2})$, which will be used in Step 4 to prove (4.11). Let $\Omega_i = \{x \in \mathbb{R}^4 : 2^{-i-1} \leq |x| \leq 2^{-i}\}$, $S_i = \{x : |x| = 2^{-i}\}$, $i = 1,2,3,\ldots$. Using arguments similar to the ones involved in the proof of Theorem 4.1, we construct Landau gauges over Ω_i which match over S_i. More precisely: Let A^i_μ denote the restriction of A_μ over Ω_i. Assume (4.16). Then the following properties hold (see Lemma 4.6 of [14]):

(a) $\partial_\mu A^i_\mu = 0$

(b) $\partial_s A^i_s = 0$ over S_i and S_{i-1}

(c) $\displaystyle\int_{S_i} A^i_r = \int_{S_{i-1}} A^i_r = 0$

(d) $A^i_s\big|_{S_i} = A^{i+1}_s\big|_{S_i}$, and the two gauges from Ω_i and Ω_{i+1} induced on S_i agree.

(e) $|A^i(x)| \leq c2^{-i}\|F\|_\infty \leq c\,\varepsilon\,2^i$

(f) $\displaystyle(4 - c^2\varepsilon^2)\int_{\Omega_i} |A^i|^2 \leq \int_{\Omega_i} |F|^2$

Step 4. From (4.16) and Step 3, we obtain by integration by parts

$$4(1 - c\varepsilon)\int_{D_{1/2}} |F(x)|^2 \leq \int_{S_{1/2}} |F(x)|^2 \tag{4.17}$$

Scaling from $D_{1/2}$ to U with radius r, and applying (4.17) we obtain

$$4\big(1 - c\varepsilon(r)\big) \int_{|x| \le r} |f(x)|^2 \le r \int_{|x| = r} |F(x)|^2 \tag{4.18}$$

where

$$\varepsilon(r) = \max_{|x| \le r} \{|x|^2 \, |F(x)|\}$$

Let

$$f(r) = \int_{|x| \le r} |F(x)|^2$$

then (4.19) becomes

$$4\big(1 - c\varepsilon(r)\big) \, f(r) \le r \, f'(r) \tag{4.19}$$

Inequality (4.19) easily yields (after an iterative integration) estimate (4.11). This completes the proof of Theorem 4.2, by earlier remarks. Corollary 4.3 is obtained from Theorem 4.2 by performing an inversion $y = R \dfrac{x}{|x|^2}$.

REFERENCES

1. A.M. Polyakov, Phys. Lett. B59, 82 (1975).
2. A.A. Belavin et al., Phys. Lett. B59, 85 (1975).
3. V. de Alfaro, S. Fubini, and G. Furlan, Phys. Lett. B65, 1963 (1976).
4. R. Jackiew, C. Nohl, C. Rebbi, Phys. Rev. D15, 1642 (1977).
5. T. Glimm and A. Jaffe, Phys. Lett. B73, 167 (1978).
6. T. Jonsson, O. McBryan, F. Zirilli, and J. Hubbard, Commun. Math. Phys. (1979).
7. G. 't Hooft, Phys. Rev. Lett. 37, 8 (1976).
8. C. Callan, R. Dashen, and D. Gross, Phys. Lett. B65, 354 (1976).
9. G. 't Hooft, Phys. Rev. D14, 3432 (1976).
10. M.F. Atiyah, N.J. Hitchin, and I.M. Singer, "Self-duality in four-dimensional Riemannian Geometry," Proc. Roy. Soc. London, Ser. A, 362 (1978) 425-461.
11. C. Callan, R. Dashen, and D. Gross, Phys. Rev. D.
12. J. Glimm and A. Jaffe, Phys. Rev. D18 (1978) 463.
13. C. Loewner and L. Nirenberg, "Partial Differential Equations Invariant under Conformal and Projective Transformations," Contributions to Analysis, Academic Press (1974), pp. 245-272.
14. K. Uhlenbeck, "Removable Singularities in Yang-Mills Fields," Bull. Am. Math. Soc. 1 (1979) 579-581, and preprint.
15. J.P. Bourgiugnon, H.B. Lawson, and J. Simons, "Stability and Gap Phenomena for Yang-Mills Fields," Proc. Nat. Acad. Sci. U.S.A. (to appear).
16. A. Schwarz, Phys. Lett. B 167, 172-174 (1977).
17. R. Hartshorne, Commun. Math. Phys. 59 (1978) 1-15.
18. V. Drinfeld and Yu Manin, Funct. Anal. i Prilozhen. 12:2 (1978) 81; Uspehi Math. Nauk 33:3 (1978) 241.
19. M. Atiyah, N. Hitchin, V. Drinfeld, and Yu Manin, Phys. Lett. 65A (1978) 185.
20. G. 't Hooft, "On the Phase Transition Towards Permanent Quark Confinement," Nucl. Phys. B138 (1978) 1, and "A Property of Electric and Magnetic Flux in non-Abelian Gauge Theories," preprint (1979).
21. R.H. Fowler, Quart. J. Math. 2 (1931) 259-288.
22. J. Cervero, L. Jacobs, and C. Nohl, Phys. Lett. 69B (1977) 351.
23. M. Obata, J. Diff. Geom. 6 (1971) 247-258.
24. B. Gidas, W.M. Ni, and L. Nirenberg
25. C.B. Morrey, Multiple Integrals in the Calculus of Variations, Springer-Verlag, N.Y. (1966).

EXISTENCE OF STATIONARY STATES IN NONLINEAR SCALAR FIELD EQUATIONS

H. BERESTYCKI, P.L. LIONS

Université Paris VI, Analyse Numérique, (55-65,5E)
4 Place Jussieu, 75230 Paris Cedex 05, France.

Abstract : We report on some recent results concerning existence of solutions for nonlinear scalar field equations that lead to semilinear elliptic boundary value problems in $\mathbb{R}^N$. Such problems arise in a wide variety of contexts in physics (solitons in non-linear Klein-Gordon or Schrödinger equations, euclidean scalar fields, statistical mechanics, cosmology, nonlinear optics etc..). Existence of a ground state and of infinitely many bound states is proved under assumptions which are "nearly optimal", using a variational technique. Other methods of resolution are also presented. Some results on bifurcation from the essential spectrum are derived for this class of problems. A generalization of the existence results for systems of equations is also provided here. Lastly, in the appendix, we present some numerical computations emphasizing some qualitative properties of these equations.

INTRODUCTION

In this paper, we present some results concerning the following problem : find u in $H^1(\mathbb{R}^N)$ (1) solution of

$$-\Delta u = g(u) \quad \text{in} \quad \mathbb{R}^N \quad , \tag{1}$$

here g is some given nonlinearity which satisfies $g(0) = 0$, and of course we look for a solution $u \neq 0$. The fact that u is in $H^1(\mathbb{R}^N)$ means, in particular, that u, in some weak sense, is zero at infinity and thus it represents some kind of "boundary condition". In conclusion, we want to solve

$$-\Delta u = g(u) \quad \text{in} \quad \mathbb{R}^N \quad , \quad u \in H^1(\mathbb{R}^N) \quad , \quad u \neq 0 \quad . \tag{2}$$

C. Bardos and D. Bessis (eds.), Bifurcation Phenomena in Mathematical Physics and Related Topics, 269–292.

This kind of problem arises in many contexts in physics : for example in nonlinear euclidean field equations, in the study of false vacuum states in statistical mechanics, in cosmology...[2] Another application is the following : let us consider some nonlinear Klein-Gordon equation (NLKG in short)

$$\phi_{tt} - \Delta\phi = f(|\phi|^2)\,\phi \qquad \text{for} \quad (t,x) \text{ in } \mathbb{R}\times\mathbb{R}^N \quad . \tag{3}$$

If we look for solitary waves in (3), and more precisely, if we look for solutions of the form

$$\phi(t,x) = e^{i\omega t}\,u(x) \qquad (\text{"standing waves"})$$

then u satisfies

$$-\Delta u = f(u^2)\,u + \omega^2 u \qquad \text{in} \quad \mathbb{R}^N \qquad ;$$

that is, some equation of the form (1).

The same remark applies for nonlinear Schrödinger equations (NLS in short) :

$$i\phi_t - \Delta\phi = f(|\phi|^2)\,\phi \qquad \text{for} \quad (t,x) \text{ in } \mathbb{R}\times\mathbb{R}^N \quad . \tag{4}$$

Let us also remark that travelling waves of (3) give rise to problems similar to (2) (take $\phi(t,x) = u(x-ct)$, $c \in \mathbb{R}^N$, $|c| < 1$).

Finally, solutions of (2) represent steady states of the nonlinear heat equation :

$$\frac{\partial u}{\partial t} - \Delta u = g(u) \qquad \text{in} \quad [0,\infty)\times\mathbb{R}^N \quad , \tag{5}$$

and this kind of equations arise in many contexts (in nuclear reactors, chemistry, population dynamics ...).

We will give here some conditions on g under which the existence of a positive, "ground-state" solution, and of infinitely many solutions of (2) (with more and more "energy") is proved. These conditions appear to be also almost necessary. The proofs of the main results (concerning (2)) are not included here and the reader is referred to [5],[6],[9].

Sections 1, 2 and 3 are devoted to three different methods to solve (2) : a variational method, a "local" one and an O.D.E. approach. In Section 4 we present some result in view of a bifurcation analysis of problem (2). Finally we consider in Section 4 the extension of problem (2) to systems. In the appendix, some numerical considerations together with some qualitative properties are given.

1. A VARIATIONAL APPROACH

1.1. Main results

Let $N \geq 3$.

Theorem 1.1 : *Let* g *be a continuous function from* $\mathbb{R}_+$ *into* $\mathbb{R}$ *satisfying* $g(0) = 0$ *and*

$$\overline{\lim_{s \to +\infty}} \frac{g(s)}{s^\ell} \leq 0 \quad , \quad \text{with } \ell = \frac{N+2}{N-2} \quad ; \tag{6}$$

$$\overline{\lim_{s \to 0_+}} \frac{g(s)}{s} < 0 \quad ; \tag{7}$$

$$G(\zeta) = \int_0^\zeta g(s)\, ds > 0 \quad , \quad \text{for some } \zeta > 0 \; . \tag{8}$$

Then, there exists u *solution of* (2), *satisfying* :

$$u \in C^2(\mathbb{R}^N) \quad , \quad u \text{ is radial} : \; u(x) = u(|x|) \quad ; \tag{9}$$

$$u \text{ is decreasing} : \; u(x) < u(y) \; , \; \text{if } |x| > |y| \; ; \quad u(x) > 0 \text{ for } x \text{ in } \mathbb{R}^N \tag{10}$$

$$|D^\alpha u(x)| \leq C\, e^{-\delta |x|} \quad \text{for every } x \text{ in } \mathbb{R}^N \text{ and } |\alpha| \leq 2 \quad , \tag{11}$$

for some $C, \delta > 0$.

Furthermore, if v *is any solution of* (2) *in* $L^\infty_{loc}(\mathbb{R}^N)$ *and such that* $G(v) \in L^1(\mathbb{R}^N)$ *then we have* :

$$0 < S(u) \leq S(v)$$

where

$$S(w) = \int_{\mathbb{R}^N} \left\{ \frac{1}{2} |\nabla w|^2 - G(w) \right\} dx \quad \text{and} \quad G(t) = \int_0^t g(s)\, ds \quad .$$

Remark 1.1 : Conditions (6)-(8) are almost necessary as we will see in section 1.2.

Remark 1.2 : Less general results concerning the existence of a positive solution of (2) were previously obtained in Refs.[23], [29],[16],[13] (where, essentially, only the case of $g(s) = \lambda |s|^p s - \mu s$ is treated by methods which do not extend to the general case). The first general results were obtained by W. Strauss[30], but are totally included in Theorem 1.1 (and the method of [30] does not extend to the general situation). Theorem 1.1 is announced in [5], and proofs of it can be found in [6],[9].

The last statement about u is an easy adaptation (and a rigorous one) of a result in [17].

Remark 1.3 : Let us indicate that (7) can be replaced by

$$\overline{\lim_{t\to 0_+}}\ \frac{G(t)}{t^2} < 0 \ , \text{ or } \overline{\lim_{s\to 0_+}}\ g(s)/s^{\ell} \leq 0 \ . \tag{7'}$$

If we assume (7) and (8), we may replace (6) by (6') : if for every $s > \zeta$ satisfying (8), we have $g(s) \geq 0$, then (6) holds.

We give now a result concerning the existence of infinitely many solutions :

Theorem 1.2 : *Under the assumptions of* Theorem 1.1, *and if we assume in addition that* g *is odd, then there exists a sequence* $(u_n)_{n\geq 1}$ *of solutions of (2) satisfying (9) and (11) and such that*

$$S(u_n)\ , \int_{\mathbb{R}^N} |\nabla u_n|^2\, dx\ , \int_{\mathbb{R}^N} G(u_n)\, dx \uparrow +\infty \quad \text{as } n \uparrow +\infty .$$

Remark 1.4 : This result is announced in [5], and proved in [6]. It generalizes previous results of [29],[13] and [30].

Remark 1.5 : A difficult mathematical question is to decide whether the assumption about the oddness of g is really needed or not (we conjecture it is not and we have a few results in this direction, see [11]). Nevertheless, it seems to be an ethical question since in all physical situations g is odd (see the preceding Introduction for example).

Remark 1.6 : The case when $N = +1$ is settled in [9], we shall discuss later on the case when $N = +2$.

1.2. The"almost necessary"character of conditions (6)-(8).

We will first show that (8) is necessary. A simple proof is to use the so-called Pohozaev identity (see [25],or [4]) we recall now :

Prop.1.1 [25],[4] : *If* u *is a solution of (2), which is in* $L^{\infty}_{loc}(\mathbb{R}^N)$ *and such that* $G(u) \in L^1(\mathbb{R}^N)$, *then we have :*

$$(N-2)\int_{\mathbb{R}^N} |\nabla u|^2\, dx = 2N\int_{\mathbb{R}^N} G(u)\, dx \quad . \tag{12}$$

Remark 1.7 : *In particular, from (12) we deduce*

$$S(u) = \frac{1}{N}\int_{\mathbb{R}^N} |\nabla u|^2\,dx = \frac{2}{(N-2)}\int_{\mathbb{R}^N} G(u)\,dx \quad .$$

Thus, if $u \not\equiv 0$, $S(u) > 0$ *and for some* x *in* $\mathbb{R}^N$, $G(u(x)) > 0$. *This proves that (8) is necessary.*

Remark 1.8 : The rigorous proof of Prop.1.1 is given in [6]. Actually, in this generality it seems to be new. We only give below some hints on the proof. Let us now give a heuristical derivation of (12) based upon a scaling argument : we denote by $u_\sigma(x) = u(x/\sigma)$. Remarking that (2) is "equivalent" to $S'(u) = 0$, we find :

$$\frac{d}{d\sigma} S(u_\sigma)\Big|_{\sigma=+1} = 0 \quad .$$

But a simple change of variables give :

$$S(u_\sigma) = \sigma^{N-2}\int_{\mathbb{R}^N} \frac{1}{2}|\nabla u|^2\,dx - \sigma^N \int_{\mathbb{R}^N} G(u)\,dx \quad .$$

Proof of Prop.1.1 : First, let us remark that since $u \in L^\infty_{loc}(\mathbb{R}^N)$, by well-known results (see [1] for example), $u \in W^{2,p}_{loc}(\mathbb{R}^N)$ (for every $p < \infty$). On the other hand, since $\nabla u \in L^2(\mathbb{R}^N)$ and $G(u) \in L^1(\mathbb{R}^N)$, there exists a sequence $R_k \underset{k\to\infty}{\longrightarrow} \infty$ such that

$$R_k \int_{\partial B_{R_k}} \left\{|\nabla u|^2 + |G(u)|\right\} dS \underset{k\to\infty}{\longrightarrow} 0 \;{}^{(3)} \quad . \qquad (13)$$

Now, we multiply (2) by $\sum_i x_i \frac{\partial u}{\partial x_i}$ and we integrate by parts on B_{R_k} . An easy computation shows that

$$(N-2)\int_{B_{R_k}} |\nabla u|^2\,dx - 2N\int_{B_{R_k}} G(u)\,dx$$

is a boundary term which goes to 0 as $k \to \infty$, because of (13). This proves (12). (For more details see [6]).

Remark 1.9 : One may give another argument to prove (8) is necessary : indeed if u is a radial solution of (2) of class $C^2(\mathbb{R}^N)$ then we have (we again denote $u = v$, where $v(r) = u(x)$ and $r = |x|$)

$$\left.\begin{aligned} &-u'' - \frac{N-1}{r}u' = g(u) \quad \text{for} \quad r \geq 0 \\ &u'(0) = 0 \quad , \quad \lim_{r\to\infty} u(r) = 0 \quad . \end{aligned}\right\} \qquad (14)$$

Multiplying by $u'(r)$, and integrating between 0 and R we obtain

$$G(u(0)) - G(u(R)) = (N-1)\int_0^R (u'(s))^2 \frac{1}{s}\,ds + \frac{1}{2}(u'(R))^2 .$$

This implies that $u'(r) \underset{r\to\infty}{\longrightarrow} 0$ since u is bounded and

$$G(u(0)) = (N-1)\int_0^\infty \frac{1}{s}(u'(s))^2\,ds .$$

Thus, as soon as $u \not\equiv 0$, $G(u(0)) > 0$ and (8) is necessary.

We now prove that (7) is almost necessary : this will follow from the following Lemma and Remark 1.10 :

Lemma 1.1 : *Let* g *be a continuous function from* $\mathbb{R}_+$ *into* $\mathbb{R}$ *such that* $g(0) = 0$ *and*

$$\underline{\lim}_{t\to 0_+} \frac{g(t)}{t} > 0 . \tag{15}$$

If u *is radial, of class* $C^2(\mathbb{R}^N)$ *and satisfies*

$$-\Delta u = g(u) \ \ in\ \mathbb{R}^N , \quad u \geq 0 \ \ in\ \mathbb{R}^N , \quad \lim_{|x|\to\infty} u(x) = 0$$

then $u \equiv 0$.

Remark 1.10 : Another way to see that (7) is almost necessary is to argue as follows. Assume that g is C^2 in a neighborhood of 0 and that $g'(0) > 0$. Then any solution u of (2) satisfies

$$-\Delta u + \left\{g'(0) - \frac{g(u)}{u}\right\} u = g'(0)\,u \quad \text{in } \mathbb{R}^N ,$$

and the "potential" $q(x) = g'(0) - \frac{g(u(x))}{u(x)}$ satisfies

$$|q(x)| \leq C|u(x)| \quad \text{for } |x| \text{ large, provided } u(x) \underset{|x|\to\infty}{\longrightarrow} 0 .$$

Thus, as soon as, in dimension $N = 3$ for example, $u(x) = O(1/|x|)$ this implies a contradiction : $u \equiv 0$, because of a well-known result of T. Kato [18]. Remark that, if u is radial and belongs to $H^1(\mathbb{R}^3)$, by [30] or [6] the asymptotic condition on u is satisfied and, again, this proves there exists no solution of (2).

Proof of Lemma 1.1 [4] : The solution u obviously satisfies (14). Now, let us introduce

$$v(r) = u(r)\, r^{(N-1)/2} .$$

An easy computation shows that v satisfies

$$v'' = -g(u)\, r^{(N-1)/2} + \frac{(N-1)(N-3)}{4}\frac{v}{r^2} \quad \text{for} \quad r > 0 .$$

Now, $g(u(r)) \geq \Theta\, u(r)$ if $r \geq r_o$, for some $\Theta > 0$.

Thus, if r is large : $r \geq r_1$, we have $v'' \leq 0$. And v' is nonincreasing, and converges to some limit α . If α were negative, this would imply v negative for large r and this contradicts $u \geq 0$.

Thus $v' \searrow_{r \nearrow +\infty} \alpha \geq 0$. And this implies : $v \nearrow_{r \nearrow +\infty} \beta \geq 0$.

If β were positive, this would imply :

$$v'' \leq -\frac{\theta}{2}\beta \quad \text{for} \quad r \geq r_2 \quad ;$$

and v' would be negative for large r . Thus $\beta = 0$ and $v \equiv 0$ on $[r_1, \infty)$. This implies $u \equiv 0$.

Remark 1.11 : Assumption (15) (or $g'(0) > 0$, as in Remark 1.10) is not exactly the negation of (7). Essentially, the only remaining case is the limit case : $g'(0) = 0$. Then, the existence questions are much more complex and very different phenomena may appear, depending on the structure of g (see [6] and [12] for the study of this case).

We now show why a condition like (6) has to be assumed. In order to understand it, let us just consider some examples :

Example 1.1 : Let us consider $g(s) = \lambda|s|^{p-1}s - ms$, with $\lambda, m > 0$ and $p > 1$. Conditions (7) and (8) are obviously satisfied, and (6) is satisfied if and only if

$$p < \frac{N+2}{N-2} = \ell \quad .$$

But, for $p \geq \ell$, there is no solution of (2) in $L^{p+1}(\mathbb{R}^N) \cap L^\infty_{loc}(\mathbb{R}^N)$ indeed if u satisfies (2) and if $u \in L^{p+1}(\mathbb{R}^N) \cap L^\infty_{loc}(\mathbb{R}^N)$, by Prop.1.1, and integration by parts, we have

$$(N-2)\int_{\mathbb{R}^N} |\nabla u|^2 \, dx = (N-2)\int_{\mathbb{R}^N} u\, g(u)\, dx = 2N \int_{\mathbb{R}^N} G(u)\, dx \quad .$$

$$\text{Thus } \left\{\frac{2N}{p+1} - (N-2)\right\}\lambda\int_{\mathbb{R}^N} |u|^{p+1} dx = \left\{N - (N-2)\right\} m \int_{\mathbb{R}^N} |u|^2 dx \quad .$$

Since the right-hand side is nonnegative, and the left-hand side is nonpositive, this equality implies $\int_{\mathbb{R}^N} |u|^2 dx = 0$ and $u \equiv 0$.

We conclude this section with a few examples which illustrate Theorems 1.1 and 1.2 .

Example 1.2 : Let $g(s) = \lambda|s|^{p-1}s - \mu|s|^{q-1}s - ms$, with $\lambda, \mu, m > 0$, $p, q > 1$. Condition (6) is satisfied if and only if we have

$$p < \max(q, \ell) \tag{16}$$

and

$$\text{if } p<q \text{ , then (8) holds : i.e. } \exists\zeta>0 \text{ such that } G(\zeta)>0 \qquad (8')$$

Remark that in this case $(p<q)$ this condition is explicit in terms of λ,μ,m,p,q and can be written

$$\text{if } p<q \text{ , then } \lambda>\lambda^*(\mu,m,p,q) \qquad (8'')$$

where λ^* is explicit and independent of N :

$$\lambda^* = (\tfrac{m}{2})^a \ \mu^b \ (q-1)(p+1)(q-p)^{-a} \ (p-1)^{-b}(q+1)^{-b}$$

where $a = \dfrac{q-p}{q-1}$ and $b = \dfrac{p-1}{q-1}$.

Notice also that in the special case $g(t) = \lambda|t|^{p-1}t - \mu|t|^{q-1}t - mt$ with $\lambda,\mu,m>0$, $1<p<q$, then $\lambda>\lambda^*$ *is a necessary and sufficient condition* for the existence of a solution to (2). Indeed if $\lambda\leq\lambda^*$ then G is non-positive everywhere and by Prop.1.1, this implies the necessary character of the assumption.

Example 1.3 : Let $g(s) = \lambda \sin s + \mu s$ (this type of nonlinearity is inspired from Sine-Gordon type models). The assumption of Theorems 1.1 and 1.2 are satisfied obviously if and only if we have :

$$\lambda \quad < \quad -\mu \quad < \quad 0 \quad .$$

Again one can check on that simple example that the given condition is rather sharp since if N=3 the condition is also necessary and if $N\geqslant 4$ it is almost necessary : there just appear, in the limit case $\lambda=-\mu<0$ solutions which are zero at infinity in a weak sense (not necessarily in $H^1(\mathbb{R}^N)$ and do not satisfy (11) (see [6])).

1.3. Ideas of the proofs

A variational method to prove Theorem 1.1 is to consider the following auxiliary problem : find u solution of

$$\left\{\begin{array}{l} \displaystyle\operatorname{Inf}_{\int_{\mathbb{R}^N} G(u)\,dx = +1} \int_{\mathbb{R}^N} \frac{1}{2}|\nabla u|^2 dx \\ u\in H^1(\mathbb{R}^N) \ , \ G(u)\in L^1(\mathbb{R}^N) \end{array}\right. = \left\{\begin{array}{l} \displaystyle\operatorname{Inf}_{V(u)=+1} T(u) \\ u\in H^1(\mathbb{R}^N) \ , \ G(u)\in L^1(\mathbb{R}^N) \end{array}\right. \qquad (17)$$

where $T(u) = \int \frac{1}{2}|\nabla u|^2$ and $V(u) = \int G(u)$.

Suppose this problem is solved, that is, we know there exists a solution u of (17), then there exists some Lagrange multiplier such that

$$T'(u) \quad = \quad \theta \ V'(u)$$

that is

$$-\Delta u = \theta\, g(u) \quad \text{in} \quad \mathcal{D}'(\mathbb{R}^N) \quad .$$

Actually, this point has to be justified rigorously, but we shall not discuss these technicalities here, see [6],[9].

Since $V(u) = +1$, $\theta \neq 0$ and one can deduce from Prop.1.1 that $\theta > 0$. Now, rescaling u, namely considering

$$\tilde{u}(x) = u(x/\sqrt{\theta})$$

we have

$$-\Delta\tilde{u} = g(\tilde{u}) \quad \text{in} \quad \mathcal{D}'(\mathbb{R}^N) \quad ;$$

and (2) is solved.

In conclusion, Theorem 1.1 is a consequence of the following result :

Prop.1.2 : *Under the assumptions of* Theorem 1.1, *there exists* u *solution of* (17), *which is radial and which satisfies :*

$$u \textit{ is nonincreasing} : u(x) \leq u(y) \text{ , if } |x| \geq |y| \text{ ; } u(x) \geq 0 \text{ on } \mathbb{R}^N \tag{10'}$$

Remark 1.12 : We shall not prove Prop.1.2 here, the reader is referred to [5],[6] or [9].

Remark 1.13 : It is easy to see that the method above fails when N=2 , since in this case, we should have for every solution u of (2) :

$$V(u) = \int_{\mathbb{R}^N} G(u)\, dx = 0 \quad ,$$

because of Prop.1.1. In addition, if we still want to minimize T over the set {V=0} , then the infinum is obviously 0 and we are not able to avoid 0 . Nevertheless, other methods (than variational) are available when N=2 , cf. Sections 2 and 3.

Remark 1.14 : Instead of (17), one could consider some dual variational problem : find u maximizing

$$\begin{cases} \displaystyle\sup_{T(u) = +1} V(u) \\ u \in H^1(\mathbb{R}^N) \text{ , } G(u) \in L^1(\mathbb{R}^N) \end{cases} \tag{18}$$

And remark that, now, the constraint $T(u) = +1$ is a smooth one :

u is supposed to lie on the sphere of $\widehat{\mathcal{D}}^{1,2}(\mathbb{R}^N) = \{u \in L^{\frac{2N}{N-2}}(\mathbb{R}^N)$, $\nabla u \in L^2(\mathbb{R}^N)\}$ $(N \geq 3)$. This remark plays an important role in the proof of Theorem 1.2 (5).

Outline of the proof of Theorem 1.2 : In a first step, we want to find critical values and critical points of V over the sphere

$$M = \{T(u) = +1 \ , \ u \in \widehat{\mathcal{D}}^{1,2}(\mathbb{R}^N)\} \quad ;$$

(T is a Hilbert norm over $\widehat{\mathcal{D}}^{1,2}(\mathbb{R}^N)$, see [20]).

And this will be achieved by a proper generalization of Liusternik-Schnirelman type results [22] or Krasnoselski variational methods [19]. Let us just recall that an even C^1 functional over a sphere in some n-dimensional space has at least n pairs of distinct critical points. This can be generalized (with some compactness assumptions) to infinite dimensional Hilbert spaces : see [22] [19],[3],[27]. Using a convenient new critical point result, we prove (see [5],[6]) that under the assumptions of Theorem 1.2, there exists a sequence $(v_k)_{k\geq 1}$ of critical points of $V_{|M}$ satisfying :

$$\begin{cases} \theta_k V'(v_k) = T'(u_k) \quad \text{or} \quad -\Delta v_k = \theta_k \, g(v_k) \ , \ \theta_k > 0 \\ V(v_k) = b_k > 0 \quad \text{and} \quad b_k \underset{k \nearrow \infty}{\searrow} 0 \ . \end{cases}$$

But Prop.1.1 implies that we have, in addition,

$$\frac{2N}{N-2}\,\theta_k b_k = +1 \quad .$$

Now, if we introduce $u_k = v_k(x/\sqrt{\theta_k})$, $(u_k)_{k\geq 1}$ satisfies the properties listed in Theorem 1.2. In particular, one has

$$S(u_k) = \frac{1}{N}\,(b_k)^{-\frac{(N-2)}{2}}\left(\frac{N-2}{2N}\right)^{\frac{(N-2)}{2}}$$

. For a complete proof, the reader is referred to [6].

2. A LOCAL METHOD

We want to investigate now the following (natural) approach to (2) : first, solve, instead of (2) :

$$-\Delta u_R = g(u_R) \ \text{in} \ B_R \ , \quad u_R = 0 \ \text{on} \ \partial B_R \ , \quad u_R \not\equiv 0 \qquad (2.R)$$

and then pass to the limit as $R \to +\infty$.

Solving (2.R) is more or less well known (see for example [2],[3],[15],[14]) but the passage to the limit presents essentially two difficulties : first we need to have uniform (in R) bounds on the solutions u_R , second we need to make sure that u_R does not converge as $R \to \infty$ to 0 or to any other solution of (1) "which is not zero at infinity" (for example constants β such that $g(\beta)=0$). These two difficulties are overcome in [7] (see also [8]) at least for positive solutions and require new results on problems like (2.R). Let us illustrate our methods on two model cases.

Theorem 2.1 : *(The bounded case). Let* $g(s)=\lambda|s|^{p-1}s-\mu|s|^{q-1}s-ms$ *with* $\lambda,\mu,m>0$; $1<p<q$. *Then if (8) is not satisfied, (2.R) has no solutions for all* R. *On the other hand, if (8) is satisfied, then there exists a "minimal radius"* $R_o>0$ *such that we have*

i) for all $R<R_o$, *there is no solution of (2.R) ;*

ii) if $R\geq R_o$, *there exists a maximum solution* $\bar{u}_R$ *of (2.R) which is radial and satisfies*

$$\bar{u}_R>0 \text{ in } B_R\ ,\ \bar{u}_R \text{ is decreasing in } B_R\ ,\ \bar{u}_R\in C^2(\bar{B}_R)\ . \tag{19}$$

Furthermore, we have, if we extend $\bar{u}_R$ *by* 0 *in* $\mathbb{R}^N\setminus B_R$

$$\bar{u}_R(x)\ <\ \bar{u}_{R'}(x) \quad \text{if} \quad R<R' \tag{20}$$

$$\bar{u}_R(x)\nearrow\beta \quad \text{as} \quad R\nearrow+\infty\ ,\ \text{for every } x \text{ in } \mathbb{R}^N \tag{21}$$

where β *is the maximum positive solution of* $g(\zeta)=0$.

iii) if $R>R_o$, *there exists another solution* u_R *of (2.R) which is distinct from* u_R , *radial and satisfies (19) and*

$$u_R(\xi_R)\ <\ u_{R_0}(\xi_R)\quad ,\ \text{for some } \xi_R \text{ in } B_{R_o} \tag{22}$$

Furthermore, if we extend u_R *by* 0 , *there exists a subsequence* u_{R_k} *which converges (locally in every Sobolev space) to a solution* u *of (2) satisfying (9), (10), (11).*

Theorem 2.2 : *(The superlinear case). Let* $g(s)=\lambda|s|^{p-1}s-\mu s$ *with* $\lambda,\mu>0$; $1<p$. *Then if* $N\geq 3$ *and if* $p\geq\ell=(N+2)/(N-2)$, *(2.R) has no solutions for all* R . *On the other hand, if* $p<\ell$ *(when* $N\geq 3$*), for all* $R_o>0$ *there exists a positive solution* u_R *of (2.R) for all* R *which is radial and which satisfies (19) and :*

$$u_R(\xi_R)\ <\ u_{R_o}(\xi_R)\quad ,\ \text{for some } \xi_R \text{ in } B_{R_o} \text{ and for } R>R_o$$

$$\|u_R\|_{L^\infty(B_R)}\ \leq\ \text{Constant (independent of R)}\ ,\ \text{for } R>R_o\ . \tag{23}$$

Furthermore, if we extend u_R *by* 0 , *there exists a subsequence* u_{R_k} *which converges (locally in every Sobolev space) to a solution* u *of (2) satisfying (9), (10), (11).*

These two results are proved in [7] (see also [8]) and are

only special cases of the results proved there. By this local approach, we are able to treat a class of nonlinearities g almost as general as in Theorem 1.1 and this enables us to treat the case when N=2 (see also Section 3).

Remark 2.1 : These two results clearly show a difference between two types of nonlinearities (bounded in Theorem 2.1 : g^+ is bounded on $\mathbb{R}_+$; unbounded or superlinear in Theorem 2.2). Indeed in the first case there is a minimal radius R_o and for $R > R_o$ there are (at least) two positive solutions, on the other hand in the second case there is no minimal radius and we conjecture there exists a unique positive solution for all R (this uniqueness result is partially proved in [10], with the restriction R large).

Remark 2.2 : The terminology "minimal radius" is motivated by Physics arguments and such an R_o seems to be related to some form of Heisenberg uncertainty principle (see [28] for a detailed explanation and [32],[33] for estimates concerning R_o).

Remark 2.3 : Part *ii)* of Theorem 2.1 illustrates one of the difficulties encountered in this local approach : indeed $\bar{u}_R$ converges as $R \to \infty$ to β which is a solution of (1) but *not* of (2). Thus, we need to select another positive solution of (2.R) (which does not converge to β) : the method is then to prove the existence of a positive solution u_R of (2.R) which satisfies (22) : then, because of (20), $u_R \neq \bar{u}_R$ and (22) will prevent u_R to converge to β (see[7] for the detailed arguments). The existence of such a solution is obtained by an easy extension of [26], given in [7] and [8].

In Theorem 2.2, (22) is also useful to derive (23) (see [7]). Remark that (23) is a new estimate (actually in [7] all positive solutions of (2.R) are estimated), which is obviously fundamental to be able to pass to the limit as $R \to \infty$.

Remark 2.4 : In [7], we also explain how this local approach can be useful in the numerical computation of a positive solution of (2).

3. AN O.D.E. METHOD

The method studied here is to look directly for radial solutions of (2), thus for solutions of the following ordinary differential equation :

$$\begin{cases} -u'' - \dfrac{N-1}{r} u' = g(u) \quad \text{for} \quad r \geq 0 \\ u'(0) = 0 \;, \quad \lim_{r \to \infty} u(r) = 0 \end{cases} \tag{24}$$

Now, for any ζ in $\mathbb{R}$ we consider the solution $u(\zeta,r)$ of the following Initial Value Problem (IVP in short) :

$$\begin{aligned} -u'' - \frac{N-1}{r} u' &= g(u) \quad \text{for} \quad r \geq 0 \\ u(0) = \zeta \quad &, \quad u'(0) = 0 \quad . \end{aligned} \tag{25}$$

Therefore to find a solution u of (24) is equivalent to determine ζ such that $u(\zeta,r) \underset{r\to\infty}{\to} 0$.

This method is developed in [12] for positive solutions of (24) that is solutions of

$$\begin{aligned} -u'' - \frac{N-1}{r} u' &= g(u) \quad \text{for} \quad r \geq 0 \quad , \\ u'(0) = 0 \quad , \quad u(r) > 0 \quad &\text{for} \quad r \geq 0 \quad , \quad \lim_{r\to\infty} u(r) = 0 \end{aligned} \tag{24'}$$

We will not give any precise result concerning the existence of a solution to (24') and we refer to [12] for some simple O.D.E. argument allowing to treat a wide class of nonlinearities g . To conclude this section, we make a few remarks :

Remark 3.1 : The method works when N=2 , see [12].

Remark 3.2 : A general open question is to prove that the ground state (of Theorem 1.1) is unique (up to translations), or, more generally, that the solution of (24') is unique. The only partial result existing is for $g(u) = u^3 - u$ with N=3 (see [16]) and it is proved by O.D.E. techniques.(6)

Remark 3.3 : It seems reasonable to compute a positive solution of (2) by looking at (24') : one computes various solutions $u(\zeta,r)$ of (25) by standard methods for a discrete set of ζ and one selects the "good" solution. Unfortunately, this method is not very efficient since the dependence in ζ is quite unstable near the solution : some calculations obtained by R. Glowinski and Bristeaux at the I.R.I.A. are included in the appendix, showing qualitative properties of solutions $u(\zeta,r)$ of (25).

4. A BIFURCATION ANALYSIS

We just want in this section to prove a simple result concerning bifurcation in the following problem :

$$-\Delta u = g(u) - \lambda u \quad \text{in} \quad \mathbb{R}^N \quad , \quad u \in H^1(\mathbb{R}^N) \quad ; \tag{2'}$$

where we assume that $g(t) = o(t)$ as $t \to 0$. If we look for bifurcation from the trivial branch $(\lambda,0)$ of solutions of (2'), it is reasonable to look near (0,0) since $\lambda = 0$ is the supremum of the essential spectrum of $-\Delta$ in $\mathbb{R}^N$ (and such a critical value seems to play an important role in these kinds of bifurcation problems, see [24] for example).

Before giving a simple result, let us make a further remark : suppose we have bifurcation from (0,0) then, since (2') is invariant under translations and rotations, the bifurcating solutions do not form some curve or even some at most countable set of curves. To avoid this multiplicity phenomena, we seek solutions in $H^1_r(\mathbb{R}^N)$ which is the subspace of $H^1(\mathbb{R}^N)$ of all radial functions. We will denote also

$$\hat{\mathcal{D}}^{1,2}(\mathbb{R}^N) = \left\{ u \in L^{\frac{2N}{N-2}}(\mathbb{R}^N) \quad , \quad \nabla u \in L^2(\mathbb{R}^N) \right\}$$

equipped with its Hilbert scalar produce $\int_{\mathbb{R}^N} \nabla u.\nabla v \, dx$, and $X = \hat{\mathcal{D}}^{1,2}_r(\mathbb{R}^N) = \left\{ u \in \hat{\mathcal{D}}^{1,2}(\mathbb{R}^N),\ u \text{ is radial} \right\}$.

We have the following result in the case when $N \geq 3$:

Theorem 4.1 : *Let* $g \in C^1(\mathbb{R},\mathbb{R})$ *such that* $g(0) = g'(0) = 0$, g *is odd and* g *satisfies* :

$$\frac{G(t)}{t^{2+\varepsilon}} \text{ increases for } t \geq 0 \text{ and is positive for large } t \ , \text{ for some } \varepsilon > 0 \tag{26}$$

$$\lim_{t \to \infty} \frac{g(t)}{t^{\ell}} = 0 \quad , \quad \ell = \frac{N+2}{N-2} \quad . \tag{27}$$

Then we have for all $\lambda > 0$:

i) there exists a sequence $(u^n_\lambda)_{n\geq 1}$ *of solutions of* (2) *satisfying* (9) *and* (11) *such that* :

$$\|u^n_\lambda\|_X \underset{n\nearrow+\infty}{\nearrow} +\infty \ .$$

ii) furthermore, (u^n_λ) *satisfies, for each fixed* $n \geq 1$,

$$u^n_\lambda \underset{\lambda \to 0}{\to} 0 \quad \text{in } X \ .$$

In particular there is bifurcation from (0,0) *in the problem* (2').

Remark 4.1 : Part *i*) of Theorem 4.1 is included in Theorem 1.2. The requirement that g is odd is used in order to prove *i*), if we do not assume that g is odd, we can prove the existence for all $\lambda > 0$ of a positive solution u_λ which converges to 0 in X as $\lambda \to 0$: this gives bifurcation from (0,0) of the positive solutions.

We have included the statement (i) in the theorem above, since we consider it as a first step in a conjecture we make : we conjecture that there exists a sequence of curves $\mathcal{C}_n$ emanating from (0,0), each point (λ, u_λ^n) on $\mathcal{C}_n$ being characterized as follows :

$\pm u_\lambda^n$ is the unique radial solution with (n-1) nodes ($\pm$ means that the uniqueness is true except for a change of sign).

We have obtained in that direction some very partial results, see [10].

Remark 4.2 : Assumption (26) is purely technical and other more general conditions could be given, but we think some assumption on g is needed since in some cases (see [6]), there exist non-trivial solutions for $\lambda=0$ and there, bifurcation may only be "vertical" (i.e. in the slice $\lambda=0$ of $\mathbb{R}\times X$), or may not exist at all. (A more precise description in this "zero mass" case is an open question).

Remark 4.3 : It is worth noting that bifurcation occurs in the space X but may not occur in the space $H^1(\mathbb{R}^N)$: indeed a simple scaling argument shows that bifurcation occurs in $H^1(\mathbb{R}^N)$ with $g(t) = |t|^{p-1}t - t$ only for some values of p (between 1 and 6) but occurs in X for all values of p (between 1 and 6). Bifurcation in the space $H_r^1(\mathbb{R}^N)$ for $g(t) = |t|^{p-1}t - t$ is considered in [31].

Proof of Theorem 4.1 : In [5], we show that for all $\lambda > 0$ there exists solutions $(v_\lambda^n)_{n\geq 1}$ of

$$-\Delta v_\lambda^n = \theta^n [g(v_\lambda^n) - \lambda v_\lambda^n] \text{ in } \mathbb{R}^N, \quad v_\lambda^n \in H_r^1(\mathbb{R}^N)$$

$$\int |\nabla v_\lambda^n|^2 = +1 \quad , \quad 0 < b_\lambda^n = \int G(v_\lambda^n) - \frac{\lambda}{2}(v_\lambda^n)^2 \underset{n \nearrow +\infty}{\searrow} 0 \quad .$$

Furthermore b_λ^n are known explicitly and it is easy to prove

$$b_\lambda^n \underset{\lambda \searrow 0}{\nearrow} +\infty \quad ;$$

and by Prop.1.1, we have $\frac{2N}{N-2}\,\theta_\lambda^n\, b_\lambda^n = +1$.

Now, let us define $u_\lambda^n(x) = v_\lambda^n(x/\sqrt{\theta_\lambda^n})$: $(u_\lambda^n)_{n\geq 1}$ is a sequence of solutions of (2) satisfying (9) and (11), and we have

$$\int |\nabla u_\lambda^n|^2 = \left(\theta_\lambda^n\right)^{\frac{N-2}{2N}} = \left(\frac{N-2}{2N}\,(b_\lambda^n)^{-1}\right)^{\frac{N-2}{2N}} \underset{n \nearrow +\infty}{\nearrow} +\infty \quad ;$$

and $\int |\nabla u_\lambda^n|^2 \underset{\lambda \searrow 0}{\searrow} 0$.

Therefore, the theorem is proved.

5. SYSTEMS

We want in this section to present (without any proofs) results concerning the following extension of (2) : find $u = (u_1,\ldots,u_m)$ solution of

$$\begin{aligned} -\Delta u_i &= g_i(u) \quad \text{in } \mathbb{R}^N , \quad u_i \in H^1(\mathbb{R}^N) , \text{ for } 1 \le i \le m \\ u &\not\equiv 0 \end{aligned} \tag{28}$$

where we assume that the nonlinearities g_i derive from some potential G :

$$\frac{\partial G(t)}{\partial t_i} = g_i(t) \quad , \text{ for all } t \text{ in } \mathbb{R}^m \quad .$$

Again (28) is motivated by problems from Mathematical Physics somewhat similar to those described in the Introduction : for example, existence of solitary waves in coupled non-linear Klein Gordon equations and (or) non-linear Schrödinger equations.

For the proofs (and generalizations) of the following results, the reader is referred to [11]. Let $N \ge 3$:

Theorem 5.1 : Let $g_i(t)$ be continuous functions from $\mathbb{R}^m$ into $\mathbb{R}$ such that $g_i(0) = 0$ and satisfying

$$\lim_{|t|\to\infty} \frac{g_i(t)}{|t|^\ell} = 0 \quad , \quad \text{with } \ell = \frac{N+2}{N-2} , \quad \text{for } 1 \le i \le m \tag{29}$$

$$g_i \text{ is differentiable at } 0 \text{ and } \left.\frac{\partial g_i}{\partial t_i}\right|_{t=0} < 0 \; ; \tag{30}$$

$$\text{there exists } G \in C^1(\mathbb{R}^m, \mathbb{R}) \; : \; \frac{\partial G}{\partial t_i} = g_i \text{ and } G(\zeta) > 0 \; , \text{ for some } \zeta \text{ in } \mathbb{R}^m \quad . \tag{31}$$

Then, there exists a solution $u = (u_1,\ldots,u_m)$ of (28) where u_i $(1 \le i \le m)$ satisfies (9) and (11) ; and such that

$$0 \le S(u) \le S(v)$$

for all solutions v of (28) in $L^\infty_{loc}(\mathbb{R}^N)$ satisfying $G(v) \in L^1(\mathbb{R}^N)$, where

$$S(v) = \int_{\mathbb{R}^N} \left\{ \sum_i \frac{1}{2} |\nabla v_i|^2 - G(v) \right\} dx$$

If we assume in addition that G is even, then there exists a sequence $(u_n)_{n\ge 1}$ of solutions of (28) satisfying (9), (11) and such that

$$S(u_n) \;;\; \int \sum_i \frac{1}{2} |\nabla u_{n,i}|^2 \;;\; \int G(u_n) \underset{n \nearrow +\infty}{\nearrow} +\infty \; .$$

Remark 5.1 : As in Section 1.2, it is possible to show that these assumptions are almost necessary, provided one assumes (31) (which is very natural in many situations from the physical point of view). For more details, the reader is referred to [11].

Remark 5.2 : For some special nonlinearities, it is possible to show that a ground state solution $u = (u_1, \ldots, u_m)$ satisfies necessarily :

$$u_i(x) > 0 \text{ in } \mathbb{R}^m \quad ; \quad u_i(x) < u_i(y) \text{ if } |x| > |y| \quad .$$

Remark 5.3 : In another context, existence results for a special system of the form (28) are proved in [21].

APPENDIX

In this appendix, we present some numerical computations based upon the O.D.E. method described in Section 3. These computations were obtained by R. Glowinski and Bristeaux at the I.R.I.A.(7) using standard schemes for solving the IVP (25) (see Section 3). The case considered here is N=3, $g(t)=t^3-\mu t^5-t$ with $\mu=1/16$ and various solutions of (25) are given corresponding to different ζ between 0 and $\beta=.97$ ($\alpha=.26$ is the other positive solution of $g(\zeta)=0$). Representations of these solutions are given in figures 1 to 8.

We just want to comment these graphs in a heuristical way and to make some conjectures on (25) (which seem to be natural). We start with $\zeta = 0$ and the corresponding solution $u \equiv 0$. As we increase ζ up to $\alpha = .26$, we find a solution which increases first, reaches a maximum above α and converges to α as $r \to \infty$ (with oscillations) remaining positive (see Fig. 1). Then for $\zeta = \alpha$, we have $u \equiv \alpha$; and for ζ in (α, ζ_1) the solution first decreases down to a positive minimum below α and then converges to α as $r \to \infty$ (with oscillations) remaining positive (see Figs.2&3). The value $\zeta = \zeta_1$ gives the ground state (or positive solution of (2)), an approximation of ζ_1 is $\zeta = .57$ (see Fig. 4) : that is a positive and decreasing solution.

For $\zeta > \zeta_1$, solutions of (25) cut the axis $u = 0$. More precisely for ζ in (ζ_1, ζ_2) solutions decrease down to a negative mini-

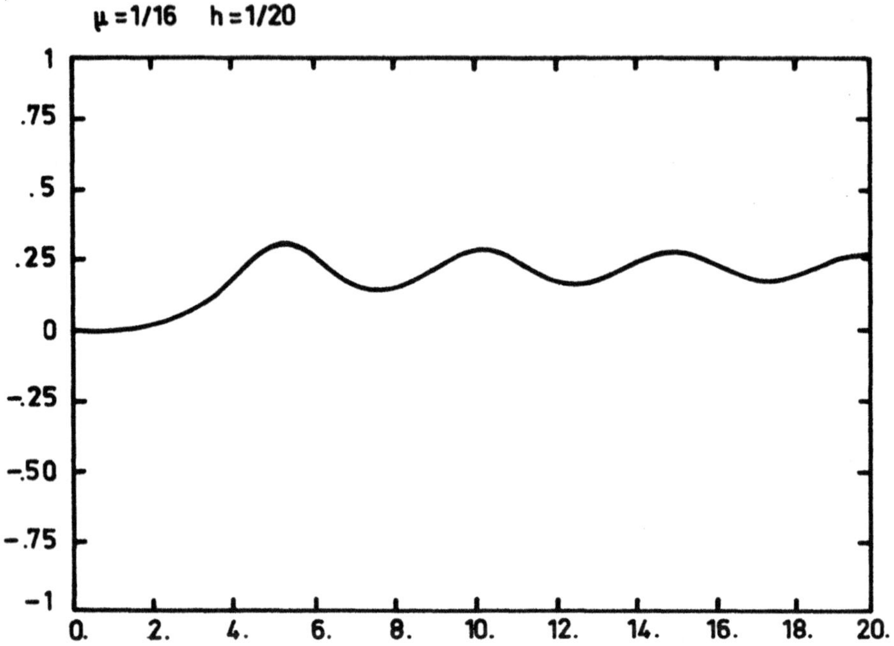

FIGURE 1

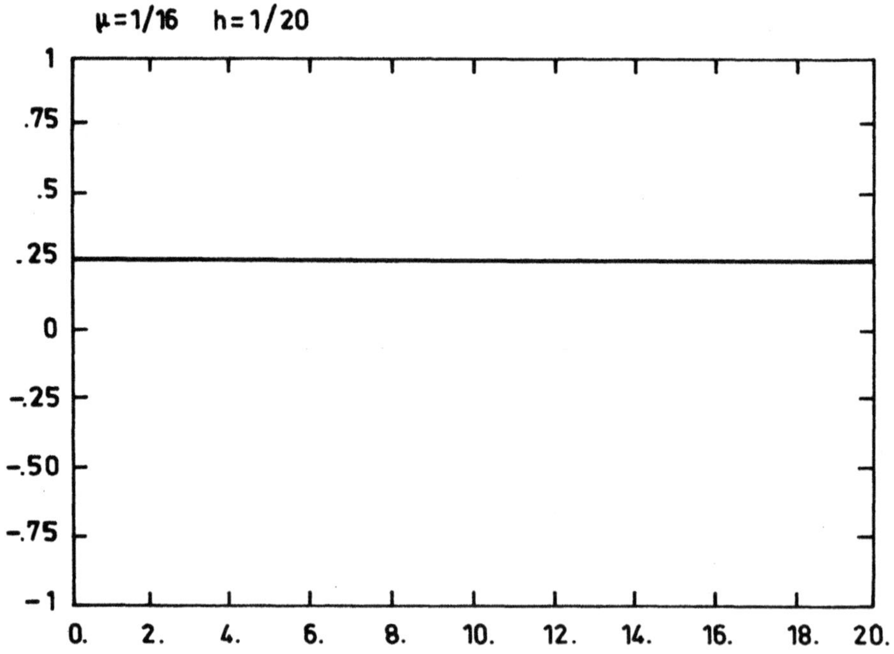

FIGURE 2

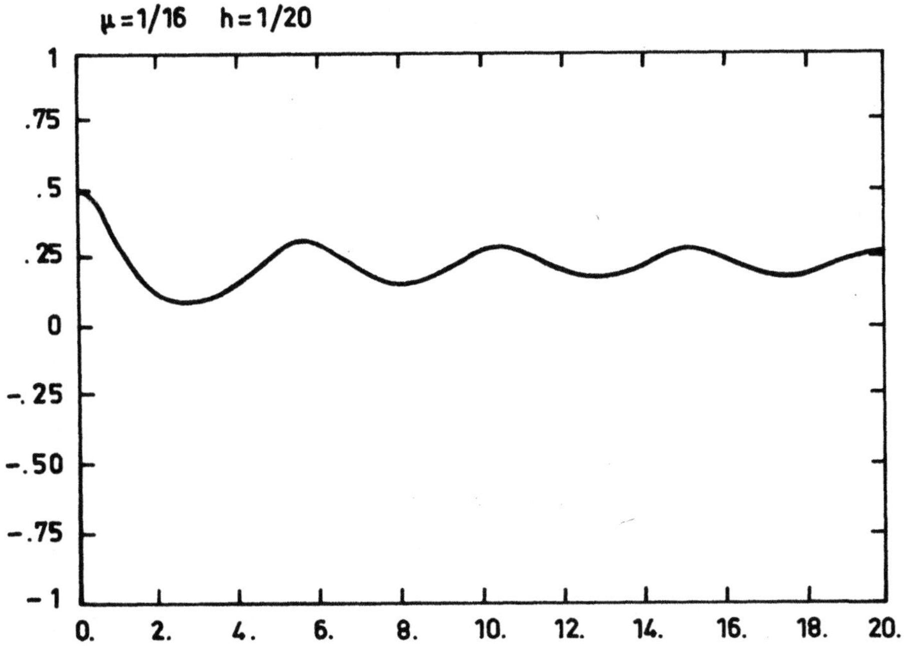

FIGURE 3

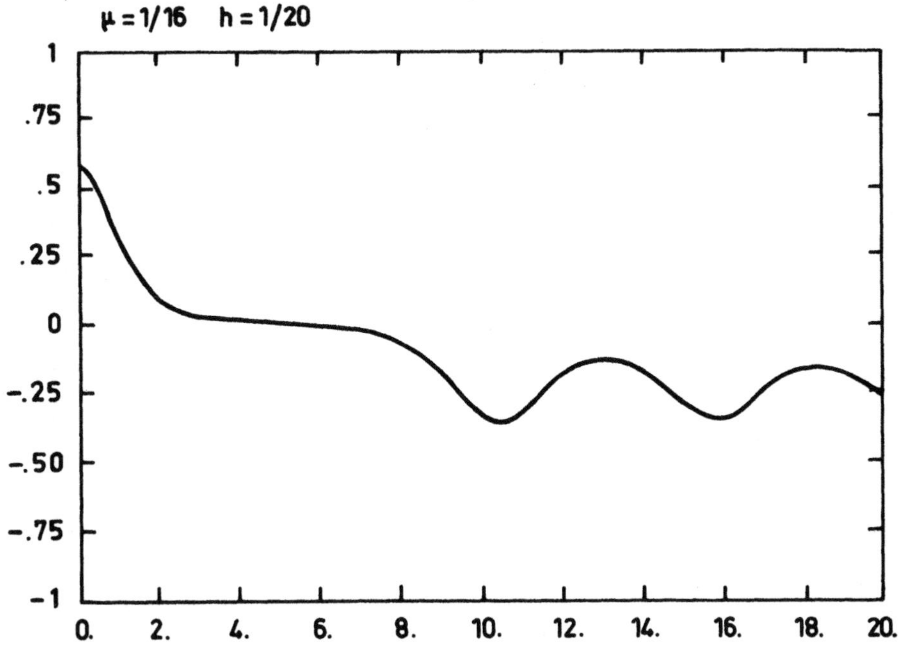

FIGURE 4

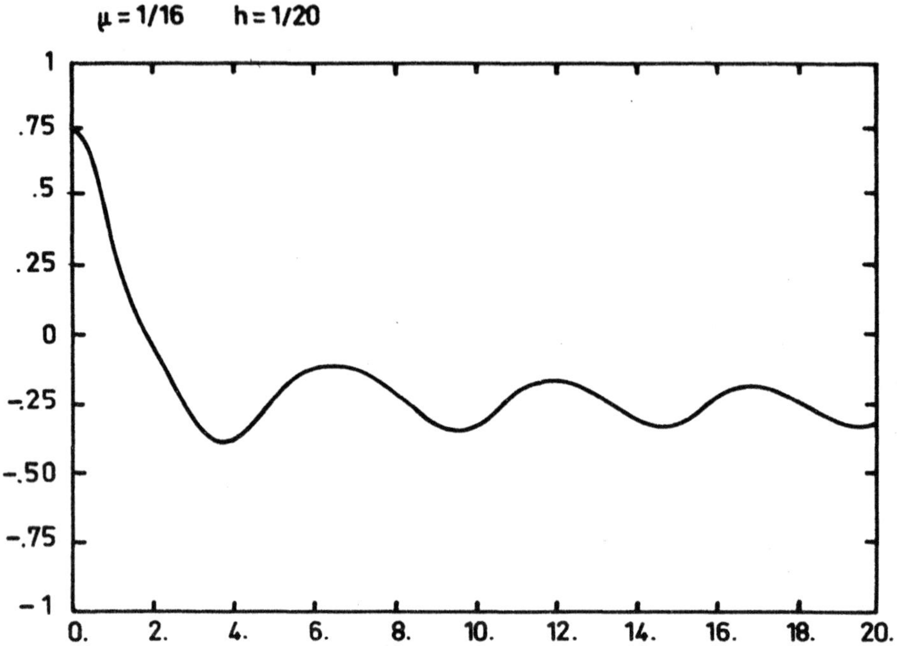

FIGURE 5

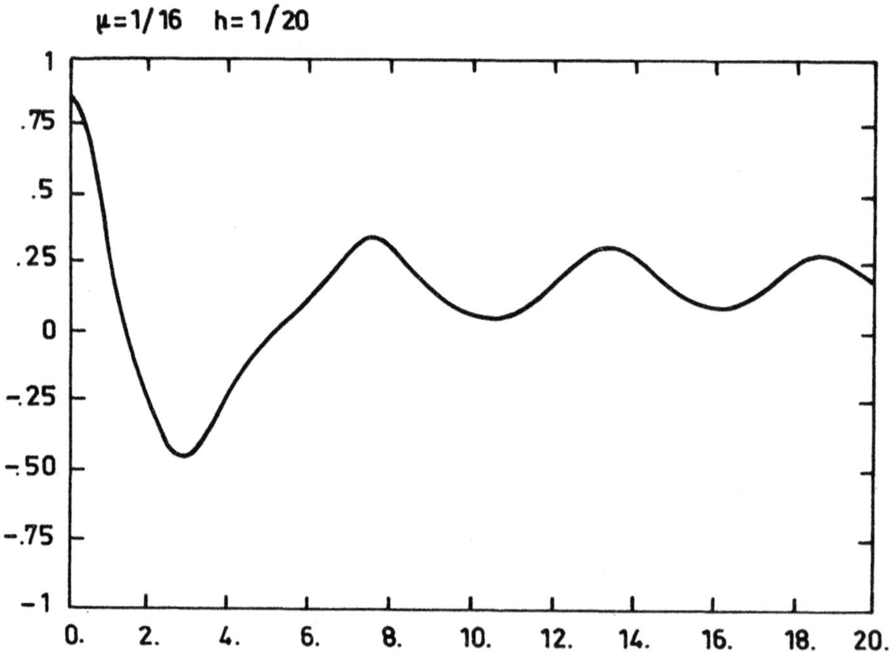

FIGURE 6

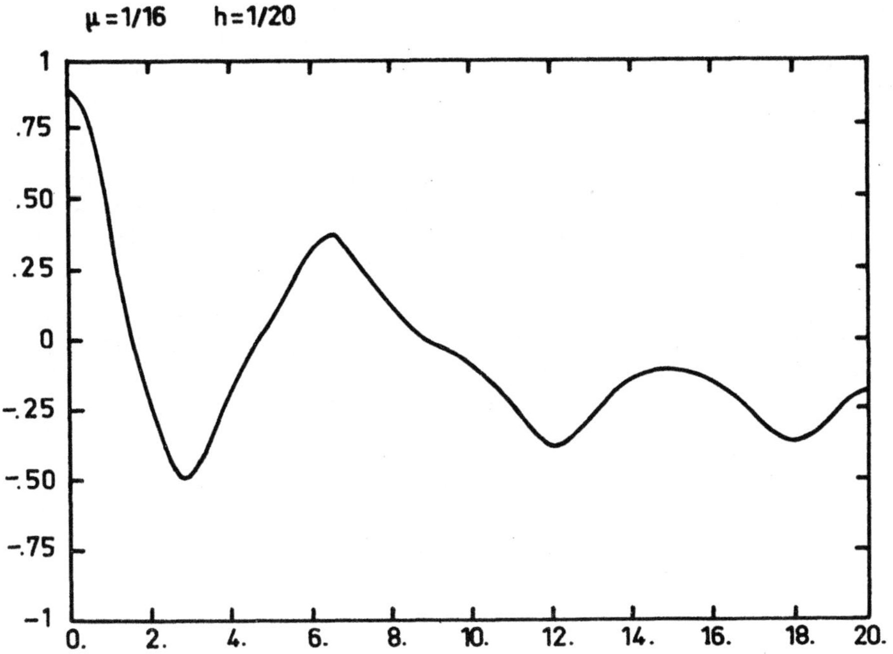

FIGURE 7

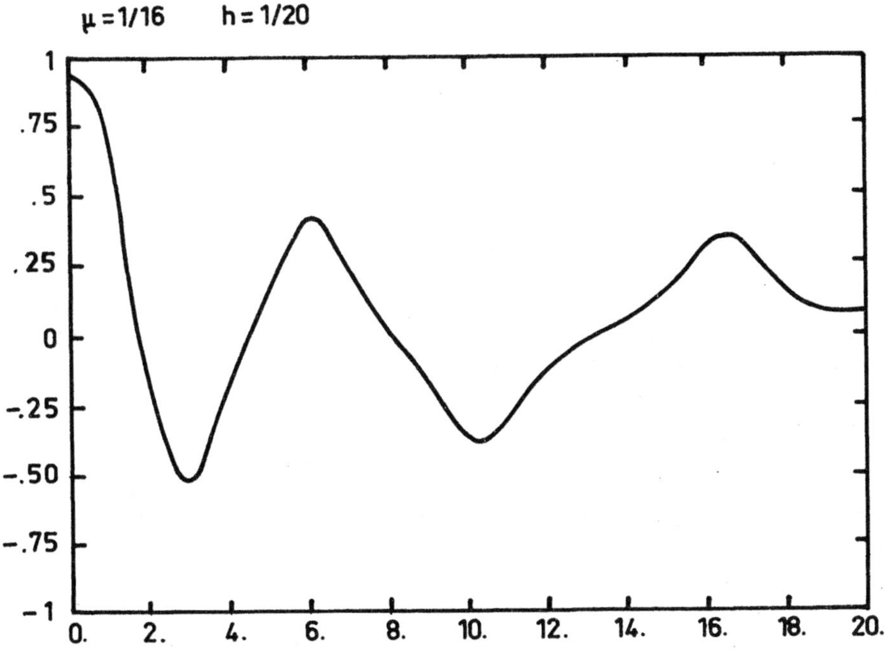

FIGURE 8

mum and then have the same behaviour (with a change of sign) as solutions for ζ in (α,ζ_1). The value $\zeta = \zeta_2$ gives the first non-positive solution of (2) (first "excited state") with exactly one node. For $\zeta > \zeta_2$, solutions of (25) have at least two nodes and increasing ζ more and more nodes appear with the same qualitative changes as between α and ζ_2 (see Figs.5 to 8). We have thus a sequence ζ_n such that $\alpha < \zeta_n < \beta, \zeta_n \nearrow \beta$ as $n \nearrow +\infty$ and the corresponding solutions u_n of (25) are solutions of (2) and are characterized by their number of nodes : u_n has (n-1) nodes. Finally for $\zeta > \beta$, the solution explodes in a finite time.

Of course, all of this is just numerical evidence and should be proved rigorously. We would like also to point out that these computations (for example of ζ_1) are somewhat difficult to make due to the instability of solutions of (2) with respect to ζ (see Table where the solution of (25) remains positive, decreasing for large r and finally cuts the axis). In conclusion, for numerical purposes, it seems more convenient to use the method described in [7] , based upon the local method given in Section 2. (One can find numerical computations of the ground state, in [7], which do converge rapidly).

NOTES

1 - $H^1(\mathbb{R}^N)$ is the Sobolev space of functions v satisfying $\int_{\mathbb{R}^N} |v|^2\,dx$, $\int_{\mathbb{R}^N} |\nabla v|^2\,dx$.

2 - The reader is referred to [6] for more detailed explanations.

3 - $B_R = \{\xi \in \mathbb{R}^N , |\xi| < R\}$.

4 - This argument is taken from [12].

5 - For the definition and basic properties of the space $\hat{\mathcal{D}}^{1,2}(\mathbb{R}^N)$, the reader is referred to [20].

6 - Very recently, other particular classes of nonlinearities g for which there is uniqueness of the ground state are obtained in Berestycki, Lions & Peletier (to appear).

7 - Rocquenquourt (France).

ACKNOWLEDGMENTS

The authors would like to thank Professors W.A. Strauss and H. Brézis for calling their attention on this type of problems. They are also indebted to Professor R. Glowinski and Mrs Bristeaux (IRIA) for the numerical examples produced in this appendix.

REFERENCES

[1] Agmon, S., Douglis, A. and Nirenberg, L. : (1959) *Comm. Pure Appl. Math.* 12, pp.623-727.
[2] Amann, H. : (1976) *SIAM Review* 18, pp.620-709.
[3] Ambrosetti, A. and Rabinowitz, P.H. : (1973) *J. Func. Anal.* 14, pp.349-381.
[4] Anderson, D. and Derrick, G. : (1970) *J. Math. Phys.* 11, pp.1336-1346 and (1971) *J. Math. Phys.* 12, pp.945-952.
[5] Berestycki, H. and Lions, P.L. : (1978) *Comptes Rendus Acad. Sci. Paris* 287, pp.503-506 and 288, pp.395-398.
[6] Berestycki, H. and Lions, P.L. :"Existence of solutions for non-linear scalar field equations". To appear.
[7] Berestycki, H. and Lions, P.L. : "A local approach to the existence of positive solutions of semilinear problems in $\mathbb{R}^N$. To appear.
[8] Berestycki, H. and Lions, P.L. : "Somme applications of the method of super and subsolutions. To appear in "*Journées de Villetaneuse*" (1979).
[9] Berestycki, H. and Lions, P.L. : "Existence of a ground state in nonlinear equations of the type Klein-Gordon". To appear in "*Variational Inequalities*", ed. Gianessi, Cottle and Lions, (1979) J. Wiley, New York.
[10] Berestycki, H. and Lions, P.L. : to appear
[11] Berestycki, H. and Lions, P.L. : to appear
[12] Berestycki, H., Lions, P.L. and Peletier, L.A. :"An O.D.E. approach to the existence of positive solutions of semi-linear problems in $\mathbb{R}^N$. To appear.
[13] Berger, M.S., (1972) *J. Func. Anal.* 9, pp.249-261.
[14] Brézis, H. and Turner, R.E.L. : (1977) *Comm. P.D.E.* 2(6), pp.601-614.
[15] Coffman, C.V. : (1969) *J. Analyse Math.* 22, pp.391-419.
[16] Coffman, C.V. : (1972) *Arch. Rational Mech. Anal.* 46, pp.81-95.
[17] Coleman, S., Glazer, V. and Martin, A. : (1978) *Comm. Math. Phys.* 58, pp.211-221.
[18] Kato, T. : (1959) *Comm. Pure Appl. Math.* 12, pp.403-425.
[19] Krasnoselski, M.A. : "*Topological methods in the theory of nonlinear integral equations*" (1964) MacMillan, New York.
[20] Lions, J.L. : "*Problèmes aux limites dans les équations aux dérivées partielles*", (1962) Presses de l'Université de Montréal.
[21] Lions, P.L. : "The Choquard equation and related questions" To appear.
[22] Liusternik, L.A. and Schnierelman, L.G. : "*Topological methods in the calculus of variations*", (1934) Hermann, Paris.
[23] Nehari, Z. : (1963) *Proc. Roy. Irish Acad.* 62, pp.117-135.
[24] Nicolaenko, B. (1977) in "*Applications of Bifurcation Theory*", P.H. Rabinowitz ed., Academic Press (New York).

[25] Pohozaev, S.I., (1965) *Sov. Math. Dokl.* 5, pp. 1408-1411.
[26] Rabinowitz, P.H. : (1973-74) *Indiana Univ. Math. J.* 23, pp.173-186.
[27] Rabinowitz, P.H. : "Variational methods for nonlinear eigenvalue problems", in "*Eigenvalues of nonlinear problems*", (1974) Rome : C.I.M.E ; Edizioni Cremonese.
[28] Rañada, A.F. and Vasquez, L. : "Kinks and the Heisenberg uncertainty principle", to appear in *Phys. Rev.*
[29] Ryder, G.H. : (1967) *Pacific J. Math.* 22, pp.477-503.
[30] Strauss, W. : (1977) *Comm. Math. Phys.* 55, pp.149-162.
[31] Stuart, C.A. : To appear.
[32] Vasquez, L. : (1977) *Lett. Nuovo Cimento*, 19 N.15, pp.561-564.
[33] Vasquez, L. : (1977) *Lett. Nuovo Cimento*, 19 N.1, pp.37-40.

IV

SOLITONS IN PHYSICS AND IN ALGEBRAIC GEOMETRY

SOLITONS : INVERSE SCATTERING THEORY AND ITS APPLICATIONS

R.K. Bullough†

NORDITA, Blegdamsvej 17
DK-2100, København Ø, Denmark

ABSTRACT. Five lectures on solitons are presented. The first summarises the discovery of the inverse scattering method for solving the initial value problem of the Korteweg de Vries equation by Kruskal and colleagues: the polynomial conserved densities of the KdV are introduced. The second lecture treats Bäcklund transformations and conserved densities, especially for the sine-Gordon equation, and extends the inverse scattering method to the 2 x 2 scheme of Zakharov and Shabat and of Ablowitz, Kaup, Newell and Segur. It is shown that all AKNS-ZS systems represent surfaces of constant negative Gaussian curvature. The geometrical analysis is used to derive non-local conserved densities for the sine-Gordon equation. In the third and fourth lectures the canonical structure of these systems is exhibited: all AKNS-ZS systems of the types considered are infinite dimensional completely integrable Hamiltonian systems. The AKNS-ZS scattering transformation is shown to be a canonical transformation. New canonical co-ordinates expressed in terms of scattering data are found. These are used for the semi-classical quantisation of these systems. The sine-Gordon equation is quantised explicitly and its eigen spectrum found. The relation of this to the eigen spectrum of the spin-$\frac{1}{2}$ x-y-z model of statistical mechanics and the massive Thirring model of massive fermions is sketched. In the final lecture the double sine-Gordon equations $u_{xx}-u_{tt} = \pm (\sin u + \frac{1}{2}\lambda \sin \frac{1}{2} u)$ which arise in resonant non-linear optics and in the theory of spin waves in ^{3}He below 2.6 mK are mentioned. The optical problem (+ ve sign) is used as a vehicle for singular perturbation theory about the sine-Gordon equation.

† On leave from Department of Mathematics, UMIST, P.O. Box 88, Manchester M60 1QD, U.K.

C. Bardos and D. Bessis (eds.), Bifurcation Phenomena in Mathematical Physics and Related Topics, 295–349.

1. INTRODUCTION: DISCOVERY OF THE INVERSE METHOD

These lectures are concerned with applications of soliton theory to certain problems in non-linear physics. However, as befits the mathematical character of this meeting there is a strong mathematical content in what I have to say and the physical applications I shall consider will be a bit incidental. I have in any case discussed elsewhere (1,2,3,4,5,6) a range of applications of soliton theory in physics which includes problems in non-linear optics and laser physics, plasma theory, solid state physics, low temperature physics, statistical mechanics, and particle physics. A book (7) devoted to the inverse scattering method of solving non-linear wave equations is to appear shortly: this book contains a number of contributions on the mathematics of the inverse method but also considers the applications of that theory to some particular physical problems and the interested reader is therefore referred to that material also.

For similar reasons I shall not repeat here the content of my first lecture as this was presented at the School. The material is already published (4,5). For present purposes that lecture made essentially one point — namely that there is a distinction between a solitary wave and a soliton. In drawing that distinction one can arrive at a working definition of a soliton.

Many equations like the Korteweg-de Vries equation (KdV)

$$u_t = 6u\,u_x - u_{xxx} \tag{1.1}$$

or the "ϕ-four" (or u-four) equation

$$u_{xx} - u_{tt} = \pm(u-u^3) \tag{1.2}$$

have solitary wave solutions. A solitary wave solution of (1.1) is

$$u = -2\eta^2\ \mathrm{sech}^2\left\{\eta(x-4\eta^2 t)\right\} \tag{1.3}$$

in which η is a parameter: the precise parametric form arises because the KdV is invariant under the scale transformation $u(x,t) \to \eta^2 u(\eta x,\eta^3 t)$. A solitary wave solution of (1.2) (+ ve sign) is

$$u = \pm\ \mathrm{sech}\left\{\frac{1}{\sqrt{2}}\ \frac{(x - Vt)}{\sqrt{1 - V^2}}\right\} \tag{1.4}$$

A 'kink' solution of (1.2) (- ve sign) of permanent profile or shape is

$$u = \pm \tanh \left\{ \frac{1}{\sqrt{2}} \frac{x - Vt}{\sqrt{1-V^2}} \right\} \tag{1.5}$$

A 'kink' solution of the sine-Gordon equation (s-G)

$$u_{xx} - u_{tt} = \sin u \tag{1.6}$$

is (the ± in (1.5) and (1.7) distinguish 'kinks' and 'antikinks', respectively)

$$u = 4 \tan^{-1} \exp \pm \left\{ \frac{x - Vt}{\sqrt{1 - V^2}} \right\} \tag{1.7}$$

However, only (1.3) and (1.7) are solitons. These have the important property that they do not interact in collision with other similar solitons. Such a property could not have been expected of the solutions of non-linear wave equations like (1.1) and (1.6). Indeed one can interpret the collision of two solutions of the forms (1.3) or (1.7) as one in which one solution of high speed (larger η^2 or larger V) "passes through" the one with smaller speed. I shall take this as the characteristic property of a soliton. It is important really only because it is also characteristic of solutions of a large class of non-linear wave or evolution equations which can be solved by an inverse scattering method. A non-linear evolution equation (NEE) is of the form $u_t = K[u]$ in which $K[u]$ is a non-linear functional of u and its derivatives u_x, u_{xx}, u_{xxx}, etc. The KdV (1.1) has this form: the s-G in light-cone coordinates

$$u_{xt} = \sin u \tag{1.8}$$

has this form in the dependent variable u_x (the metric is (-1,1), otherwise the right side is $-\sin u$).

Since this is a fairly mathematically minded meeting it should in any case be sufficient to launch straight away into the inverse scattering method itself. Different authors have different approaches to the method (see for example (8)). To my mind one of the most direct is that actually found by Kruskal and colleagues (9) in the first place. I give here an account of the discovery of the method for solving the KdV as I have heard it directly from Kruskal (10). Naturally any errors or misunderstandings in the presentation of this account are entirely mine. The material is substantially that I presented in (7).

In the period 1955-65 Kruskal was concerned with the recurrences exhibited by one-dimensional anharmonic lattices — the Fermi-Pasta-Ulam (FPU) problem (11). This led Zabusky and Kruskal (12) to study the equation obtained for a cubic non-linearity in the continuum limit

$$y_{tt} = y_{xx}(1 + \epsilon y_x) \quad . \tag{1.9}$$

The parameter ϵ scaling the nonlinearity $\sim 1/10$. By looking at the Riemann invariants of the linear problem $y_{tt} = y_{xx}$, namely

$$u = \tfrac{1}{2}(y_x + y_t) = O(1), \quad v = \tfrac{1}{2}(y_x - y_t) = O(\epsilon)$$

they reduced (1.9) to the "simple wave"

$$u_\tau + u\,u_x = 0 \qquad (\tau = 4\epsilon\,(t-x)). \tag{1.10}$$

Characteristics of this equation are $\dfrac{dx}{d\tau} = u = \text{constant}.$

Intersection of these suggests that the system shocks. It may be integrated by the hodograph transformation to yield the solution $u = u(x-ut)$ which shocks (the speed u is large where the disturbance u is large). An element of dispersion was next added by returning to the discrete lattice model: the result was

$$y_{tt} = y_{xx}(1 + \epsilon y_x) + \frac{\Delta x^2}{12}\, y_{xxxx}, \tag{1.11}$$

where Δx was the lattice spacing and from this

$$u_\tau + u\,u_x + \delta^2 u_{xxx} = 0 \tag{1.12}$$

where $\delta^2 \equiv \Delta x^2/12\epsilon$. For $\Delta x = 1/64$, $\epsilon \sim 1/10$, $\delta \sim 0.022$.

Numerical solutions of (1.12), which is a Korteweg de Vries equation, showed (12) that, under periodic boundary conditions, harmonic initial data developed a smooth jump approaching the simple wave shock for the chosen small value of δ^2. Thereafter the profile broke into spikes the collision properties of which caused Zabusky and Kruskal (12) to call them solitons. In order to characterize the solution across the jump Kruskal and co-workers then looked for local conservation laws. These take the form

$$\frac{\partial T}{\partial t} + \frac{\partial X}{\partial x} = 0 \tag{1.13}$$

in which T is a conserved density and X is a flux. Typically $\int_{-\infty}^{\infty} T\,dx$ is a constant of the motion by appeal to the boundary conditions on u. Roughly u is exponentially vanishing at $x = \pm\infty$.

Equation (1.12) is already in the form (1.13) with density $T^1 = u$. However, $T^2 = u^2$ and $T^3 = 2u^3 - u_x^{\ 2}$ are also conserved densities of the KdV

$$u_t + u\,u_x + u_{xxx} = 0\,. \tag{1.14}$$

Miura, Gardner and Kruskal (13) found 10 such <u>polynomial</u> conserved densities for (1.14) and proved that there was an infinity of non-trivial polynomial conserved densities T^r. Each term $\Pi\, u_j^{\,a_j}$ in T^r (where u_j is the x-derivative of order j) has "rank" $r = \sum_j (1 + \tfrac{1}{2}j)\, a_j$: $T^{10} = \frac{1}{10} u^{10} - 36 u^7 u_1^{\ 2} \ldots + \frac{419904}{1255} u_8^{\ 2}$ has rank 10 and contains 32 terms (13).

FPU also looked at a lattice with quartic rather than cubic anharmonicity. This leads to $y_{tt} = y_{xx}(1 + \epsilon y_x^{\ 2})$ and thence to the modified KdV

$$v_t + v^2 v_x + v_{xxx} = 0 \tag{1.15}$$

This also has conservation laws. Miura (14) found the transformation $u = v^2 + \sqrt{-6}\, v_x$ connecting these. The $\sqrt{-6}$ is innocuous and may be scaled away. The Miura transformation

$$u = v^2 + v_x \tag{1.16}$$

is a part of the "Bäcklund transformation" found subsequently by Wahlquist and Estabrook (15) scaled in this form to connect the KdV and modified KdV equations in the forms (1.1) and

$$v_t - 6v^2 v_x + v_{xxx} = 0\,. \tag{1.17}$$

We look at Bäcklund transformations and conserved densities again in § 2. The reader should have no problem in seeing here at least that scaling of x, t and possibly u changes the KdV from (1.1) to (1.12) to (1.14); that similar scalings are possible for the modified KdV; but that these two equations need some non-linear transformation like (1.16) to map between them.

Next we note that the Riccati transformation $v = \partial(\log\psi)/\partial x$

linearises $v^2 + v_x$ and from (1.16) $u = \psi_{xx}/\psi$. The Galilean transformation

$$x' = x - Vt \ , \ t' \to t \ , \ u \to u + V$$

leaves e.g. $u_t - u\,u_x + u_{xxx} = 0$ invariant. Hence $u \to u + \lambda = \psi_{xx}/\psi$ and the Schrödinger eigenvalue problem

$$-\psi_{xx} + u\,\psi = -\lambda\,\psi \tag{1.18}$$

has emerged. Furthermore $u = \psi_{xx}/\psi - \lambda$ into the KdV (1.1) yields (9)

(a) $\lambda_t = 0$, i.e. the eigenvalues are constants of the motion for the bound states ($\lambda > 0$) even though u depends on t ;

(b) $$\psi_t + \psi_{xxx} - 3(u-\lambda)\psi_x = C\psi + D\psi \int^x \frac{dx'}{\psi^2} \tag{1.19}$$

and (9) D vanishes for $\lambda > 0$, whilst C vanishes through the normalisation of the bound state eigenfunctions. From the property $u \to 0$ as $|x| \to \infty$ the eigenfunction ψ_n belonging to $\lambda_n > 0$ takes the form

$$\psi_n \sim c_n(t) \exp(-k_n x) \quad \text{for} \quad x \to \infty \tag{1.20}$$

($k_n = \lambda_n^{\frac{1}{2}}$) and from (1.19) $c_n(t) = c_n \exp 4k_n{}^3 t$.

For $\lambda = -k^2 < 0$, a solution of (1.18) for large $|x|$ is a linear combination of $\exp \pm ikx$. If

$$\psi \sim \exp - ikx + b \exp ikx, \quad x \to \infty \tag{1.21a}$$

$$\psi \sim a \exp - ikx \qquad , \quad x \to -\infty \tag{1.21b}$$

$a(k,t)$ is a transmission coefficient and $b(k,t)$ a reflexion coefficient and $|a|^2 + |b|^2 = 1$. Since the spectrum for $\lambda < 0$ is continuous λ can be chosen so that $\lambda_t = 0$ here also.

Equations (1.21a) and (1.21b) in (1.18) yield $D = 0$, $C = 4ik^3$ and two equations which easily give

$$a(k,t) = a(k,o)$$

$$b(k,t) = b(k,o) \exp(8ik^3 t). \tag{1.22}$$

The key point now is that the scattering data a, b, c_n and λ_n are sufficient to allow reconstruction of u at any time t. Let $K(x,y)$ for $y \geq x$ be the solution of the Gel'fand-Levitan equation (16,17)

$$K(x,y) + B(x+y) + \int_x^\infty K(x,z)\, B(y+z)\, dz = 0 \qquad (1.23a)$$

with

$$B(\xi) \equiv \frac{1}{2\pi} \int_{-\infty}^{\infty} b(k) \exp(i k \xi)\, dk + \sum_n c_n^2 \exp k_n \xi \,. \qquad (1.23b)$$

Then

$$u(x,t) = 2 \frac{\partial}{\partial x} K(x,x) \,. \qquad (1.23c)$$

The procedure for solving the KdV equation (1.1) is therefore (i) map the initial data $u(x,0)$ into scattering data

$$S = \left\{ b\,(k,0),\ c_n(0) \text{ and } \lambda_n (n = 1,\ldots, N) \right\} ;$$

(ii) compute the time evolution of the scattering data as indicated above;

(iii) solve the Gel'fand-Levitan equation (1.23a) and calculate $u(x,t)$ $(t > 0)$.

The reader may care to check, for example, that for a single c_n and λ_n the solution is the 1- soliton solution (1.3). Note that this is a reflexionless potential, i.e. $b(k,t) = 0$; η in (1.3) is exactly the quantity $\lambda_n^{\frac{1}{2}}$, that is the bound state eigenvalues determine the speeds and amplitudes of the solitons.

In the next lecture we look at the Bäcklund transformations and conserved quantities associated with other NEE's and generalise the inverse scattering method to a 2×2 scattering problem.

2. MATHEMATICAL PROPERTIES OF SOME INTEGRABLE NEEs.

The inverse method for the KdV equation introduced in the last lecture solves the initial and boundary value problem $u(x,0) = f(x)$; $u(x,t)$, u_x, u_{xx} etc. $\to 0$, $|x| \to \infty$ for that equation. Such an equation is an "integrable system" in the technical sense that it is an infinite dimensional completely integrable Hamiltonian system (18). We explain more precisely what this means in the third lecture presented in § 4. In this § 2 we introduce other systems which have the same integrable character. We start with the Bäcklund transformation for the sine-Gordon equation (1.8), the s-G.

This transformation involving a single real parameter k connects two functions $u(x,t)$ and $u'(x,t)$ by

$$u'_x = u_x + 2k \sin \tfrac{1}{2} (u+u')$$

$$u'_t = -u_t + 2k^{-1} \sin \tfrac{1}{2} (u-u') . \tag{2.1}$$

The integrability condition $u'_{xt} = u'_{tx}$ implies both

$u_{xt} = \sin u$ and $u'_{xt} = \sin u'$. Thus the transformation (2.1)

is an auto-Bäcklund transformation (aBT) mapping a solution u of the s-G (1.8) into a second solution. Auto-BTs of similar character exist for the KdV, modified KdV and non-linear Schrödinger equations for example (7). The non-linear Schrödinger equation (NLS) is (19)

$$i u_t = u_{xx} + 2u |u|^2 \tag{2.2}$$

for a complex field $u(x,t)$. All these equations are integrable systems.

A BT which is not an auto-BT is

$$u'_x = k^{-1} \sin (u'-u) \tag{2.3a}$$

$$u'_t = u_t + k \sin u' . \tag{2.3b}$$

This maps a solution u of $u_{xt} = \sin u$ to a solution u' of

$$u'_{xt} = \left\{1 - k^2 u'^2_x\right\}^{\frac{1}{2}} \sin u' . \tag{2.4}$$

It can be used to find an infinity of conserved densities for the s-G (20,21).

We write

$$u' = \sum_{n=0}^{\infty} f_n(u)\, k^n \tag{2.5}$$

and find from (2.3a) that

$$u' = u + ku_x + k^2 u_{xx} + k^3(u_{xxx} + \frac{1}{3!} u_x^3) + k^4(u_{xxxx} + u_x^2 u_{xx}) + \ldots\ . \tag{2.6}$$

A local conservation law for (2.4) is

$$\left\{(1 - k^2 u'_x)^{\frac{1}{2}}\right\}_t - k^2(\cos u')_x = 0\ . \tag{2.7}$$

If (2.6) is inserted into (2.7) we get, after dropping constants and constant factors, the expression (2.8) for the conserved density in (2.7)

$$\begin{aligned} &u_x^2 + k(2u_x u_{xx}) + k^2(u^2_{xx} + 2u_x u_{xxx} + \frac{1}{4} u_x^4) \\ &+ k^3 (2u_x u_{xxxx} + 2u_{xx} u_{xxx} + 2u_x^3 u_{xx}) \\ &+ k^4 (u^2_{xxx} + 2u_{xx} u_{xxxx} + 2u_x u_{xxxxx} + \frac{1}{8} u_x^6 + \\ &+ 3 u_{xxx} u_x^3 + \frac{13}{2} u_x^2 u_{xx}^2) + \ldots\ . \end{aligned} \tag{2.8}$$

Since k is arbitrary the coefficients of each term k^r are conserved separately. Perfect differentials like $2u_x u_{xx}$ are trivially conserved since they may be moved from the density T to the flux X in any local conservation law of the form (1.13). Modulo certain perfect differentials we then find that a non-trivial set of conserved densities for the s-G is therefore

$$T^2 = \tfrac{1}{2} u_x^2,\quad T^4 = \tfrac{1}{2} u_{xx}^2 - \frac{1}{8} u_x^4,\quad T^6 = \tfrac{1}{2} u^2_{xxx} - \frac{5}{4} u_{xx}^2 u_x^2 + \frac{1}{16} u_x^6,$$

$$T^8 = \tfrac{1}{2} u^2_{xxxx} - \frac{7}{4} u^2_{xxx} u_x^2 + \frac{7}{8} u_{xx}^4 + \frac{35}{16} u_{xx}^2 u_x^4 + \frac{5}{128} u_x^8\ ,$$

$$T^{10} = \ldots\ . \tag{2.9}$$

Notice that these densities are again polynomials in u_x,

u_{xx}, etc. Every term $\prod_j u_j^{a_j}$ in T^r (with u_j the j-th partial derivative) has the same rank r now defined by $r = \sum_j j\, a_j$.

It is proved (21) that there is an infinite set of such non-trivial polynomial conserved densities T^r for each even integer r . A Bäcklund transformation can be used similarly to find the infinity of polynomial densities for the KdV described in § 1 (22).

We show later that the set (2.8) does not exhaust the important conserved densities of the s-G: there is an additional important infinite set of densities which is neither polynomial nor local. However polynomial densities are rather typical of the integrable systems and in the case of the KdV, modified KdV or NLS equations the polynomial densities exhaust the important densities. In practice this is easy to see only via the inverse scattering method. The method as described for the KdV in § 1 is not applicable to the modified KdV, NLS or s-G equations. The simplest scattering problem capable of solving these equations is 2 × 2 . This is the AKNS (23)- generalised Zakharov-Shabat (19) scattering problem. I introduce it now.

If we set $\Gamma = \tan\frac{1}{4}(u+u')$ in the aBT (2.1) then

$$\Gamma_x = \tfrac{1}{2} u_x (1 + \Gamma^2) + k\Gamma$$

$$\Gamma_t = k^{-1}\Gamma \cos u - (2k)^{-1}(1 - \Gamma^2)\sin u \ . \qquad (2.10)$$

The Riccati transformation $\Gamma = v_2 v_1^{-1}$ now yields

$$\begin{bmatrix} v_{1,x} \\ v_{2,x} \end{bmatrix} = \begin{bmatrix} -\frac{1}{2}k & -\frac{1}{2}u_x \\ +\frac{1}{2}u_x & +\frac{1}{2}k \end{bmatrix} \begin{bmatrix} v_1 \\ v_2 \end{bmatrix} \qquad (2.11)$$

$$\begin{bmatrix} v_{1,t} \\ v_{2,t} \end{bmatrix} = \frac{-1}{2k} \begin{bmatrix} \cos u & \sin u \\ \sin u & -\cos u \end{bmatrix} \begin{bmatrix} v_1 \\ v_2 \end{bmatrix} \qquad (2.12)$$

This combines a scattering problem (2.11) with a 'time evolution equation' (2.12) actually used by AKNS (24) to solve the s-G equation: the problem is exactly their form of the problem if

$k \equiv 2i\zeta$ and ζ is the eigenvalue of the scattering problem. AKNS (23) generalised the problem to scattering and time evolution equations

$$\hat{L}v = \zeta v \;, \quad \hat{L} = \begin{bmatrix} i\partial/\partial x & -iq \\ ir & -i\partial/\partial x \end{bmatrix} \tag{2.13}$$

$$\hat{A}v = v_t \;, \quad \hat{A} = \begin{bmatrix} A & B \\ C & -A \end{bmatrix} \tag{2.14}$$

in which v is a 2-component column vector and A, B, C, q, r are chosen so that (compare § 1), under $\zeta_t = 0$,

$$\hat{L}_t = [\hat{A}\,,\,\hat{L}] \tag{2.15}$$

read as an operator equation (the right side is the commutator) is the required pair of evolution equations

$q_t = K_1[q,r]$, $r_t = K_2[q,r]$ (in which the K_i are functionals of q,r).

To fix ideas we note that (2.15) component by component follows from the integrability condition $v_{tx} = v_{xt}$ and $\zeta_t = 0$. Moreover if $k = 2i\zeta$, $r = -q = \frac{1}{2}u_x$ and $A = \frac{i}{4\zeta}\cos u$, $B = C = \frac{i}{4\zeta}\sin u$, (2.13) and (2.14) are (2.11) and (2.12) whilst (2.15) is the s-G. One finds

$$[\hat{A}\,,\,\hat{L}] = -\frac{\sin u}{2\zeta}\begin{bmatrix} 0 & 1 \\ 1 & 0 \end{bmatrix}\begin{bmatrix} \partial/\partial x & \frac{1}{2}u_x \\ \frac{1}{2}u_x & -\partial/\partial x \end{bmatrix}$$

$$= i\,\frac{\sin u}{2}\begin{bmatrix} 0 & 1 \\ 1 & 0 \end{bmatrix} = i\begin{bmatrix} 0 & \frac{1}{2}u_{xt} \\ \frac{1}{2}u_{xt} & 0 \end{bmatrix} = \hat{L}_t \tag{2.16}$$

We now have an important result: the system (2.13) and (2.14) together with the condition $\zeta_t = 0$ is equivalent to the NEE's (2.15). The general form (2.15) as a single scalar operator for the KdV equation was first given by Lax (24) and pairs of operators like $\hat{L}$ and $\hat{A}$ are called 'Lax pairs'. The AKNS scheme is somewhat unusual as a Lax pair (7) in that in Lax's original approach both $\hat{L}$ and $\hat{A}$ were differential operators: in the AKNS scheme $\hat{A}$ depends on the eigenvalue ζ but of course (see (2.16)) the NEE does not. All the integrable systems seem to arise as integrability conditions on pairs of operators like $\hat{L}$ and $\hat{A}$.

Of course the equivalence of (2.13) and (2.14) with $\zeta_t = 0$ to the NEE (2.15) is not helpful unless as in § 1 they can be solved for $q(x,t)$ and $r(x,t)$. This can be done and the method is another 'inverse method'.

For given initial data $q(x,0)$, $r(x,0)$ we can solve the direct scattering problem (2.13). It is convenient to introduce the idea of scattering data. Consider the two Jost function solutions of (2.13) $\psi(x,t)$ and $\bar{\psi}(x,t)$. Set $\xi \equiv \mathrm{Re}\,\zeta$, $\eta = \mathrm{Im}\,\zeta$. Since $q, r \to 0$ as $x \to +\infty$, (the boundary condition — actually q, q_x, q_{xx}, etc. $\to 0$,

r, r_x, r_{xx} etc. $\to 0$ is imposed)

$$\psi \sim \begin{bmatrix} 0 \\ 1 \end{bmatrix} \exp\ i\xi x, \qquad \bar{\psi} \sim \begin{bmatrix} 1 \\ 0 \end{bmatrix} \exp\ -i\xi x \tag{2.17}$$

as $x \to +\infty$. Correspondingly there is a pair of Jost functions $\bar{\phi}$, ϕ asymptotic to the same forms as $x \to -\infty$. Since there can be only one linearly independent pair we have

$\phi = a\,\bar{\psi} + b\,\psi$, $\bar{\phi} = \bar{b}\,\bar{\psi} + \bar{a}\,\psi$ where $a,b,\bar{a},\bar{b}$ depend only on ξ

(and possibly t). Two solutions of (2.13) with row vectors

$\tilde{u} = (u_2,u_1)$, $\tilde{v} = (v_2,v_1)$ (the transposed form is chosen this way

to define the natural inner product $< u,v > \equiv \int (u_2\, v_1 + u_1\, v_2)\, dx$

of the theory — see especially (25)) have Wronskian $W(u,v) = u_1 v_2 - u_2 v_1$ independent of x . It is easy to show

$a = W(\phi,\psi)$, $b = -W(\phi,\bar{\psi})$, $\bar{a} = W(\bar{\phi},\bar{\psi})$, $\bar{b} = W(\bar{\phi},\psi)$ and $a\bar{a} - b\bar{b} = 1$.

(Our choice of signs in the asymptotic behaviour

$\bar{\phi} \sim \begin{bmatrix} 0 \\ +1 \end{bmatrix} \exp\ i\xi x$ as $x \to -\infty$ differs from that of (23) so

that the minus sign multiplies $b\bar{b}$ in the normalisation condition. Note that $|a|$ and $|b|$ are unbounded.) The quantities a and b can be analytically continued into the upper half ζ-plane, $\bar{a}$ and $\bar{b}$ into the lower half ζ-plane. From the asymptotic behaviour of ϕ, ψ and $\bar{\psi}$, a^{-1} is the transmission coefficient and $ba^{-1} \equiv w$ is the reflexion coefficient for the wave starting as $\bar{\psi}$ at $x \to +\infty$. The set $\{a, \bar{a}, b, \bar{b}\}$ (of which one quantity is redundant) constitutes a part of the scattering data.

The Jost solutions of (2.13) are unbound scattering states. In addition (as for the Schrödinger operator) there are bound states in general. These occur at the roots of $a(\zeta) = 0$ in the upper half of the complex ζ-plane: here $\phi = b(\zeta)\psi$, there is no 'input' wave, and the transmitted and reflected waves become a single bound state wave function. Typically the roots ζ_j of $a(\zeta) = 0$, which are the bound state eigenvalues, have a finite imaginary part and both ϕ and ψ are exponentially damped as $|x| \to \infty$. Similar remarks apply to the roots of $\bar{a}(\zeta) = 0$ in the lower half ζ-plane. Note that it is essential for bound states that $\hat{L}$ is <u>not</u> Hermitean ($r \neq q^*$): otherwise the eigenvalues ζ are real. It is this non-Hermitean character which influences the choice of inner product (see remark above).

It is convenient (see(7,23,25)) to write $b(\zeta_j) = i\, k_j\, \dot{a}\,(\zeta_j)$; $\bar{b}(\bar{\zeta}_j) = -i\, \ell_j\, \dot{\bar{a}}\,(\bar{\zeta}_j)$ $(\dot{f}\,(\zeta_j) \equiv (df/d\zeta)_{\zeta=\zeta_j})$. The set S of scattering data is then defined by

$$S = \left\{ \zeta_j,\ \bar{\zeta}_j,\ k_j,\ \ell_j,\ a(\xi),\ \bar{a}(\xi),\ w(\xi),\ \bar{w}(\xi)\ ; \right.$$
$$\left. \xi \in R\ ;\ j = 1,2,\dots, M;\ \ s = 1,2,\dots, N \right\}. \tag{2.18}$$

It is not the only possible choice (7) but it is one choice complete in the sense that S can be inverted by a Marchenko equation to yield $q(x,t)$ and $r(x,t)$.

Although $q = r^*$ is not an interesting case other symmetries can and do occur: for example if $r = -q$, $M = N$, $\bar{a}(\zeta) = a(-\zeta)$, $-\bar{b} = b(-\zeta)$, $\bar{\zeta}_k = -\zeta_k$, $\bar{b}(\bar{\zeta}_k) = -b(\zeta_k)$; if $r = -q^*$, $\bar{a}(\zeta) = a^*(\zeta^*)$, $\bar{b}\,(\zeta) = b^*\,(\zeta^*)$, $\bar{\zeta}_k = \zeta_k{}^*$, $\bar{b}(\bar{\zeta}_k) = \{b(\zeta_k)\}^*$. In the case of the s-G both conditions obtain since $q = -r$ and $q = -\frac{1}{2}u_x$ is real. The bound state eigenvalues therefore arise on the imaginary

ζ-axis ($\zeta_k = i\eta_k$, $\bar{\zeta}_k = -i\,|\bar{\eta}_k|$) or they arise as pairs $(\zeta_k, -\zeta_k^*)$ with $(\bar{\zeta}_k, -\bar{\zeta}_k^*) = (\zeta_k^*, -\zeta_k)$. The bound state eigenvalues $\zeta_k = i\,\eta_k$ determine the speeds and steepness of the kink solutions (1.7) associated with each eigenvalue (the kink is $4\tan^{-1}\exp\{-(k_j x + k_j^{-1}t)\}$ with $k_j = 2i\zeta_j$ for the s-G (1.8) in light cone co-ordinates) the pairs $(\zeta_k, -\zeta_k^*)$ determine the breathers; the continuous part of the spectrum with eigenvalues ξ determines a 'background' or 'radiation' content of the solution about which we have more to say later.

From the scattering data at $t = 0$ one can find the scattering data at the current time $t > 0$ very much as in the KdV Schrödinger case of § 1. But now $\hat{A}$ describes all this. Conditions on A, B, C in $\hat{A}$ are $B, C \to 0$, $A \to \Omega(\zeta)$ as $|x| \to \infty$; for example $A = \frac{-1}{2k} = -\frac{1}{4i\zeta}$ and $B = C = 0$ when $u = 0 \pmod{2\pi}$, in (2.12). Thus $\Omega(\zeta) = -\frac{1}{4i\zeta}$ for the s-G (1.8) and this is related to the linearised dispersion relation $\omega(k)$ of that equation by $\Omega(\zeta) = \frac{-i}{2}\,\omega(-2\zeta)$ (25). The dispersion relation $\Omega(\zeta)$, and thereby $\omega(k)$, plays an important role in the class of integrable equations as we show in the next lecture presented in § 4. In particular for the AKNS scattering data S we have here for $x \to \infty$ that

$$(\hat{A} - \Omega(\zeta)\,\hat{I})\phi \sim \begin{bmatrix} 0 & 0 \\ 0 & -2\Omega(\zeta) \end{bmatrix} \begin{bmatrix} a(\xi)e^{-i\xi x} \\ b(\xi)e^{i\xi x} \end{bmatrix} = \begin{bmatrix} a_t(\xi)e^{-i\xi x} \\ b_t(\xi)e^{i\xi x} \end{bmatrix} . \tag{2.19}$$

We can always add $-\Omega(\zeta)\,\hat{I}$ ($\hat{I}$ is the unit matrix) to $\hat{A}$ in (2.14) since this does not change the NEE (2.15). Indeed this choice is actually forced upon us by the need to satisfy the boundary conditions like (2.17) for the Jost functions. By this choice a is independent of t , i.e. <u>a is a constant of the motion</u>, and $b(\xi,t) = b(\xi,0)e^{-2\Omega(\xi)t}$. A similar argument for the bound states yields $b(\zeta_k,t) = b(\zeta_k,0)e^{-2\Omega(\zeta_k)t}$. Thus for the s-G $b(i\eta_k,t) = b_o e^{-t/2\eta_k}$ (with $\eta_k > 0$). Notice, that equations

(2.19) can be trivially integrated depends critically on the fact that $q, r \to 0$ as $x \to \infty$; in (2.13) q, r alone depend on t since the eigenvalues are by choice time independent; the precise simplification in (2.19) is achieved by making it so (23), but the vanishing of q and r asymptotically underpins the whole of the inverse scattering method and restricts soluble problems to the real line $-\infty < x < \infty$.

As in the KdV Schrödinger problem the crux of the argument is now to infer the potentials q, r at time t from the scattering data S at time t . This is done through the generalised Gel'fand-Levitan-Marchenko equations

$$\bar{\underset{\sim}{K}}(x,y) + \begin{pmatrix}0\\1\end{pmatrix} F(x+y) + \int_x^{\infty} \underset{\sim}{K}(x,z)\, F(z+y)\, dz = 0$$

$$\underset{\sim}{K}(x,y) + \begin{pmatrix}1\\0\end{pmatrix} \bar{F}(x+y) + \int_x^{\infty} \bar{\underset{\sim}{K}}(x,z)\, \bar{F}(z+y)\, dz = 0 \qquad (2.20a)$$

where (compare (1.23))

$$F(x) = (2\pi)^{-1} \int_{-\infty}^{\infty} w(\xi,t)e^{i\xi x}\, d\xi - i \sum_{j=1}^{M} k_j(t)e^{i\zeta_j x}$$

$$\bar{F}(x) = (2\pi)^{-1} \int_{-\infty}^{\infty} \bar{w}(\xi,t)e^{-i\xi x}\, d\xi + i \sum_{s=1}^{N} \ell_s(t)e^{-i\bar{\zeta}_s x} \qquad (2.20b)$$

and

$$r(x,t) = -2\bar{K}_2(x,x,t) \;;\; q(x,t) = -2K_1(x,x,t) \qquad (2.20c)$$

($\underset{\sim}{K}$ and $\bar{\underset{\sim}{K}}$ are column vectors with components K_1, K_2 and $\bar{K}_1$, $\bar{K}_2$.)

A derivation of this Marchenko equation for inverting the scattering problem (2.13) is given in Ref. (25). We shall use this paper, simply as a matter of convenience, as a general reference for these lectures. It will be clear already that the work of that paper owes much to previous work (especially (9,19,23)). A variety of inverse scattering problems are solved in (26). The inverse problem was already important in physics before its solution was applied to soliton theory since scattering potentials must usually be inferred from observed asymptotic scattering data. It is remarkable that the solution of the Schrödinger inverse problem could be seized upon and exploited by Kruskal and co-workers (9) in the fashion described in § 1 to solve the KdV equation.

As an example of the solution of (2.20) consider the single soliton problem which has

$F(x) = -ik_o\, e^{-2\Omega(i\eta)t}\, e^{-\eta x} = \bar{F}^*$. We have to solve

$$K_1(x,y) + i\,k(t)\,e^{-\eta(x+y)} + i\,k(t)\int_x^\infty \bar{K}_1(x,z)e^{-\eta(z+y)}\,dz = 0$$

$$\bar{K}_1(x,y) - i\,k(t)\int_x^\infty K_1(x,z)e^{-\eta(z+y)}\,dz = 0 \quad . \tag{2.21}$$

The solution is obtained by setting $K_1(x,y) = f(x)e^{-\eta y}$ and to satisfy (2.21) we then need

$$f(x) = i\,k(t)e^{-\eta x}\,[1 - \tfrac{1}{4}\,k^2(t)\,\eta^{-2}\,e^{-4\eta x}]^{-1} \,. \tag{2.22}$$

Thus

$$q(x,t) = -2K_1(x,x,t) = -2i\,k(t)e^{-2\eta x}[1 - \tfrac{1}{4}\,k^2\eta^{-2}e^{-4\eta x}]^{-1}$$

$$= 2\eta\,\mathrm{sech}\,[2\Omega(i\eta)t + 2\eta x + \delta] \tag{2.23}$$

in which η and δ can be treated as free parameters fixed by $q(x,0)$ ($\delta = \log -i\,k_o/2\eta$ and $k_o = k(i\eta,t=0)$ will be purely imaginary). The reader may care to find the 2-soliton solution similarly. Notice amazingly enough that the 1-soliton solution will always be a sech with speed $-\Omega(i\eta)/\eta$ and only the function $\Omega(i\eta) = \Omega(\zeta)$ relates the result to the original NEE. This leads us to expect that the most general equation soluble by the inverse scattering method in the AKNS-ZS formulation is determined by $\Omega(\zeta)$ alone. In fact one finds that a fairly general pair of NEEs solvable this way is (23,25)

$$\frac{\partial}{\partial t} \hat{\sigma}\, s(x,t) + 2\Omega(L^+)\, s(x,t) = 0 \tag{2.24a}$$

$$L^+ = \frac{1}{2i}\begin{bmatrix} \frac{\partial}{\partial x} - 2r\int_{-\infty}^{x} dx'\, q & 2r\int_{-\infty}^{x} dx'\, r \\ -2q\int_{-\infty}^{x} dx'\, q & \frac{-\partial}{\partial x} + 2q\int_{-\infty}^{x} dx'\, r \end{bmatrix}, \quad s = \begin{pmatrix} r \\ q \end{pmatrix}, \quad \hat{\sigma} = \begin{bmatrix} 1 & 0 \\ 0 & -1 \end{bmatrix} \tag{2.24b}$$

providing $\Omega(\zeta)$ is a simple ratio of entire functions. The reader may wish to express the s-G in the form (2.24) and derive (1.8).

In Ref. (25) it is shown, as we have already mentioned, that $\Omega(\zeta)$ plays a fundamental role in the description of NEEs as Hamiltonian systems. We describe this Hamiltonian formalism in the next lecture in § 4. In Ref. (25) two extensions of the theory are made: one allows $\Omega(\zeta)$ to depend on time t ; the other adds more time-like variables so that $\Omega \to \Omega(\zeta,\vec{y},t)$. The reader is referred to Ref. (25) for this work, especially the Appendix 3.

In the case of the s-G, $\Omega(\zeta) = -1/4\, i\, \zeta$ and $\Omega(i\eta) = (4\eta)^{-1}$. Thus (2.23) becomes

$$u_x = 4\eta \operatorname{sech}\left\{(2\eta)^{-1} t + 2\eta x + \delta\right\}$$

$$u = 4 \tan^{-1} \exp\left\{(2\eta)^{-1} t + 2\eta x + \delta\right\} \tag{2.25}$$

This is the 2π - kink solution (1.7) in light cone co-ordinates. The N-kink solution of the s-G (27) can be obtained similarly: the breathers come from pairs of eigenvalues $(\zeta,-\zeta^*)$.

There is a valuable method for finding conserved densities which uses the scattering problem (2.13). This method generalises that given by Zakharov and Shabat (19). The argument starts from the fact that $a(\zeta)$ and $\log a(\zeta)$ are constants of the motion (5,6, 7,25,28). From $a(\zeta) = W(\phi,\psi)$, $a(\zeta) \sim \phi_1\, e^{i\zeta x}$ as $x \to \infty$ (ϕ_1 is the 1-component of the Jost function ϕ). From (2.13)

$\Phi \equiv \log(\phi_1 \exp i\zeta x)$ satisfies (19,25)

$$2i\zeta\Phi_x = \Phi_x{}^2 - qr + q\frac{\partial}{\partial x}(q^{-1}\Phi_x). \tag{2.26}$$

As $|\zeta| \to \infty$,

$$\Phi_x \sim \sum_{n=1}^{\infty} \mathcal{H}_n (2i\zeta)^{-n} \quad . \tag{2.27}$$

So from (2.26)

$$\mathcal{H}_{n+1} = q\frac{\partial}{\partial x}(q^{-1}\mathcal{H}_n) + \sum_{j+k=n} \mathcal{H}_j \mathcal{H}_k \, . \tag{2.28}$$

Also as $x \to \infty$, $\log a(\zeta) \sim \Phi$ and since $a(\zeta) \to 1$ as $|\zeta| \to \infty$

$$\log a(\zeta) \sim \sum_{n=1}^{\infty} \zeta^{-n}(2i)^{-n}\int_{-\infty}^{\infty}\mathcal{H}_n\,dx = \sum_{n=1}^{\infty}\zeta^{-n} H_n \, . \tag{2.29}$$

Since $\log a(\zeta)$ is a constant of the motion the

$H_n \equiv (2i)^{-n}\int_{-\infty}^{\infty}\mathcal{H}_n\,dx$ are also constants of the motion. The

first few $\mathcal{H}_n$ prove to be (5,25)

$(2i)^{-1}\mathcal{H}_1 = -(2i)^{-1} qr$, $(2i)^{-2}\mathcal{H}_2 = (8)^{-1}(q r_x - q_x r)$,

$(2i)^{-3}\mathcal{H}_3 = -(8i)^{-1}(r_x q_x + q^2 r^2)$, $(2i)^{-4}\mathcal{H}_4 = (32)^{-1} \times$

$$\times \; (q_x r_{xx} - q_{xx} r_x - 3r^2 q q_x + 3q^2 r r_x) \, ,\ldots. \tag{2.30}$$

If we set $q = -r = -\frac{1}{2}u_x$, the $\mathcal{H}_{2n}$ vanish and the $\mathcal{H}_{2n+1}$ coincide with the T^{2n+2} of (2.9) (upto numerical factors).

We show in § 4 that we need in general a further set of

conserved quantities. The set (2.30) is a set of polynomial conserved densities. For the s-G there is a second set which is obtained in this case by expanding log $a(\zeta)$ about $\zeta = 0$. Although we have derived a density $\{1 - \cos u(x,t)\}$ this way (5,25), and this acts as a Hamiltonian for the s-G (see § 4), the integrations for higher rank densities became too complicated to handle through (2.26). We have recently devised a different approach (29) and I describe this now. The approach interprets every solution $u(x,t)$ of an NEE (2.15) as a surface. It is a remarkable and elegant fact that every AKNS-ZS system (2.15) describes a surface of constant negative Gaussian curvature if the metric is properly chosen.

3. GEOMETRY OF THE AKNS-ZS SYSTEMS

The material in this section is due to my colleague R. Sasaki (29,30). I shall use the language of exterior differential forms (31) since this allows a simple and transparent presentation. The symbols I use are "wedge" ($\wedge$) and d : d has the properties (31)

(i) $d(\omega + \eta) = d\omega + d\eta$ for any p-forms ω,η;

(ii) $d(\lambda \wedge \mu) = d\lambda \wedge \mu + (-1)^{(\deg \lambda)} \lambda \wedge d\mu$ for any p-form λ (degree $\lambda = p$) and q-form μ

(iii) for each ω, $d(d\omega) = 0$;

(iv) for each function f, $df = \sum_i \frac{\partial f}{\partial x^i} dx^i$.

In the case of the co-ordinate x^i, dx^i is the usual dx^i, and $d^2x^i = 0$. This property (iii) is the Poincaré Lemma or integrability condition (31,33). To see this consider a monomial p-form $\lambda = a\, dx^p$ where dx^p symbolizes any product of p 1-forms dx^i. Then by (iv) (31)

$$d(d\omega) = d \sum_i \frac{\partial a}{\partial x^i} dx^i\, dx^p$$

$$= \sum_{i \neq j} \frac{\partial^2 a}{\partial x^i \partial x^j} dx^j\, dx^i\, dx^p$$

$$= \frac{1}{2} \sum_{i \neq j} \left(\frac{\partial^2 a}{\partial x^i \partial x^j} - \frac{\partial^2 a}{\partial x^j \partial x^i} \right) dx^j\, dx^i\, dx^p$$

$$= 0 \ .$$

The statement is just that of the equality of the mixed second derivatives (the integrability conditions). The argument has used $d^2x^i = 0$. And it has also used a property of $\wedge$.

The properties of $\wedge$ are

(i) $\lambda \wedge \mu$ is distributive

(ii) $\lambda \wedge (\mu \wedge \nu) = (\lambda \wedge \mu) \wedge \nu$ (associative)

(iii) $\mu \wedge \lambda = (-1)^{pq} \lambda \wedge \mu$ for any p-and q-forms λ, μ.

In the argument above $dx^j dx^i$ denotes $dx^j \wedge dx^i$ (it is customary to omit the $\wedge$ here) and the argument uses the property (iii).

Notice that the interpretation as $dx^j \wedge dx^i$ is sensible: consider (for A,B,....,G functions of x,y,z)

$$(Adx + Bdy + Cdz) \wedge (Edx + Fdy + Gdz)$$

$$= (BG - CF)dy\, dz + (CE - AG)dz\, dx + (AF - BE)dx\, dy$$

which is the 'vector product' of the two 'vectors'. Notice that for any λ of odd degree $\lambda \wedge \lambda = -\lambda \wedge \lambda = 0$: this was used above for e.g. $dx \wedge dx = 0$. Other natural properties are discussed by Flanders (31).

The AKNS scattering and time evolution equations (2.13) and (2.14) are equivalent to the completely integrable Pfaffian system

$$d\,v = \Omega\, v, \quad v = \begin{pmatrix} v_1 \\ v_2 \end{pmatrix}, \quad \Omega = \begin{pmatrix} \omega_1 & \omega_2 \\ \omega_3 & -\omega_1 \end{pmatrix} . \tag{3.1}$$

The matrix Ω of 1-forms has nothing to do with $\Omega(\zeta)$ which appeared in § 2. It is a traceless matrix consisting of a one-parameter family of 1-forms $(\omega_1, \omega_2, \omega_3)$ in the independent variables (x,t), the dependent variable u, and its derivatives; the parameter is chosen conveniently as $\eta = -i\zeta$ in which ζ is the eigenvalue of the scattering problem (2.13). The integrability of equation (3.1) requires that the following 2-form, written in wedge notation, vanishes:

$$\Theta \equiv d\Omega - \Omega \wedge \Omega = 0. \tag{3.2}$$

This follows from $d^2v = 0 = d\Omega v - \Omega \wedge dv$ (see property (ii) of d). All we are doing here is appealing to the integrability condition $v_{tx} = v_{xt}$ and (3.2) is simply an alternative expression of the NEE itself, (2.15). There may be subtleties about 'factorising' out the eigenvectors v (7): (2.16) suggest a problem at $\zeta = 0$ (7).

The proof that solutions of NEEs like (2.15) represent surfaces of constant negative curvature with respect to a particular metric rests on the observation that (3.2) has the same form as the fundamental equations of such a surface. To see this consider the following: at each point P of the surface one has a tangent plane T_P which is spanned by a local basis $\underline{e}_1$, $\underline{e}_2$ (say) for which one can always choose

$$\underline{e}_i \cdot \underline{e}_j = \delta_{ij} \quad . \tag{3.3}$$

The 'structure equations' are (31)

$$\begin{aligned} dP &= \sigma^1 \underline{e}_1 + \sigma^2 \underline{e}_2 \\ d\underline{e}_1 &= \omega \underline{e}_2 \\ d\underline{e}_2 &= -\omega \underline{e}_1 \end{aligned} \tag{3.4}$$

in which σ^1 and σ^2 are 1-forms; ω is the 'connection' 1-form. The first equation states that P is confined to the surface; the remaining two equations follow from (3.3).

The integrability condition in this case is $d^2P = 0$ from which follows

$$\begin{aligned} d\sigma^1 &= \omega\sigma^2 \\ d\sigma^2 &= -\omega\sigma^1 \end{aligned} \tag{3.5}$$

The Gaussian curvature K is defined by

$$d\omega = -K\, \sigma^1 \wedge \sigma^2 \tag{3.6}$$

For $K = -1$ precisely the choice of Ω

$$\Omega \equiv \begin{bmatrix} -\frac{1}{2}\sigma^2 & \frac{1}{2}(\omega + \sigma^1) \\ \frac{1}{2}(-\omega + \sigma^1) & \frac{1}{2}\sigma^2 \end{bmatrix} \tag{3.7}$$

(for example) satisfies (3.2) exactly. Pseudo-spherical surfaces have as a matter of definition Gaussian curvature $K = -1$: thus (3.2) with (3.7) is a pseudo-spherical surface. The metric is given by (31)

$$ds^2 = (\sigma^1)^2 + (\sigma^2)^2 \tag{3.8}$$

An early example of a surface with constant negative curvature is Beltrami's surface of revolution obtained by spinning the tractrix

$$y = \sqrt{a^2 + x^2} - a \log \left\{\frac{a + \sqrt{a^2 + x^2}}{x}\right\}$$

about the y-axis. This surface has curvature $-a^{-2}$. The s-G actually arose in the first place in the study of surfaces of constant negative curvature carried on before 1883 (7). The 'surfaces of Enneper' associated with different solutions of the s-G were catalogued by Steurwald in 1936 (7). These contained both the kink and breather solutions as well as the 2-kink solution. The elegance of the situation that the result for the s-G extends to all AKNS systems (2.15) is obvious. But the result also provides important insights.

The aBT is a mapping of one pseudospherical surface into another via a gauge transformation (see (3.14) below and Ref. (30)). An isometry is a mapping of the surface into itself preserving scalar products and distances. For the 2-surface of constant negative curvature the number of isometries is maximal (7,30). The isometry groups are SL(2,R) and SU(1,1) and this 'explains' for example, the emergence of the Lie algebras of these groups in the 'prolongation structures' of AKNS systems (7,32).

We can use the same geometrical approach to construct the conservation laws. We first consider certain Riccati forms of the AKNS system (2.13) and (2.14). We introduce variables

$$\Gamma_1 = v_2/v_1 \quad \text{and} \quad \Gamma_2 = v_1/v_2 \tag{3.9}$$

to rewrite the Pfaffian (3.1) as either

$$0 = \varepsilon_1 \equiv d\Gamma_1 - \omega_2 + 2\Gamma_1 \omega_1 + \Gamma_1^2 \omega_3 \tag{3.10a}$$

or

$$0 = \varepsilon_2 \equiv d\Gamma_2 - \omega_3 + 2\Gamma_2 \omega_1 + \Gamma_2^2 \omega_2 \quad . \tag{3.10b}$$

From the integrability condition (3.2) the exterior derivatives of 1-forms ε_1 and ε_2 can now be calculated to be

$$\begin{aligned} d\varepsilon_1 &= 2\varepsilon_1 \wedge (\omega_1 + \Gamma_1\omega_2) \\ d\varepsilon_2 &= 2\varepsilon_2 \wedge (-\omega_1 + \Gamma_2\omega_3) \end{aligned} \tag{3.11}$$

and these are necessary and sufficient conditions for the ε_i to be completely integrable. From closure $d^2 \varepsilon_i = 0$ of each of equations (3.11) it follows that the 1-forms

$$\begin{aligned} \delta_1 &\equiv \omega_1 + \Gamma_1\omega_3 \\ \delta_2 &\equiv -\omega_1 + \Gamma_2\omega_3 \end{aligned} \tag{3.12}$$

are closed, that is

$$d\delta_1 = d\delta_2 = 0 \ . \tag{3.13}$$

These abstract expressions are actually the conservation laws. They are one-parameter families of conservation laws because of the dependence on the eigenvalue $\zeta = i\eta$. However by expanding these in powers of η^{-1} and η we get precisely the conserved quantities of the s-G for example. The calculation is facilitated by noting that equations (3.1) and (3.2) are form invariant under the 'gauge' transformation

$$\begin{aligned} &v \rightarrow v' = Bv \ , \quad \Omega \rightarrow \Omega' = B\Omega B^{-1} + dB B^{-1} \\ &\Theta \rightarrow \Theta' = B\Theta B^{-1} \end{aligned} \tag{3.14}$$

where B is an arbitrary (space-time dependent) 2 × 2 matrix with determinant unity. The condition guarantees that the Wronskian $W(\phi,\psi)$ defined in § 2 by

$$W(\phi,\psi) = \phi_1 \psi_2 - \phi_2 \psi_1 \tag{3.15}$$

in terms of solutions ϕ , ψ of (3.1) is invariant under the gauge transformation. Since $a(\zeta) = W(\phi,\psi)$ with ϕ , ψ Jost solutions, it follows that $a(\zeta)$ is gauge invariant. One can also demonstrate (29) that suitably gauge transformed quantities Γ_i satisfy the same Riccati equations (3.10) and that the corresponding gauge transformed δ_i are gauge invariant upto complete derivatives

$$\delta_1' = \delta_1 - d\,\beta_1$$

$$\delta_2' = \delta_2 - d\,\beta_2 \quad . \tag{3.16}$$

Finally one shows (29) that because (3.1) has

$\omega_1 = \eta\, dx + A\, dt$, $\omega_2 = q\, dx + B\, dt$, $\omega_3 = r\, dx + C\, dt$ (compare the AKNS system (2.13) with (2.14) against (3.1)) the x-"component" of equation (3.10a) and equation (2.26) are identical providing one identifies $\Gamma_1 \equiv q^{-1}\,\Phi_x$. From $\delta_1 \equiv \omega_1 + \Gamma_1\omega_2$ and the similar expression for δ_2 there follows

$$\log a(\zeta) = \int_{-\infty}^{\infty} (\delta_1 - d\,\beta_1) = \int_{-\infty}^{\infty} (\delta_2 - d\,\beta_2) \tag{3.17}$$

where the β_i are functions depending on the chosen gauge: typically they do not matter since they may be integrated out and the result vanishes at both limits by the boundary conditions. Thus by expanding the δ_i about points μ in the complex ζ-plane one gets the same conservation laws as are found by expanding $\log a(\zeta)$ as in § 2.

For the s-G in particular the facility of the method depends on the best choice of gauge (29). The procedure is to symmetrize Ω so that (29)

$$\omega_1 = \frac{i}{4}\,(u_x\, dx - u_t\, dt)$$

$$\omega_2 = \tfrac{1}{2}\,\eta'\, e^{-iu/2}\, dx + \frac{1}{2\,\eta'}\, e^{iu/2}\, dt$$

$$\omega_3 = \tfrac{1}{2}\,\eta'\, e^{iu/2}\, dx + \frac{1}{2\,\eta'}\, e^{-iu/2}\, dt \tag{3.18}$$

where $\eta' = 2\eta$ $(=-2i\zeta)$. We drop the primes on η in the following.

The metric is

$$ds^2 = (dx)^2 + 2\cos u\, dx\, dt + (dt)^2$$

which follows from

$$\sigma^1 = \cos\frac{u}{2}\,.\,(dx + dt)$$

$$\sigma^2 = \sin\frac{u}{2}\,.\,(dx - dt)$$

$$\omega = \tfrac{1}{2}\,(u_x\, dx - u_t\, dt) \tag{3.19}$$

and the possible choice (compare (3.7))

$$\Omega = \begin{bmatrix} \frac{i}{2}\,\omega & \frac{1}{2}(\sigma^1 - i\sigma^2) \\ \frac{1}{2}(\sigma^1 + i\sigma^2) & -\frac{i}{2}\,\omega \end{bmatrix} \tag{3.20}$$

together with the scale transformation $x \to x' = \eta x$, $t \to t' = \eta^{-1}t$ (η is η not η' here)

Because $\omega_1{}^* = -\omega_1$, $\omega_2{}^* = \omega_3$ from (3.18)

$$\Gamma_1 = \Gamma_2{}^*$$

(* denotes complex conjugate). Then

$$\delta_1 = \delta_2{}^* \tag{3.21}$$

Thus we need treat only Γ_1 and δ_1. We omit suffixes on these. The Riccati equation (3.10a) is equivalent then to

$$\Gamma_x - \tfrac{1}{2}\eta\, e^{iu/2} + \frac{i}{2}\, u_x\Gamma + \tfrac{1}{2}\eta\, e^{-iu/2}\,\Gamma^2 = 0 \tag{3.22a}$$

$$\Gamma_t - \tfrac{1}{2}\eta^{-1} e^{-iu/2} - \frac{i}{2}\, u_t\Gamma + \tfrac{1}{2}\eta^{-1} e^{-iu/2}\,\Gamma^2 = 0\,. \tag{3.22b}$$

The procedure now is to solve for Γ in the forms

$$\Gamma = \sum_{k=0}^{\infty} \gamma_k\, \eta^{-k} \quad \text{and} \quad \Gamma = \sum_{k=0}^{\infty} \tilde{\gamma}_k\, \eta^{k} \tag{3.23}$$

and the γ_k ($\tilde{\gamma}_k$) are determined by equating powers of η^{-1}(η).

Consistency of this procedure is guaranteed by the complete integrability of (3.10a).

There is one case of symmetry when Γ in inverse powers of η is put into (3.22a) and Γ in powers of η is put into (3.22b). In this case

$$\tilde{\gamma}_k = \gamma_k \, (u \to -u, \; u_x \to -u_t, \; u_{xx} \to -u_{tt}) \tag{3.24}$$

and we have to determine only the γ_k's . This case is exploiting what is essentially the following consideration (29): since $T^2 = \frac{1}{2} u_x^{\;2}$ satisfies the conservation law for the s-G

$$(\tfrac{1}{2} u_x^{\;2})_t - (1-\cos u)_x = 0, \tag{3.25}$$

and the s-G is invariant under interchange of x and t, a conserved density is also (1-cos u). From T^4 we find the density $-\frac{1}{2} u_t^{\;2} \cos u$ by a similar argument, and an infinite sequence of non-trivial conservation laws for the s-G develops. However, as shown in Ref. (29), this sequence is a property of the s-G alone: for the s-G the quantities certainly are conserved quantities but they are not the important ones. The important ones are obtained by making the two expansions (3.23) on (3.22a) alone: this equation represents the gauge transformed scattering problem (2.13).

The results are (29) the polynomial sequence (2.9) and a new set

$$\bar{\mathcal{H}}_1 = -(1-\cos u) ,$$

$$\bar{\mathcal{H}}_3 = \sin u \int_{-\infty}^{x} \cos u \, dx_1 \int_{-\infty}^{x_1} \sin u \, dx_2 + \cos u \int_{-\infty}^{x} \sin u \, dx_1 \int_{-\infty}^{x} \sin u \, dx_2 \; ,$$

$$\bar{\mathcal{H}}_5 = \; ..... \tag{3.26}$$

These are manifestly not polynomial and except for $\bar{\mathcal{H}}_1$ are non-local. We make use of all of these conserved densities in the next lecture presented in § 4 .

4. CANONICAL STRUCTURE OF THE INTEGRABLE NEEs.

Let us consider the two AKNS potentials q , r as canonical variables. Since each depends on the running variable x they each represent a continuous infinity of such variables. There is a doubly infinite number of degrees of freedom and the Hamiltonian system is an infinite dimensional Hamiltonian system.

From the $\mathcal{H}_n$ of (2.30) and the Hamiltonians

$H_n = (2i)^{-n} \int_{-\infty}^{\infty} \mathcal{H}_n \, dx$ we find for $n = 3$ and $H = 8\,H_3$

$$q_t = -8\,\frac{\delta H_3}{\delta r} = i^{-1}\,(-q_{xx} + 2q^2 r) \tag{4.1a}$$

$$r_t = 8\,\frac{\delta H_3}{\delta q} = -i^{-1}\,(-r_{xx} + 2q\,r^2)\ . \tag{4.1b}$$

The symbol $\delta H_3/\delta q$, for example, is the "functional derivative". The functional derivative of $\int_{-\infty}^{\infty} A[q,r]dx$ in which $A[q,r]$ is a functional of q, r, q_x, r_x, etc. is defined by either of the equivalent alternative expressions

$$\delta A = \int_{-\infty}^{\infty} \left[\frac{\delta A}{\delta q}\,\delta q + \frac{\delta A}{\delta r}\,\delta r \right] dx\ , \tag{4.2a}$$

so that for example

$$\frac{dA}{dt} = \int_{-\infty}^{\infty} \left[\frac{\delta A}{\delta q}\, q_t + \frac{\delta A}{\delta r}\, r_t \right] dx\ ,$$

or

$$\lim_{\varepsilon \to 0} \frac{d}{d\varepsilon} \int_{-\infty}^{\infty} A[q + \varepsilon v(x), r]\, dx = \int_{-\infty}^{\infty} \frac{\delta A}{\delta q}\, v(x)\, dx\ . \qquad (4.2b)$$

If $q = -r^*$, (4.1b) is the NLS (2.2) and (4.1a) is its complex conjugate. From $H = -16\, H_4$

$$q_t = -\frac{\delta H}{\delta r} = -\left\{ q_{xxx} - 6\, rq\, q_x \right\} . \qquad (4.3)$$

and, if $q = -r$, this is the modified KdV

$$q_t + 6\, q^2 q_x + q_{xxx} = 0 \qquad (4.4)$$

Formally we can also set $r = -1$ (although this does not satisfy the boundary conditions on r) to get the KdV (1.1) with $u \to -u$. We can also choose $H = -16\sigma H_4 + 8\rho H_3$ (σ,ρ constants) and $r = -\gamma q^*$ to get

$$iq_t = i\sigma q_{xxx} + 6i\gamma\sigma |q|^2 q_x + \rho q_{xx} + 2\rho\gamma q |q|^2\ . \qquad (4.5)$$

This is the Hirota equation (33) with $2\gamma\sigma = \alpha$, $2\rho\gamma = \delta$, $\alpha\rho = \sigma\delta$. Clearly we can choose any linear combination of the H_n to obtain interesting NEEs in q and r. In fact all of these NEEs fall into the AKNS-ZS scheme (25,28). And indeed all the coefficients of the H_n can depend on time without fundamentally changing this situation.

Notice also that for $q_t = -16\delta H_{2n+2}/\delta r$ ($n = 1,2,\ldots$) and $r = -q$ we obtain an infinite sequence of generalised modified KdV equations with (4.4) as first member; for $r = -q^*$, $r_t = -8\,\delta H_{2n+1}/\delta q$ ($n = 1,2,\ldots$) is an infinite sequence of NLS equations. The infinite sequence of KdV equations obtained formally through $q_t = -16\delta\, H_{2n+2}/\delta r$ and $r = -1$ was first discovered by Lax (24). In any of the sequences each member has the same infinite set of conserved densities; each member proves to be soluble by the inverse method with scattering problem (2.13); and each member has the same eigen spectrum. The flows described by these NEEs commute (34). This is so because all the conserved densities are 'in involution'. We explain what we mean by this once we have defined a Poisson bracket below.

Perhaps the most striking result so far is the absence of the s-G (1.8). The $\mathcal{H}_{2n+1}$ coincide with the T^{2n+2} and for $q = -r = -\frac{1}{2}u_x$ these are actually momentum densities for an infinite sequence of s-G equations. The Hamiltonian densities are obtained from the sequence (3.26) obtained by expanding log $a(\zeta)$ about $\zeta = 0$. One apparent difficulty is that $\mathcal{H}_1$ depends on the single set of co-ordinates u or u_x. Formally however we can write (25)

$$\overline{H}_1 = \frac{-m^2}{4\gamma_0} \int_{-\infty}^{\infty} \left\{ \cos \left(-2 \int_{-\infty}^{x} q\, dx'\right) + \cos(-2\gamma_0 p) - 2 \right\} dx \tag{4.6a}$$

$$P_1 = \frac{1}{2\gamma_0} \int_{-\infty}^{\infty} dx \,(q^2 + \gamma_0{}^2\, p_x{}^2) \tag{4.6b}$$

for total energy and momentum respectively. We have done several things here: $\overline{H}_1$ is scaled by the factor $\frac{m^2}{4\gamma_0}$ in which m is a mass and γ_0 is a dimensionless coupling constant. We have introduced canonical co-ordinates $q = -\frac{1}{2}u_x$, and $p = \gamma_0{}^{-1} \int_{-\infty}^{x} q\, dx' = -\frac{1}{2}\gamma_0{}^{-1} u$. It is easily checked that the equations of motion derived from each of Hamilton's equations are

$$-q_t \;\left(= \tfrac{1}{2}u_{xt}\right) = -\frac{\delta H_1}{\delta p} = \tfrac{1}{2} m^2 \sin u \tag{4.7a}$$

$$-p_t \;\left(= \tfrac{1}{2}\gamma_0{}^{-1} u_t\right) = \frac{\delta \overline{H}_1}{\delta q} = \frac{m^2}{2\gamma_0} \int_{-\infty}^{x} \sin u\, dx' \tag{4.7b}$$

so that both equations are $u_{xt} = m^2 \sin u$. For $m = 1$ this is of course precisely (1.8). Notice that γ_0 does not appear in the equations of motion (essentially because q does not involve γ_0 but the canonical momentum $p = -\frac{1}{2}\gamma_0{}^{-1} u$ does: a simple argument is $L = \int_{-\infty}^{\infty}(p\dot{q} - \mathcal{H})dx = \frac{1}{4\gamma_0}\int_{-\infty}^{\infty} u\, u_{xt}\, dx - \overline{H}_1$, $\frac{\delta L}{\delta \dot{q}} = -\frac{1}{2\gamma_0} u$).

Notice also that directly from (3.26)

$$u_{xt} = \frac{\partial}{\partial x}\frac{\delta \overline{H}_1}{\delta u_x} = \sin u \tag{4.8}$$

after changing the sign of $\overline{\mathcal{H}}_1$. We discuss this form later noting here that if u_x is q (upto a factor) $\frac{\delta \overline{H}_1}{\delta p} \to \frac{\partial}{\partial x}\frac{\delta \overline{H}_1}{\delta q}$. In a similar way to (4.8) we find from $-\overline{\mathcal{H}}_3$ in (3.26) that

$$u_{xt} = -\tfrac{1}{2}\sin u \left(\int_{-\infty}^{x} \sin u\, dx\right)^2 + \cos u \int_{x}^{\infty} \cos u\, dx_1 \int_{-\infty}^{x_1} \sin u\, dx_2 , \tag{4.9}$$

and in this way one can develop a sequence of generalised s-G equations which is the required analogue of the sequences of modified KdV, KdV and NLS equations just discussed.

One can check, for example, that the linearised form of (4.9) is $u_{xxxt} = -u$ and its linearised dispersion relation is $\omega = k^{-3}$ (to find the linearised form of (4.9) one needs the result demonstrated in Ref. (29) that $\int_{-\infty}^{\infty} \cos u\, dx_1 \int_{-\infty}^{x_1} \sin u\, dx_2 = 0$ for all members of the s-G hierarchy: it is obviously true for the s-G since it is $\int_{-\infty}^{\infty} \cos u\, u_t\, dx = -\frac{d}{dt}\int_{-\infty}^{\infty} \sin u\, dx = -\frac{d}{dt}[u(\infty,t)_t - u(-\infty,t)_t] = 0$).
One can also check that a solution of (4.9) is $u = 4\tan^{-1}\exp[2\Omega(i\eta)t + 2\eta x + \delta]$ with $\Omega(\zeta) = -\frac{i}{2}\omega(-2\zeta)$ with $\omega(k) = k^{-3}$. These are just the properties needed of the second member of a hierarchy whose first member has $\omega = k^{-1}$ as we explain below. It is already clear from what we have just said that the single kink solution has the correct from: the form of this was derived at (2.23).

From (4.2a) it is plain that there is a natural Poisson bracket

$$\{A,B\} = \int_{-\infty}^{\infty} \left\{ \frac{\delta A}{\delta q} \frac{\delta B}{\delta r} - \frac{\delta A}{\delta r} \frac{\delta B}{\delta q} \right\} dx \qquad (4.10)$$

When B is H,

$$\{A, H\} = \int_{-\infty}^{\infty} \left\{ \frac{\delta A}{\delta q} q_t + \frac{\delta A}{\delta r} r_t \right\} dx = \frac{dA}{dt} \qquad (4.11)$$

and when A is a constant of the motion $\{A, H\} = 0$. If any of the set of constants of the motion H_n is a possible Hamiltonian, $\{H_n, H_m\} = 0$. Such a set is said to be in involution.

The argument just used had identified $q(x,t)$, the potential in (2.13), as a canonical co-ordinate, and the other potential $r(x,t)$ as its canonical momentum. This is the situation for (4.1b); and it is also so for the NLS where $r = -q^*$. It is also so for (4.5), but it is true for the modified KdV (4.4) only because $r = -q$ is imposed after evaluating the equations of motion. It is possible to set $r = -q$ before doing this (25) and indeed this is what is done for the s-G equation (4.7): here $q = -r = -\frac{1}{2} u_x$ and $p = \int_{-\infty}^{x} q\, dx'$ (we choose $\gamma_0 = 1$ for simplicity at the moment). The proper choice of Poisson bracket can be motivated by the remark by (4.8) that $\delta/\delta r$, that is $\delta/\delta p$, is replaced by $\frac{\partial}{\partial x}\frac{\delta}{\delta q}$. Thus with an integration by parts on the second term in (4.10) (and a factor one half) we choose

$$\{A, B\} = \int_{-\infty}^{\infty} \left(\frac{\delta A}{\delta q} \frac{\partial}{\partial x} \frac{\delta B}{\delta q} \right) dx \qquad (4.12)$$

a form first found by Gardner (35) for the KdV. If B is H, $q_t = \frac{\partial}{\partial x}\frac{\delta H}{\delta q}$ and $\{A, H\} = \frac{dA}{dt}$. If $A = \int q(x',t)\, \delta(x-x')dx' = q(x,t)$, $q_t = \{A, H\} = \frac{\partial}{\partial x}\frac{\delta H}{\delta q}$, etc. as required.

There is an interpretation of this equation of motion which is instructive. Compare the prescription for one degree of freedom

$$\begin{bmatrix} \dot{q} \\ \dot{p} \end{bmatrix} = \begin{bmatrix} 0 & 1 \\ -1 & 0 \end{bmatrix} \begin{bmatrix} \partial H/\partial q \\ \partial H/\partial p \end{bmatrix} = J \text{ grad } H. \tag{4.13}$$

In this, J is a linear map taking phase space to phase space. It connects the contravariant vector field $(\dot{q},\dot{p})$ with a covariant one (grad H). It is isometric (preserves inner products), and this is the key point, is skew. Evidently the motion $q_t = \frac{\partial}{\partial x}\frac{\partial H}{\partial q}$ replaces J by the skew symmetric operator $\partial/\partial x$.

Phase space is symplectic: it carries a closed skew symmetric differential 2-form ϖ (31). In terms of arbitrary local co-ordinates x^i $(i = 1,2)$ in phase space

$$\varpi = \omega_{ij}\, dx^i{}_\wedge dx^j = dq_\wedge dp . \tag{4.14}$$

In the second form the co-ordinates are canonical $(x^1,x^2) = (q,p)$ and $2\omega_{ij}$ is just the matrix J. We call the form ϖ the 'symplectic form': it is invariant (co-ordinate free) as we show below and closed $d\varpi = 0$. It appeared first in soliton theory (18) as the bilinear form ω which is associated with ϖ (25). With the symplectic form ϖ defined above, $\omega = \delta_1 q\, \delta_2 p - \delta_2 q\, \delta_1 p$ in which $\delta_1 q$ and $\delta_2 q$ ($\delta_1 p$ and $\delta_2 p$) are independent variations in q (p). The correspondence between the symplectic form and the bilinear form ω is 1:1 and we may work with either (25).

Consider ω : as a heuristic motivation for considering it take Hamilton's principle for the single pair (q,p) of canonical co-ordinates

$$\delta \int (p\, dq - H\, dt) = 0 \tag{4.15}$$

In the canonically transformed set $(\bar{p}\ ,\ \bar{q}\ )$ with Hamiltonian $\bar{H}$

$$\delta \int \bar{p}\, d\bar{q} - \bar{H}\, dt) = 0\ , \tag{4.16}$$

and so for some function F

$$\delta \int (p\, dq - H\, dt - \bar{p}\, d\bar{q} + \bar{H}\, dt + dF) = 0 \tag{4.17}$$

Since for conservative systems the <u>values</u> of H and $\bar{H}$ are necessarily the same

$$\delta\, [p\, dq - \bar{p}\, d\bar{q} + dF] = 0$$

and

$$\delta p\, dq - \delta\bar{p}\, d\bar{q} + p\, \delta(dq) - \bar{p}\, \delta(d\bar{q}) + \delta(dF) = 0\ ; \tag{4.18}$$

but since δp and dq can be interpreted as independent arbitrary variations we also have that

$$dp\,\delta q - d\bar{p}\,\delta\bar{q} + p\,d(\delta q) - \bar{p}\,d(\delta\bar{q}) + d(\delta F) = 0 \tag{4.19}$$

so that

$$dp\,\delta q - \delta p\,dq = d\bar{p}\,\delta\bar{q} - \delta\bar{p}\,d\bar{q} \tag{4.20}$$

and ω is invariant under the transformation, i.e.

$$\omega = \bar{\omega}\quad . \tag{4.21}$$

For many degrees of freedom we sum $dp_i\delta q_i - \delta p_i dq_i$. For continuously many degrees of freedom we integrate. This heuristic argument suggests that ω defined by

$$\omega = \int_{-\infty}^{\infty} [\delta_1 q(x)\delta_2 p(x) - \delta_2 q(x)\delta_1 p(x)]\, dx \tag{4.22}$$

is invariant under a canonical transformation $\{p(x),\ q(x)\} \to \{\bar{p}\ ,\bar{q}\}$ (in which some at least of the $\{\bar{p}\ ,\bar{q}\}$ could be discrete. Conversely reversing the argument to (4.21) if ω is invariant the transformation is canonical. The symplectic form associated with ω is

$$\Omega = \int_{-\infty}^{\infty} (d\,q \wedge d\,p)\ dx \tag{4.23}$$

If and only if Ω is invariant the transformation is canonical.

The next point is that the transformation to and from scattering data S given by (2.18) is actually a canonical transformation: the proof is given in (25) and demonstrates invariance of the symplectic forms (4.22) or (4.23). The value of this fact is that everything becomes much simpler written in terms of scattering data as we shall show below. We sketch the proof that the transformation to and from scattering data S is canonical here now.

If r and q in (2.13) are independent we can use them as p and q in ω ; if r and q are dependent, and we consider especially $r = -q$, we can choose q and $\int_{-\infty}^{x} q\,dx' = p$. We develop the theory for this case of linear dependence $r = -q$. For although the independent case is in some ways simpler (25), the dependent case includes that of the s-G as we have noted. The steps of the proof in (25) are to establish a connection between variations in scattering data and consequential variations in q and p : the connection is established via a generalised AKNS-ZS Marchenko equation. In this way the symplectic form Ω in dq and dp can be expressed in terms of new canonical co-ordinates expresses in terms of scattering data. It is shown (25), in the case $r = -q$, $q = q$, $p = +\gamma_0^{-1} \int_{-\infty}^{x} q\,dx'$, that, under the scattering transform (2.13), the symplectic form defined on the manifold with canonical co-ordinates $q(x)$, $p(x)$ transforms into

$$\varpi = \sum_{j=1}^{M} dq_j \wedge dp_j + \int_{-\infty}^{\infty} d\xi\, (dQ(\xi)_\wedge dP(\xi)) \tag{4.24}$$

on the manifold with canonical co-ordinates

$$\{p_j,\ q_j,\ P(\xi),\ Q(\xi),\ j = 1, \ldots, M;\ \xi \in R\}\ . \tag{4.25}$$

The complex canonical co-ordinates are defined in terms of scattering data by

$$p_j = \gamma_0^{-1} \log \zeta_j\ ,\ q_j = 2 \log b_j,\ j = 1, \ldots, M;$$

$$P(\xi) = (2\pi\, \xi\, \gamma_0)^{-1} \log \{a(\xi)\, \bar{a}(\xi)\},\ Q(\xi) = \arg b(\xi). \tag{4.26}$$

The momenta and Hamiltonians corresponding to the H_n in (2.29) and (2.30) and the $\bar{H}_m$ obtained by expanding $\log a(\zeta)$ about an ordinary point $\zeta = \mu$ can be obtained by reconstructing $\log a(\zeta)$ in terms of scattering data through Cauchy's theorem (25) and using (4.26). We find (25,28)

$$a(\zeta) = \prod_{j=1}^{M} \left(\frac{\zeta-\zeta_j}{\zeta+\zeta_j}\right) \exp - \frac{\zeta}{2\pi i} \int_{-\infty}^{\infty} \frac{d\xi}{(\xi^2-\zeta^2)} \log\,(1-w(\xi)\bar{w}(\xi)),$$
$$\mathrm{Im}\zeta > 0\ . \tag{4.27}$$

Thus the expansion of $\log a(\zeta)$ in inverse powers of ζ as $|\zeta| \to \infty$ yields (with p_j written for $\gamma_0 p_j$)

$$H_{2m+1} = -2 \sum_{j=1}^{M} \frac{\exp\{-(2m+1)\, p_j\}}{(2m+1)} - i\gamma_0 \int_{-\infty}^{\infty} d\xi\, \xi^{2m+1}\, P(\xi) \tag{4.28}$$

whilst expansion about the ordinary point $\zeta = \mu$ yields (for $m \geq 1$)

$$\bar{H}_m(\mu) = \sum_{j=1}^{M} -(m)^{-1} \{(\exp(-p_j)-\mu)^{-m} + (-1)^{m+1}(\exp(-p_j)+\mu)^{-m}\}$$
$$- \frac{\gamma_0}{2i} \int_{-\infty}^{\infty} d\xi\, P(\xi)\, \xi \left\{(\xi-\mu)^{-(m+1)} + (-1)^{m+1}(\xi+\mu)^{-(m+1)}\right\} \tag{4.29}$$

Notice that the canonical co-ordinates (4.26) have the character of action-angle variables (although the q_j are not precisely angle variables). The action-angle variable character is reflected in the fact that both (4.28) and (4.29) depend only on the momenta p_j, $j = 1, \ldots, M$; $P(\xi)$, $\xi \in R$; whilst these quantities are constants of the motion. We can actually integrate the equations of motion therefore and connect the evolution of the $Q(\xi)$ in particular with the evolution of the scattering data $b(\xi,t)$ to see that $\Omega(\zeta) = 4\, i\, \zeta^{2n+1}$ from (4.28) and

$$\Omega(\zeta) = -\frac{1}{8i}\,\zeta\left\{(\zeta-\mu)^{-(m+1)} + (-1)^{m+1}(\zeta+\mu)^{-(m+1)}\right\} \text{ from (4.29).}$$

Thus to get the equations of motion (in terms of scattering data) for the modified KdV we need to take H_3 from (4.28), whilst for the s-G we need to set $\mu = 0$ and take $\overline{H}_1(0)$ from (4.29). It is then clear for example that the family of s-G's has $\Omega(\zeta) = \frac{1}{2}(-2i\zeta)^{-(2n+1)}$, $n = 0, 1, 2, \ldots$ and Hamiltonians (we are scaling (4.28) and (4.29) by an extra factor $(-2i)^{-(2n+1)}\gamma_0^{-1}$)

$$\overline{H}_{2n+1}(0) = \left(\frac{(-2i)^{-(2n+1)}}{\gamma_0}\right)\left[\left(\frac{-2}{2n+1}\right)\left(\sum_{j=1}^{N}\zeta_j^{-(2n+1)} + \sum_{k=1}^{L}\{\zeta_k^{-(2n+1)} - \zeta_k^{*-(2n+1)}\}\right) - i^{-1}\gamma_0\int_{-\infty}^{\infty} P(\xi)\,\xi^{-(2n+1)}\,d\xi\right] \tag{4.30}$$

and "momenta"

$$P_{2n+1} = \left(\frac{(-2i)^{-(2n+1)}}{\gamma_0}\right)\left[\left(\frac{-2}{2n+1}\right)\left(\sum_{j=1}^{N}\zeta_j^{(2n+1)} - \sum_{k=1}^{L}\{\zeta_k^{(2n+1)} - \zeta_k^{*(2n+1)}\}\right) - i\gamma_0\int_{-\infty}^{\infty} d\xi\, P(\xi)\,\xi^{2n+1}\right] \tag{4.31}$$

The Hamiltonians (4.30) expressed in terms of scattering data are consistent with the Hamiltonians obtained from the densities (3.26) which are expressed as functionals of $u(x,t)$: the "Hamiltonians" $\overline{H}_{2n}$ all vanish, whilst the $\overline{H}_{2n+1}$ yield equations of motion which have the correct linearised dispersion relations $\Omega(\zeta)$ and the kink solutions associated with (2.23). Notice that in (4.30) and (4.31) we have divided the bound state eigenvalues into the two groups ζ_j $(1 \le j \le N)$ and the group of pairs $(\zeta_k, -\zeta_k^*)$ $(1 \le k \le L)$: the former are associated with the kinks — $\zeta_j = i\,\eta_j$, $\eta_j > 0$; and the latter with the breathers. These possibilities arise from the symmetry $r = -q$ as we noted in § 2. Evidently $N + 2L = M$ the total number of bound states.

Notice that by working in terms of scattering data and identifying $\Omega(\zeta)$ one finds which families of evolution equations are "integrable". Notice that we now know that all the integrable systems which we have found this way are infinite dimensional completely integrable Hamiltonian systems. As a matter of definition 'completely integrable' can be taken to mean that in the case of 2n degrees of freedom there are n constants of the motion 'in involution'. In our case n is infinite but the system is demonstrably completely integrable in that we can proceed directly to carry that integration through. Notice that the constants of the motion are the infinite set of momentum variables p_j, $j=1, \ldots, M$; $P(\xi)$, $\xi \in R$. Notice that these are trivially in involution, that is their Poisson brackets necessarily vanish. Notice also that from (4.30) $\{\overline{H}_{2n+1}, \overline{H}_{2m+1}\} = 0$ because the $\overline{H}$ do not depend on the $(q_j, Q(\xi))$; and notice that $\{\zeta_j, \zeta_k\} = 0$; and so on.

Completely integrable systems are easy to quantise: one can replace Poisson brackets by commutators — so called semi-classical quantisation. Evidently every eigen state of energy is a simultaneous eigen state of momentum and all the other operators in involution with H. In the case of the s-G itself the quantisation goes as follows: from (4.30) and (4.31) with a mass m included in $\overline{H}_1$

$$\overline{H}_1 = \frac{-m^2 i}{2\gamma_0}\left\{2i\sum_{j=1}^{N}\eta_j^{-1} - 2\sum_{k=1}^{L}(\zeta_k^{-1}-\zeta_k^{*-1}) + i\gamma_0\int_{-\infty}^{\infty} d\xi\, \xi^{-1} P(\xi)\right\}$$

$$P = \frac{2i}{\gamma_0}\left\{-2i\sum_{j=1}^{N}\eta_j - 2\sum_{k=1}^{L}(\zeta_j-\zeta_1^{*}) - i\gamma_0\int_{-\infty}^{\infty} d\xi\, \xi\, P(\xi)\right\} . \tag{4.32}$$

These expressions can be written in the forms

$$P = \sum_{j=1}^{N} p_j + \sum_{k=1}^{L} \hat{p}_k + \int_{-\infty}^{\infty} d\xi\, p(\xi)\, P(\xi)$$

$$H = \sum_{j=1}^{N} h_j + \sum_{k=1}^{L} \hat{h}_k + \int_{-\infty}^{\infty} d\xi\, h(\xi)\, P(\xi) . \tag{4.33}$$

The p_j now defined through (4.32) and (4.33) are not the p_j

defined in (4.26) but (4.24) shows the change is canonical. However, $P(\xi)$ (the old momentum associated with the continous part of the spectrum of the scattering operator $\hat{L}$) is now reinterpreted as a momentum density and new momentum variables $p(\xi) = 2\xi$ have been introduced. This way there is now a "particle" interpretation for all of (4.32) as we shall see. But for the continuous part of the spectrum we view the change as formal: $P(\xi)$ (and $Q(\xi)$) will be kept as canonical coordinates preserving (4.24) as well as the equations of motion.

From (4.32) and (4.33)

$$h_j p_j = 4m^2\gamma_0^{-2}, \quad \hat{h}_k\hat{p}_k = 16m^2\gamma_0^{-2}\sin^2\theta_k, \quad h(\xi)p(\xi) = m^2, \qquad (4.34)$$

in which $\theta_k = \arg\zeta_k$ is an internal momentum of the breather labelled by k. If we define new energies and momenta by

$$\begin{aligned} h(\xi) &= \tfrac{1}{2}\{h'(\xi) + p'(\xi)\} \\ p(\xi) &= \tfrac{1}{2}\{h'(\xi) - p'(\xi)\} \end{aligned} \qquad (4.35)$$

and define similar quantities for h_j, p_j and $\hat{h}_k$, $\hat{p}_k$, then if we define a new mass $m' = 2m$, we find, after dropping the primed notation, that

$$\begin{aligned} h_j &= [4m^2\gamma_0^{-2} + p_j^2]^{\frac{1}{2}}, \quad \hat{h}_k = [16m^2\gamma_0^{-2}\sin^2\theta_k + \hat{p}_k^2]^{\frac{1}{2}} \\ h(\xi) &= [m^2 + p(\xi)^2]^{\frac{1}{2}} . \end{aligned} \qquad (4.36)$$

When these expressions are put into $\overline{H}_1$ this Hamiltonian of the s-G is that of a collection of free relativistic particles of masses m, $2m\gamma_0^{-1}$, and $4m\gamma_0^{-1}\sin\theta$. In terms of a corrected coupling constant $\gamma_0' \equiv 4\gamma_0$ these masses become m, $8m\gamma_0^{-1}$, and $16m\gamma_0^{-1}\sin\theta$. These results agree with the rest masses of the sine-Gordon kinks, breathers and the mass of the field m associated with $u_{xx}-u_{tt} = m^2\sin u$: the reason for changing γ_0 to γ_0' becomes clearer below.

From the requirement that the symplectic form remains closed under the canonical transformation one can see that before the transformation of (4.35) the canonical pairs of co-ordinates for solitons and breathers in (4.32) are

$$4\gamma_0^{-1}\eta_j, (2\eta_j)^{-1}\log|b_j| \;;\; 8\gamma_0^{-1}|\zeta_k|, (2|\zeta_k|)^{-1}\log|b_k| \qquad (4.37a)$$

and

$$\gamma_0^{-1}\theta_k, 4\arg b_k \qquad (4.37b)$$

for the internal co-ordinates of the breathers. Only the last of these occupy a compact phase-space (the other two have $p \geq 0$ and $q \geq 0$). Since $0 \leq \theta_k \leq \pi/2$, $0 \leq \arg b_k \leq 2\pi$, the phase space volume is $4\pi^2 \gamma_0^{-1}$ for each pair of internal breather co-ordinates. With $\hbar = 1$, semi-classical quantisation indicates that if there are N quantum states this volume is also $2\pi N$ so that $N = 2\pi \gamma_0^{-1}$. Thus θ (for each breather) is discretely quantised:

$$\theta \rightarrow \theta_n = \frac{1}{4} n \gamma_0, \; n = 1,2\ldots, N, \tag{4.38}$$

approximately and N is that largest integer $\leq 2\pi \gamma_0^{-1}$. This is very close to $2\pi \gamma_0^{-1}$ if N is large. From the result for the mass of the breather, remembering the changes in γ_0 and m ($\gamma_0 \rightarrow \gamma_0' = 4\gamma_0$ and the prime is dropped) one finds that the breather spectrum is discretely quantised with levels

$$M_n = \frac{16m}{\gamma_0} \sin \frac{n\gamma_0}{16}, \; n = 1,2,\ldots, N, \tag{4.39}$$

a result first obtained by Fadeev (36) via the canonical formalism and by Dashen, Hasslacher and Neveu (37) by path integral techniques (WKB semi-classical quantisation extended however to include renormalisation of the coupling constant and probably exact).

After the transformation (4.35) the phase space available to the other canonical variables of the solitons is RxR, and canonical quantisation yields

$$[q_j, p_k] = i\,\delta_{jk}, \; [\hat{q}_j, \hat{p}_k] = i\,\delta_{jk} \tag{4.40}$$

whilst, for quantisation, the co-ordinates $P(\xi)$, $Q(\xi)$ are assumed irrelevant[1]. This quantisation will now be equivalent, via invariance of the Poison brackets, to the quantisation

$$[u,u_t] = i\,\gamma_0\,\delta(x-x') \tag{4.41}$$

of the s-G $u_{xx} - u_{tt} = m^2 \sin u$. For from the Poisson brackets $\{p,q\} = \{-\frac{1}{2}\gamma_0^{-1}u, -\frac{1}{2}u_x\}$ and $x \rightarrow x + t$, $t \rightarrow x - t$, one finds $m' = 2m$ (used already above) and $\delta(x-x') = \frac{1}{4}\{\gamma_0^{-1}u,u_t\}$; thus, quantising $[u,u_t] = 4\gamma_0 i\delta(x-x') = \gamma_0' i\delta(x-x')$. In terms of γ_0' called γ_0 and the 'laboratory co-ordinates' x,t of the Lorentz covariant s-G one thus has (4.41).

The beautiful mass spectrum (4.39) has stimulated a lot of work in field theory: some of it is summarised in the first article in the book (7). A number of non-trivial field theories in 1 + 1 dimensions have been solved at semi-classical level or

---

[1] For a further remark see 'Note in proof' at the end of the refs.

better: I cite here only the work on the non-linear σ-models (7, 38) and remark that the S-matrix of the s-G has been calculated (39) and the same mass spectrum (4.39) found. One of the earlier results was the equivalence found by Coleman (40) between the massive Thirring model and the quantised sine-Gordon equation. Since the sine-Gordon particles are bosons (see the commutation relations (4.40) and (4.41)) and the massive Thirring model describes massive fermions the result is striking enough. Luther in the book (7) discusses this boson-fermion equivalence and shows that assumptions of locality and Lorentz covariance are sufficient (essentially) to determine this unique field theory which has the s-G and massive Thirring representations. The theory also has other representations in statistical mechanics — namely on a discrete lattice the model is the spin-$\frac{1}{2}$ x-y-z model (41)

$$H_s = -\sum_{j=1}^{N} J_\alpha S_j^\alpha S_{j+1}^\alpha$$

$$[S_j^\alpha, S_{j'}^\beta] = i\,\varepsilon^{\alpha\beta\gamma}\, S_j^\gamma\, \delta_{jj'} \qquad (4.42)$$

the spectrum of which has been calculated (41). This model maps onto a fermion model via the Jordan-Wigner transformation to fermion operators (see article 12 in (7)) and becomes essentially the massive Luttinger model (42). Thereafter one passes to a continuum limit renormalising coupling constants or operators to do so and introduces a bose representation of the fermi field operators which has the same 2-point correlation function (it is in this sense of equivalence under correlation functions that the bose and fermi operators are equivalent: they are not identical but this equivalence is what is needed). In this way one reaches the Lorentz covariant quantised s-G. The point now is that in the continuum limit all of these models have the spectrum (4.33) with the Lorentz covariant contributions (4.36)! From this it would be possible to find the eigenvalue spectrum of the spin-chain model (4.42) (at least in the continuum limit). In fact the argument is pursued the other way (7,41) and the spin-chain spectrum is used to confirm 'exactly' that the spectrum (4.33) is the eigen spectrum of the quantised s-G.

There seems to be a fruitful field of investigation in statistical mechanics for connections of this type. There seem to be connections between the theory of Painlevé transcendentals and Ising models (43) and between Painlevé transcendentals and inverse scattering (44). The c-number s-G with the continuous breather spectrum h_k of (4.36) and the soliton and 'phonon' contributions h_j and $h(\xi)$ is of interest both in the theory of linear conductors like TTF-TCNQ (45) and as a model system of non-linear excitations in statistical mechanics (46,47). Notice again that this classical system is formally just a relativistic gas of independent particles, solitons, breathers and phonons - although viewed as a density of states $P(\xi)$ depends on the initial data (trace the argument back through the scattering problem!) and in fact we have retained $P(\xi)$ here as the canonical momentum.

Within these terms results for the partition function of the grand canonical ensemble have been found (46): for example the soliton contribution is $\sum_{N_s=0}^{\infty} e^{\beta\mu_s N_s} Z_s^{N_s} = \exp\{e^{\beta\mu_s} 2MK_1(M\beta)\}$

in which μ_s is a soliton chemical potential, $\beta = (k_B T)^{-1}$ and

$Z_s = 2MK_1(M\beta)$ with $M = 8m\,\gamma_0^{-1}$ and $K_\nu = \frac{\pi}{2\sin\nu\pi}[I_{-\nu}(x) - I_\nu(x)]$

is a modified Bessel function with $I_\nu(x) = i^{-\nu}J_\nu(ix)$. However the results are still incomplete and in particular the breather contributions do not yet agree with those found by the transfer integral technique (47): the low temperature expansion is cubic in T. No doubt these problems will soon be resolved. The direct calculation of the partition function for the relativistic gas (4.33) with (4.36) seems the natural way to proceed. This relativistic gas result illustrates the power of working in terms of the scattering data. We have no physical example yet of a direct application of the quantised s-G in 1 + 1 dimensions but we have investigated two (3,4,6,7,48): one of these is the propagation of ultra-short optical pulses (1,2,7) for which the electromagnetic field is quantised in the first instance: the other is the problem of spin waves in the A-phase of ^{3}He below 2.6 mK (1,2,3,4). In the last lecture presented in § 5 following we look at a c-number problem in short optical pulse propagation as a vehicle for perturbation theory. This concerns the 'double sine-Gordon' equations

$$u_{xx} - u_{tt} = \pm\,(\sin u + \tfrac{1}{2}\sin\tfrac{1}{2}u) \qquad (4.43)$$

in the case when the +ve sign is taken. It is interesting to note that the case of the -ve sign occurs in the description of spin waves in the B-phase of ^{3}He (1,2,3,4). So far it has not been possible to treat this case of -ve sign by such perturbation theory.

5. THE DOUBLE SINE-GORDON EQUATION IN NON-LINEAR OPTICS AND SINGULAR PERTURBATION THEORY FOR IT.

The sine-Gordon equation

$$u_{tt} + u_{xt} = \alpha \sin u \tag{5.1}$$

governs resonant so called "sharp-line" optical pulse propagation in a medium with a non-degenerate resonant optical transition (1,2,3,4,5,6,7). The electric field envelope is $\mathcal{E} = u_t$. In the case of 5-fold degenerate transitions the right side of (5.1) becomes $\alpha(\sin u + \frac{1}{2} \sin \frac{1}{2} u)$ (α is a number, dimensions $(\text{frequency})^{-2}$, proportional to the number density n of atoms: in typical experiments n is 'small' $\sim 10^{11} cc^{-1}$). The independent variables are transformed $\sqrt{\alpha}\, t \to t$, $\sqrt{\alpha}\,(t-2x) \to x$ and one reaches the 'double sine-Gordon' equation

$$u_{xx} - u_{tt} = \sin u + \tfrac{1}{2} \lambda \sin \tfrac{1}{2} u \tag{5.2}$$

where $\lambda = 1$. This double s-G can of course be expressed in terms of light cone co-ordinates. But neither in this evolution equation form nor in the form (5.2) is it an integrable system: it has neither an aBT (49) nor an infinity of polynomial conserved densities (21). Only single kink solutions are known in analytical terms (see the article 3 in (7)): one of these is a kink of zero jump — essentially a breather. My colleagues and I have investigated this solution in terms of perturbation theory about the integrable s-G (7,50).

The boundary conditions for this breather are (7,50) $u \to 2\pi$ when $|x| \to \infty$. In its rest frame the solution may be put into the form

$$u = 4 \tan^{-1} \left(\sqrt{\frac{\lambda}{4-\lambda}} \cosh \tfrac{1}{2} \sqrt{4-\lambda}\, x \right) \quad . \tag{5.3}$$

This represents an unstable bound pair of, roughly speaking, a 2π-kink and 2π-antikink of the s-G. For initial data corresponding to a compression of the spacing of the kink and antikink the solution (5.3) goes into oscillation much as a breather of the s-G does (the breather of the s-G oscillates even in its rest frame — only the external co-ordinates are at rest: the $\sin\theta$ in the breather mass is the frequency of the s-G breather's oscillations (7)). The Figure 1 shows the result of direct numerical integration of (5.2) for $\lambda = 1$ in this oscillatory regime. For data corresponding to an extension of the spacing of the kink-antikink pair the system unbinds as shown in Figure 2.

The s-G is integrable as we have seen at length. It is natural to regard the double s-G as a perturbation for small λ. In fact we shall see that λ need not in fact be small. A convenient starting point is a 2-kink solution of the s-G which may be written in terms of scattering data as

$$u = 4 \tan^{-1} \left[\frac{1 - \left(\frac{\zeta_1-\zeta_2}{\zeta_1+\zeta_2}\right)^2 \frac{\gamma_1}{2\zeta_1} \frac{\gamma_2}{2\zeta_2} e^{i\,2(k_1+k_2)x}}{\frac{\gamma_1}{2\zeta_1} e^{i\,2k_1x} + \frac{\gamma_2}{2\zeta_2} e^{i\,2k_2x}} \right] \tag{5.4}$$

with

$$k_i = \tfrac{1}{2} (\zeta_i - \frac{1}{4\zeta_i}) , \quad i = 1,2 \ . \tag{5.5}$$

In this γ_1 and γ_2 are the residues of the reflection coefficient $w = b/a$ continued into the upper half ζ-plane and evaluated at the complex eigenvalues ζ_1 and ζ_2 respectively. The combination of (5.5) arises essentially because we are working with the s-G in Lorentz covariant rather than evolution equation form (7).

For the s-G the evolution of the scattering data is rather trivial as we have seen. For (5.2) however it is (7,50,51)

$$\zeta_{i,t} = - \frac{\lambda}{8\gamma_i \mathring{a}_i^{\,2}} \int_{-\infty}^{\infty} \sin \tfrac{1}{2} u \cdot [\phi_1{}^2+\phi_2{}^2]_{\zeta_i} \, dx, \ (i = 1,2) \tag{5.6a}$$

$$\gamma_{i,t} = i2\gamma_i\omega_i - \frac{\gamma_i \ddot{a}_i}{\dot{a}_i} \zeta_{i,t} - \frac{\lambda}{8\dot{a}_i^{\,2}} \int_{-\infty}^{\infty} \sin \tfrac{1}{2} u \cdot \frac{\partial}{\partial\zeta} [\phi_1{}^2+\phi_2{}^2]_{\zeta_i} \, dx \tag{5.6b}$$

$$\left(\frac{b}{a}\right)_t = i\,2\,\omega \left(\frac{b}{a}\right) - \frac{\lambda}{8a^2} \int_{-\infty}^{\infty} \sin \tfrac{1}{2} u \cdot [\phi_1{}^2+\phi_2{}^2]_\zeta \, dx \tag{5.6c}$$

$$\omega = \tfrac{1}{2} (\zeta + \frac{1}{4\zeta}) , \ \gamma_i = \frac{b_i}{\dot{a}_i} \ . \tag{5.6d}$$

where $\dot{a}_i = \left[\frac{da}{d\zeta}\right]_{\zeta=\zeta_i}$ as earlier. The expression $(\phi_1{}^2 + \phi_2{}^2)_\zeta$ denotes the squared eigenfunction belonging to the eigenvalue ζ of the s-G under the scattering transform (2.13) with (2.14) worked however in Lorentz covariant co-ordinates (see article 3 in (7) for more details and reference to a relevant article by Kaup). Notice how the equations (5.6) go over to the equations (2.19) for the evolution of scattering data when $\lambda = 0$: the complication is then only that of the Lorentz covariant form. Even so it should be realised that equations (5.6) are potentially exact: in fact in evolution equation form they are an exact consequence of transcribing u_{xt} via the scattering transform (2.13) (7,25,51).

The equations simplify if we discard the contributions of the continuous part of the spectrum (and therefore (5.6c)) on the assumption that they are negligible. This seems intuitively plausible if we start with (5.4) which has purely discrete scattering data and λ is taken sufficiently small. How small this must be is not clear; but we indicate the few failures of this approximation shortly.

The expression (5.4) represents a breather of the s-G if $\zeta_1 = -\zeta_2^*$, $\gamma_1 = -\gamma_2^*$. The identifications

$$\zeta_1 = \tfrac{1}{2} e^{i\phi} \qquad \gamma_1 = \tan\phi\, e^{i\psi} \tag{5.7}$$

ensure that (5.4) is symmetrical about the origin in its rest frame. If we put $\phi = \cos^{-1}\sqrt{\lambda/4}$ and $\theta = \psi - \phi = 0$ in (5.4) we get the rest frame solution (5.3): equation (5.3) can be given a Lorentz "boost" to satisfy (5.2) and that boosted solution travels without distortion (i.e. is of permanent profile). Perturbation theory investigates departures from this motion for initial data not leading to (5.3).

The time evolution of (5.4) is given by (5.6) with continuum contributions (continuous part of the spectrum contributions) neglected. This was first analysed by my colleague A.L. Mason: provided the breather solution remains valid the equations resulting are

$$C_t = \frac{\lambda}{2}\tan Q \cdot \frac{J(a)+a}{1+a}, \quad a = \cos^2\theta\tan^2\phi,$$

$$J = \frac{a}{\sqrt{1+a}}\tanh^{-1}\frac{a}{\sqrt{1+a}} \quad ;$$

$$Q_t = \sin^2 Q - \left(\frac{\lambda}{4} - C\right)\cos^2 Q$$

$$Q = \tan^{-1}(\tan\theta\cos\phi), \quad C = \cos^2\phi\ . \tag{5.8}$$

These equations show that for initial conditions $\theta = 0$,

$0 < \phi < \cos^{-1}\sqrt{\lambda/4}$ the resulting motion is oscillatory: for

$\phi > \cos^{-1}\sqrt{\lambda/4}$, ϕ eventually reaches $\pi/2$. At this point, the solution becomes a rapidly separating kink and antikink. The breather form is now inappropriate and the analysis has to be repeated using kink parameters. A suitable parametrisation is now

$$\zeta_1 = \frac{i}{2}\, e^{-q} \;,\; \frac{\gamma_1}{2\zeta_1} = \coth q\; e^{p} \tag{5.9}$$

the equations of motion being

$$P_t = P^2 + C - \frac{\lambda}{4} \;,\; P = \coth p \sinh q,\; C = -\sinh^2 q$$
$$C_t = \frac{\lambda}{4}\, P\, \frac{J(a) + a}{1 + a} \;,\; a = \coth^2 q \;\sinh^2 p \;. \tag{5.10}$$

The equations (5.10) are really of the same form as (5.8) and this fact enables a smooth match to be made across the different regions of parametrisation (7).

My colleagues (50) have numerically integrated (5.8) and (5.10) for the initial conditions

$$\theta = 0 \;,\; \cos^2\phi = \frac{1}{4}\,\lambda(1-\epsilon)^{-1}$$

with $\epsilon = 0.1$ or -0.1 and $\lambda = 1.0$. When $\epsilon = -0.1$ the motion is oscillatory as shown in Figure 3. For $\epsilon = 0.1$ break up occurs (Figure 4) and this requires the change of parametrisation when C becomes negative. A comparison of Figure 4 with Figure 2 shows good agreement for break-up in this no-continuum approximation. However, eventually truncation errors cause the parametrisation to break down due to the asymptotic properties of the hyperbolic functions in (5.10). For the oscillatory regime agreement is almost perfect until u first reaches its maximum amplitude. Subsequently, a slight periodic emission of ripples may be discerned representing energy being pumped into the continuum. The net result is to shorten the period of the breather-like oscillation. We have tried to remove this discrepancy by taking some account of the growth of the continuum. The procedure is then extremely complicated, however, and seems unlikely to yield a set of equations with the relative simplicity of (5.8) and (5.10). This situation is indicative of the state of perturbation theoretical studies at the present time. It is noteworthy that in all other respects agreement between theory and direct numerical integration is almost perfect even for the large value $\lambda = 1$.

The reader is referred to article 3 of (7) where a second solution of (5.2), the 4π -"wobbler" is analysed in perturbation theoretical terms. The reader is also referred there for references to the form of perturbation theories based on (5.6) used here - and especially for reference to the pioneering work of Kaup and Newell in this connection. Note that in effect the equations (5.6) are also derived in (25) (cf. eqns. (3.11) in (25)) essentially by inverting the expressions for the variations of the AKNS potentials q,r found in terms of variations in the scattering data. This result is of course exact and illustrates the essentially exact character of equations (5.6).

6. ADDITIONAL REMARKS

The following comments on the material of the lectures were stimulated by discussions at the School and it may be helpful to record them here.

First of all notice (see above eqn. (1.20)) that the inverse method exploits the boundary condition $u \to 0$ as $|x| \to \infty$. The time evolutions (1.20) and (1.21) are obtainable in consequence only because of this. The same remarks apply to the time evolution (2.19) of the AKNS-ZS scattering data. The importance of the boundary conditions $q \to 0$, $r \to 0$ as $|x| \to \infty$ is emphasized below (2.19).

Second the KdV flow is (see (a) below (1.18)) isospectral, i.e. $\lambda_t = 0$. The condition $\zeta_t = 0$ is imposed on the AKNS system (2.13) and (2.14) so that the evolution equation is (2.15). But conversely (2.15) is isospectral since (for v belonging to ζ)

$$\hat{L}_t v = [\hat{A}, \hat{L}]\, v = \zeta v_t - \hat{L}\, v_t, \text{ so } \zeta v_t = (\hat{L}v)_t = \zeta v_t + \zeta_t v.$$

Originally Lax (24) chose $\hat{A}$ to be a skew symmetric differential operator. If $\hat{A}$ is skew the solution $\hat{U}$ of $\hat{U}_t = \hat{A}\,\hat{U}$ is unitary ($\hat{U}$ is the time evolution operator) and $\hat{L}_t = [\hat{A}, \hat{L}]$ implies (7) that $\hat{U}^{-1}\,\hat{L}\,\hat{U}$ is independent of t. From this it follows easily that $\zeta_t = 0$, i.e. the flow is isospectral. It is this aspect of the 'Lax pair' formulation which could be exploited by Moser (for matrices $\hat{A}$ and $\hat{L}$) to solve the many-body problem with x^{-2} interactions described by Calogero in his second lecture.

For evolution equations the important point is that certain evolution equations together with certain scattering problems are isospectral. The transform into scattering data S through the scattering problem $\hat{L}$ is a spectral transform: but the evolution of the scattering data is determined by the flow, i.e. by the particular NEE, and, this may be isospectral or it may be not. When it is isospectral it is simple. But the expressions (5.6) for example show that it is in general very complicated. Equation (3.11c) in (25) shows that the relation for AKNS systems

$$\zeta_{j,t} = -i\,k_j \int_{-\infty}^{\infty} dx\,(\psi_2{}^2\, q_t - \psi_1{}^2\, r_t)_{\zeta_j}$$

is exact for any flows q_t, r_t and scattering problem $\hat{L}v = \zeta v$. For the integrable systems the expression reduces to the isospectral condition $\zeta_{j,t} = 0$ (23,51).

Notice (see above (2.19)) that conditions on B, C in $\hat{A}$ are B, C $\to 0$ as $|x| \to \infty$. Without these the time evolution of the scattering data is not simple enough to be effective. Notice that $A \to \Omega(\zeta)$ as $|x| \to \infty$ and that $\Omega(\zeta)$ consequentially completely determines the form of the integrable evolution equations. This is stressed in the lectures and is made clear by the form of (2.24a) whilst (and cf.(25)) the whole of § 4 is concerned with constructing Hamiltonians flows which (cf. above (4.30)) are identified through $\Omega(\zeta)$.

It is worth noting that, just as (2.24a) contains the s-G, it also contains the next member of the s-G hierarchy corresponding to $\Omega(\zeta) = 4i\ \zeta^{-3}$. It seems to me as easy to find the flow (4.9) and the conservation laws (3.26) by the route I have used as it is to invert $(L^+)^{-3}$ in (2.24a) (or to multiply through by $(L^+)^3$) and to separately find the conservation laws. (This remark is made in reply to a criticism by Calogero.)

In further comment on the inverse method, notice that u in (1.3) is negative. Thus the potential u really is attractive and the solitons correspond to the bound states of u. Of course u in (1.1) could be made positive, in which case $u_t + 6u\,u_x + u_{xxx} = 0$. But the form (1.1) must be used for the inverse method. For $u < 0$, (1.3) is a reflexionless potential with exactly one bound state of the Schrödinger eigenvalue problem with 'energy' eigenvalue $-\lambda_n$: in the corresponding ζ-plane $(-\lambda_n)^{\frac{1}{2}} = i\eta$ $(\eta > 0)$. Notice from (1.23b) that, when there are no bound states (all $c_n = 0$) but there is then reflexion $(b(k) \neq 0)$, $b(\xi)$ is the Fourier transform of $b(k)$. From this one can see (7,23,25) that for u small enough (linearised theory) the whole inverse method is just the usual method of the Fourier transform on x (incidentally the Fourier transform is a canonical transformation). Thus the inverse or spectral transform method is a generalisation of the Fourier transform method for linear partial differential equations, and especially linear evolution equations, to NEEs. Of course it is important to the effectiveness of the method that the isospectral flows are particularly simple and that the Marchenko equation is a linear integral equation. Thus the method reduces the solution of NEEs to a sequence of linear steps.

Third in comment on the geometrical theory of § 3 notice that the 'gauge transformation' (3.14) interprets Ω as a gauge field of Yang-Mills type (cf. Stora's lectures); Θ is a sort of curvature field (corresponding to $F_{\mu\nu}$ in quantum electrodynamics). The integrability condition (3.2) is thus one of vanishing curvature. Notice that B in (3.14) is in the group of gauge transformations, but Ω is in the algebra. From (3.2) one has immediately that $d\omega_1 = \omega_2 \wedge \omega_3$, $d\omega_2 = 2\omega_1 \wedge \omega_2$, $d\omega_3 = -2\omega_1 \wedge \omega_3$ and formally these are defining relations (Maurer-Cartan relations (52)) for an SL(2,R) or SL(2,C) Lie algebra. This really explains the SL(2,R) structure (see the remark above (3.9)). Notice that the AKNS systems in general contain two potentials q and r (or q and q*) so that the pseudo-spherical surface is in general complex and the algebra is SL(2,C): for the s-G $q = -r = -r^*$ and the algebra is SL(2,R).

Fourthly concerning Bäcklund transformations the following remarks may be helpful. The choice of a sufficiently comprehensive useful definition of a Bäcklund transformation is perhaps still an open question. A working definition for present purposes is that it relates the solution u of one NEE to the solution u' of another or the same NEE by

$$u'_x = f_1(u, u_x, u_t, u')$$

$$u'_t = f_2(u, u_x, u_t, u')$$

The transformation from u to u' is a generalisation of a contact transformation in which f_1 and f_2 are extended to include dependence on the a priori unknown new solution u': f_1 and f_2 might also depend on x,t, explicitly but this has not proved helpful; obviously, depending on the order of the non-linear equation, there are extensions relating second derivatives u'_{xx}, etc. to f_1, f_2 and f_3 depending on u', u'_x, u'_t, on u, and on the derivatives of u to order two. A different approach to BTs is via the geometry of § 3. In this formulation the BT is a gauge transformation mapping one pseudo-spherical surface to another (30). The specific form of the BT is not gauge invariant although particular gauge transformations can apparently be chosen to embrace the known forms of BTs (30).

The most useful BTs are invertible. The Miura transformation (1.16a) maps a solution v of (1.17) to a solution u of (1.1). But it is easily checked (and cf. (14)) that (1.16a) does not necessarily imply that, if u satisfies (1.1), v satisfies (1.17). However the two equations of the BT overcome this difficulty. One finds that the integrability condition $u'_{xt} = u'_{tx}$ on the BT

$$u'_x = -2u - 2u'^2$$

$$u'_t = 8u^2 + 4nu'^2 + 2u_{xx} - 4u_x u'_x$$

(this BT contains the Miura transformation $u = \frac{1}{2}(u'_x + u'^2)$), leads to

$$2(u_{xxx} + 12u u_x + u_t) + 2u' (u'_{xxx} - 6u'^2 u'_x + u'_t) = 0.$$

Thus u is a solution of the KdV $\Longleftrightarrow$ u' is a solution of the modified KdV (including the solution u' = 0).

Notice that I state in § 5, by (5.2), that the double s-G

$u_{xt} = \sin u + \frac{1}{2}\lambda \sin \frac{1}{2} u$ has neither an aBT nor an infinity of polynomial conserved densities (p.c.d.s.). This follows from the theorems (49,21) that $u_{xt}=F(u)$ has an aBT and an infinity of p.c.d.s. if and only if $F''(u) + \alpha^2 F(u) = 0$ for some α. From these theorems it also follows that $u_{xt} = u^2$ has neither an aBT nor an infinity of p.c.d.s. Barnsley in his lectures derives $u_{xt} = u^2$ from his isotropic Boltzmann equation and indicates how to find $u_{xt} = u^n$. Thus we conclude that $u_{xt} = u^n$ is not an integrable system in the sense described in these lectures. And Barnsley's method of solving the initial value problem for this family of equations is a distinct, exciting and altogether new method relying on the properties of the Laguerre polynomials.

In drawing connections with material in other lectures at the School I should refer at least to Stora's remarks on the non-linear σ-models (see my brief reference to these above (4.42)) and to the remarks of the two Chudnovskys on the Painlevé transcendentals (my reference to these below (4.42) is in connection with Ising models but a connection with inverse scattering is mentioned: they arise through similarity solutions of the integrable systems: a similarity solution of the s-G for example is scale invariant and depends on the product xt).

Finally those more expert and aware of the fundamental work of Wahlquist and Estabrook (53) on the 'prolongation structures' associated with NEEs will realise that the geometrical arguments of § 3 provide a natural basis for interpreting the pseudopotentials which arise in that prolongation structure theory. A preliminary investigation in this direction has been made by my colleague Sasaki (53). Moreover the arguments of § 3 appear to generalise to arbitrarily large n x n matrix scattering problems with which are associated the Lie algebras SL(n,R) or SL(n,C). P.c.d.s for the 3 x 3 problem have been derived this way by Sasaki (54). The 3 x 3 scattering problem includes that for the '3-wave interaction' solved with different degrees of completeness in 1973 and 1976 (55,56).

REFERENCES

1. Bullough, R.K.: 1978, 'Solitons' in Physics Bulletin, February 1978 pp. 78-82.

2. Bullough, R.K.: 1977, 'Solitons' in 'Interaction of Radiation and Condensed Matter' Vol. 1 IAEA-SMR-20/51 (International Atomic Energy Agency: Vienna, 1977) pp. 381-469.

3. Bullough, R.K. and Caudrey, P.J.: 1978, 'The multiple sine--Gordon equations in non-linear optics and in liquid ^{3}He' in 'Nonlinear evolution equations solvable by the spectral transform' F. Calogero ed. (Pitman: London, 1978) pp. 180-224.

4. Bullough, R.K.: 1978, 'Solitons in Physics' in 'Nonlinear Equations in Physics and Mathematics' NATO Advanced Study Institute Series. A.O. Barut ed. (D. Reidel Publ. Co.: Dordrecht, Holland, 1978) pp. 99-141.

5. Bullough, R.K. and Dodd, R.K.: 1977, 'Solitons. I. Basic Concepts. II. Mathematical Structures' in 'Synergetics. A Workshop' Proc.Intl.Workshop on Synergetics at Schloss Elmau, Bavaria, May 1977. H. Haken ed. (Springer-Verlag: Heidelberg, 1977) pp. 92-119. 'Solitons in Physics and Mathematics' in 'Structural Stability: A symposium in honour of René Thom' W. Güttinger ed. (Springer-Verlag: Heidelberg, 1979).

6. Bullough, R.K., Kitchenside, P.W., Jack, P.M. and Saunders, R.: 1979, 'Solitons in laser physics' Physica Scripta. In press.

7. 'Solitons' R.K. Bullough and P.J. Caudrey eds. (Springer-Verlag: Heidelberg). In the press: to appear 1979.

8. Degasperis, A.: 1978, 'Solitons, Boomerons, Trappons' in 'Nonlinear evolution equations solvable by the spectral transform' F. Calogero ed. (Pitman: London, 1978) pp. 97-126.

9. Gardner, C.S., Greene, J.M., Kruskal, M.D. and Miura, R.M.: 1967, Phys. Rev. Lett. 19, pp. 1095-1097.

10. Kruskal, M.D.: 1978, 'The birth of the soliton' in 'Nonlinear evolution equations solvable by the spectral transform' F. Calogero ed. (Pitman: London, 1978) pp. 1-8.

11. Fermi, E., Pasta, J.R. and Ulam, S.M. 'Studies of Nonlinear Problems' Vol. 1. Los Alamos Rept. LA-1940 (May 1955); 'Collected Works of E. Fermi' Vol. 2 (Univ. of Chicago Press, 1965) pp. 978-88.

12. Zabusky, N. and Kruskal, M.D.: 1965, Phys. Rev. Lett. 15, pp. 240-243.

13. Miura, R.M., Gardner, C.S. and Kruskal, M.D.: 1968, J. Math. Phys. 9, pp. 1204-1209.

14. Miura, R.M.: 1968, J. Math. Phys. 9, pp. 1202-1204.

15. Wahlquist, H.D. and Estabrook, F.B.: 1975, J. Math Phys. 16, pp. 1-7.

16. Gel'fand, I.M. and Levitan, B.M.: 1951, Izv. Akad. Nauk. SSSR, Ser. Mat. 15 p. 309. Translated in American Math. Soc. Translations (AMS: Providence, Rhode Island, 1955) Ser. 2 Vol. 1, p. 253.

17. Kay, I., Courant Institute of Math. Sciences, New York Univ. Rept. No. EM-74, 1955; Kay, I. and Moses, H.E.: 1956, Il Nuovo Cim. 3 pp. 276-304.

18. Zakharov, V.E. and Fadeev, L.D.: 1971, Funkt. Anal. i Ego Prilozh. 5 p. 18.

19. Zakharov, V.E. and Shabat, A.B.: 1973, Zh. Eksp. Teor. Fiz. (USSR) 64 p. 1627 (:1973, Soviet Physics JETP 37, p. 823).

20. Kruskal, M.D.: 1974, 'The Korteweg de Vries equation and related evolution equations' in Lect. Appl. Math. 15 A.C. Newell ed. (AMS: Providence, Rhode Island, 1974) pp. 61-83.

21. Dodd, R.K. and Bullough, R.K.: 1977, Proc. Roy. Soc. (London) A 352, pp. 481-503.

22. Satsuma, J.: 1974, Prog. of Theor. Phys. 52, p. 1396 L.

23. Ablowitz, M.J., Kaup, D.J., Newell, A.C. and Segur, H.: 1973, Phys. Rev. Lett. 31, pp. 125-127: 1974, Studies in Appl. Maths. 53, pp. 249-315.

24. Lax, P.D.: 1968, Comm. Pure Appl. Maths. 21, pp. 467-490.

25. Dodd, R.K. and Bullough, R.K.: 1979, 'The Generalised Marchenko Equation and the Canonical Structure of the A.K.N.S.-Z.S. Inverse Method' Physica Scripta. In press.

26. Chadan, K. and Sabatier, P.C.: 1977, 'Inverse Problems in Quantum Scattering Theory' (Springer-Verlag: Heidelberg, 1977).

27. Hirota, R.: 1972, J. Phys. Soc. Japan 33, pp. 1459-1463; Caudrey, P.J., Gibbon, J.D., Eilbeck, J.C. and Bullough, R.K.: 1973, Phys. Rev. Lett. 30, pp. 237-238; Ablowitz, M.J., Kaup, D.J., Newell, A.C. and Segur, H.: 1973, Phys. Rev. Lett. 30, pp. 1262-1264; Caudrey, P.J., Eilbeck, J.C. and Gibbon, J.D.: 1974, J. Inst. Maths. Applics. 14, pp. 375-386.

28. Flaschka, H. and Newell, A.C.: 1975, 'Integrable systems of non-linear evolution equations' in 'Dynamical Systems Theory and Applications' Springer Lecture Notes in Physics. J. Moser ed. (Springer-Verlag: Heidelberg, 1975).

29. Sasaki, R. and Bullough, R.K.: 1979, 'Non-local conserved quantities for the sine-Gordon equation: geometrical basis' To be published.

30. Sasaki, R.: 1979, 'Geometrisation of soliton equations' Phys.Lett. 71A, pp. 390-393. 'Soliton equations and pseudo--spherical surfaces' Nuclear Physics. In press.

31. Flanders, H.: 1963, 'Differential Forms' (Academic Press: New York, 1963).

32. Dodd, R.K. and Gibbon, J.D.: 1978, Proc. Roy. Soc. (London) A 359. pp. 411-433.

33. Hirota, R.: 1973, Journ. Math. Phys. 14, pp. 805-809; Dodd, R.K. and Bullough, R.K.: 1975, Lett. al Nuovo Cim. 13, pp. 313-318.

34. Lax, P.D.: 1975, Comm. Pure Appl. Maths. 28, pp. 141-187.

35. Gardner, C.S.: 1971, J. Math. Phys. 12, pp. 1548-1551.

36. cf. e.g. Korepin, V.E. and Fadeev, L.D.: 1975, Teoreticheskaya i Matematischeskaya Fiz. 25, p. 147.

37. Dashen, R.F., Hasslacher, B. and Neveu, A.: 1975, Phys. Rev. D 11, pp. 3424-3450.

38. Pohlmeyer, K.: 1976, Comm. Math. Phys. 46, pp. 207-221.

39. Zamolodchikov, A.B.: 1977, Comm. Math. Phys. 55, pp. 183-186: 1977, JETP Lett. 25, pp. 468-471: 1977, ITEP (Moscow) Preprint (ITEP 12). Also see refs. [1.94],[1.95],[1.96] of Ref. 7.

40. Coleman, S.: 1975, Phys. Rev. D 11, pp. 2088-2097.

41. Baxter, R.J.: 1971, Phys. Rev. Lett. 26, pp. 832-833; Johnson, J.D., Krinsky, S. and McCoy, B.M.: 1973, Phys. Rev. A 8, pp. 2526-2547. For the connection with the quantised s-G see Luther, A.: 1976, Phys. Rev. B 14, pp. 2153-2159.

42. Luttinger, J.M.: 1963, J. Math. Phys. 4, pp. 1154-1162; Mattis, D.C. and Lieb, E.H.: 1965, J. Math. Phys. 6, pp. 304-312; Theuman, A.: 1967, J. Math. Phys. 8, pp. 2460 -2467.

43. Tracey, C.A.: 1978, 'Painlevé Transcendents and Scaling Functions of the Two-Dimensional Ising Model' in 'Nonlinear Equations in Physics and Mathematics' NATO Advanced Study Institute Series. A.O. Barut ed. (D. Reidel Publ. Co.: Dordrecht, Holland, 1978) pp. 221-237.

44. Ablowitz, M.J., Ramani, A. and Segur, H.: 1978, 'A connection between nonlinear evolution equations and ordinary differential equations of P-type. I' Preprint January 1979.

45. Rice, M.J.: 1978, 'Charge Density Wave Systems: The ϕ-Particle Model' in 'Solitons and Condensed Matter Physics' A.R. Bishop and T. Schneider eds. (Springer-Verlag: Heidelberg, 1978) pp. 246-253.

46. Timonen, J.: 1979, private communication.

47. Krumhansl, J.A. and Schrieffer, J.R.: 1975, Phys. Rev. B 11, pp. 3535-3545; Guyer, R.A. and Miller, M.D.: 1978, Phys. Rev. A 17, 1205-1217.

48. Bullough, R.K. and Caudrey, P.J.: 1978, 'Optical Solitons and Their Spin Wave Analogues in ^{3}He' in 'Coherence and Quantum Optics IV' L. Mandel and E. Wolf eds. (Plenum: New York, 1978) pp. 767-780.

49. Dodd, R.K. and Bullough, R.K.: 1976, Proc. Roy. Soc. (London) A 351, pp. 499-523.

50. Kitchenside, P.W., Mason, A.L., Bullough, R.K. and Caudrey, P.J.: 1978, 'Perturbation Theory of the Double Sine-Gordon Equation' in 'Solitons and Condensed Matter Physics' A.R. Bishop and T. Schneider eds. (Springer-Verlag: Heidelberg, 1978) pp. 48-51.

51. Kaup, D.J.: 1976, SIAM J. Appl. Math. 31, pp. 121-133.

52. Crampin, M.: 1978, Phys. Lett. 66 A, pp. 170-172.

53. Wahlquist, H.D. and Estabrook, F.B.: 1975, J. Math. Phys. 16, pp. 1-7, and subsequent papers.

54. Sasaki, R.: 1979, 'Pseudopotentials for the general AKNS system' Phys. Lett. To appear.

55. Sasaki, R.: 1979, 'Geometric approach to soliton equations' Preprint.

56. Zakharov, V.E. and Manakov, S.V.: 1973, JETP Lett. 18, pp. 243-245.

57. Kaup, D.J.: 1976, Studies in Applied Mathematics, 55, pp. 9-44.

(<u>Note in proof</u>: concerning the quantisation of the sine-Gordon equation it is usually assumed that the continuous modes of the scattering problem with apparent classical energy $(m^2 + p(\xi)^2)^{\frac{1}{2}}$ do not contribute to the quantised eigenspectrum. For small enough γ_0 (namely the "classical limit") these energies are given by the lowest excitation of the breather — for note from (4.39) that for small γ_0 and $n = 1$, $M_1 \approx m$. The analogy appears to be between the linearised quantised Klein-Gordon equation with harmonic oscillator eigenvalues labelled by a continuous wavenumber ξ and the "<u>breather</u>" eigenvalues (4.39) of the sine-Gordon equation.)

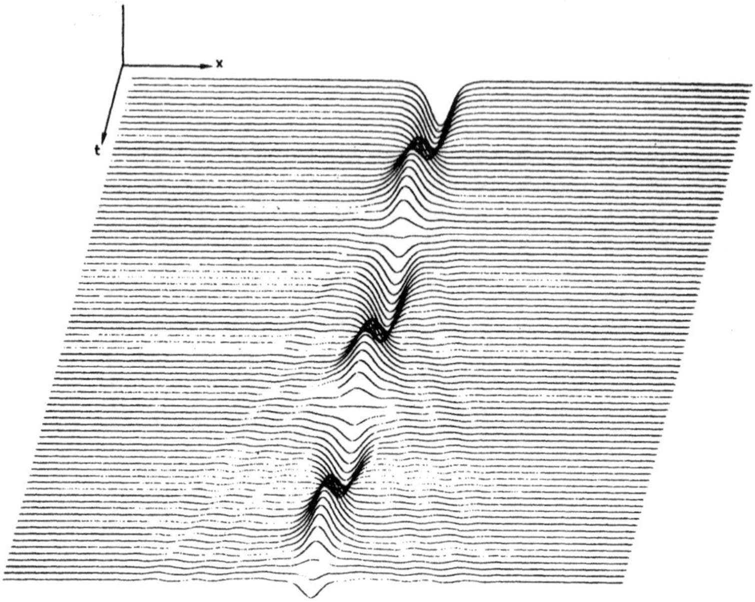

Figure 1. A numerical integration of the double sine-Gordon equation (5.2) in the oscillatory regime. Note the small emission of radiation associated with the continuum. (Time t runs towards the observer.)

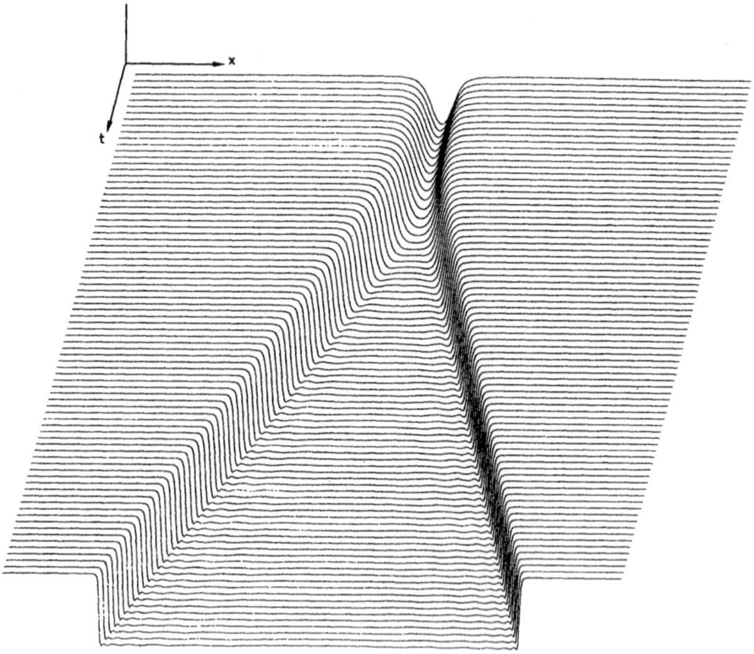

Figure 2. The break-up regime.

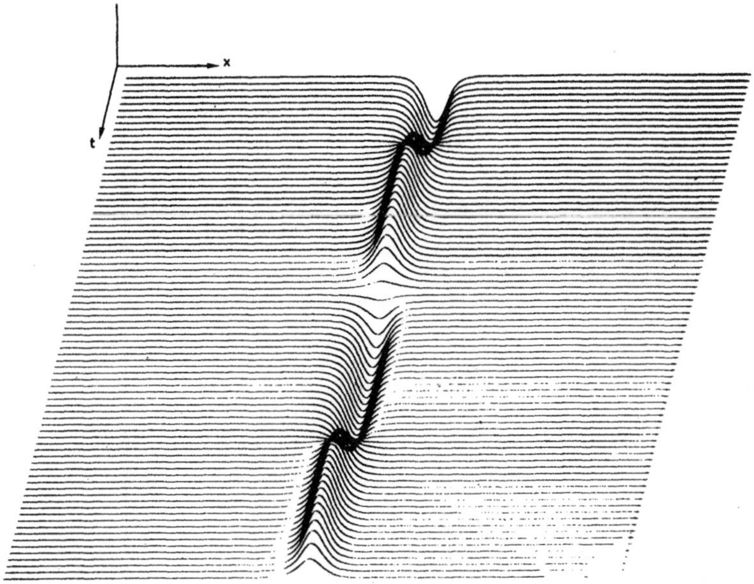

Figure 3. The oscillatory regime described by equation (5.8). Note the longer period of oscillation and the absence of radiation.

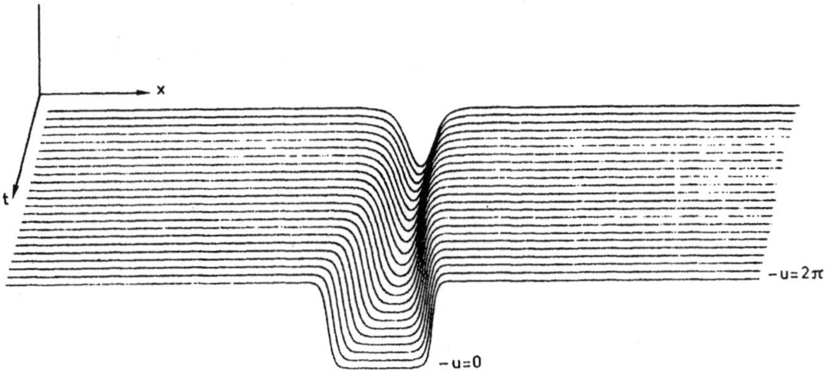

Figure 4. Break-up described by equations (5.8) and (5.10).

NEW RESULTS ON THE NONLINEAR BOLTZMANN EQUATION

Michael F. Barnsley

Dept. of Mathematics
Georgia Tech
Atlanta, GA 30332

Giorgio Turchetti

Istituto di Fisica
University of Bologna
Italy

Recent developments in the theory of the nonlinear Boltzmann equation are described. It is shown how the Abel transform relates two different models, how the method of a priori estimates provides existence theorems, and how the asymptotic behavior of solutions is quantized. The pure solutions of the considered equations are discussed, and finally a general family of solutions is given.

§1. INTRODUCTION

In these two lectures we both review and add to some recent progress towards the general solution of the nonlinear Boltzmann equation. The new advances grow out of the work of Krook and Wu [1,2], Bobylev [3], and Tjon and Wu [4], and they have been presented in various papers by Barnsley and Cornille [5,6] and Barnsley and Turchetti [7,8]. Results appearing here which have not been reported elsewhere include the discussion of the existence question using a priori estimates, given in Section 3.

We consider the relaxation of nonequilibrium distribution functions to Maxwellian form [9]. A strong motivation is the possibility of discovering initial distributions, of a type reproducible in the laboratory, which will evolve in such a way that the high energy tail of the velocity distribution will, at certain typically high velocities and for a certain interval in time occurring a while after initiation, display significant enhancement with respect to both the initial distribution and the final steady state Maxwellian distribution. A parallel image is of a bath tub full of water, which in its initial state has some warm patches and some cold patches. In the evolution of this

C. Bardos and D. Bessis (eds.), Bifurcation Phenomena in Mathematical Physics and Related Topics, 351–370.

'distribution' towards its final uniformly tepid state, is it possible that a hot spot could occur? Of course, in the case of Boltzmann's equation we are talking about quite a different system and the analogy must in no way be stretched. Possible modifications of the Boltzmann distribution function away from a Maxwell distribution can significantly change calculated values of certain gas-phase reaction rates, at given temperature. We have in mind the physically important case of controlled thermonuclear fusion of a confined hydrogen plasma.

It has recently been brought to our attention that some of the results described here have been obtained independently by Ernst [10].

§2. TWO BOLTZMANN EQUATIONS AND THE RELATION BETWEEN THEM

Following Krook and Wu [1,2], and Tjon and Wu [4], we consider an infinite spatially homogeneous and isotropic gas composed of a single species which interacts only through binary collisions. The microscopic state of the system at time t is specified by a distribution function $nf(\underline{v},t) = nf(v,t)$ where v is the velocity variable, $v = |\underline{v}|$, and n is the constant number density. Then $f(\underline{v},t)$ satisfies the Boltzmann equation

$$\frac{\partial f}{\partial t}(\underline{v},t) + \int [f(\underline{v},t)f(\underline{w},t) - f(\underline{v}',t)f(\underline{w},t)]A d\underline{v}' d\underline{w}' d\underline{w} = 0 \qquad (2.1)$$

where the third term provides the contribution from collisions between particles with initial velocities $\underline{v}'$ and $\underline{w}'$ to produce particles with final velocities $\underline{v}$ and $\underline{w}$.

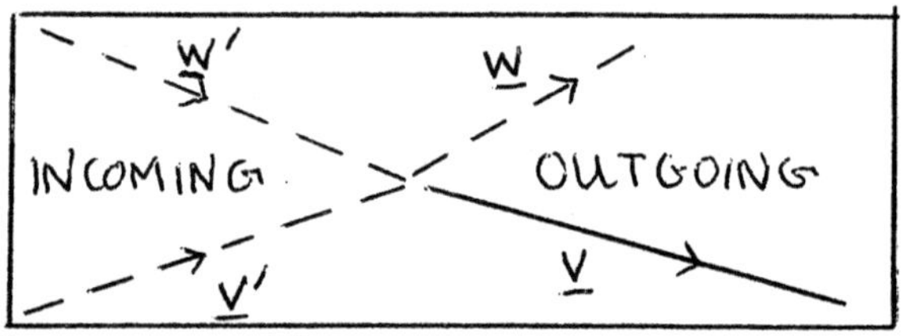

The collision matrix A is such that the normalization and the average energy are conserved quantities. For the isotropic model studied here A is characterized by an elastic cross section

$$\sigma = K/|\underline{v}'-\underline{w}'| \quad , \quad K \text{ a constant} , \qquad (2.2)$$

and (2.1) can be reexpressed in the form

$$\frac{\partial f}{\partial t}(v,t) + f(v,t) = \int_0^\infty w^2 dw \int_0^\pi d\chi \, \sin\chi \int_0^{2\pi} d\varepsilon f(v',t) f(w',t). \qquad (2.3)$$

In this equation $v' = |\underline{v}'|$ and $w = |\underline{w}'|$ are complicated functions of $v = |\underline{v}|$ and $w = |\underline{w}|$: we have

$$v'^2 = \frac{1}{2}(v^2+w^2) - \frac{1}{2}(v^2-w^2)\cos\chi + |\underline{v}\times\underline{w}| \sin\chi \cos\varepsilon , \qquad (2.4)$$

$$w'^2 = \frac{1}{2}(v^2+w^2) = \frac{1}{2}(v^2-w^2)\cos\chi + |\underline{v}\underline{x}\underline{w}|\sin\chi\cos\varepsilon\ , \tag{2.5}$$

where χ and ε are angles which describe the scattering geometry, see [2] for example. The units for velocity v and time t are

$$\text{unit velocity} = (K_B T/m)^{1/2}, \text{ and unit time} = 4\pi nK, \tag{2.6}$$

where K_B is Boltzmann's constant, m is the molecular mass, and T is the kinetic temperature defined by

$$4\pi\int_0^\infty v^4 f(v,t)dv = 3K_B T/m\ . \tag{2.7}$$

The problem is to solve (2.3) for all t > 0 subject to any given initial condition

$$f(v,0) = f_0(v)\quad,\quad 0 \le v < \infty \tag{2.8}$$

which satisfies the normalization conditions

$$4\pi\int_0^\infty v^2 f_0(v)dv = 1\ ,\ 4\pi\int_0^\infty v^4 f_0(v)dv = 3\ . \tag{2.9}$$

Physically $f_0(v)$ should be positive but mathematically the general case is of interest. We will refer to the model Boltzmann equation formulated above as the Original Representation.

With the aid of two linear integral operations L_1 and L_2 the Original Representation is transformed into an attractive form which we will call the Tjon-Wu Representation, as illustrated in the following scheme:

ORIGINAL REPRESENTATION

$$\frac{\partial f}{\partial t} + f = \text{collision integral}$$

$$f(v,0) = f_0(v)$$

$$4\pi\int_0^\infty v^2 f_0(v)dv = 1$$

$$4\pi\int_0^\infty v^4 f_0(v)dv = 3$$

$$F(x,t) = \frac{4\pi}{\sqrt{2}} f(v,t),\ x = \frac{1}{2}v^2$$

? ←--→

TJON-WU REPRESENTATION

$$\frac{\partial \bar{F}}{\partial t}(x,t) + \bar{F}(x,t) = \int_x^\infty \frac{dx'}{x'}\int_0^{x'} \bar{F}(x'-x'',t)\bar{F}(x'',t)dx''$$

$$\bar{F}(x,0) = \bar{F}_0(x) = L_2^{-1}L_1 f_0(v)$$

$$\int_0^\infty \bar{F}_0(x)dx = 1$$

$$\int_0^\infty x^2\bar{F}_0(x)dx = 1$$

$$\bar{F}(x,t) = \bar{f}(v,t),\ x = \frac{1}{2}v^2$$

$L_1 f = G$ ↘ ↗ $L_2^{-1}G = \bar{F}$

$$\frac{\partial^2}{\partial\xi\partial t}(\xi G) + \frac{\partial}{\partial\xi}(\xi G) = G^2$$

$$G(\xi,0) = G_0(\xi) = L_1 f_0(v)$$

INTERMEDIATE REPRESENTATION

The two linear transformations act according to

$$G(\xi,t) = L_1 f = 4\pi\int_0^\infty L_1(\xi,v)f(v,t)dv , \tag{2.10}$$

where the kernel is

$$L_1(\xi,v) = (\pi/2\xi)^{1/2} v \exp[\xi v^2/2]\mathrm{erf}[\xi^{1/2}v] , \tag{2.11}$$

and

$$G(\xi,t) = \int_0^\infty \exp[-\xi x]\bar{F}(x,t)dx \tag{2.12}$$

so that $\bar{F}$ is the inverse Laplace transform of G. G is referred to as the <u>Normalized Moment Generating Function</u> and is discussed in detail in [6]. We will not be concerned with this function here, being instead interested in the direct connection between the Original Representation and the Tjon-Wu Representation. Furthermore, for brevity we will treat L_1 and L_2 in a purely formal fashion; a discussion of their domains and ranges is given in [5].

An important point concerning the Tjon-Wu Representation is that not only is it a much easier problem to think about, but also it is in its own right a Blotzmann equation corresponding to a two dimensional model with diffusive scattering, see [4].

To discover in a formal manner the direct connection between the Original Representation and the Tjon-Wu Representation, we merely observe as in [7] that the relationship between the moments

$$M_n = 4\pi\int_0^\infty v^{2n+2} f(v,t)dv, \text{ and } \bar{M}_n = \int_0^\infty x^n \bar{F}(x,t)dx , \tag{2.13}$$

is

$$\bar{M}_n = \frac{2^n (n!)^2}{(2n+1)!} M_n, \quad n=0,1,2,\cdots . \tag{2.14}$$

This can be seen by expanding the kernels of L_1 and L_2. Moreover, we note that

$$\frac{2^{2n+1}(n!)^2}{(2n+1)!} = \int_0^1 \frac{x^n}{\sqrt{(1-x)}} dx , \quad n=0,1,2,\cdots . \tag{2.15}$$

We now make use of the formal observation that if

$$\alpha_n = \int_0^\infty x^n A(x)dx \quad \text{and} \quad \beta_n = \int_0^\infty x^n B(x)dx \tag{2.16}$$

then

$$\alpha_n \beta_n = \int_0^\infty x^n C(x)dx \quad \text{for} \quad n=0,1,2,\cdots , \tag{2.17}$$

where

$$C(x) = \int_0^\infty \frac{A(y)B(x/y)}{y}\, dy \ . \tag{2.18}$$

Thus we uncover that the direct connection between the Original Representation and the Tjon-Wu Representation is provided by the Abel Transform, or fractional derivative operator [11]. Defining

$$F(x,t) = \frac{4\pi}{\sqrt{2}} f(v,t) \quad \text{where} \quad x = \frac{1}{2} v^2 \ , \tag{2.19}$$

we have

$$\left\{ \begin{array}{l} \bar{F}(x,t) = \int_x^\infty \frac{F(z,t)}{\sqrt{z-x}}\, dz \\ \\ F(x,t) = -\frac{1}{\pi} \frac{d}{dx} \int_x^\infty \frac{\bar{F}(z,t)}{\sqrt{z-x}}\, dz \end{array} \right\} . \tag{2.20}$$

We see that, in this case, the Abel Transform relates a model Boltzmann equation with another seemingly simpler one.

Concerning positivity we observe that the Abel Transform carries positive distributions into positive distributions, while its inverse at least takes decreasing distributions back into positive distributions.

Secure in the knowledge that we have a straightforward mechanism which relates the Original Representation with the Tjon-Wu Representation, we will for the most part work in the latter frame.

In the following table the two known closed form solutions of the studied Boltzmann equation are given. Note that, at least formally, the H-Theorem applies in both representations which means that all normalized positive initial distributions should tend to the stationary solution as $t \to \infty$. In fact a much more general statement than this is rigourously true, as will become apparent.

ORIGINAL REPRESENTATION	TJON-WU REPRESENTATION
$F(x,t) = \frac{1}{\sqrt{\pi}} \exp[-x]$	$\bar{F}(x,t) = \exp[-x]$
Stationary Solution [9]	
$F(x,t) = \frac{\exp[-x/\tau]}{\sqrt{\pi}\,\tau^{3/2}} \left(\frac{5\tau-3}{2\tau} + \frac{(1-\tau)x}{\tau^2} \right)$	$\bar{F}(x,t) = \frac{\exp[-x/\tau]}{\tau^2} \left(2\tau - 1 + \frac{(1-\tau)x}{\tau} \right)$
Similarity Solution [1,3]	
$\tau = 1 - \exp[-(t_0+t)/6]$, $t_0 > 0$, $x = v^2/2$.	

§3. EXISTENCE USING A PRIORI ESTIMATES

Here we indicate how theorems on uniqueness, existence, and smoothness can be established with the aid of some simple a priori estimates.

We will drop the bar from $\bar{F}(x,t)$ in the Tjon-Wu Representation for the time being. Let us assume that the problem

$$\left.\begin{aligned} &\frac{\partial F}{\partial t}(x,t) + F(x,t) = \int_x^\infty \frac{dx'}{x'} \int_0^{x'} dx'' F(x'-x'',t)F(x'',t) = S_x[F,F], \\ &F(x,0) = F_0(x) \geq 0 \ \forall\ 0 \leq x < \infty \text{ with } \int_0^\infty F_0(x)dx = 1, \text{ and } F_0(x) \text{ continuous;} \end{aligned}\right\} \quad (3.1)$$

possesses a solution $F(x,t)$ such that

$$N(t) = \int_0^\infty F(x,t)dx \qquad (3.2)$$

exists for all $t \geq 0$. We can rewrite (3.1) as

$$\frac{\partial}{\partial t}\{e^t F(x,t)\} = e^{-t} S_x[e^t F, e^t F] \qquad (3.3)$$

which indicates that $F(x,t)$ remains positive. Upon integrating (3.1) with respect to x and using (3.2) we obtain the Normalization Equation

$$\left.\begin{aligned} &\frac{dN(t)}{dt} + N(t) = N(t)^2, \\ &N(0) = 1. \end{aligned}\right\} \qquad (3.4)$$

On looking at the solutions of the differential equation here as functions of the initial condition, we see that there are three types: (i) $N(0) > 1 \Rightarrow N(t) \to +\infty$ for some finite $t > 0$, (ii) $N(0) = 1 \Rightarrow N(t) = 1 \ \forall\ t > 0$; (iii) $N(0) < 1 \Rightarrow N(t) \to 0$ as $t \to \infty$. Thus, in an obvious sense $N(t) \equiv 1$ is an unstable solution while $N(t) \equiv 0$ is a stable solution.

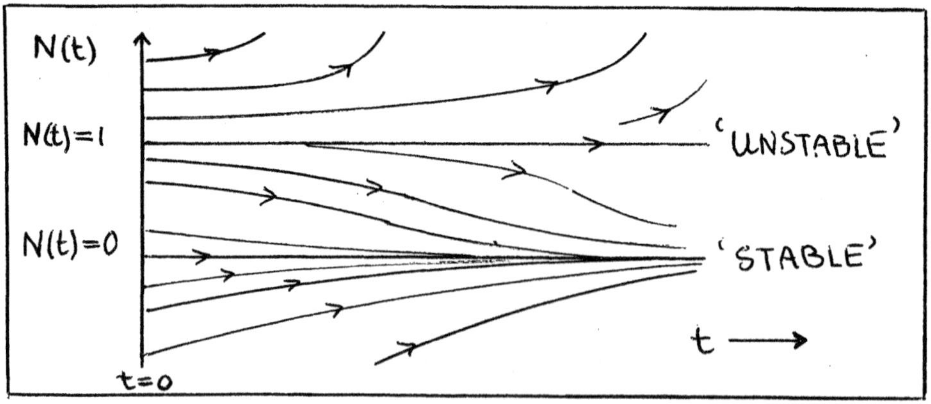

This situation indicates why it is difficult to formulate numerical schemes for solving (3.1); see however [4, 12].

We now observe that $\underline{N}_1(t) = e^{-t}$ constitutes a lower solution for (3.4), i.e.

$$\left.\begin{aligned} \frac{d}{dt}\underline{N}_1(t) + \underline{N}_1(t)^2 &= 0 \le \underline{N}_1(t)^2 , \\ \underline{N}_1(0) &\le N(0) . \end{aligned}\right\} \tag{3.5}$$

Hence it is easy to prove that the sequence $\{\underline{N}_n(t)\}_{n=1}^{\infty}$ of monotone iterates defined by

$$\left.\begin{aligned} \frac{d}{dt}\underline{N}_{n+1}(t) + \underline{N}_{n+1}(t) &= \underline{N}_n(t)^2 , \\ \underline{N}_{n+1}(0) &= 1 \quad , \quad n=1,2,3,\cdots \end{aligned}\right\} \tag{3.6}$$

constitutes a monotone increasing sequence of functions which is uniformly convergent to 1 on any finite interval $t \in [0,T]$.

We now consider the sequence of iterates $\{F_n(x,t)\}_{n=1}^{\infty}$ defined by

$$F_1(x,t) = F_0(x)e^{-t} ,$$

$$\left.\begin{aligned} \frac{\partial}{\partial t}\{e^t F_{n+1}(x,t)\} &= e^t S_x[F_n,F_n] \\ F_n(x,0) &= F_0(x) \; , \; n=1,2,3,\cdots \end{aligned}\right\} \tag{3.7}$$

Lemma 1. $\{F_n(x,t)\}_{n=1}^{\infty}$ is a monotone increasing sequence of functions for $(x,t) \in [0,\infty)\times[0,\infty)$. Each $F_n(x,t)$ is continuous and integrable in x for $0 \le x < \infty$, and is infinitely differentiable in t.

Proof. To see that the sequence is increasing we proceed inductively, noting that

$$\left.\begin{aligned} \frac{\partial}{\partial t}\{e^t(F_{n+1}(x,t)-F_n(x,t))\} &= e^t S_x[F_n-F_{n-1},F_n+F_{n-1}] \\ (F_{n+1}(x,0)-F_n(x,0)) &= 0 . \end{aligned}\right\} \tag{3.8}$$

Here is it implicitly assumed that each $S_x[F_n,F_n]$ exists. To see that this is true we first observe, again by induction, that each $F_n(x,t)$ is of the form

$$F_n(x,t) = F_0(x)e^{-t} + \sum_{m=0}^{M(n)} \theta_{n,m}(t)F_{n,m}(x) \tag{3.9}$$

for some finite integer M(n), where each $F_{n,m}(x)$ has the structure

$$F_{n,m}(x) = S_x[S_x[\cdots S_x[F_0(x),F_0(x)]\cdots],S_x[\cdots S_x[F_0(x),F_0(x)]\cdots]] \tag{3.10}$$

and where each $\theta_{n,m}(t)$ is a positive infinitely differentiable function of $t \in [0,\infty)$. Thus, it suffices to prove (†) that if $G(x)$ is positive and continuous on $0 \le x < \infty$ with $\int_0^\infty G(x)dx = 1$ then the same is true for $S_x[G,G]$. To prove (†) we observe that the integrand in $S_x[G,G]$ is positive and we have the two majorizations (i) $\forall x > 0$,

$$S_x[G,G] \le \frac{1}{x}\int_0^\infty dx'\int_0^{x'} G(x'-x'')G(x'')dx'' = \frac{1}{x} \tag{3.11}$$

and (ii), since $G(x)$ is continuous $\exists\ 0 < c < \infty$ such that

$$G(x) \le c \quad \text{for all} \quad 0 \le x \le 1/c \ ; \tag{3.12}$$

from which we find that for all $0 \le x \le 1/c$

$$S_x[G,G] \le S_0[G,G] = \{\int_0^{1/c}\frac{dx'}{x'} + \int_{1/c}^\infty \frac{dx'}{x'}\}\int_0^{x'} G(x'-x'')G(x'')dx'' \le 2c \tag{3.13}$$

Hence $S_x[G,G] \le \text{Min}\{\frac{1}{x},2c\}$ for all $x \in [0,\infty)$, from which it follows that $S_x[G,G]$ exists and is continuous for all $x \in [0,\infty)$. Moreover, integration by parts is now justified and yields $\int_0^\infty S_x[G,G]dx = 1$, which completes the proof of (t), and thus the proof of the Lemma. ■

Remark: If $F_0(x) \in C^1$ with $\frac{dF_0}{dx}(x) \le 0$ for $x \in [0,\infty)$, then it is straightforward to show that the same is true for $F_n(x,t)$.

Lemma 2. For given $\varepsilon > 0$ the sequence $\{F_n(x,t)\}_{n=1}^\infty$ converges uniformly to a continuous function $\tilde{F}(x,t)$, $x \in [\varepsilon,\infty)$. $\tilde{F}(x,t)$ is integrable over $0 \le x < \infty$, with $\int_0^\infty \tilde{F}(x,t)dx = 1$.

Proof. Since by Lemma 1 each $F_n(x,t)$ is integrable in x, we can define $\underline{N}_n(t) = \int_0^\infty F_n(x,t)dx$. But then on integration of (3.7) we find that the sequence $\{\underline{N}_n(t)\}_{n=1}^\infty$ are precisely the sequence of monotone iterates given by (3.6). In particular, for all positive integers p and q we must have

$$1 \ge \underline{N}_p(t) - \underline{N}_q(t) \ge 0 \quad , \quad \forall\ t \in [0,\infty). \tag{3.14}$$

We now use the observation that if $0 \le x_0 \le x \le \infty$ then

$$\frac{\partial}{\partial t}\{e^t[F_{n+1}(x,t)-F_n(x,t)]\} = e^t S_x[F_n-F_{n-1},F_n+F_{n-1}]$$
$$\le e^t S_{x_0}[F_n-F_{n-1},F_n+F_{n-1}] \le \frac{\partial}{\partial t}\{e^t[F_{n+1}(x_0,t)-F_n(x_0,t)]\} \tag{3.15}$$
$$\text{for} \quad n=2,3,4,\cdots \ .$$

This implies

$$0 \le F_{n+1}(x,t) - F_n(x,t) \le F_{n+1}(x_0,t) - F_n(x_0,t) \ \text{if} \ 0 \le x_0 \le x \le \infty . \tag{3.16}$$

But if for some $x_1 > 0$ the sequence $\{F_n(x_1,t)\}_{n=1}^{\infty}$ is divergent, then there must exist integers p and q such that $F_p(x_1,t) - F_q(x_1,t) > \frac{1}{x_1}$, and (3.16) then implies that

$$F_p(x,t) - F_q(x,t) > \frac{1}{x_1} \quad \text{for all} \quad 0 \le x \le x_1 . \tag{3.17}$$

Integrating the latter, we obtain

$$\underline{N}_p(t) - \underline{N}_q(t) > 1 \tag{3.18}$$

which contradicts (3.14). Thus we conclude that $\{F_n(x,t)\}$ converges for all $(x,t) \in (0,\infty)\times[0,\infty)$. The rest of the theorem now follows with the aid of (3.16) and Dini's Theorem. ∎

If we now define $\tilde{F}(x,t) = \lim_{n\to\infty} F_n(x,t)$ then, using the device of rewriting (3.1) as an integral equation, we can prove the following theorem.

Theorem. $\tilde{F}(x,t)$ is a solution of (3.1). It is nonnegative for all $(x,t) \in (0,\infty) \times [0,\infty)$.

Remark: If $F_0(x) \in C^1$ with $\frac{dF_0}{dx}(x) \le 0$ for all $x \in [0,\infty)$, then one can show that $\tilde{F}(x,t)$ is non-increasing in x for all $t \ge 0$.

Following the above lines it is possible to establish uniqueness in a wide variety of situations. We do not present further results here because our intention has been to indicate the kind of method we use, and because we believe that these will be provided more succinctly by some more sophisticated approach. Note that the constructive theory which follows provides existence for classes of initial conditions which are not treatable by methods of the above type.

§4. A QUANTIZATION ARGUMENT, AND DISCOVERY OF PURE SOLUTIONS.

We continue our examination of the Tjon-Wu equation

$$\frac{\partial F}{\partial t}(x,t) + F(x,t) = \int_x^{\infty} \frac{dx'}{x'} \int_0^{x'} dx'' F(x'-x'',t) = S_x[F,F] \tag{4.1}$$

subject to an initial condition

$$F(x,0) = F_0(x) \quad \text{where} \quad \int_0^{\infty} F_0(x)dx = \int_0^{\infty} xF_0(x)dx = 1. \tag{4.2}$$

Assuming that an integrable solution to this problem exists, we obtain from (4.1) the normalization equations

$$\overset{\circ}{N}(t) + N(t) = N(t)^2 \quad , \quad \text{and} \quad \overset{\circ}{E}(t) + E(t) = N(t)E(t) \tag{4.3}$$

the first of which was discussed in Section 3, and where $E(t) = \int_0^\infty xF(x,t)dx$ corresponds to average energy. Then the initial conditions (4.2) imply $N(t) = E(t) = 1$ for all $t \geq 0$, which expresses conservation of particle number and energy. The stationary solution of (4.1) is e^{-x}.

Thus we seek solutions $F(x,t)$ such that (i) $F(x,t) \to e^{-x}$ as $t \to \infty$, (ii) $\int_0^\infty F(x,t)dx = 1$, and (iii) $\int_0^\infty xF(x,t)dx = 1$, for all $t \geq 0$. To begin, we observe that for non-negative integers N and M,

$$S[x^M e^{-x}, x^N e^{-x}] = P_{N+M}(x)e^{-x} , \tag{4.4}$$

where $P_{N+M}(x)$ is a polynomial in x of degree N+M. Hence we look for solutions in the form

$$F(x,t) = \{\sum_{n=0}^\infty c_n(t)x^n/n!\}e^{-x} \tag{4.5}$$

where (i) $c_n(t) \to \delta_{0,n}$ (=1 if n=0, =0 if n≠0), (ii) $\sum_{n=0}^\infty c_n(t) = 1$ and (iii) $\sum_{n=0}^\infty (n+1)c_n(t) = 1$ for all $t \geq 0$. Substituting (4.5) into (4.1) and equating coefficients of powers of x provides

$$\overset{\circ}{c}_0(t) + c_0(t) = \sum_{N=0}^\infty \sum_{M=0}^\infty \frac{c_N(t)\, c_M(t)}{(N+M+1)} \tag{4.6.0}$$

$$\overset{\circ}{c}_n(t) + c_n(t) = \overset{\circ}{c}_{n-1}(t) + c_{n-1}(t) - \frac{1}{n}\sum_{M+N=n-1} c_M(t)c_N(t) \tag{4.6.n}$$

$$\text{for } n=1,2,3,\cdots .$$

Instead of asking "How does $F_0(x)$ evolve?" we ask "In what manner can $F(x,t) \to e^{-x}$?".

In this spirit we make an <u>Ansatze</u> for the asymptotic behaviour of the $c_n(t)$'s,

$$c_n(t) = \delta_{0,n} + a_n e^{-bt} + \text{Higher Order (H.O.)}. \tag{4.7}$$

Then the equations (4.6) provide

$$1 + (1-b)a_0e^{-bt} = 1 + 2\sum_{n=1}^\infty a_n e^{-bt}/n + \text{H.O.} \tag{4.8.0}$$

$$(1-b)a_n e^{-bt} = (1-b-2/n)a_{n-1}e^{-bt} + \text{H.O.} \tag{4.8.n}$$

$$\text{for } n=1,2,3,\cdots .$$

Assuming that the H.O. terms are negligible as $t \to \infty$, the latter equations imply

$$a_n = a_0 \prod_{m=1}^{n} \left(1 - \frac{2}{(1-b)m}\right) \text{ for } n=1,2,3,\cdots . \quad (4.9)$$

But this implies that the normalization term $\sum_{n=0}^{\infty} a_n$ is divergent unless*

$$b = b_N = N/(N+2) \quad \text{for some integer } N=1,2,3,\cdots , \quad (4.10)$$

in which case

$$a_n = (-1)^n \binom{N+1}{n} a_0 \quad n=0,1,2,\cdots \quad (4.11)$$

and $\sum_{n=0}^{\infty} a_n$ correctly vanishes. Here $\binom{\Gamma}{s} = \frac{\Gamma!}{s!(\Gamma-s)!}$ when $0 \le s \le \Gamma$, $= 0$ when $\Gamma < s$. Moreover, in this case (4.8.0) and (iii) are also satisfied.

To see that we are on the right track, we observe that the similarity solution can be expanded

$$F(x,t) = e^{-x/\tau}\left(\frac{2\tau-1}{\tau^2} + \frac{(1-\tau)x}{\tau^3}\right) = \sum_{n=0}^{\infty} C_n^{sim}(t) x^n e^{-x}/n! \quad (4.12)$$

with, for appropriately chosen constant a,

$$\left.\begin{aligned} C_0^{sim}(t) &= 1 - a^2 e^{-t/3} - 2a^3 e^{-t/2} + \text{H.O.} \\ C_1^{sim}(t) &= \quad + 2a^2 e^{-t/3} + 6a^3 e^{-t/2} + \text{H.O.} \\ C_2^{sim}(t) &= \quad - a^2 e^{-t/3} - 6a^3 e^{-t/2} + \text{H.O.} \\ C_3^{sim}(t) &= \quad + 2a^3 e^{-t/2} + \text{H.O.} \\ C_4^{sim}(t) &= \quad \text{H.O.} \end{aligned}\right\} \quad (4.13)$$

We see that the quantization argument is true in at least one case. Note that here there is a superposition of quantized leading behaviours. It is tempting to look for a family of Pure Solutions characterized by time dependendences which are power series in e^{-bt}, for some fixed value of the quantization parameter in (4.10).

With this in mind we fix N and make the Ansatze

$$F(x,t) = F^N(x,t) = \sum_{n=0}^{\infty} C_n^N(t) x^n e^{-x}/n! \quad (4.14)$$

where $C_n^N(t) = \sum_{m=0}^{\infty} d_m^N (-1)^n \binom{(N+1)m}{n} (ae^{-bt})^m$, a = constant. (4.15)

Upon substituting into (4.6), and equating coefficients of powers of ae^{-bt} we obtain

$$[(N+1)n+1][nb-1]d_n^N + \sum_{p+q=n} d_p^N d_q^N = 0 \ , \ n=0,1,2,\cdots \ . \tag{4.16}$$

The first equation here implies $d_0^N = 0$, or 1, so we choose $d_0^N = 1$ to obtain a nontrivial solution. The second equation then leaves d_1^N arbitrary, so we take $d_1^N = 1$, the arbitrariness being absorbed into the constant a. The remaining equations can then be solved iteratively according to

$$d_n^N = -(N+2) \sum_{p=1}^{n-1} d_p^N d_{n-p}^N / [N(N+1)n^2-(N^2+2N+2)n+(N+2)] \tag{4.17}$$
$$\text{for } n=2,3,4,\cdots \ .$$

This specifies the values $\{d_n^N\}_{n=0}^{\infty}$ uniquely, and our proposed Pure Solution $F^N(x,t)$ is prescribed up to the arbitrary constant a.

Theorem: For fixed $N \in \{1,2,3,\cdots\}$, and any real or complex value for a such that $|a| < (N^2+N-1)/(N+2)$, $F^N(x,t)$ is an entire function in x for all $t \geq 0$. It satisfies the Tjon-Wu equation (4.1), and obeys the conservation conditions $\int_0^\infty F^N(x,t)dx = \int_0^\infty xF^N(x,t)dx = 1$ for all $t \geq 0$. Moreover, as $t \to \infty$, $F^N(x,t) \to e^{-x}$.

Proof. More details can be found in [5]. Briefly, from the iterative relations (4.17) one shows that

$$\left|d_n^N\right| \leq \frac{1}{n-1} \frac{N+2}{N^2+N-1} \quad \text{for all} \quad \begin{array}{l} N=1,2,3,\cdots \ , \\ n=2,3,4,\cdots \ ; \end{array} \tag{4.18}$$

from which one deduces that the function $D^N(z) = \sum_{n=0}^{\infty} d_n^N z^n$ has radius of convergence $R \geq (N^2+N-1)/(N+2)$. One now reexpresses $F^N(x,t)$ in the form

$$F^N(x,t) = \sum_{\ell=0}^{\infty} d_\ell^N L_{\ell(N+1)}(x)(ae^{-bt})^\ell e^{-x} \tag{4.19}$$

where $L_k(x)$ denotes the Laguerre polynomial of order zero and degree k. The proof is completed by comparing (4.19) with the generating function

$$G(x,s) = \sum_{k=0}^{\infty} s^k L_k(x) = \frac{e^{-xs/(1-s)}}{(1-s)} \ , \tag{4.20}$$

whose series expansion is absolutely convergent for all $|s| < 1$. ∎

The nature of the pure solutions is readily investigated by direct calculation using (4.19). Two typical structures appear, depending upon the sign of a and the parity of N. One structure exhibits infinitely many nodes along the x-axis, distributed almost periodically as indicated in Figure 1. As t increases the waves move in the direction of increasing x, leaving behind them a function which converges to e^{-x}. The other structure is positive for $x \in [0,x_0]$, some $x_0 > 0$, after which it has a negative tail. x_0 increases with increasing t, again leaving behind it a function which converges to e^{-x}, see Figure 2. In both cases $F^N(x,t)$ is 'non physical' in that it displays negative values at all finite times. This does not detract from its mathematical interest!

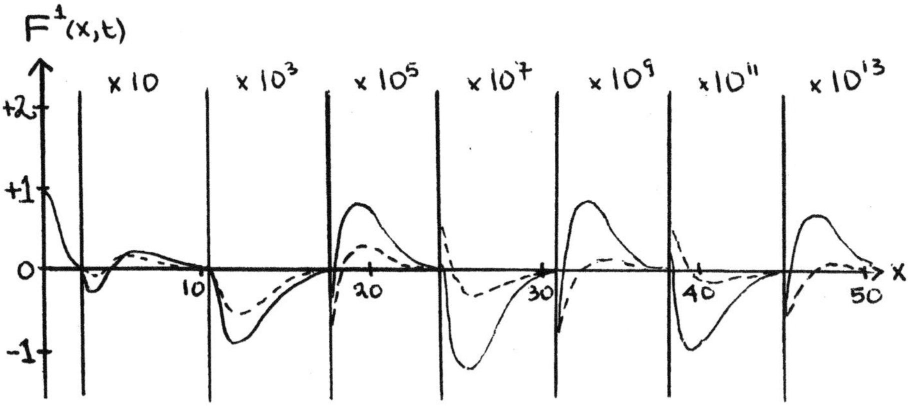

Figure 1. Exact $F^1(x,t)$, a=0.6, for t=0 indicated unbroken, and for t=0.5 indicated by the dashed curve. As t increases the "wavelength" grows while the "amplitude" decreases.

The analytic structure of the pure solutions can be probed with the aid of Baker-Gammel Padé approximants, [13]. The point is that the generating function $D^N(z) = \sum_{n=0}^{\infty} d_n^N z^n$ is well approximated in a significant neighborhood of the origin by near diagonal rational fraction Padé approximants [13]. This is because the series $\{d_n^N\}_{n=0}^{\infty}$ is very nearly a finite sum of geometrical series, as evidenced by the fact that its Hadamard determinents become rapidly very small with increasing order, relative to the size of the d_n^N's involved. It is interesting in this context to note that $D^N(z)$ satisfies the ordinary differential equation

$$\left.\begin{aligned} \frac{(N+1)N}{(N+2)} z^2 \frac{d^2D^N}{dz^2}(z) - z\frac{dD^N}{dz}(z) + D^N(z)(D^N(z)-1) = 0, \\ D^N(0) = \frac{dD^N}{dz}(z)\Big]_{z=0} = 1. \end{aligned}\right\} \tag{4.21}$$

To obtain approximants for $F^N(x,t)$, one begins by forming a rational fraction Padé approximant for $D^N(z)$, say

$$D^N(z) \simeq [(M+1)/M]_{D^N} = 1 + z + \sum_{i=1}^{M} \frac{v_i}{(1-\varepsilon_i z)} . \tag{4.22}$$

The expression on the right here denotes the partial fraction expansion of the approximant, where the v_i's and ε_i's are

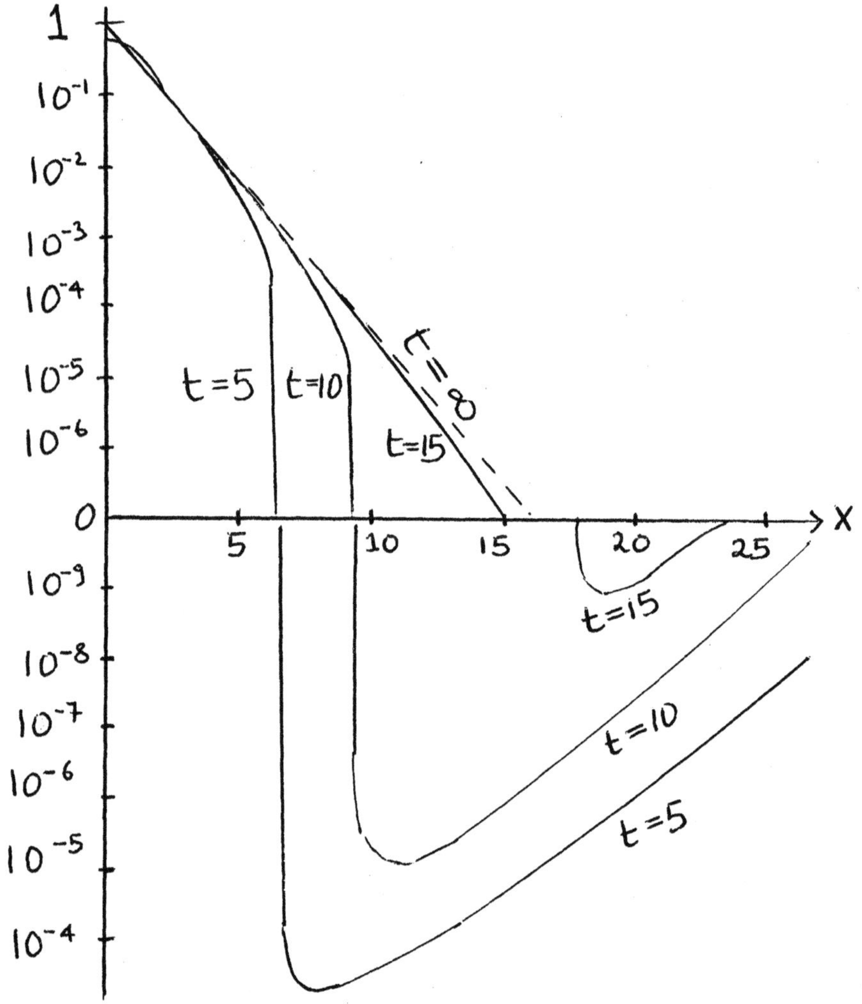

Figure 2. Evolution of $F^1(x,t)$ in time in the case $a = -1$. The vertical scale is logarithmic both in the positive direction and in the negative direction. In this case there is only a single node which moves towards $x = \infty$ as $t \to \infty$.

appropriate complex constants. Upon combining (4.22) with the generating function (4.20) and the expression (4.19) we are led to the generalized Padé approximant

$$F^N(x,t) \simeq F_M^N(x,t) \quad \text{for} \quad xe^{-bt} \text{ not too large,} \tag{4.23}$$

where

$$F_M^N(x,t) = (1-\sum_{i=1}^{M} v_i)e^{-x} + (1-\sum_{i=1}^{M} v_i\varepsilon_i)L_{(N+1)}(x)e^{-x}ae^{-bt}$$
$$+ e^{-x}\sum_{i=1}^{M} v_i G^N(x,\varepsilon_i ae^{-bt}) , \tag{4.24}$$

with

$$G^N(x,z) = \frac{1}{(N+1)}\sum_{m=0}^{N} G\left(x,\ e^{\frac{2m\pi i}{N+1}} z^{\frac{1}{N+1}}\right) \tag{4.25}$$

In particular, the essential structure of the approximant is seen to be carried by G^N because F_M^N consists of a superposition of such terms. (We mean this in the same sense that one can say that the essential structure of a Stieltjes function $\int_0^\infty \frac{dQ(u)}{(1+ux)}$ is carried by for example $(1+x)^{-1}$.) As an illustration, we present the approximant

$$F_1^1(x,t) = e^{-x}(4-2L_2(x)ae^{-t/3}-3G^1(x,-ae^{-t/3})), \tag{4.26}$$

where

$$G^1(x,-ae^{-t/3}) = \frac{\exp\left(\frac{-xae^{-t/3}}{(1+ae^{-t/3})}\right)}{(1+ae^{-t/3})}\left(\cos\left(\frac{x\sqrt{ae^{-t/3}}}{1+ae^{-t/3}}\right) + \sqrt{ae^{-t/3}}\sin\left(\frac{x\sqrt{ae^{-t/3}}}{1+ae^{-t/-}}\right)\right).$$

The following features of the approximant are noted: (i) $F_1^1(x,t) \to e^{-x}$ as $t\to\infty$, (ii) $\int_0^\infty F_1^1(x,t)dx = 1$ for all $t \geq 0$, (iii) $\int_0^\infty xF_1^1(x,t)dx = 1$ for all $t \geq 0$, (iv) for $a > 0$, $F_1^1(x,t)$ has an almost periodic character in x for all finite t, (v) for $a < 0$, however small, the approximant blows up at some finite negative t.

§5. MIXTURES OF PURE SOLUTIONS TO PROVIDE GENERAL SOLUTIONS

We begin with the idea of mixing two pure solutions $F^M(x,t)$ and $F^N(x,t)$ with $M \neq N$. We seek a solution $F^{M,N}(x,t)$ which, at low asymptotic order, appears as a superposition of pure solutions, namely as $t \to \infty$

$$F^{M,N}(x,t) = F^M(x,t) + F^N(x,t) - e^{-x} + \text{H.O.} \tag{5.1}$$

Thus we make the Ansatze

$$F^{M,N}(x,t) = \sum_{n=0}^{\infty} c_n(t) x^n e^{-x}/n! \tag{5.2}$$

where

$$c_n(t) = \sum_{p,q=0}^{\infty} d_{p,p}(-1)^n \binom{(M+1)p+(N+1)q}{n} \left(a^{(M)} e^{-b_M t}\right)^p \left(a^{(N)} e^{-b_N t}\right)^q. \tag{5.3}$$

Then upon substituting into equations (4.6), and equating coefficients of $\left(e^{-b_M t}\right)^p \left(e^{-b_N t}\right)^q$, for each p and q, we find that (5.2) formally satisfies (4.1) provided that

$$\Big((M+1)p + (N+1)q + 1\Big)\Big(pb_M + qb_N - 1\Big) d_{p,q} = - \sum_{\substack{r+u=p \\ s+v=q}} d_{r,s} d_{u,v} . \tag{5.4}$$

From the latter we readily discover that we must choose $d_{0,0} = 1$ for a nontrivial solution, and that we can take $d_{1,0} = d_{0,1} = 1$ without loss of generality. All of the remaining $d_{p,q}$'s, for $p \geq 1$ and $q \geq 1$, can then be uniquely determined in a recursive manner. Changing the order of summation in (5.2) we thus derive the formal two-parameter family of solutions

$$F^{M,N}(x,t) = \sum_{y \in Y} \theta_y(t) L_y(x) e^{-x} \tag{5.5}$$

where

$$\theta_y(t) = \sum_{\{p,q \geq 0 \,|\, (M+1)p+(N+1)q=y\}} d_{p,q} \left(a^{(M)} e^{-b_M t}\right)^p \left(a^{(N)} e^{-b_N t}\right)^q, \tag{5.6}$$

with

$$Y = \{y \,|\, y=(M+1)p + (N+1)q;\ p,q \text{ nonnegative integers}\}. \tag{5.7}$$

Under various different assumptions on the constants $a^{(M)}$ and $a^{(N)}$ it can be shown that the series (5.5) is absolutely convergent for all $t \geq 0$ and provides a bona fide solution of (4.1). Here we merely observe that in the case $a^{(M)} = 0$, $F^{M,N}(x,t)$ reduces to the pure solution $F^N(x,t)$, and that in the case $M = 1$, $N = 2$, $a^{(1)} = -a^2$, and $a^{(2)} = -2a^3$ where a is a constant, $F^{1,2}(x,t)$ reduces to the similarity solution (4.12) of Bobylev and Krook and Wu.

More generally, as described in [5], we can follow the above lines to form mixtures of any number of pure solutions. The results are equivalent to the those obtained via the Hilbert space formulation of the next section.

§6. GENERAL SOLUTION IN A HILBERT SPACE

Let h be the Hilbert space of real valued functions $f(x)$, defined on $0 \le x \le \infty$, corresponding to the symmetric inner product

$$<f_1,f_2> = \int_0^\infty f_1(x)f_2(x)e^x dx \ . \tag{6.1}$$

Then we assume that the solution $F(x,t)$ of (4.1) belongs to h for all $t \ge 0$, so that it admits a representation

$$F(x,t) = \sum_{p=0}^{\infty} \theta_p(t)\mathcal{Q}_p(x) \ , \tag{6.2}$$

where

$$\mathcal{Q}_p(x) = e^{-x}L_p(x) \qquad p=0,1,2,\cdots \tag{6.3}$$

constitutes a complete orthonormal basis set for h. If we now project the equation (4.1) along $\mathcal{Q}_p$ for $p=0,1,2,\cdots$ we find

$$\theta_0(t) = 1 \text{ for all } t, \text{ since } N(t) = 1 \ , \tag{6.4}$$

$$\theta_1(t) = 0 \text{ for all } t, \text{ since } E(t) = 1 \ , \tag{6.5}$$

and

$$\mathring{\theta}_p(t) + \frac{p-1}{p+1}\theta_p(t) = \frac{1}{(p+1)} \sum_{k=1}^{p-1} \theta_k(t)\theta_{p-k}(t) \qquad p=2,3,4,\cdots \tag{6.6}$$

The later set of equations can be solved successively subject either to (i) initial conditions

$$\theta_p(0) = \alpha_p = <F_0,\phi_p> \quad p=2,3,4,\cdots \tag{6.7}$$

or else subject to (ii) final conditions

$$\lim_{t\to\infty} e^{\frac{Nt}{N+2}} \theta_{N+1}(t) = a^{(N)} \ , \quad N=1,2,3, \quad , \tag{6.8}$$

where the $a^{(N)}$'s are constants to be prescribed. (i) is discussed in [6]. The $a^{(N)}$'s in case (ii) correspond to the components of the pure solutions involved in the general solution of the form (6.2). In [5] it is shown how the requisite $a^{(N)}$'s can be determined successively in terms of the initial conditions α_p, $p=2,3,\cdots$.

The result of solving (6.6) subject to (ii) can be expressed as follows: Let $\mathbb{R}^\infty$ denote the set of infinite dimensional vectors with real elements. Put

$$\underline{b} = \frac{1}{3}, \frac{2}{4}, \frac{3}{5}, \cdots \qquad , \underline{b} \in \mathbb{R}^\infty , \tag{6.9}$$

$$\underline{a} = (a^{(1)}, a^{(2)}, a^{(3)}, \cdots), \ \underline{a} \in \mathbb{R}^\infty , \tag{6.10}$$

$$\underline{N} = (2,3,4,\cdots) \qquad , \underline{N} \in \mathbb{R}^\infty , \tag{6.11}$$

Furthermore, let $V \subset \mathbb{R}^\infty$ denote the set of infinite dimensional vectors

$$\underline{v} = (p,q,r,\cdots) \quad , \quad \underline{v} \in V \tag{6.12}$$

whose components are nonnegative integers, only finitely many of which are different from zero. Then

$$\theta_p(t) = \sum_{\{\underline{v}\in V:\underline{N}\cdot\underline{V}=p\}} d_{\underline{v}} e^{-\underline{b}\cdot\underline{V}t} \underline{a}^{\cdot\underline{v}} \tag{6.13}$$

where the numbers $d_{\underline{v}}$ are determined recursively from

$$(\underline{N}\cdot\underline{V}+1)(1-\underline{b}\cdot\underline{v})d_{\underline{v}} = \sum_{\substack{\underline{v}_1+\underline{v}_2=\underline{v}\\ \underline{v}_1,\underline{v}_2\in V}} d_{\underline{v}_1} d_{\underline{v}_2} \quad , \quad \underline{v} \in V , \tag{6.14}$$

together with the definitions

$$d_{0,0,0,\cdots} = d_{0,0,\cdots,0,1,0,\cdots} = 1 . \tag{6.15}$$

These latter expressions are the generalizations of (4.10) and (5.4). We have used the notations

$$\underline{r}\cdot\underline{s} = \sum_{i=1}^{\infty} r_i s_i \text{ and } \underline{r}^{\cdot\underline{s}} = \begin{cases} 1 & \text{if } \underline{s} = 0 \\ \prod_{\{i|r_i\neq 0\}} r_i^{s_i} & \text{otherwise.} \end{cases} \tag{6.16}$$

The connection between the present formalism and that given earlier is readily seen. If we choose all of the $a^{(M)}$'s zero except $a^{(N)}$, then (6.2) reduces to $F^N(x,t)$, while if we choose two of the $a^{(M)}$'s nonzero it becomes a mixture of two pure solutions. More generally any $F_0(x) \in h$ determines uniquely the values of the $a^{(N)}$'s such that $F(x,0) = F_0(x)$, and thus $F(x,t)$ defined by (6.2) constitutes, formally at least, a general family of solutions of (4.1).

The formal solutions derived here are substantiated as bona fide solutions when the assumption that $F(x,t)$ belongs to h is justified. Thus we would like to know general conditions under which the implication '$F_0(x) \in h \Rightarrow F(x,t) \in h$ for all $t \geq 0$,' is true. So far we have obtained the following results.

(α) If either all of the $a^{(N)}$'s have the same sign, or else if they alternate in sign, then $|\theta_p(t)|$ is a monotone decreasing function of t for all p, and the solution remains in h if it starts in it. For example, this condition is satisfied if $F_0(x) = e^{-x}\{1+\beta L_{2M}(x)\}$, where β is a parameter which can be chosen so that $F_0(x) \geq 0$ for all $x \geq 0$.

(β) If $|a^{(N)}|^{1/N} < 1$ for all large N, then F(x,t) belongs to h for all $t \geq 0$.

(γ) If $\int_0^\infty F_0(x) e^{x/2} dx < \infty$ and $F_0(x) \in h$, then F(x,t) remains in h at least for some finite time.

(δ) All numerical calculations which we have done to date indicate that regardless of the initial condition, the solution always arrives in h after a short time, and remains in it thereafter. Based on these observations we conjecture that $F_0(x) \in h \Rightarrow F(x,t) \in h$ for all $t > 0$.

Upon applying the inverse Abel transformation (2.20) to the above results, we find that the solution $\bar{F}(x,t) = \sum_{n=0}^{\infty} \theta_p(t) L_p(x) e^{-x}$ in the Tjon-Wu Representation (we've put the bar back on F) becomes $F(x,t) = \sum_{n=0}^{\infty} \theta_p(t) L_p^{(1/2)}(x) e^{-x}$ in the Original Representation. Thus all of our analysis works equally well with respect to (2.3) with initial conditions normalized as in (2.9).

In practice, we have been able to calculate F(x,t) and $\bar{F}(x,t)$ for a wide variety of initial distributions. These calculations are based directly on the Laguerre expansion formulas, and the series are found to converge reasonably fast for a large range of x-values. The range of convergence for a given order of truncation can be dramatically increased with the aid of a convergence acceleration technique based on the use of Padé approximants, as described in [8] .

Acknowledgement

The authors thank C. Bardos, D. Bessis, and T. T. Wu for helpful discussions.

REFERENCES

[1] M. Krook and T. T. Wu, Phys. Rev. Lett. 36(1976), 1107.

[2] M. Krook and T. T. Wu, Physics of Fluids 20(1977), 1589.

[3] A. V. Bobylev, Soviet Physics Doklady 20(1976), 822.

[4] J. A. Tjon and T. T. Wu, Phys. Rev. A.

[5] M. F. Barnsley and H. Cornille, Proc. Roy. Soc. Lond. to appear.

[6] M. F. Barnsley and H. Cornille, J. Math. Phys. to appear.

[7] M. F. Barnsley and G. Turchetti, Phys. LHrs. (1979).

[8] M. F. Barnsley and G. Turchetti, Nuovo Cimento LHrs. (1979).

[9] J. C. Maxwell, Scientific Papers edited by W. D. Niven, (Dover, New York, 1890), Vol. 2, p. 37 ff.

[10] M. H. Ernst, Phys. Letters (1979) 69A, 390; Ibid. 70A, 183.

[11] I. M. Gelfand and G. E. Shilov, Generalized Functions, (Academic, 1964), Vol. 1, p. 115 ff.

[12] J. A. Tjon, Phys. Letters(1979) 70A, 369.

[13] G. A. Baker Jr., 'The Padé approximant method and some related generalizations'. In 'The Padé Approximant in Theoretical Physics' (Eds. G. A. Baker, Jr., and J. L. Gammel), (New York: Academic Press, 1965).

SOLVABLE MANY-BODY PROBLEMS AND RELATED MATHEMATICAL FINDINGS (AND CONJECTURES)

F. Calogero

Dept. of Applied Maths., Queen Mary College, London
*Istituto di Fisica, Università di Roma, 00185 Roma,
Istituto Nazionale di Fisica Nucleare, Sezione di Roma

This is a terse survey of recent results and ideas.

The purpose and scope of this paper is to survey selectively certain recent results on classical one-dimensional many-body problems, focusing on related mathematical developments. Since the field is vast, we are able to provide only a flair of these topics. The reader interested in pursuing these matters will find more details in the review paper [1] and especially in the original references quoted in [1] and/or below.

We start from the classical one-dimensional n-body problem characterised by the hamiltonian

$$H = \frac{1}{2} \sum_{j=1}^{n} (p_j^2 + q_j^2) + \sum_{j>k=1}^{n} (q_j - q_k)^{-2} , \qquad (1)$$

that has played a key role in this field [2]. Here, and always in the following, a prime appended to a summation symbol indicates that the singular term in the sum must be omitted.

The hamiltonian (1) describes n (unit-mass) point particles on the line, acted upon by an external harmonic potential and interacting pairwise via a "centrifugal" inverse-square repulsive potential. The corresponding equations of motion read

$$\ddot{q}_j + q_j = 2 \sum_{j=1}^{n} (q_j - q_k)^{-3} , \quad j = 1, 2, \ldots, n. \qquad (2)$$

There is clearly a unique (up to permutations) equilibrium configuration,

*permanent address

C. Bardos and D. Bessis (eds.), Bifurcation Phenomena in Mathematical Physics and Related Topics, 371–384.

$$q_j(t) = x_j, \quad j = 1, 2, \ldots n, \tag{3}$$

the set of n numbers x_j being the solution of the system of nonlinear algebraic equations

$$x_j = 2 \sum_{k=1}^{n}{}' (x_j - x_k)^{-3}, \quad j = 1, 2, \ldots n. \tag{4}$$

Below we show that these numbers x_j have a remarkable mathematical significance, being just the n zeros of the Hermite polynomial of order n, [3]

$$H_n(x_j) = 0, \quad j = 1, 2, \ldots, n. \tag{5}$$

The small oscillations of the system (7) around its equilibrium configuration (3) can be investigated in the standard fashion, namely setting

$$q_j(t) = x_j + \varepsilon\, \xi_j(t), \quad j = 1, 2, \ldots n, \tag{6}$$

in (2), expanding in ε and keeping only linear terms. There obtains for the vector

$$\xi(t) = (\xi_1(t), \xi_2(t), \ldots, \xi_n(t)) \tag{7}$$

the equation

$$\ddot{\xi}(t) + B^2(t) = 0, \tag{8}$$

the real symmetrical matrix B^2, of order n, being defined as

$$B^2 = 1 + 6 A^{(2)} \tag{9}$$

$$A^{(2)}_{jk} = \delta_{jk} \sum_{\ell=1}^{n}{}' (x_j - x_\ell)^{-4} - (1 - \delta_{jk})(x_j - x_k)^{-4}. \tag{10}$$

The notation here anticipates a property of the matrix B^2, namely that it is the square of another, simpler, matrix (see below). In any case the physics of the problem clearly implies that all the eigenvalues of B^2 are positive (and different), since the system (2) is obviously stable. We therefore term b_m^2 the eigenvalues of B^2 and introduce the corresponding eigenvectors $v^{(m)}$ writing

$$B^2 v^{(m)} = b_m^2 v^{(m)}, \quad m = 1, 2, \ldots n. \tag{11}$$

The small oscillations of the system (2) are then given by the standard decomposition in normal modes:

$$\xi_j(t) = \sum_{m=1}^{n} [c_m \cos(b_m t) + s_m \sin(b_m t)]\, v_j^{(m)}, \quad j = 1, 2, \ldots n. \tag{12}$$

By convention we assume all the b_m's to be positive; and we show below that they are given by the extremely simple formula

$$b_m = 1 + n - m, \quad m = 1, 2, \ldots, n. \tag{13}$$

(Note that we could just as well have written, at this stage, $b_m = m$ in place of $b_m = 1 + n - m$; the second form is preferable for notational consistency with the definition of the eigenvectors $v^{(m)}$ in (11); see below.) This implies that the small oscillations of the system (2) around its equilibrium configuration (3) are completely periodic, with period 2π (for any initial condition; since (13) and (12) clearly imply

$$\xi_j\,(t + 2\pi) = \xi_j\,(t)\ , \quad j = 1, 2, \ldots, n, \tag{14}$$

for any choice of the constants c_m and s_m).

Actually, not only the small oscillations of the system (2) are completely periodic with period 2π, but the general solution of (2) is itself completely periodic with this period; a property that underscores the very special nature of the hamiltonian (1). Indeed this property had been originally conjectured [2a] on the basis of the solution of the quantal problem for the system characterised by this hamiltonian; its validity is clearly implied by the following result [2c, d, e, f], that provides a fairly explicit solution for the initial value problem of (2): the solution of (2) characterised by the initial values $q_j(o)$, $\dot{q}_j(o)$, is given by the n eivenvalues of the hermitian matrix of order n

$$\begin{aligned} M_{jk}(t) &= \delta_{jk}\,[q_j(o)\cos(t) + \dot{q}_j(o)\sin(t)] + \\ &\quad + i\,(1-\delta_{jk})\sin(t)\,[q_j(o) - q_k(o)]^{-1} \end{aligned} \tag{15}$$

namely

$$M(t)\,w^{(m)}(t) = q_m(t)\,w^{(m)}(t), \quad m = 1, 2, \ldots, n, \tag{16}$$

where we have evidently denoted by $w^{(m)}(t)$ the eigenvectors of $M(t)$.

The interested reader is referred, for a proof of this result, to the original papers [2c, d, e, f] (for a more pedestrian version of the proof see [1]). Note that the initial condition for the coordinates q_j (i.e., the identification of $q_j(o)$) follows trivially from (15) and (16), while the initial condition for the momenta (i.e., the identification of $\dot{q}_j(o)$ from (15) and (16)) is less obvious, since it clearly also reflects the time dependence of the eigenvectors.

Another remarkable property of the system (2) refers to its relationship to the system of equations

$$\ddot{r}_j = 2 \sum_{j=1}^{n}{}' (r_j - r_k)^{-3} \quad , \quad j = 1, 2, \ldots, n, \tag{17}$$

that clearly obtain from the hamiltonian

$$H = \tfrac{1}{2} \sum_{j=1}^{n} p_j^2 + \sum_{j>k=1}^{n} (r_j - r_k)^{-2} \quad , \tag{18}$$

that differs from (1) due to the elimination of the external harmonic potential. This is itself a remarkable system, whose solution is known in rather explicit form, being given by the result [2]: the solution of (17) characterised by the initial values $r_j(o)$, $\dot{r}_j(o)$, is given by the n eigenvalues of the hermitian matrix of order n,

$$N_{jk}(t) = \delta_{jk} \left[r_j(o) + \dot{r}_j(o)\, t \right] + i\,(1 - \delta_{jk})\, t \left[r_j(o) - r_k(o) \right]^{-1}, \tag{19}$$

namely

$$N(t)\, y^{(m)}(t) = r_m(t)\, y^{(m)}(t), \quad m = 1, 2, \ldots, n, \tag{20}$$

where we have indicated by $y^{(m)}(t)$ the eigenvectors of $N(t)$.

It can be easily shown from these formulae that the trajectories of (17) have the property

$$\{ r_j(t) - [p_j^{(\pm)}\, t + a_j^{(\pm)}]\}_{t\to\pm\infty} = 0, \quad j = 1, 2, \ldots, n, \tag{21}$$

with

$$p_j^{(+)} = p_j\,(+\infty) = p_{n+1-j}^{(-)} \equiv p_{n-1-j}(-\infty), \quad j = 1, 2, \ldots, n, \tag{22a}$$

$$a_j^{(+)} = a_{n+1-j}^{(-)} \quad , \quad j = 1, 2, \ldots, n, \tag{22b}$$

where we have labeled the particles according to their ordering on the line. Also this property had been originally discovered in the quantal context [2a].

The remarkable relation between the solutions of (2) and (17) (mentioned above) is given by the simple formulae [4]

$$q_j(t) = \cos(t) \quad r_j\,[\mathrm{tg}(t)] \qquad j = 1, 2, \ldots, n, \quad (23a)$$

$$r_j(t) = (1 + t^2)^{1/2}\, q_j \quad [\mathrm{artcg}\ (t)]\ , \quad j = 1, 2, \ldots, n. \quad (23b)$$

These formulae are implied by a comparison of (15) and (19); but can also be trivially verified by direct insertion in (2) and (17); indeed there is a larger class of equations for which similar relationships hold [4].

Let us return now to the equilibrium configuration (3) of (2). We may remark incidentally that (23b) and (3) imply that a special solution of (17) is provided by the explicit formula

$$r_j(t) = (1 + t^2)^{1/2} \quad x_j, \quad j = 1, 2, \ldots, n. \qquad (24)$$

Moreover (11) and (13) imply that the matrix B^2, defined by (9) and (10), has the eigenvalues m^2, $m = 1, 2, \ldots, n$; while (3) and (15) clearly imply that the hermitian matrix L, of order n, defined by

$$L_{jk}(t) = \delta_{jk} \cos\,(t)\, x_j + i\,(1-\delta_{jk}) \sin\,(t)(x_j-x_k)^{-1} \qquad (25)$$

has the (t-independent !) eigenvalues x_j, $j=1, 2, \ldots, n$.

These results display remarkable properties of the numbers x_j corresponding to the equilibrium configuration (3) of the system (2). The interest of these properties is highlighted by the identification of these numbers with the n zeros of the Hermite polynomial of order n; a result that was already mentioned above and that shall now be proved. But let us first call attention to the remarkable appearance of Hermite polynomials in connection with the classical dynamical system characterised by the hamiltonian (1), whose relationship to a (higher dimensional) pure oscillator problem (namely, a problem characterised by the hamiltonian (1) without the last term) is evident from (15) (for more details see [2c, d, e, f]); while of course Hermite polynomials also play a fundamental role in the solution of the quantal pure oscillator problem.

In order to prove the result mentioned above we now introduce a novel approach, itself of some interest: the relationship between many-body problems and the motion of zeros of solutions of partial differential equations [5].

Consider the partial differential equation

$$\psi_{tt} + \tfrac{1}{2} [\psi_{xx} - 2 x \psi_x + 2 n\psi] = 0, \quad \psi \equiv \psi(x, t) \quad . \tag{26}$$

This equation admits solutions that are polynomials of degree n in x. It is convenient to represent such a solution in the following two forms:

$$\psi(x, t) = \prod_{j=1}^{n} [x - q_j(t)] \quad , \tag{27}$$

$$\psi(x, t) = 2^{-n} H_n(x) + \sum_{m=1}^{n} c_m(t) H_{n-m}(x) \tag{28}$$

where of course the H_n's are the Hermite polynomials, namely the solutions of the ordinary differential equation

$$H''_m(x) - 2x H'_m(x) + 2m H_m(x) = 0 \tag{29}$$

with normalisation

$$\lim_{x\to\infty} [H_m(x) / (2 x)^m] = 1 \; . \tag{30}$$

Note that the simultaneous validity of the two formulae (27) and (28) implies a nonlinear bijective mapping between the quantities $q_j(t)$ and $c_m(t)$. On the other hand (26) and (27) imply for the zeros $q_j(t)$ of $\psi(x, t)$ the equations of motion

$$\ddot{q}_j + q_j = \sum_{k=1}^{n}{}' (1 + 2 \dot{q}_j \dot{q}_k)(q_j - q_k)^{-1} \; , \; j = 1, 2, \ldots n, \tag{31}$$

while (26), (28) and (29) imply for the coefficients $c_m(t)$ the equations

$$\ddot{c}_m + m c_m = 0 \; , \quad m = 1, 2, \ldots, n \quad . \tag{32}$$

Clearly the latter equations are immediately solved

$$c_m(t) = c_m(o) \cos(m^{1/2} t) + \dot{c}_m(o) m^{-\frac{1}{2}} \sin(m^{1/2} t), \quad m = 1, 2, \ldots, n, \tag{33}$$

and through these formulae and the nonlinear mapping mentioned above it is therefore also possible to solve the equations of motion (31), that may be interpreted as the equations of motion of a many-body problem (with "velocity-dependent" forces). Note that, since the equations of motion (32) are clearly hamiltonian, and since the mapping between the variables q_j and c_m, being a point transformation, is clearly canonical, the equations of motion (31) are also hamiltonian (but it is not obvious how to write them in hamiltonian form for general n).

The equations of motion (31) are merely one instance of a large class of dynamical systems that can be solved by the technique we have just outlined; the interested reader is referred to [5] for a more extended treatment. Here we focus attention only on the equilibrium configuration of (31),

$$q_j(t) = x_j, \quad j = 1, 2, \ldots, n, \tag{34}$$

with

$$x_j = \sum_{k=1}^{n}{}' (x_j - x_k)^{-1} \quad , \quad j = 1, 2, \ldots n \tag{35}$$

It can be shown, by simple algebraic manipulations [6] (see below), that (35) is equivalent to (4). On the other hand (27), (28) and (33) clearly imply (5) (the equilibrium configuration (34) corresponds to a t-independent $\psi = 2^{-n} H_n(x)$; and the x_j's are the zeros of this ψ).

Having proved that the solutions of (35) and (4) are the zeros of the Hermite polynomial of order n, see (5), let us proceed and investigate the behaviour of the system (31) in the neighbourhood of this equilibrium configuration. Thus we set

$$q_j(t) = x_j + \varepsilon \zeta_j(t) \tag{36}$$

in (31), and linearise in ε. There obtains for the vectors

$$\zeta(t) = (\zeta_1(t), \zeta_2(t), \ldots, \zeta_n(t)) \tag{37}$$

the equation of motion

$$\ddot{\zeta}(t) + (1 + A)\, \zeta(t) = 0, \tag{38}$$

with the real symmetrical matrix A, of order n, defined by

$$A_{jk} = \delta_{jk} \sum_{\ell=1}^{n}{}' (x_j - x_\ell)^{-2} - (1-\delta_{jk})(x_j - x_k)^{-2} . \tag{39}$$

But we know from (33) that the general solution of (31) contains only the frequencies $m^{1/2}$, $m = 1, 2, \ldots n$; we may therefore conclude that the eigenvalues of the matrix A are the integers $0, 1, 2, \ldots n-1$,

$$A\, v^{(m)} = (n-m)\, v^{(m)} \quad , \quad m = 1, 2, \ldots , n. \tag{40}$$

This is another remarkable property of the zeros of the Hermite polynomials. The connection of (40) to (11) and (13) is provided by the matrix formula

$$B^2 = (1 + A)^2 \tag{41}$$

implying

$$A^{(2)} = \frac{1}{6} A(A+2) \tag{42}$$

with $A^{(2)}$ defined by (10). As for (41), it can be viewed as a corollary of the following result [7], valid for two dynamical systems characterised by the two hamiltonians

$$H^{(s)} = \tfrac{1}{2} \sum_{j=1}^{n} p_j^2 + V^{(s)}(q_1, q_2, \ldots, q_n), \quad s = 1, 2, \tag{43}$$

with

$$V^{(s)}(q_1, q_2, \ldots, q_n) = \tfrac{1}{2} \sum_{k=1}^{n} [\partial V^{(s)}(q_1, q_2, \ldots q_n)/\partial q_k]^2 + \\ + \text{const} . \tag{44}$$

Let the system characterised by the hamiltonian $H^{(1)}$ possess the equilibrium configuration $q_j(t) = x_j$, and let $\omega_m^{(1)}$, $m = 1, 2, \ldots, n$, be the circular frequencies of the small oscillations of this system around this equilibrium configuration; then the system characterised by the hamiltonian $H^{(2)}$ has the same equilibrium configuration, its small oscillations around this equilibrium configuration have the same normal modes as those of $H^{(1)}$ and have circular frequencies

$$\omega_m^{(2)} = [\omega_m^{(1)}]^2 , \quad m = 1, 2, \ldots, n \tag{45}$$

The proof of this result is easy. Since (44) implies

$$\partial V^{(2)}/\partial q_j = \sum_{k=1}^{n} [\partial^2 V^{(1)}/\partial q_j \, \partial q_k][\partial V^{(1)}/\partial q_k], \tag{46}$$

the equation determining the equilibrium configuration of $H^{(1)}$,

$$\partial V^{(1)}/\partial q_j \Big|_{q=x} = 0, \quad j = 1, 2, \ldots, n , \tag{47}$$

also implies

$$\partial V^{(2)}/\partial q_j \Big|_{q=x} = 0, \quad j = 1, 2, \ldots, n; \tag{48}$$

hence the coincidence of the equilibrium configurations. Moreover the matrices whose eigenvalues and eigenvectors determine the frequencies and normal modes of the small oscillations are given by

$$W_{jk}^{(1)} = \partial^2 V^{(1)}/\partial q_j \, \partial q_k \Big|_{q=x} , \tag{49}$$

and clearly differentiating (46) and using (47) there obtains the matrix relation

$$W^{(2)} = [W^{(1)}]^2 , \tag{50}$$

q.e.d.

This result is applicable to our case, since the choices

$$V^{(1)}(q) = \tfrac{1}{2} \sum_{j=1}^{n} q_j^2 - \sum_{j>k=1}^{n} \log (q_j - q_k) , \tag{51}$$

$$V^{(2)}(q) = \tfrac{1}{2} \sum_{j=1}^{n} q_j^2 + \sum_{j>k=1}^{n} (q_j - q_k)^{-2} , \tag{52}$$

satisfy (44) (with const $= \frac{1}{2}\, n(n-1)$), as can be verified with a little labour, and they yield the equations of motion

$$\ddot{q}_j + q_j = \sum_{k=1}^{n}{}' \, (q_j - q_k)^{-1} , \quad j = 1, 2, \ldots n, \tag{53}$$

$$\ddot{q}_j + q_j = 2\sum_{k=1}^{n}{}' \, (q_j - q_k)^{-3} , \quad j = 1, 2, \ldots n . \tag{54}$$

The second of these coincides with (2). The first differs from (21) due to the absence of "velocity-dependent" contributions in the r.h.s.; but such contributions are quadratic, and are therefore ignorable for all results relevant to small oscillations (the linearised equations around equilibrium are identical for (53) and (31)).

This concludes our analysis of mathematical results for the zeros x_j of the Hermite polynomial $H_n(x)$, results obtained as a by-product of the study of certain integrable many-body problems on the line. Let us summarise our findings:

(i) The x_j's satisfy the system of nonlinear equations (35) and (4). (Actually (35) was discovered almost a cuntury ago by

Stieltjes [8]; (4) is more recent [3]; and several other analogous equations are collected in [9]).

(ii) The matrix A, see (39), has the first n nonnegative integers as eigenvalues, see (40).

(iii) The matrix $A^{(2)}$, see (10), is related to A, see (39), by (41).

(iv) The matrix $L(t)$, see (25), has the (t-independent) eigenvalues x_m,

$$L(t)u^{(m)}(t) = x_m\, u^{(m)}(t), \qquad m = 1, 2, \ldots, n. \tag{55}$$

These results, and others, can also be obtained, more directly, by a different technique [9]. For instance, to obtain (40) it is sufficient to introduce the function

$$v_j^{(m)}(z) = (z-x_j)^{-2}\, H_{m-1}(z)/H_n(z), \quad j = 1, 2, \ldots n; \quad m = 1, 2, \ldots, n \tag{56}$$

This is a rational function of the complex variable z, with $n-1$ simple poles (at $z = x_k$, $k \neq j$) and one triple pole (at $z = x_j$); and it vanishes faster than $|z|^{-1}$ as $|z|\to\infty$ in all directions of the complex z plane. Hence the sum of all its residues vanishes. This condition yields precisely (40) (using (29)), and it identifies moreover the eigenvectors $v^{(m)}$:

$$v_j^{(m)} = [2^{n-m}(n-1)!]^{1/2}\, H_{m-1}(x_j)/\{[n(m-1)!]^{1/2}\, H_{n-1}(x_j)\}, \quad j = 1, 2, \ldots n;\ m = 1, 2, \ldots, n. \tag{57}$$

Here we have also introduced normalisation constants, so that the (real) eigenvectors $v^{(m)}$ satisfy the orthonormality and completeness relations

$$\sum_{j=1}^{n} v_j^{(m)}\, v_j^{(\ell)} = \delta_{\ell m}\,, \tag{58}$$

$$\sum_{m=1}^{n} v_j^{(m)}\, v_k^{(m)} = \delta_{jk}\,. \tag{59}$$

By analogous techniques it can be shown [7], [10] that the two matrices $A^{(\pm)}$, of order n, defined by

$$A^{(\pm)}_{jk} = \delta_{jk} \sum_{\ell=1}^{n}{}' (x_j - x_\ell)^{-1} \quad \mp (1-\delta_{jk})(x_j - x_k)^{-1} \tag{60}$$

or, equivalently (see (35)),

$$A^{(\pm)}_{jk} = \delta_{jk}\, x_j \quad \mp (1-\delta_{jk})(x_j - x_k)^{-1} , \tag{60b}$$

act as raising and lowering operators on the eigenvectors (57):

$$A^{(-)} v^{(m)} = [2(m-1)]^{1/2} \quad v^{(m-1)}, \quad m = 1, 2, \ldots, n, \tag{61a}$$

$$A^{(+)} v^{(m)} = [(2m)^{1/2} - \delta_{nm} (2n)^{1/2}], \quad m = 1, 2, \ldots n, \tag{61b}$$

and satisfy the following matrix equations

$$[A^{(\pm)}, A] = \pm A^{(\pm)} , \tag{62}$$

$$A = n-1-\tfrac{1}{2} A^{(+)} A^{(-)} , \tag{63}$$

$$[A^{(+)} A^{(-)}] = 2(J-1) , \tag{64}$$

with the matrix J having all elements equal to unity,

$$J_{jk} = 1, \quad j = 1, 2, \ldots, n; \quad k = 1, 2, \ldots, n. \tag{65}$$

Of course these equations, as well as (42) and the trace formulae implied by (40) and (55), namely

$$\mathrm{tr}\,[A^p] = \sum_{m=1}^{n-1} m^p, \quad p = 1, 2, 3, \ldots , \tag{66}$$

$$\mathrm{tr}\,[L^p] = \sum_{j=1}^{n} x_j^p , \quad p = 1, 2, 3, \ldots , \tag{67}$$

yield additional sets of nonlinear equations satisfied by the x_j's.

It is appealing to conjecture that each of these sets of nonlinear equations is also sufficient to identify uniquely (up to permutations, that are hereafter ignored, and up to some other obvious ambiguity, see below) the numbers x_j, and therefore also the Hermite polynomial $H_n(x)$ that is of course essentially determined once its zeros are assigned. In some case this is known to be the case; for instance it is obvious on physical grounds that the equilibrium configuration for the two hamiltonians (43) with (51) and (52) is unique, and therefore, at least in the real field, both (35) and (4) must have a unique solution (indeed it has been proved that (35) has a unique solution even in the complex domain [8], [11]). In other cases, however, this has not yet been proved. In particular the conjecture [5], [12] that the requirement that the matrix A of order n defined, in terms of n *a priori* arbitrary numbers x_j, by

$$A_{jk} = \delta_{jk} \sum_{\ell=1}^{n}{}' (y_j-y_\ell)^{-2} - (1-\delta_{jk})(y_j-y_k)^{-2} \tag{68}$$

(see (39)), have the first n nonnegative integers as eigenvalues, implies that the y_j's are, up to a translation, the n zeros of the Hermite polynomial of order n,

$$H_n\,(y_j + y_o) = 0, \quad j = 1, 2, \ldots, n \quad , \tag{69}$$

has not yet been proved (except for n=2 and n=3 [13]). This conjecture is appealing, since its validity would allow basing the entire theory of Hermite polynomials on an elegant diophantine requirement. We consider it plausible, although the following result [9a] points in the opposite direction: the requirement that the matrix K, defined in terms of the n *a priori* arbitrary numbers y_j by the formula

$$K_{jk} = \delta_{jk}\, y_j \sum_{\ell=1}^{n}{}' (y_j-y_\ell)^{-1} + (1-\delta_{jk})\, y_j\, (y_j-y_k)^{-1}, \tag{70}$$

have the first n nonnegative integers as eigenvalues, implies no restriction on the y_j's, since for any choice of these numbers (all different, of course) the matrix K does indeed have the first n nonnegative integers as eigenvalues.

In this lecture we have concentrated, for the sake of simplicity, on the case of Hermite polynomials. Let us end by mentioning that analogous results exist for all the classical polynomials (Hermite, Laguerre, Jacobi) and also for Bessel functior [5], [9], [14]. The form that some of the results for the classical polynomials take as their order n is (appropriately) made to diverge have also been studied [15], and this has led to the discovery of an interesting class of singular integral operators having integer eigenvalues and polynomial eigenfunctions [16]; but this should be the subject of a separate lecture.

References

[1] F. Calogero: Integrable many-body problems. In Nonlinear equations in physics and mathematics (A.O.Barut, ed.), Reidel, 1978, pp.3-53.

[2](a) F. Calogero: Solution of the one-dimensional N-body problem with quadratic and/or inversely quadratic pair potentials. J.Math.Phys.12, 419-436 (1971). (b) J. Moser: Three integrable hamiltonian systems connected with isospectral deformations. Adv. in Math. 16, 197-220 (1975). (c) M.A. Olshanetzky and A.M. Perelomov: Explicit solution of the Calogero model in the classical case and geodesic flows on symmetric spaces of zero curvature. Lett.Nuovo Cimento 16, 33-39 (1976). (d) M.Adler: Some finite dimensional integrable systems. Proceedings of the Conference on the Theory and Applications of Solitons (H.Flaschka and D.W.McLaughlin, eds.), Rocky Mountain Mathematics Consortium (Arizona State University), Tempe, 1978; Some finite dimensional integrable systems and their scattering behaviour. Commun.Math.Phys. 55, 195-230 (1977). (e) H.Airault, H.P.McKean and J.Moser: Rational and elliptic solutions of the Korteweg-de-Vries equation and a related many-body problem. Commun.Pure Appl.Math 30, 95-148 (1977). (f) D.Kazhdan, B.Kostant and S.Sternberg: Hamiltonian group actions and dynamical systems of Calogero type. Commun.Pure Appl.Math. 31, 481-507 (1978).

[3] F. Calogero: Equilibrium configuration of the one-dimensional N-body problem with quadratic and inversely quadratic pair potentials. Lett.Nuovo Cimento 20 ,251-253 (1977).

[4] A.M. Perelomov: Simple relation between certain dynamical systems. Commun.Math.Phys. 63, 9-11 (1978).

[5] F. Calogero: Motion of poles and zeros of special solutions of nonlinear and linear partial differential equations, and related "solvable" many-body problems. Nuovo Cimento 43B, 177-241 (1978).

[6] F. Calogero: On the zeros of Hermite polynomials. Lett. Nuovo Cimento 20, 489-490 (1977).

[7] A.M. Perelomov: Equilibrium configurations and small oscillations of some dynamical systems. Ann.Inst.H. Poincaré A28, 407-415 (1978).

[8] G. Szëgo: Orthogonal polynomials. Amer.Math.Soc.Coll. Publ.XXIII. Providence, R.I. 1939, subsect. 6.7.

[9](a) S.Ahmed, M.Bruschi, F.Calogero, M.A.Olshanetsky and A.M. Perelomov: Properties of the zeros of the classical polynomials and of the Bessel functions. Nuovo Cimento 49B, 173-199 (1979). (b) M.Bruschi and F. Calogero: Eigenvectors of a matrix related to the zeros of Hermite polynomials. Lett.Nuovo Cimento 24, 601-604 (1979).

[10] M.Bruschi: On the algebra of certain matrices related to the zeros of Hermite polynomials. Lett.Nuovo Cimento 24, 509-600 (1979).

[11] P. Sabatier: On the solutions of an infinite system of nonlinear equations. Lett.Nuovo Cimento 21, 41-44 (1978); M.E. Muldoon: An infinite system of equations characterising the zeros of Bessel functions. Lett.Nuovo Cimento 23, 447-448 (1978).

[12] F. Calogero: On the zeros of the classical polynomials. Lett.Nuovo Cimento 19, 505-508 (1977).

[13] F. Calogero: unpublished.

[14] M.Bruschi: On the algebra of certain matrices related to the zeros of Laguerre polynomials. Lett.Nuovo Cimento 25, 417-423 (1979).

[15] F. Calogero and A.M. Perelomov: Asymptotic density of the zeros of Hermite polynomials of diverging order, and related properties of certain singular integral operators. Lett. Nuovo Cimento 23, 650-652 (1978); Asymptotic density of the zeros of Laguerre and Jacobi polynomials. Lett.Nuovo Cimento 23, 653-656 (1978).

[16] F. Calogero: Singular integral operators with integral eigenvalues and polynomial engenfunctions. Nuovo Cimento 51B, 1-14 (1979) and Erratum (to be published); Integral representation and generating function for the polynomials $U_n^{(\alpha,\beta)}(x)$. Lett. Nuovo Cimento 24, 595-600 (1979); D.Atkinson (to be published).

RIEMANN MONODROMY PROBLEM, ISOMONODROMY DEFORMATION EQUATIONS AND COMPLETELY INTEGRABLE SYSTEMS

D.V. CHUDNOVSKY

DPh.T. CEN-Saclay - 91190 Gif-sur-Yvette, France

and

Dept. of Mathematics, Columbia University, New-York
NY 10027, USA

ABSTRACT

These lectures serve as an introduction into the Riemann-Hilbert monodromy problem. Our main aim is to relate known completely integrable systems with isomonodromy deformation. We describe Schlesinger isomonodromy deformation equations for Fuchsian linear differential equations and their connection with Painlevé transcendents. Moreover it is shown that all the one-dimensional classical completely integrable systems, connected with commuting matrix differential operators, can be represented as simplified Schlesinger systems. We generalize isomonodromy deformation equations for the case of two-dimensional systems. This way arise Painlevé type equations in one space and one time situation. The last part of the lectures is devoted to Riemann boundary value problem. It is explained, how using boundary problem for analytic function on a Riemann surface of finite genus one can solve classical multidimensional isospectral and isomonodromy deformation equations. Examples of so called Bakes's functions are given.

INTRODUCTION

We shall try to present how completely integrable systems in one, two and several dimensions arise from the Riemann problem on Riemann surface. What is understood here by Riemann problem ?

First of all there is Riemann-Hilbert problem concerning determination of an equation with the given monodrony group. On the other

C. Bardos and D. Bessis (eds.), Bifurcation Phenomena in Mathematical Physics and Related Topics, 385–447.

hand there is the boundary value Riemann problem for analytic functions in the complex plane or on Riemann surface. It is known that the boundary value problem is one of the ways to solve monodromy Riemann problem. We consider these Riemann problems as a natural substitute for the inverse scattering method in the sense, that this way certain non linear equations are reduced to linear integral ones. While the situation in one dimension is clear (see §5 for complete description), in two dimensions it is more or less clear, in truly multidimensional situation (especially in four dimensions) the complete analysis still requires time for investigation.

Basic facts and elementary introduction to the isomonodrony theory are given in what follows. Very interesting connections with automorphic functions, Fuchsian groups and quasi-conformal mappings will not be discussed, though they are very deep and exist.

Some examples of applications of the Riemann problem and isomonodromy deformations :

1) In recent time attention was attracted to isomonodromy deformations due to the use in statistical mechanics. Mac Coy, Tracy, Wu [49] and then Jimbo, Miwa, Sato [62] and Abraham demonstrated that n-point correlation functions for two dimensional Ising models are expressed explicitly in terms of Painlevé transcendents and other non linear equations of the isomonodromy deformation.

2) Generally speaking Sato [62] and his collaborators developed a two-dimensional theory of linear differential equations with regular singularities in order to show that solutions of isomonodromy equations are expressed in terms of classical operators of field theory. Sato's series of papers [63] is now one of the bestsellers among preprint (or "samizdat") publications.

3) For the last 50 years the Riemann problem, Riemann boundary value problem and corresponding singular integral equations have been intensively studied by hundreds of pure and applied mathematicians. They have applied this to the elasticity theory and to the problems of aero - and hydrodynamics. The most complete treatises of this subject were written by Vekua [67] and Muskhelishvili [53] and Plemelj [59] .

4) Very important is the relation of the Riemann boundary value problem and scattering problems in the multi-dimensional case. First of all there are very deep papers of Kawai and Kashivara relating monodromy problems and general properties of S-matrix which are not yet completely understood. On the other hand even for

such simple equations as three dimensional Schrödinger the best studies of the direct and inverse scattering problems of Jost and Newton have been based completely on the Riemann problem [43] .

Organization of lectures

These lectures are organized as follows : We define classical Riemann monodromy problem in §1. In appendix to §1 we formulate solutions of Riemann problem as a function of all the parameters involved. This relates our studies with those of Lappo-Danilevsky [46] and Sato et al. [62] . More detailed exposition of this subject see in [24],[13] , [26] . In [26] the Riemann notion of the class of functions (having the same monodromy group) is reproduced. In order to make the exposition self-contained we include in §2 connection between the Riemann monodromy problem and the corresponding Riemann boundary value problem together with description of local monodromy for linear differential equations. Schlesinger equations are considered in §3 for the second order linear differential equations. In this case, following basically R. Fuchs [30],we obtain all the Painlevé transcendents [57] . The complete list of them with interrelations is presented in the appendix to §3. Also in §3 we deduce these and more general transcendents as self-similarity solutions of two dimensional completely integrable systems In §4 Schlesinger system is derived and it's limiting version - simplified Schlesinger system is presented as a model one dimensional completely integrable system. Russian chain (example 4.3) is obtained as a particular case of the simplified Schlesinger system with one movable singularity. Another class of one dimensional completely integrable systems is furnished to us by the condition of commutativity of two matrix ordinary linear differential operators. This class of systems (see, e.g. [6]) is defined in §5. We show in §5 that these (isospectral deformation) systems are one-to-one equivalent to the simplified Schlesinger systems. For example, Russian chains are equivalent to stationary Korteweg de Vries (KdV) equations [16] - [18] . The equations like KdV belong to a general class of two dimensional isospectral deformation equations. These equations are defined in §6 via Zakharov-Mikhailov-Shabat scheme as consistency condition of two linear eigenvalue problems with rational dependence on spectral parameter. Examples of field theory equations are presented in §6 and in §7 we generate infinite component and operator analogs of all known two dimensional classical completely integrable systems. Among them the main place belongs to infinite component and operator non-linear Schrödinger equations (Russian chain is their stationary subsystem). Here §7 provides an introduction to new approach of inverse scattering for Sturm-Liouville problem and possible quantization of two-dimensional completely integrable field theories.

The general idea of isomonodromy deformation and equations, differential with respect to spectral parameter, enables us to present in §8 natural two dimensional analogs of the Painlevé transcendents. These equations appear here for the first time. In conclusion of §§1-8 we try to summarize briefly the results and we are not afraid to be optimistic about future progress of isomonodromy studies.

The lectures are concluded by §9, where deep underlying analytic and algebraic -geometric structure of actual solutions is touched. In order to solve the conditions of consistency of two linear problems we need the Riemann boundary value problem on certain Riemann surfaces. The resulting object, so called Baker function, can be constructed this way and classical expressions [3],[55] ,[51] for hyperelliptic Riemann surfaces are presented. In the references we try to keep good balance between recent and ancient progress in this field (made sometimes independently).

We thank to the organizers of the Cargèse Conference and especially to D. Bessis for cordial hospitality and excellent possibility for work and preparation of these lectures.

1. CLASSICAL RIEMANN PROBLEM AND THE MONODROMY GROUP IN THE ONE-DIMENSIONAL CASE.

1.1. Riemann monodromy problem for $\mathbb{C} \cup \{\infty\}$ is always formulated for Fuchsian linear differential equation (i.e. for linear differential equations with regular singularities only).

So we start with matrix differential equation having the simplest singularities (so-called canonical or normal form) :

$$\frac{d\bar{y}}{dx} = \sum_{i=1}^{n} \frac{A_i}{x - a_i} \bar{y} , \qquad (1.1)$$

where $\bar{y} = {}^t(y_1,\ldots,y_m)$ and A_i are $m \times m$ constant matrices : $i = 1,\ldots,n$.

The system (1.1) is the subject of all our discussions and applications. Also from the point of view of Riemann problem (1.1) is the best example for all the construction.

If we turn to an arbitrary system of linear differential equations the general Riemann problem is formulated as follows.

Let

$$\frac{d\bar{y}}{dx} = A(x)\bar{y} \quad , \quad \bar{y} = {}^t(y_1,\ldots,y_m) \quad , \qquad (1.2)$$

be a system of equations with a matrix A(x) with rational coefficients. Let Y = Y(x) be a fundamental matrix of (1.2) and let

$$\{a_1,\dots,a_n\}$$

be the set of poles of A(x). In general the fundamental matrix Y(x) is a multivalued function having $a_1,\dots,a_n$ and $a = \infty$ as its branch points and

$$Y(x) \rightarrow Y(x)\,M(\gamma)$$

when prolonged along γ (look at Figure 1).

Here $M(\gamma) \in GL(m,C)$ and $M(\gamma)$ depends only on the homotopy class of γ. Then, of course

$$M(\gamma_1\gamma_2) = M(\gamma_1)\,M(\gamma_2) \quad ,$$

so the set $\mathfrak{M} = \{M(\gamma)\}$ is a group-monodromy group of the equation (1.2).

This group is generated by the matrices $M_i = M(\gamma_i) : i = 1,\dots,n,$ where γ_i is a clockwise circuit around a_i, which does not contain other singular points inside, and we have

$$M_1\dots M_n M = I. \tag{1.3}$$

<u>Classical Riemann problem.</u> Given branch points $a_1,\dots,a_n \in \mathbb{C}$ and matrices $M_1,\dots,M_n \in GL(m,\mathbb{C})$ find a linear differential equation (1.2) or, better, (1.1), whose monodromy group $\mathfrak{M}$ coincides with the group generated by $M_1,\dots,M_n$.

We also naturally demand in the <u>classical</u> Riemann problem for the fundamental matrix Y(x) of (1.2) to be at most regularly singular at the points $a_1,\dots,a_n$. This means that

$$Y(x) = H_i(x)(x - a_i)^{-L_i} \quad : \quad i = 1,\dots,n, \quad , \tag{1.4}$$

(natural parameter near ∞ is 1/x). In (1.4) $H_i(x)$ is an invertible holomorphic matrix at $x = a_i$ and L_i (the exponent) is a constant matrix such that

$$e^{2\pi i L_j} = M_j \quad : \quad j = 1,\dots,n. \tag{1.5}$$

The reason that this problem appears as the "Riemann-Hilbert" problem is very simple : Riemann in 1857[60] was working with the system of functions satisfying the linear differential equation with a given monodromy group $\mathfrak{M}$. The only thing that was not

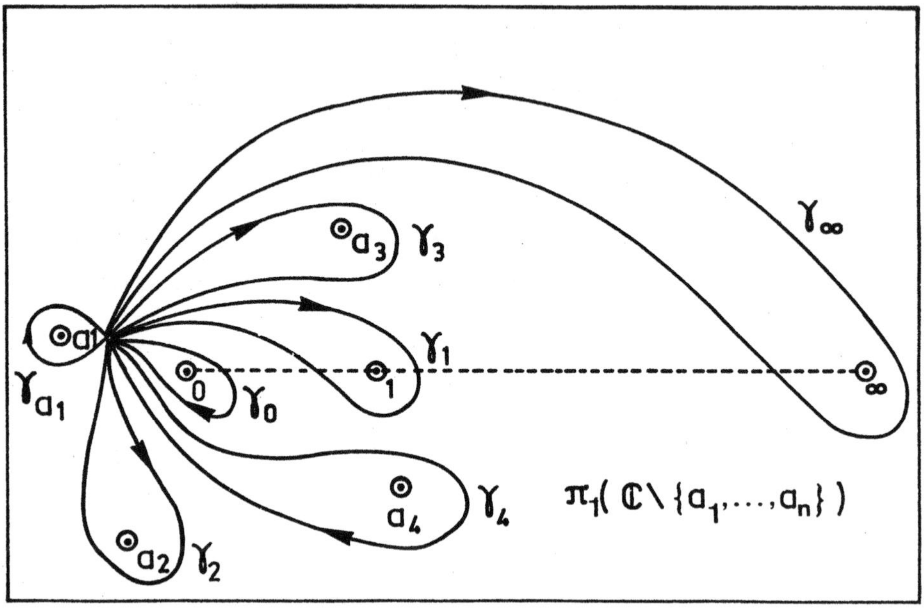
γ_∞
a_3 γ_3
a_1
γ_{a_1}
γ_1
0 γ_0
1
∞
a_4 γ_4
$\pi_1(\mathbb{C} \setminus \{a_1, \ldots, a_n\})$
a_2 γ_2

Figure 1.

proved by Riemann was the existence of the system of m functions analytic outside of $\{a_1,\ldots,a_n\}$ and having a given monodromy group $\mathfrak{M}$.

It is unclear whether Riemann knew about the problem of existence or not. In any case, Riemann computed the dimensions of the set of Fuchsian linear differential equations with given singularities and the number of possible groups (more precisely the dimension of the possible monodromy matrices). On the basis of such computations he claimed that there are infinitely many systems (of functions) belonging to a given class (i.e. having a given monodromy group).

Hilbert proposed the proof of this statement as problem N°21 in his list and in 1912 gave [41] the first serious solution for m = 2 on the basis of Riemann boundary value problem.

Since Riemann and Hilbert this problem was solved in different ways by Plemelj [59], Birkhoff [4] (reduction to the singular integral equations) ; by Röhrl [61] (for an arbitrary Riemann surface using fibre bundles) ; by Lappo-Danilevsky [46], [29] (using series expansions in hyperlogarithms).

From our present point of view the most important was the reduction of the linear Riemann problem to the non-linear "completely integrable" system of equations : Schlesinger's equations [64], - [66].

<u>"Schlesinger's Theorem".</u> [64] Let $Y = Y(x_o;x)$ be the fundamental matrix solution of (1.1) ; $Y(x_o;x_o) = I$. The necessary and sufficient conditions for $A_i : i = 1,\ldots,n$ as the function of the parameters $a_1,\ldots,a_n,x_o$ to have the fixed monodromy group of $Y(x)$ is the following completely integrable system of total differential equations

$$dA_j = -\sum_{i\neq j} [A_j,A_i]\ d\log\frac{a_j - a_i}{x_o - a_i} \quad : \quad j = 1,\ldots,n. \qquad (1.6)$$

This Schlesinger system can be written in a classical form [64]-[66]-[62] :

$$\frac{\partial A_j}{\partial a_i} = [A_j,A_i]\left(\frac{1}{a_j - a_i} - \frac{1}{x_o - a_i}\right), \quad j \neq i$$

$$\frac{\partial A_i}{\partial a_i} = -\sum_{j\neq i} [A_i,A_j]\frac{1}{a_i - a_j},$$

$$\frac{\partial A_i}{\partial x_o} = \sum_{j\neq i} [A_i,A_j]\frac{1}{x_o - a_j} \quad : \quad i = 1,\ldots,n. \qquad (1.7)$$

This system of equation is the source of "classical" completely integrable systems like, e.g. stationary KdV equation.

There is, of source, a natural temptation to identify the deformation equations of the types (1.7) with the Lax type of isospectral deformation. This is not the case at all ! The deformation equation (1.7) are reduced in the most interesting cases to the equations, differential with respect to a spectral parameter as it will be explained below.

APPENDIX TO § 1

BACKGROUNDS AND THE ANALYTIC CONSTRUCTION OF THE SCHLESINGER's SYSTEM.

1.2. Let now $a_1,\dots,a_n$ and x_o be distinct points on $P\mathbb{C}^1$ and let $L_1,\dots,L_n$ be $m \times m$ matrices satisfying the natural condition

$$e^{2\pi i L_1}\dots e^{2\pi i L_n} = 1 \ . \tag{1.8}$$

We consider the following precise version of the Riemann problem :
Find a matrix $Y(x)$ with the properties

a) $Y(x)$ is a multi-valued analytic matrix on $P\mathbb{C}^1\setminus\{a_1,\dots,a_n\}$;

b) $Y(x) = H_i(x).(x-a_i)^{-L_i}$ at $x = a_i (i = 1,\dots,n)$, where $H_i(x)$ is an invertible holomorphic matrix at $x = a_i$;

c) $\det Y(x) \neq 0$ for $x \neq a_1,\dots,a_n$;

d) $Y(x_o) = 1$.

Such a matrix $Y(x)$ is unique (but exists not always). Lappo-Danielevsky [42] and Sato et al.[63] have proved that for sufficiently small $|L_i|$: $i = 1,\dots,n$ (excluding $i = \infty$) such a matrix

$$Y = Y(x_o;x;{}^{a_1,\dots,a_n}_{L_1,\dots,L_n})$$

exists and can be written as a series in L_i : $= 1,\dots,n$.

Let us suppose now that $a_i \neq \infty$: $i = 1,\dots,n$, then the function $Y(y;x;{}^{a_1,\dots,a_n}_{L_1,\dots,L_n})$ gives us simultaneously : 1) the Fuchsian linear differential system with the monodromy group, generated by $\{e^{2\pi i L_1},\dots,e^{2\pi i L_n}\}$ and 2) the Schlesinger system of the equations (1.7) for the coefficients $A_1,\dots,A_n$ of (1.1).

We can write this for a function

$$Y = Y(y;x;{}^{a_1\ldots a_n}_{L_1\ldots L_n})$$

as a linear total differential equation

$$dY = \Omega Y \quad , \tag{1.9}$$

$$\Omega = \sum_{i=1}^{n} A_i \, d \log \frac{x - a_i}{y - a_i} =$$

$$= \sum_{i=1}^{n} A_i \left(\frac{d(x - a_i)}{x - a_i} - \frac{d(y - a_i)}{y - a_i} \right)$$

where

$$A_i = A_i(y;{}^{a_1\ldots a_n}_{L_1\ldots L_n}) = - H_i(a_i) L_i H_i(a_i)^{-1} : i = 1,\ldots,n \tag{1.10}$$

are matrices independent of x satisfying

$$\sum_{i=1}^{n} A_i = 0.$$

The equation (1.9) means that Y as a function of x satisfies the Fuchsian system of linear ordinary differential equations

$$\frac{dY}{dx} = \sum_{i=1}^{n} \frac{A_i}{x - a_i} Y.$$

Now the coefficients A_i : $i = 1,\ldots,n$ as functions of $Y = x_o$ and $\bar{a} = (a_1,\ldots,a_n)$ satisfy the Schlesinger's equations (1.6) :

$$dA_j = - \sum_{j \neq i} [A_j, A_i] \, d \log \frac{a_i - a_j}{a_i - y} \quad : j = 1,\ldots,n.$$

Sato, Miwa and Jimbo in a series of papers [62], [63] , describe a quantum field theory approach of the construction of $Y(y;x;{}^{a_1\ldots a_n}_{L_1\ldots L_n})$ and the solution of the Riemann problem. They represent Y and Y_i in terms of classical field theory operators and then apply such representations to the explicit expressions for the n-th correlation functions in the two-dimensional Ising model. It should be noted, however, that the series expansions for $Y(y,x;{}^{a_1,\dots,a_n}_{L_1,\dots,L_n})$ proposed by Sato, Miwa, Jimbo are basically the same as in the papers of Lappo-Danilevsky.

2. RIEMANN BOUNDARY VALUE PROBLEM AND LOCAL MONODROMY.

Hilbert was the first to propose the method for investigation and solution of the monodromy problem using the reduction to the singular integral equations (1901 - 1912) [41].

This reduction of the monodromy problem to linear integral equation is based on another Hilbert observation : it is possible to reduce monodromy problem to the Riemann boundary value problem in the complex plane.

This reduction was established by Hilbert [41] for the case of the linear differential equation of the second order (the case of two linearly independent functions). Later Plemelj (1908 - 1912) [59] proposed the general solution, extended by Birkhoff [4] using the matrix Riemann boundary value problem.

The reduction looks as follows. Let

$$z_1,\dots,z_n$$

be given points in $\mathbb{C} \cup \{\infty\}$ - positions of the regular singularities - and

$$A_1,\dots,A_n$$

be given (monodromy) matrices of sizes $m \times m$.

We consider the closed contour C (analytic in $\mathbb{C}$) connecting cyclically the points $z_1,\dots,z_n$.

We want to construct the system of functions $\bar{f} = (f_1,\dots,f_m)$ that are analytic (or, say, meromorphic) in the whole complex plane outside of C in such a way that the values of $\bar{f}$ on the outside part of C_i : $i = 1,\dots,n$ and the values of $\bar{f}$ on the internal part of C_i are related by a given linear homogeneous relation with given complex coefficients.

Precisely, we want to construct $\bar{f}^+ = (f_1^+,\dots,f_n^+)$ and $\bar{f}^- = (f_1^-,\dots,f_n^-)$ such that $\bar{f}^+$ is analytic outside of C and $\bar{f}^-$ is analytic inside C (possibly meromorphic) and their limit values

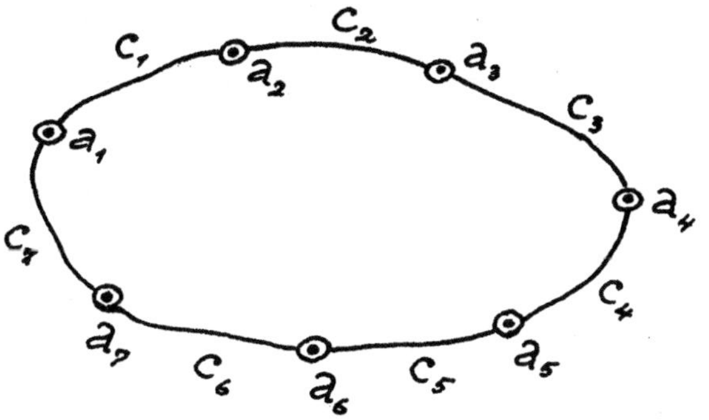

Figure 2.

$\bar{f}^{+}_{\lim}, \bar{f}^{-}_{\lim}$ on C are continuous everywhere on C and on the part C_i of C between z_i and z_{i+1} : $i = 1,\ldots,n$ $(z_{n+1} = z_1)$, we have

$$\bar{f}^{+}_{\lim} = A_i \, \bar{f}^{-}_{\lim} \quad \text{on } C_i \ : \ i = 1,\ldots,n.$$

Warning I. Though the "inverse monodromy problem" : Riemann-Hilbert problem is in principle solved, there is no effective way to solve the "direct monodromy problem" : to determine the monodromy group knowing the form of a Fuchsian linear differential equation.

The situation looks very similar to the spectral problem : there exists a very good apparatus for the solution of the inverse scattering problem, but it is not easy to solve the direct spectral and scattering problem (especially for a continuous spectrum).

Remark. The difficulties in the solution of the "direct monodromy problem " look a little strange. Indeed, you can always determine the exponents of the differential equation corresponding to any given singularity.

These exponents correspond to the so-called local monodromy of the equation. Roughly speaking, local monodromy exponents are eigenvalues of the corresponding monodromy matrices.

The knowledge of the eigenvalues of A_i : $i = 1,\ldots,n$ together with the identity

$$A_1 \ldots A_n = I$$

determines the whole monodromy group only for $n = 3$.

This is the case of three regular singularities : $\{0,1,\infty\}$

which correspond to the ordinary Gauss hypergeometric ${}_2F_1$ functions (for $(m = 2)$ or to a generalized hypergeometric function ${}_mF_{m-1}$ (exposition of the theory of hypergeometric functions see in [60],[42],[26]).

Let us define explicitly the local monodromy exponents of the differential equation

$$\frac{d^n w}{dz^n} + P_1(z)\frac{d^{n-1}w}{dz^{n-1}} + \dots + P_{n-1}(z)\frac{dw}{dz} + P_n(z)w = 0 \quad (E)$$

with rational functions $P_1(z),\dots,P_n(z)$. Let z_o be any singularity of (E), i.e. singularity of one of $P_i(x)$ (including, possibly ∞).

Let

$$w_1,\dots,w_n$$

be a fundamental set of solutions of (E). A simple closed circuit γ is drawn, not passing through any singular point of (E), but enclosing z_o in its interior.

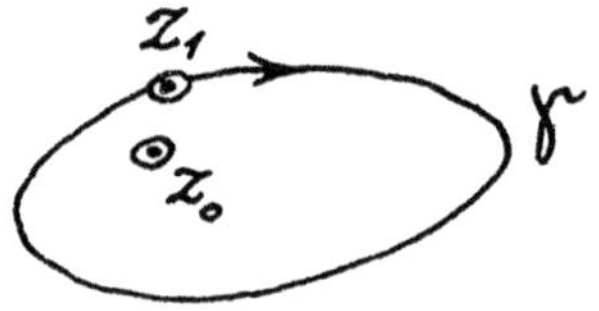

Figure 3.

Let z_1 be any point on γ and

$$W_1,\dots,W_n$$

be what $w_1,\dots,w_n$ respectively become after the variable z has described the circuit γ.

As (E) is linear, W_i can be expressed in terms of w_i,

$$\vec{W} = A\vec{w}$$

for some $n \times n$ matrix A (constant matrix). Now

$$\det(A) = \frac{\Delta(W_1,\dots,W_n)}{\Delta(w_1,\dots,w_n)} \neq 0$$

(where $\Delta(w_1,\ldots,w_n)$ is a Wronskian of $w_1,\ldots,w_n$).

The local monodromy exponents of (E) corresponding to the circuit γ are eigenvalues of the matrix A in

$$\vec{W} = A\vec{w}$$

Really, let s be an eigenvalue of A. Let $\vec{\lambda} = (\lambda_1,\ldots,\lambda_n)$ be an eigenfunction of A corresponding to s : $A\vec{\lambda} = s\vec{\lambda}$.

Then there exists a solution $u = u(x)$ of the equation (E) such that $u(x)$ becomes $s.u(x)$ after the circuit γ has been completely described once.

This solution u(x) is simply $\vec{\lambda}.\vec{w}(x) = \sum_{i=1}^{n} \lambda_i w_i(x)$.

Here comes the relation with regular singularities and exponents :

We consider, in particular, the case where the contour γ encloses only one singular point z_o and consider the multiform function - $(z - z_o)^\rho$. After one complete circuit has been described, this function becomes $e^{2\pi i\rho}(z - z_o)^\rho$.

Let exponents ρ_k be chosen so that

$$s_k = e^{2\pi i\rho_k}$$

for all eigenvalues s_k of the matrix A. Then the function

$$\phi_k(z - z_o) = (z - z_o)^{-\rho_k}.u_k(z) \quad : \quad k = 1,\ldots,n,$$

will return to its initial value after the description of a complete circuit about z_o, i.e. $\phi_k(z - z_o)$ are uniform functions of z in the domain of the z_o.

If ρ_k can be so determined that $\phi_k(0)$ is finite but non zero, the solution $u_k(z)$ is called regular.

Regular solution $u_k(z)$ has the form

$$u_k(z) = (z - z_o)^{\rho_k}\,\phi_k(z - z_o) \quad ,$$

where $\phi_k(0) \neq 0, \infty$.

Of course if n independent solutions of (E) are regular at z_0, then z_0 is said to be a regular singularity of (E).

3. ISOMONODROMY DEFORMATION FOR FUCHSIAN LINEAR DIFFERENTIAL EQUATIONS OF THE SECOND ORDER AND CORRESPONDING PAINLEVE TRANSCENDENTS I-VI (WITH THE LIST OF PAINLEVE TRANSCENDENTS).

3.1. One of the applications of the deformation equations was found by R. Fuchs (1905) see [30] , who showed that, as a particular case of deformation (Schlesinger) equation you can obtain all Painlevé equations of the kind I-VI. These Painlevé equations [57] are characterized by the following property [57],[42], [5],[7] : absence of the movable singularities. We consider all the equations of the second order

$$\frac{d^2w}{dz^2} = F(z,w,\frac{dw}{dz})$$

for the rational function F(z,w,p) in w and p and analytic in z all solutions of which have their critical points : branch points and essential singularities fixed (immovable).

Painlevé [57]and Gambier [32] (1887-1907) showed that all such equations can be reduced to the Painlevé transcendents of the kinds I-VI see also [40] . The most general is the sixth equation

$$\frac{d^2w}{dz^2} = \frac{1}{2}\left\{\frac{1}{w} + \frac{1}{w-1} + \frac{1}{w-z}\right\}\left(\frac{dw}{dz}\right)^2 - \left\{\frac{1}{z} + \frac{1}{z-1} + \frac{1}{z-x}\right\}\frac{dw}{dz} +$$

$$+ \frac{w(w-1)(w-z)}{z^2(z-1)^2}\left\{\alpha + \frac{\beta z}{w^2} + \frac{\gamma(z-1)}{(w-1)^2} + \frac{\delta z(z-1)}{(w-z)^2}\right\} \tag{VI}$$

The other equations I-V can be deduced from (VI) taking special values of $\alpha, \beta, \gamma, \delta$. We present only

$$\frac{d^2w}{dz^2} = 6w^2 + z. \tag{I}$$

Then, as it was shown by R. Fuchs [30] and Garnier (1916)[35], all of the equations I-VI are indeed deformation equations for the Fuchsian linear differential equation of the second order with regular singular points at $0, 1, \infty, z$ and an apparent singularity w.

In general, any known algebraic nonlinear equation without movable critical points can be represented as a certain isomonodromy deformation equation, though we do not know the proof of any general statement here.

We know only that any Schlesinger's isomonodromy deformation equation is an equation without movable singular points [64], [30].

3.2. As an example of the isomonodromy equations we present the way in which R. Fuchs [30] (the son of L. Fuchs) found the most general Painlevé transcendents. This was the first time that Schlesinger's equations were applied to some concrete system. It is also a very important example from the point of view of introduction of the spectral parameter. It will be very interesting to observe how the ideas of Schlesinger Fuchs-Fuchs which are similar to the iso-spectral (Lax) deformation turn out to be absolutely different.

R. Fuchs starts from the simplest equation of the second order

$$\frac{d^2y}{dx^2} = p(x)\, y \tag{3.1}$$

of the Fuchsian type. This means e.g. that p(x) is a rational function on x, but we assume p(.) to be a function of some additional parameter t.

Let us assume, however, that the monodromy group $\mathfrak{M}$ of the equation (3.1) is independent of the parameter t.

Then, according to Riemann [60] or [26] (the same arguments were used often by Fuchs [31] : 1888-1894) the fundamental solutions Y of (3.1) belong to the same class (independent of t) and for any solution y_i of (3.1) we can write

$$\frac{\partial y_i}{\partial t} = B.y_i + A\,\frac{\partial y_i}{\partial x}\ . \tag{3.2}$$

Here A and B are functions in x (and t) and for p(x) being rational in x, A and B are also rational in x.

Now, following the ideas of Schlesinger, Fuchs considers the condition of the consistence of the equations (3.1)-(3.2) :

$$\frac{\partial^2 y}{\partial t \partial x} = \frac{\partial^2 y}{\partial x \partial t}. \tag{3.3}$$

The equation (3.3) takes the form of a single non-linear partial differential equation :

$$\frac{\partial p}{\partial t} = A.\frac{\partial p}{\partial x} + 2p\,\frac{\partial A}{\partial x} - \frac{1}{2}\,\frac{\partial^3 A}{\partial x^3} . \tag{3.4}$$

It is natural and "almost necessary" to call this equation "of the Korteweg-de Vries" type.

Indeed you can put, say $p = \alpha .A$ for a constant $\alpha \neq 0$ and you obtain immediately the KdV equation

$$\alpha A_t = 3\alpha A A_x - \frac{1}{2} A_{xxx} .$$

Looking at this you can say : "it's clear that isomonodromy deformation leads to the same type of equations as isospectral deformations (inverse scattering method)".

It is not true in a trivial sense.

The equation (3.4) has nothing to do with KdV. Indeed, p(x) in (3.4) is a rational function in x (as A is) and A/p is not a constant at all. On the contrary, the main sense of (3.4) is that p(x) has prescribed poles : $\{0,1,t,\infty\}$ corresponding to the positions of regular singularities.

By the solution of (3.4) we understand the evolution of the residues of the p(x) as functions of t only.

In other words looking at (3.4) and taking into account the structure of p(x) and A(x) we find that the equation (3.4) is the nonlinear equation in t only.

The only genuine variable is t. The variable x is an auxiliary variable and is basically the spectral parameter.

For usual equations of the inverse scattering the spectral variable x enters only algebraically (see §6 for an exposition) :

$$\frac{\partial A_1(x)}{\partial t} - \frac{\partial A_2(x)}{\partial y} = [A_1(x), A_2(x)] ,$$

where $A_1(x)$, $A_2(x)$ are rational functions in the spectral parameter

$$x \equiv \lambda .$$

But in the isomonodromy case x enters the game both algebraically and as the independent variable of the differential (Schlesinger's) equations.

We can be more explicit with the structure of the equation (3.1).

Since equations with 3 regular singularities can always be reduced to the fixed singularities : $0,1,\infty$ (Gauss ${}_2F_1(a,b;c;x)$ functions), the first nontrivial case is the case of

4 singularities : $\{0,1,t,\infty\}$,

where t is a new variable.

Now we have a variable t. But in order to get an equation we need to have some object to vary it.

According to the Fuchsian theory of linear differential equations for the equation of the second order with four regular singularities, we have only one accessory parameter [40] , [42] . In other words we can vary in (3.1) at most one independent quantity. R. Fuchs chooses for this quantity another singularity of the equation (3.1) a so-called apparent singularity.

Definition 3.1. The singularity x = a of the equation (3.1) is called apparent if all the solutions of (3.1) are analytic at x = a.

Changing slightly the notations of Fuchs we denote this apparent singularity by w.

Remark. The conditions that the exponents are positive integers or zero and that no logarithmic terms appear, ensure that the singularity is apparent.

We are considering "isomonodromy" deformations. Then by Schlesinger's equations or by the "KdV-type" equation (3.4) w is a function of t.

In order to find this dependence of w on t (as a single differential equation of the second order) we write down the form of p(x) according to L. Fuchs' studies

$$p(x) = \frac{a}{x^2} + \frac{b}{(x-1)^2} + \frac{c}{(x-t)^2} + \frac{3}{4(x-w)^2} + \frac{\alpha}{x} + \frac{\beta}{x-1} + \frac{\gamma}{x-t} + \frac{\varepsilon}{x-w} \,. \tag{3.5}$$

Here a,b,c are constant (independent of x and t), $\alpha,\beta,\gamma,\varepsilon$ are functions of t (independent of x).

Now the rational function A takes the form

$$A = -\frac{x(x-1)}{x-w} \cdot \frac{t-w}{t(t-1)} \cdot \qquad (3.6)$$

Substituting (3.5) and (3.6) into (3.4) we obtain the following deformation equation for the linear differential equations of the second order with four regular singularities :

$$\frac{d^2w}{dt^2} = \frac{1}{2}\left\{\frac{1}{w} + \frac{1}{w-1} + \frac{1}{w-t}\right\}\left(\frac{dw}{dt}\right)^2 -$$

$$- \left\{\frac{1}{t} + \frac{1}{t-1} \quad \frac{1}{w-t}\right\}\frac{dw}{dt} +$$

$$+ \frac{w(w-1)(w-t)}{t^2(t-1)^2}\left\{\bar{\alpha} + \frac{\bar{\beta}t}{w^2} + \frac{\bar{\gamma}(t-1)}{(w-1)^2} + \frac{\bar{\delta}t(t-1)}{(w-t)^2}\right\}$$

Here the <u>constants</u> $\bar{\alpha},\bar{\beta},\bar{\gamma},\bar{\delta}$ can be expressed in terms of a,b,c :

$$\bar{\alpha} = 2a + 2b + 2c + k \quad , \quad k = \text{const},$$

$$\bar{\beta} = -2(a + \tfrac{1}{4}) \quad , \quad \bar{\gamma} = 2(b + \tfrac{1}{4}) \quad ,$$

$$\bar{\delta} = -2c.$$

This is nothing but Painlevé transcendent (VI) introduced before !

3.3. Now we know that the equations I-V of Painlevé and some other equations without movable singularities can be considered as the equations for similarity solutions (automodel, selfsimilar solutions) for completely integrable two dimensional systems [1] . For example, the second Painlevé equation can be considered to be an equation for similarity solutions of the modified KdV equation

$$u_t = (6u^3 - u_{xx})_x$$

Also the first Painlevé transcendent can be reconstructed from the Bousinesq equation

$$u_{tt} = (6u^2 - u_{xx})_{xx} \, ,$$

the soliton solutions $u(x-Vt)$ of which are reduced either to elliptic functions or to that of Painlevé (I).

Example. In the case of the generalized multi-component nonlinear Schrodinger

$$i\varphi_{jt} = -\varphi_{jxx} + \varphi_j \sum_{k=1}^{n} \varphi_k \psi_k ,$$

$$-i\psi_{jt} = -\psi_{jxx} + \psi_j \sum_{k=1}^{n} \varphi_k \psi_k : j = 1,\dots,n,$$

(see below in §7,7.1).

We can look for self-similarity solutions having the form

$$\varphi_i(x,t) = t^{\mu_i} u_i(xt^{-1/2}) ,$$

$$\psi_i(x,t) = t^{\lambda_i} v_i(xt^{-1/2}) : i = 1,\dots,n$$

Then we obtain such a system

$$u_{itt} = \lambda_i u_i + \frac{1}{2} tu_{it} - u_i \sum_{j=1}^{n} u_j v_j ;$$

$$v_{itt} = (1+\lambda_i)v_i - \frac{1}{2} tv_{it} - v_i \sum_{j=1} u_j v_j , \quad \text{(nR)}$$

$$i = 1,\dots,n.$$

By a reason explained below in §4 we call (nR) a "non-stationary Russian chain."

APPENDIX TO §3

3.4. THE COMPLETE LIST OF PAINLEVE TRANSCENDENTS

All the equations $(d^2w/dt^2) = F(t,w,\frac{dw}{dt})$ without movable critical points are reduced (by simple transformations) either to Ricatti equations or to elliptic function or to one of the following G types of Painlevé transcendents [57],[42],[32]:

$$\frac{d^2w}{dt^2} = 6w^2 + t ; \quad \text{(I)}$$

$$\frac{d^2w}{dt^2} = 2w^3 + tw + \alpha; \quad \text{(II)}$$

$$\frac{d^2w}{dt^2} = \frac{1}{w}\left(\frac{dw}{dt}\right)^2 - \frac{1}{t}\frac{dw}{dt} + \frac{1}{t}(\alpha w^2 + \beta) + \gamma w^3 + \frac{\delta}{w} \quad ; \tag{III}$$

$$\frac{d^2w}{dt^2} = \frac{1}{2w}\left(\frac{dw}{dt}\right)^2 + \frac{3w^3}{2} + 4tw^2 + 2(t^2 - \alpha)\, w + \frac{\beta}{w} \quad ; \tag{IV}$$

$$\frac{d^2w}{dt^2} = \left\{\frac{1}{2w} + \frac{1}{w-1}\right\}\left(\frac{dw}{dt}\right)^2 - \frac{1}{t}\frac{dw}{dt} + \frac{(w-1)^2}{t^2}\left\{\alpha w + \frac{\beta}{w}\right\} + \frac{\gamma w}{t} + \frac{\delta w(w+1)}{w-1} \quad ; \tag{V}$$

$$\frac{d^2w}{dt^2} = \frac{1}{2}\left\{\frac{1}{w} + \frac{1}{w-1} + \frac{1}{w-t}\right\}\left(\frac{dw}{dt}\right)^2 - \left\{\frac{1}{t} + \frac{1}{t-1} + \frac{1}{w-t}\right\}\frac{dw}{dt} + \frac{w(w-1)(w-t)}{t^2(t-1)^2}\left\{\alpha + \frac{\beta t}{w^2} + \frac{\gamma(t-1)}{(w-1)^2} + \frac{\delta t(t-1)}{(w-t)^2}\right\} \quad ; \tag{VI}$$

In fact all types (I)-(VI) can be deduced from (VI) alone. For this we define the following sequence of transformations [42] :

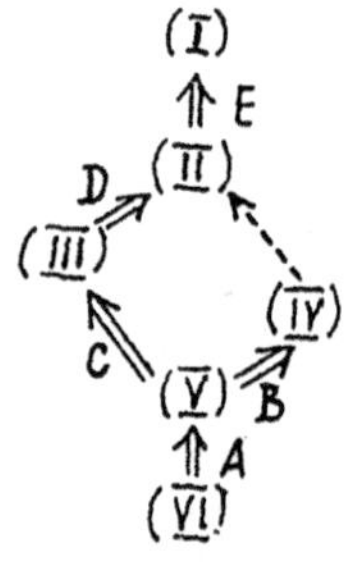

A. (VI) → (V). Replace z by $1 + \varepsilon z$, δ by $\frac{\delta}{\varepsilon^2}$, γ by $\frac{\gamma}{\varepsilon} - \frac{\delta}{\varepsilon^2}$ and let $\varepsilon \to 0$.

B. (V) → (IV). Replace w by $\varepsilon w\sqrt{2}$, z by $1 + \varepsilon z\sqrt{2}$, α by $\frac{1}{2\varepsilon^4}$, γ by $-\frac{1}{\varepsilon^4}$, δ by $-(\frac{1}{2\varepsilon^4} + \frac{\delta}{\varepsilon^2})$ and let $\varepsilon \to 0$.

C. (V) $\to$ (III). Replace w by $1+\varepsilon w$, β by $-\frac{\beta}{\varepsilon^2}$, α by $\frac{\beta}{\varepsilon^2}+\frac{\alpha}{\varepsilon}$, γ by $\gamma\varepsilon$ and δ by $\delta\varepsilon$ and let $\varepsilon\to 0$.

D.(III) $\to$ (II). Replace z by $1+\varepsilon^2 z$, w by $1+2\varepsilon w$, γ by $\frac{1}{4\varepsilon^6}$, δ by $-\frac{1}{4\varepsilon^6}$, α by $-\frac{1}{2\varepsilon^6}$, δ by $\frac{1}{2\varepsilon^6}+\frac{2}{\varepsilon^3}$, and let $\varepsilon\to 0$.

E. (II) $\to$ (I). Replace z by $\varepsilon^2 z-\frac{6}{\varepsilon^{10}}$, w by $\varepsilon w+\frac{1}{\varepsilon^5}$, α by $\frac{4}{\varepsilon^{15}}$ and let $\varepsilon\to 0$.

4. ISOMONODROMY AND SIMPLIFIED ISOMONODROMY DEFORMATION EQUATIONS RUSSIAN CHAIN.

4.1. As the particular example of "KdV type" equations in 3.2 the equation (3.4) shows there is no even trivial relation between isomonodromy and isospectral deformation equations.

There is one deep reason explaining this and one rather technical We present here both of them.

The basic reason is genuinely multidimensional character of isomonodromy deformation equations. From the algebraic point of view isomonodromy deformation is natural for any multidimensional situation see [39] . Here still unexploited are the relations with GL_n theory of multidimensional automorphic forms. On the other hand isospectral deformations, if understood in the straightforward way, have no natural multidimensional generalization. For example, on the manifolds of negative curvature the spectrum of $\Delta + U$ (Δ is Beltrami-Laplace operator and U is potential) determines uniquely both the manifold and the potential.

In other words isospectral deformations in multidimensional case can be trivial.

The technical reason is different and is connected with the form of deformation equations.

Isospectral deformation equations are the condition of consistency of two or more differential equations that are algebraic in spectral parameter λ . This is so called Zakharov-Shabat - Mikhailov general scheme, which has been given by Burchnall and Chaundy [6] in one dimensional case.

The isomonodromy equations also arise as a consistency condition of differential equations but being now differential in spectral parameter λ as well. This is more general approach than the isospectral one.

We rigorously show below in §5 that in one dimensional case

isomonodromy equations contain as a very particular case all the isospectral. We present in §4 the deduction of Schlesinger and simplified Schlesinger systems (A). These systems (A), are in fact, equivalent to one-dimensional isospectral deformation equations.

4.2. Let us start with the system of linear differential equations of the order m with n + 2 singularities $t_1,\dots,t_{n+2}$ (where we can assume $t_{n+1} = 0$, $t_{n+2} = 1$) :

$$\frac{dy_k}{dx} = \sum_{j=1}^{m} y_j \sum_{i=1}^{n+2} \frac{A^i_{jk}}{x - t_i} \quad : \quad k = 1,\dots,m. \tag{4.1}$$

We assume that the equation (4.1) is Fuschian and the monodromy group $\mathcal{J}$ of (4.1) is independent on $t_i (i = 1,\dots,n)$. Then the quantities A^i_{jk} satisfy the Schlesinger system of nonlinear non-autonomous equations [64] , [36] :

$$\frac{\partial A^j_{hk}}{\partial t_i} = \sum_{\ell=1}^{m} \frac{A^i_{h\ell}A^j_{\ell k} - A^j_{h\ell}A^i_{\ell k}}{t_j - t_i} \quad j \neq i \ ; \ h,k = 1,\dots,m \tag{4.2}$$

$$\sum_{j=1}^{n+2} \frac{\partial A^j_{hk}}{\partial t_i} = 0 \quad : \quad i = 1,\dots,n. \tag{4.3}$$

In order to obtain the system (4.2)-(4.3) we consider simply the evolution of y in t_i as

$$\frac{\partial y}{\partial t_i} = y\,\beta_i \quad : \quad i = 1,\dots,n$$

Together with the original equation (4.1)

$$\frac{\partial y}{\partial x} = ya = y\beta_o,$$

for $t_o = x$. Then the condition of consistency gives us the following equations

$$\frac{\partial \beta_i}{\partial t_j} - \frac{\partial \beta_j}{\partial t_i} = [\beta_i, \beta_j] \ . \tag{4.4}$$

Now let us look at the behaviour of y near t_j :

$$y = [(x - t_j)^r]z ,$$

where z is holomorphic near $x = t_j$. Then

$$\beta_i = z^{-1} \frac{\partial z}{\partial t_i} ,$$

and we can find that $\beta_i + a$ is a holomorphic function everywhere , i.e.

$$\beta_i = - \frac{A^i}{x - t_i} + \gamma_i \quad : \quad i = 1,\ldots,n+2,$$

where γ_i are now independent of x. We can also make in (4.1) a transformation

$$y = Yc$$

for c independent on x in such a way that

$$\frac{\partial c}{\partial t_i} = c\gamma_i \quad \text{or} \quad \frac{\partial \gamma_i}{\partial t_j} - \frac{\partial \gamma_j}{\partial t_i} = [\gamma_i,\gamma_j] \quad .$$

In other words we can assume

$$\beta_i = - \frac{A^i}{x - t_i} ,$$

and the system (4.4) is reduced to the form

$$\frac{\partial A^j}{\partial t_i} = [A^i,A^j] / t_j - t_i \quad : \quad i \neq j \quad ;$$
$$\sum_{j=1}^{m} \frac{\partial A^j}{\partial t_i} = 0 \qquad (4.5)$$

and the system (4.5) is naturally equivalent to (4.2) - (4.3). This completely integrable system depends on $m^2(n + 1)$ constants.

We still do not know too much about such systems : we do not know whether all such systems are Hamiltonian, whether they are indeed completely integrable ; whether we can solve these equations in terms of θ-functions (but not necessarily Abelian functions (see §9), etc...

There is, however, one very special case where we know a rather simple and nice answer to all such questions. This special case corresponds to the situation when we treat the Schlesinger system (4.5) in <u>the first approximation</u> in t_i. Such simplified systems were first introduced by Painlevé [57] and Boutroux [5] , who showed that Painlevé transcendences under a transformation like

$$t_i = a_i + \varepsilon t_i' \quad , \qquad w_i = \varepsilon^{-1} w_i' \quad ,$$

transform into a classical elliptic function. Then these investigations were continued by R. Garnier[35],[36](1916,1919) who considered many examples (especially m = 2 or n = 1).

In the Schlesinger system (4.5) we make the following tranformation

$$t_i \to a_i + t_i$$

$$A^i \to \varepsilon^{-1} A^i \quad : \quad i = 1,\ldots,n+2 \quad ,$$

for constants a_i, where, in particular, $a_{n+1} = a_{n+2} = 1$. If we put $\varepsilon \to 0$ then the system (4.5) takes the following form

$$\frac{\partial A^j}{\partial t_i} = \frac{[A^i,A^j]}{a_j - a_i} \quad : \quad j \neq i$$

$$\sum_{j=1}^{n+2} \frac{\partial A^j}{\partial t_i} = 0 \quad : \quad i = 1,\ldots,n \quad . \qquad \text{(A)}$$

As we shall see the system (A) is indeed completely integrable and, moreover, can be solved in Abelian integrals (of the first and second kind) corresponding to a certain algebraic curve.

Moreover, any completely integrable one-dimensional system connected with the algebra of commuting differential operators can be represented in a form (A) (see the next chapter). We call the system (A) "simplified Schlesinger equations".

4.3. Let us investigate the system (A). In (4.1) we have

$$\frac{dy_k}{dx} = \sum_{k=1}^{m} a_{hk} y_h \quad \text{for} \quad a_{hk} = \sum_{n=1}^{n+2} \frac{A^i_{hk}}{x - t_i} \quad . \qquad (4.6)$$

Now in the simplified system (A) we put

$$\overset{o}{a}_{hk} = \sum_{i=1}^{n+2} \frac{A^i_{hk}}{x - a_i} \quad : h,k = 1,\ldots,m, \quad .$$

Let

$$\varphi(x) = \prod_{i=1}^{n+2} (x - a_i) = x(x - 1) \prod_{i=1}^{n} (x - a_i),$$

and we define

$$a_{hk}^{o} = \frac{b_{hk}(x)}{\varphi(x)} , \qquad (4.7)$$

for some polynomials $b_{hk}(x)$: $h,k = 1,\ldots,m$ of degree $\leq n + 1$ in x and $b_{hk}(x)$ depending on t_i (because A^j depend on t_i).

If we put

$$b = (b_{hk}(x)) \quad h,k = 1,\ldots,m$$

then the system of the equations (A) is equivalent to

$$\frac{\partial b}{\partial t_i} = \frac{[A^i, b]}{x - a_i} \quad : \quad i = 1,\ldots,n. \qquad (4.8)$$

One of the main results concerning the system (A) has the following form :

Theorem 4.1. Let us define the following algebraic curve

$$f(x,y) = \det(b_{hk}(x) + y\delta_{hk})_{h,k=1}^{m} = 0. \qquad (4.9)$$

Then the coefficients of $f(x,y)$ are the first integrals of the system (A), i.e. for any solution A^i of (A) the coefficients of $f(x,y)$ are constants.

On the other hand, for an algebraic curve $f(x,y) = 0$ we can always find a $b_{hk}(x)$ such that $f(x,y) = 0$ can be represented in a form (4.9).

Proposition 4.2. ([36]) For any $f(x,y)$ having the form

$$f(x,y) = y^m + \sum_{i=1}^{m} y^{m-i} P_{i(n+1)}(x) \qquad (4.10)$$

for polynomials $P_{i(n+1)}(x)$ of degree $\leq i(n+1)$ there are $b_{hk}(x)$ of degree $\leq n+1$ such that

$$f(x,y) = \det(b_{hk}(x) + y\delta_{hk})_{h,k=1,\ldots,m}.$$

If the theorem 4.1 enables us to show that the system (A) is completely integrable and to integrate it in terms of Abelian functions, then proposition 4.2 gives us the possibility to reduce any classical one-dimensional completely integrable system to a simplified Schlesinger equation (A).

Example 4.3 (Garnier). Let us consider an arbitrary hyperelliptic curve of the genus $m-1$: $m \geq 2$. We can present such a curve in the most general form

$$f(x,y) = x^2 P(y) + x\,Q(y) + R(y) \tag{4.11}$$

for polynomials $P(y)$, $Q(y)$, $R(y)$ of degrees $m-2$, $m-1$ and m, respectively. Here we put $n = 1$ (i.e. singularities at $0,1,t$ but, possibly, some apparent singularities). Now we can put

$$b_{11}(x) = x^2 + C_{11} \quad ; \quad b_{hk}(x) = A_{hk}\cdot x + C_{hk} \; : \; (h,k) \neq (1,1).$$

We can always transform b in such a way that $A_{hk} = 0$ for $h > 1$ or $k > 1$. Looking at $b_{hk}(x)$ corresponding to the equation (4.11) we deduce from the equation (4.8), or the system (A), that

$$a_1(a_1 - 1)\frac{dA_{1k}}{dt_1} = -(A_{1k}a_1 + C_{1k})$$

$$a_1(a_1 - 1)\frac{dA_{h1}}{dt_1} = A_{h1}\,a_1 + C_{h1}$$

where putting

$$\tau = \frac{t_1}{a_1(a_1 - 1)} \quad ,$$

we see that

$$A_{1k} = f_k e^{-a_1\tau} \quad , \quad A_{h1} = g_h e^{a_1\tau}$$

$$C_{1k} = -f_k'\,e^{-a_1\tau} \quad , \quad C_{h1} = g_h' e^{a_1\tau}$$

(' is the derivative on τ). Next

$$a_1(a_1 - 1)\frac{dC_{hk}}{dt_1} = C_{h1}A_{1k} - A_{h1}C_{1k},$$

or for $h > 1$ or $k > 1$

$$C_{hk} = f_k g_h + C_{hk} \quad \text{and} \quad -C_{11} = f_2 g_2 + \ldots + f_m g_m .$$

From C'_{1h} and C'_{k1} we obtain

$$\begin{cases} f''_i = \lambda_i f_i + 2f_i \sum_{j=2}^{m} f_j g_j ; \\ g''_i = \lambda_i g_i + 2g_i \sum_{j=2}^{m} f_j g_j : i = 2,\ldots,m. \end{cases} \qquad (\mathcal{R})$$

Here $\lambda_i = c_{ii}$ and we can consider (c_{kk}) to be reduced to a diagonal form.

This system (especially in the case of $m = 3$) appears in many different works. E.g. in 1870 ($m = 3$) it was considered by Neumann. Since 1880 it was been the subject (for $m = 3$ again) of investigation for Russian mathematicians : Kowalevskaya [45] , Tchapligin, Kolossoff, Goriachov, so we decided to call the system $(\mathcal{R})$ the "Russian chain".

5. ALGEBRA OF COMMUTING DIFFERENTIAL OPERATORS, SIMPLIFIED SCHLESINGER SYSTEM AND THEIR CONNECTION.

Let us describe results on the algebra of commuting differential operators and corresponding completely integrable systems. We start with two linear differential operators in d/dx :

$$L_1 = \sum_{i=0}^{n} u_i \frac{d^i}{dx^i}, \quad L_2 = \sum_{j=0}^{m} v_j \frac{d^j}{dx^j},$$

where $u_n = v_m = 1$, $u_{n-1} = v_{m-1} = 0$. Then we can consider the following non-linear system of differential equations on u_i, v_j :

$$[L_1, L_2] = 0 \qquad (5.1)$$

being equivalent to a condition of consistency of a system

$$L_1\psi = \lambda\psi , \qquad L_2\psi = \mu\psi . \qquad (5.2)$$

The basic result here is

Theorem 5.1. (Burchnall-Chaundy [6]). An equation (5.1) is equivalent to an algebraic relation

$$Q(L_1,L_2) = 0$$

for a polynomial $Q(x,y)$. Here

1. For eigenfunctions ψ in (5.2) corresponding to the solution of (5.1), $Q(\lambda,\mu) = 0$;

2. The coefficients of $Q(x,y)$ are the first integrals of the system (5.1) ; they can be expressed as a differential polynomials in u_i, v_j and are constant for any given solution of (5.1).

The case of relatively prime m and n is especially simple. In this case $Q(x,y)$ has the form $Q(x,y) = x^m - y^n + \ldots$ and the space $\mathcal{L}_\lambda$ of ψ satisfying (5.2) is one-dimensional. In this case we have again

Theorem 5.2. (Buchnall-Chaundy [6], [3], 1928). In the case of a one-dimensional $\mathcal{L}_\lambda$ (the space of ψ in (5.2)), in particular, for $(n,m) = 1$ the system (5.1) is completely integrable. In this case the solutions of (5.1) (the coefficients u_i, v_j of L_1, L_2) are expressed in terms of a θ-function on $Q(x,y) = 0$.

For any fixed curve $Q(x,y) = x^m - y^n + \ldots$ the algebras $\mathfrak{A}$ of commuting differential operators $\{L_1,L_2\}$ satisfying $Q(L_1,L_2) = 0$ is naturally isomorphic to the Jacobian $Jac(Q)$ of $Q(x,y) = 0$.

For not relatively prime m and n or $\dim \mathcal{L}_\lambda = r > 1$ the description of L_1, L_2 can be effectively given in terms of algebraic vector bundles of dimension r over $Q(x,y)$.

The case $m = 2$, $n = 2k+1$ - odd number is especially interesting. The equations

$$[L_2, L_{2k+1}] = 0$$

for $L_2 = -\frac{d^2}{dx^2} + u$ have the form

$$Q(u,u_x,\ldots,u_{\underbrace{x\ldots x}_{2k+1}}) = 0$$

and are called stationary higher Korteweg-de Vries equations (k-th KdV)[47] , [48] , [34] . The reason for such a name is very simple. If we consider the Korteweg-de Vries equation

$$u_t = 6uu_x - u_{xxx}$$

then it can be written in the form of a Lax representation

$$\frac{dL_2}{dt} = [L_2,L_3]$$

for $L_3 = -4\frac{d^3}{dx^3} + 3(u\frac{d}{dx} + \frac{d}{dx}u)$. Now the KdV have infinitely many first integrals (conserved quantities)

$$I_n = \int P_n(u,u',\dots)dx$$

and the corresponding commuting Hamiltonian flows

$$u_t = \frac{\partial}{\partial x} \cdot \frac{\delta I_n}{\delta u}$$

Each of these Hamiltonian systems is equivalent to a Lax representation [47],[48],[34] :

$$\frac{dL_2}{dx} = [L_2,A_{2n+1}]$$

Now, the stationary system

$$[L_2,A_{2n+1}] = 0$$

is

$$\frac{\delta I_n}{\delta u} = c$$

and is an n-th stationary KdV equation for the function u(x).

We showed [16] - [18] that a general n-th stationary KdV equation is equivalent to the following completely integrable Hamiltonian system on n parameters $\lambda_1,\dots,\lambda_n$

$$f_i'' = \lambda_i f_i + 2f_i \sum_{j=1}^{n} f_j^2 : \quad i = 1,\dots,n \tag{5.3}$$

But the system (5.3) is simply "Russian chain" corresponding to a simplified Schlesinger equation (A) with 3 singularities : 0,1,x and of the order n+1. In this case for L_1,L_2 in (5.1) we have a hyperelliptic curve Q(x,y) :

$$Q(x,y) = x^2 - P_{2n+1}(y) ,$$

which can be reduced to a f(x,y) above.

This particular result can be generalized to a :

Theorem 5.3. For any relatively prime m and n any system (5.1) with $\mathrm{ord}(L_1) = n$, $\mathrm{ord}(L_2) = m$ can be represented as the system (A) : - "simplified Schlesinger equations". In this case the curves $Q(x,y) = 0$ and $f(x,y) = 0$ are birationally equivalent.

Moreover we can prove certain results in the case $(m,n) \neq 1$ and in the case of matrix differential operators as well.

Theorem 5.4. For any m and n the system (5.1) can be represented as the system (A).

Theorem 5.5. Let $\hat{L}_1, \hat{L}_2$ be matrix differential operators of size $k \times k$:

$$L_1 = \sum_{i=0}^{n} U_i \frac{d^i}{dx^i}, \quad \hat{L}_2 = \sum_{j=0}^{m} V_j \frac{d^j}{dx^j}$$

where $U_n = \mathrm{diag}(a_1,\dots,a_k)$; $V_m = \mathrm{diag}(b_1,\dots,b_k)$ and $U_{n-1,\alpha\beta} = 0$ if $a_\alpha = a_\beta$; $V_{m-1,\alpha\beta} = 0$ if $b_\alpha = b_\beta$.

Then the non-linear equation

$$[\hat{L}_1, \hat{L}_2] = 0$$

can be reduced to a form (A).

The order of (A) and the number of singularities of the system (A) are determined now by a genus and degree of $Q(x,y) = 0$.

Of course, the Schlesinger approach is more efficient than the classical Burchnall-Chaundy-Lax-Novikov- ... approach [6], [47], [55], since it can be generalized to great extent to an arbitrary dimension (cf. [39]).

We should notice that the additional variable x in the Schlesinger system is not a variable t (the position of the singularity). The variable t is a "real" variable in a system (A) but now play the role of a spectral parameter. Such an analogy can be continued further and we can indeed show that the system

$$\frac{d\bar{y}}{dx} = \sum_{i=1}^{n+2} \bar{y} \frac{A^i}{x - t^i}$$

can be considered as to be some auxiliary system of differential linear equations, where x is a spectral parameter for the Lax representation

$$L_1\psi = x\psi \; ; \quad L_2\psi = y\psi \; ; \quad Q(x,y) = 0$$

of the simplified Schlesinger system (A).

6. TWO-DIMENSIONAL COMPLETELY INTEGRABLE SYSTEMS AS CONDITION OF COMMUTATIVITY. RATIONAL DEPENDENCE ON SPECTRAL PARAMETER AND ISOSPECTRAL DEFORMATIONS.

6.1. The most general approach to the completely integrable system is based on the representation of the system of non linear equation as condition of commutativity of linear differential operators in partial derivatives. The first such representation belongs to Burchnall and Chaundy [6] . We present the corresponding idea along the lines of Zakharov-Shabat-Mikhailov [70] ,[71] approach. In (x,t)-space-time dimensional case, $x_1 = x$, $x_2 = t$ we also put $2\xi = t-x$, $2\eta = t+x$.

The most general commutativity condition in (x,t) have the following form

$$i\psi_\xi = U\psi \, , \quad i\psi_\eta = V\psi \qquad (6.1)$$

for (complex) operators U,V and ψ depending on ξ, η. Then (6.1) is equivalent to

$$U_\eta - V_\xi + i\,[U,V] = 0 \qquad (6.2)$$

The system (6.1)-(6.2) is gauge invariant under the transformation

$$U \to \widetilde{U} = f\,U\,f^{-1} + i\,f_\xi\,f^{-1} \quad V \to \widetilde{V} =$$

$$= f\,U\,f^{-1} + i\,f_\eta f^{-1} \; , \quad \psi \to \widetilde{\psi} = f\psi \, ,$$

for an arbitrary nonsingular operator function f.

The system (6.2) itself is trivially solvable if you don't put restrictions on U,V. It's general solution has the form

The system (6.2) becomes non-trivial if you introduce spectral parameter λ. For this we demand U and V to be rational (or, in general meromorphic) functions in λ.

We assume according to the Zakharov-Mikhailov-Shabat [70]-[71], [21] scheme that U and V, as functions of λ, have poles of fixed orders $m_1,\dots,m_n$ at fixed points $\lambda_1^o,\dots,\lambda_n^o$ of a λ-plane.

Then the system (6.2) becomes a "completely" integrable system of non-linear equations on the coefficients of U and V : i.e. on the residues of U and V at $\lambda=\lambda_j^o$. The classical Lax representation is imbedded into this scheme, if you consider both U and V to have poles at one fixed λ_o, with orders of poles at $\lambda = \lambda_o$ for U and V being equal, respectively, to the orders of operators L and A.

On the other hand, after some transformations the system (6.2) can be reduced to a certain form of Lax representation.

In the simplest case, when $U(\lambda),V(\lambda)$ have each only one pole, we put

$$U = U_o + \frac{U_1}{\lambda + 1}, \qquad V = V_o + \frac{V_1}{\lambda - 1}, \tag{6.3}$$

where U_o, U_1, V_o, V_1 are now independent of λ. If we now choose the gauge f in which

$$\tilde{U}_o = 0, \ \tilde{V}_o = 0, \ \tilde{U}_1 = A, \ \tilde{V}_1 = -B$$

we obtain very important for applications system of equations [70], [8], [21]:

$$A_\eta = \frac{i}{2}[A,B], \qquad B_\xi = -\frac{i}{2}[A,B]. \tag{6.4}$$

Equation (6.4) is equivalent to the condition of consistency for the system

$$i\psi_\xi = \frac{A}{\lambda+1}\psi, \qquad i\psi_\eta = -\frac{B}{\lambda-1}\psi \tag{6.5}$$

6.2. Another formulation of the system (6.4) as the commutativity condition is based on the Lax representation for (6.4). This formalism is due to K. Pohlmeyer, A. Neveu, L. Takhtadzhyan. We consider two operators

$$L = i\begin{pmatrix} A & 0 \\ 0 & 0 \end{pmatrix} \frac{d}{dx} + \begin{pmatrix} Q_1 & Q_2 \\ Q_3 & 0 \end{pmatrix} ,$$

$$M = i\begin{pmatrix} I & 0 \\ 0 & -I \end{pmatrix} \frac{d}{dx} + \begin{pmatrix} 0 & 2BQ_2 \\ 2Q_3B & 0 \end{pmatrix}$$

For diagonal matrix $A = \text{diag}(a_1,\dots,a_N)$, $a_i = 0 : i = 1,\dots,N$; $B = A^{-1}$ and $N \times N$ matrices Q_1, Q_2, Q_3, where

$$Q_{1jj} = 0 \quad : \quad j = 1,\dots,N \quad , \quad Q_3BQ_2 = B.$$

Then (6.4) can be reduced to the classical Lax representation

$$L_t = i[L,M] \ .$$

6.3. The system (6.4) leads to field theories, e.g. to σ models, connected with Lie groups [70] , [8] . Let us suppose that we have at any point (ξ,η) some element $g(\xi,\eta)$ of the Lie group G, where G is considered as subgroup of group of complex matrices.

We discuss the equations of the motion

$$g_{\xi\eta} = \frac{1}{2}(g_\xi g^{-1} g_\eta + g_\eta g^{-1} g_\xi) \tag{6.6}$$

and the corresponding action

$$S = \int d_\xi d_\eta \, \frac{1}{2} \, T_2\left(\frac{\partial}{\partial\xi} g \, \frac{\partial}{\partial\eta} g^{-1}\right) .$$

The equation (6.6) can be reduced to the form (6.4). We define :

$$A = ig_\xi g^{-1} \quad , \quad B = ig_\eta g^{-1} . \tag{6.7}$$

Then in the notation (6.7), our system (6.6) is equivalent to the system (6.4) :

$$A_\eta - B_\xi - i[A,B] = 0, \qquad A_\eta + B_\xi = 0.$$

From a physical (and mathematical) point of view it is reasonable to restrict ourselves to the two most important cases

$$G = SO(N) \quad \text{and} \quad G = SU(N).$$

Moreover it is necessary to consider also some reduction of the principle field g in (6.6). For example, we obtain one of the interesting systems (σ-models) under the restriction

$$g^2 = 1. \tag{6.8}$$

The system (6.6) is thereby reduced to

$$g_{\xi\eta} = \frac{1}{2}(g_\xi g g_\eta + g_\eta g g_\xi). \tag{6.9}$$

The restriction (6.8) is consistent with the equation (6.9).

6.4. Let us find to what systems of equations the systems (6.8), (6.9) can be reduced for G = SO(N), SU(N). We present g in the form

$$g = 1 - 2P_k \tag{6.10}$$

where $P_k^2 = P_k$. i.e. P_k is a projection operator

and k is the dimension of the image of P_k : the number k is unchanged in the evolution.

The most important case is <u>k = 1</u>, when we have $\mathbb{RP}^{N-1}$ or $\mathbb{CP}^{N\ 1}$. In this case the projector P_1 can be written in the form

$$(P_1)_{ij} = u_i u_j^* \quad \text{and} \quad \sum_{i=1}^{N} u_i u_i^* = 1. \tag{6.11}$$

Thus we have two σ-models [70] :

A. <u>Real $\mathbb{RP}^{N-1}$ σ-model</u>

$$\vec{u}_{\xi\eta} + (\vec{u}_\xi, \vec{u}_\eta)\vec{u} = 0 \ ; \qquad (\vec{u},\vec{u}) = 1 \ ;$$

B. <u>Complex $\mathbb{CP}^{N-1}$ σ model</u>

$$\vec{u}_{\xi\eta} + \frac{1}{2}\Big((\vec{u}_\xi,\vec{u}_\eta) + (\vec{u}_\eta,\vec{u}_\xi) + (\vec{u}_{\xi\eta},\vec{u}) - (\vec{u},\vec{u}_{\xi\eta})\Big)\vec{u}$$
$$- (\vec{u},\vec{u}_\xi)\vec{u}_\eta - (\vec{u},\vec{u}_\eta)\vec{u}_\xi + 2(\vec{u},\vec{u}_\eta)(\vec{u},\vec{u}_\xi)\vec{u} = 0$$

and $(\vec{u},\vec{u}) = 1$.

Here we put $\vec{u} = (u_1,\ldots,u_N)$ and $(\vec{a},\vec{b}) = \sum_{i=1}^{N} a_i b_i^*$ in the complex case.

7. EXAMPLES OF INFINITE COMPONENT AND OPERATOR TWO-DIMENSIONAL ISOSPECTRAL DEFORMATION EQUATIONS.

Here we present several interesting examples of two-dimensional isospectral deformation systems. These systems can be considered as infinite-component (or operator) generalization of already known classical completely integrable systems : non-linear Schrodinger, KdV, modified KdV etc.

However, infinite-component systems exhibit very important new properties. E.g. we show that the solution of the stationary (i.e. 1-dimensional) infinite-component non-linear Schrodinger contains the solution of the inverse scattering problem for general class of Sturm-Liouville problem (see [9]). On the other hand, operator equations are the natural form for considering quantum analogs of classical completely integrable systems. The exposition here is based on our papers [9] , [10], [12], [20] .

7.1. In order to show how two-dimensional completely integrable systems arise directly from Lax representation [34] ,[47] (or what is the same, from commutativity condition), we decided to take an example of infinite-component nonlinear-like Schrodinger equation [54], [9] , [20]. This system is very important from many points of view [8] , [23] , but also in the context of our lectures : it naturally contains "non-stationary" and "stationary" Russian chains (see § 3 : 3.3, § 4 : 4.3, § 5) in the finite component case.

First of all we present the multicomponent nonlinear-like Schrodinger equation. This equation in n-components $(\varphi_1,\ldots,\varphi_n)$, $(\psi_1,\ldots,\psi_n)$ has the form

$$i\varphi_j = -\varphi_{jxx} + \varphi_j \sum_{k=1}^{n} \psi_k \varphi_k \quad , \; j = 1,\ldots,n \qquad (7.1)$$

$$-i\psi_j = -\psi_{jxx} + \psi_j \sum_{k=1}^{n} \varphi_k \psi_k \; : \; j = 1,\ldots,n .$$

This system arises in a natural way from the Lax representation

$$iL_t = [L,M]$$

for the (n+1) (n+1) matrix differential operator L of order one

$$L = \begin{pmatrix} 1 & & 0 \\ & -1 & \\ 0 & & -1 \end{pmatrix} \frac{d}{dx} + \begin{pmatrix} 0 & \varphi_1 \ldots \varphi_n \\ \psi_1 & \\ \vdots & \\ \psi_n & & 0 \end{pmatrix}$$

and M being the (n+1) (n+1) matrix differential operator of order two [16] - [18]. Cf. §6 and example in §3,3.3.

7.2. An important approach in the completely integrable systems, connected with the infinite-dimensional Hamiltonian form, is based on the existence of the resolvent expansion [37], [38], [19]. This idea was realized first by J. Drach [27] and the modern exposition see in [37], [38].

We start from two equations for the resolvent for Sturm Liouville problem : $-\varphi'' + (u + \zeta)\varphi = 0$. These equations for $\bar{R}(x_1, z) \in C[[z^{-1}]]$ have the form

$$-\bar{R}''' + 4(u + \zeta)\bar{R}' + 2u'\bar{R} = 0, \quad \zeta = z^2 ;$$

$$-2\bar{R}\bar{R}'' + (\bar{R}')^2 + 4(u + \zeta)\bar{R}^2 = c(\zeta) ,$$

for $c(\zeta) \in C[[\zeta^{-1}]]$ with constant coefficients and

$$\bar{R} = \sum_{k=0}^{\infty} \bar{R}_k z^{-k} ,$$

where $\bar{R}_k = R_k[u,u',\dots]$ are differential polynomials in $u,u',\dots$. For $c(\zeta) \equiv 1$ we have a canonical resolvent

$$R(x,z) = \sum_{k=1}^{\infty} R_k[u] z^{-2k+1} \qquad (7.2)$$

Here

$$R_1 = \frac{1}{2}, \quad R_2 = \frac{1}{4}u, \quad R_3 = \frac{1}{16}(3u^2 - u''), \dots \qquad (7.3)$$

Then the equation

$$u_t = \frac{\partial}{\partial x} \cdot \frac{\delta I_n}{\delta u} \quad \text{for} \quad I_n = \int P_n[u,u',\dots]\, dx$$

is (n-2)-th KdV equation as it was introduced in [47], [34], [37] see §6. All I_m are conserved quantities for higher KdV.

7.3. The infinite component non-linear Schrodinger

The most natural way to introduce the infinite component system is to consider an index k of φ_k as a variable. We see later that this corresponds to the introduction of the spectral parameter k as the new variable.

We consider the Hilbert space $L^2(\Omega, d\mu_k)$) and $d\mu_k$ is a measure on Ω. We treat the functions $u(k)$, $\varphi(k) : k \in \Omega$ as elements of $L^2(\Omega, d\mu_k)$. Let $\lambda(k)$ be a measurable function on $(\Omega, d\mu_k)$.

In $L^2(\Omega, d\mu_k)$ there exists a natural definition of a scalar product

$$(u,v) = \int_\Omega u(k)\, v(k)\, d\mu_k ,$$

for $u(k)$, $v(k) \in L^2(\Omega, d\mu_k)$ and $|u|^2 = (u,u^*)$.

The most general system has the form

$$\begin{aligned} i\varphi_t &= -\varphi_{xx} + \varphi \int_\Omega \varphi\, \psi\, d\mu_k \\ -\, i\psi_t &= -\psi_{xx} + \psi \int_\Omega \varphi\, \psi\, d\mu_k \end{aligned} \tag{7.4}$$

for $\varphi = \varphi(x,t,k)$, $\psi = \psi(x,t,k)$. As a natural generalization of the nonlinear Schrodinger we have one equation

$$i\varphi_t = -\varphi_{xx} + \varphi \int_\Omega |\varphi|^2\, d\mu_k \tag{7.5}$$

In the stationary cases the systems (7.4)-(7.5) we write in the form

$$\begin{aligned} f_{xx} &= \lambda(k)\, f - f \int_\Omega fg\, d\mu_k \\ g_{xx} &= \lambda(k)\, g - g \int_\Omega fg\, d\mu_k , \end{aligned} \tag{7.6}$$

for $f = f(x,k)$, $g = g(x,k)$. As the generalized of the system from [16]-[18] we get

$$f_{xx} = \lambda(k)\, f - f \int_\Omega f^2\, d\mu_k . \tag{7.7}$$

First of all we must stress that φ, ψ, f, g or $d\mu_k$ are not necessarily continuous functions of k ; we demand only L_2 integrability with respect to $d\mu_k$. For example, the most interesting cases correspond to the situation., when φ or $d\mu_k$ has some singular components.

Let us present the Hamiltonian structure and conservation laws of the stationary equation (7.7).

We introduce a natural Hamiltonian

$$H = \frac{1}{2}\left\{\int_{\Omega} v^2 \, d\mu_k - \int_{\Omega} \lambda(k)u^2 \, d\mu_k + \frac{1}{2}\left(\int_{\Omega} u^2 \, d\mu_k\right)^2\right\} \qquad (7.8)$$

The Hamiltonian system which corresponds to (7.8) has the natural form

$$\begin{pmatrix} u \\ v \end{pmatrix}_x = \begin{pmatrix} 0 & 1 \\ -1 & 0 \end{pmatrix}\begin{pmatrix} \delta/\delta u \\ \delta/\delta v \end{pmatrix} H \qquad (7.9)$$

or, in shorter form, we have

$$u_{xx} = \lambda(k)\, u - u \int_{\Omega} u^2 \, d\mu_k \qquad (7.7')$$

for $u = u(x,k)$.

We have a very simple form of the conservation law for (7.7). We define $C(k)$ in the following expression

$$C(k) = 2v(x,k)^2 - 2\lambda(k)\, u(x,k)^2 + u(x,k)^2 \int_{\Omega} u(x,\ell)^2 \, d\mu_\ell$$
$$+ \int_{\Omega} \frac{(v(x,k)\, u(x,\lambda) - v(x,\ell)\, u(x,k))^2}{\lambda(k) - \lambda(\lambda)} \, d\mu_\ell$$

Theorem 7.1. For the solution u,v of (7.9) and arbitrary k,

$$\frac{dC(k)}{dx} = 0.$$

We have the following general problem : for an arbitrary potential $u(x)$ for which the corresponding Sturm-Liouville problem has sense (e.g. in the self-adjoint case and for non trivial spectrum) the solution of (7.9) which is naturally associated with $u(x)$ exists in the sense that

$$u(x) = -\int_{\Omega} u(x,k)^2 \, d\mu_k + C \qquad (7.10)$$

where $u(x,k)$ is the eigenfunction associated with k, of the Schrodinger with the potential $u(x)$.

One of the main results is the following theorem from [16] - [18], [9] .

Theorem 7.2 : For any solution u(x,k) of (7.7) let the potential u(x) be defined by (7.10).

$$u(x) = - \int_{\Omega} u(x,k)^2 d\mu_k + C_1.$$

Then for an arbitrary $n \geq 1$ there is a constant C_n such that

$$R_n[u] = (-1)^n \int_{\Omega} (\lambda(k))^{n-1} u(x,k)^2 d\mu_k + C_n \quad : n \geq 1. \quad (7.11)$$

The simplest way to prove theorem 7.2 is to look at any system $\varphi(x,k)$ of the eigenfunctions for the potential u(x) :

$$-\phi_{xx} + u(x)\phi = k\phi,$$

$$E(k) = \int_{\Omega} \frac{\phi^2(x,\ell)\, d\mu_\ell}{\lambda(\ell) - \lambda(k)}$$

Then E(k) satisfies the resolventa-like equation

$$- E_{xxx} + 4(u-k) E_x + 2u_x E = 2 \left\{ \int_{\Omega} \phi^2(x,\ell) d\mu_\ell \right\}_x \quad (7.12)$$

independently of the representation (7.10). You can obtain (7.11) from (7.10), (7.12) and the definition of a resolvent $R(x,\lambda)$.

7.4. We know already [16] - [18] that the system (7.9) together with the representation (7.10) describes all finite-band potential or solutions of the stationary KdV's in this case f(x,k) is singular.

The same representation takes place for an arbitrary C^∞-periodic potential u(x). In this case we have the same representation

$$u(x) = - \sum_{i=1}^{\infty} f_i^2(x) + C$$

For eigenvalues λ_i being nondegenerate periodic (left ends of the zonae) and $f_i(x)$ being periodic eigenfunctions corresponding to λ_i.

We have also a complete reduction of the inverse scattering problem for the system (7.9)-(7.10) in the case of u(x) rapidly decreasing on infinity.

Let us define all spectral data in this case. Let $\int_{-\infty}^{\infty} |u(x)|(1+x^2)dx < \infty$ and f(x,k) be the solution of $-f'' + uf = k^2 f$ with

$$f(x,k) \sim e^{ikx} \; : \; x \to +\infty$$

and

$$f(x,k) = a(k)\, e^{ikx} + b(k)\, e^{-ikx} \; : \quad x \to -\infty.$$

The reflection coefficient r(k) can be defined as

$$r(k) = -\frac{b^*(k)}{a(k)} .$$

It is known that u(x) may have only a finite number of bound states $-\beta_n^2 < \dots < -\beta_1^2 < 0$. Then the potential is uniquely determined by the reflection coefficient r(k), n bound states $\beta_1^2, \dots, \beta_n^2$ and n norming constants

$$c_j = \left\{ \int_{-\infty}^{\infty} f^2(x; i\beta_j)dx \right\}^{-1} \; ; \quad j = 1,\dots,n.$$

It is possible to express u(x) in terms of $f(x;k)^2$ in the form of the system (7.9)-(7.10). This was done for the first time by Kruscal, Gardner, Greene and Miura [34] for purely soliton solutions, for which $r(k) \equiv 0$ and then by Deift and Trubowitz for potential without bound states [28] .

Moreover, we have a more general representation for all $R_K[u]$:

<u>Theorem 7.3.</u> a) In the case r(k) = 0 we have

$$R_{K+1}[u] = (-1)^K \sum_{j=1}^{n} \beta_j^{2K-1} c_j f(x; i\beta_j)^2$$

b) In the case of the absence of the bound states (n=0) :

$$R_{K+1}[u] = (-1)^{K+1} \frac{i}{2\pi} \int_{-\infty}^{\infty} k^{2K-1} r(k) f^2(x;k)dk.$$

In the most general case we have

$$R_{K+1}[u] = \frac{i}{2\pi} \int_{-\infty}^{\infty} (-1)^{K+1} k^{2K-1} r(k) f^2(x,k)dk$$

$$+ \sum_{j=1}^{n} (-1)^K \beta_j^{2K-1} c_j f(x;i\beta_j)^2 : K = 0,1,2,3\ldots .$$

We would like to mention that such a representation of u(x) as (7.10) is not unique ! E.g. a finite band potential u(x) can be represented as a linear combination of any n from the 2n + 1 squares of eigenfunctions which correspond to the end of the bands.

However under a proper normalization and for a special measure, we have uniqueness and a very simple representation for a potential u(x) and higher quantities $R_K[u]$.

Main assertion 7.4 : for the case of self-adjoint L and the spectral measure $d\mu_k$ corresponding to a $L\varphi = k\varphi$ there exists a system $\varphi(x,k)$ of eigenfunctions of

$$\left(- \frac{d^2}{dx^2} + u(x)\right) \varphi = k\varphi$$

such that

$$R_K[u] = (-1)^{K+1} \int k^{K-1} \varphi^2(x,k)d\mu_k \qquad (B)$$

for all K = 1,2,3,...

In particular,

$$\frac{1}{2} = \int \varphi^2(x,k)d\mu_k ;$$

$$\frac{u}{4} = \int k\varphi^2(x,k)d\mu_k ;$$

Now the measure $d\mu_k$ and the system of eigenfunctions $\varphi(x,k)$ is fixed by the conditions (B).

7.5. Two-dimensional operator systems.

The aim of this part of the §7 is to present a procedure that enables to consider together with each classical completely integrable system its operator analogue. In particular it is related to a quantization of these classical systems.

We presented before in § 7. infinite-component generalizations of non-linear Schrodinger equation (see also [8] , [9] , [20]). That generalizations [8], [9] as can be verified, are a particular case of the present approach, where rank of the operators is one.

Now we construct only such "completely integrable" operator systems and represent them in terms of isospectral deformations. The rigorous proof of the complete integrability and the reduction to the action-angle variables (under special boundary conditions) are done using only spectral interpretation but will be published later.

Very interesting classes of solutions are connected with vector bundles over algebraic curves and can be deduced from the expressions for Baker's function of §9. Solutions connected with hyperelliptic curves of genus g are described practically in 9.6 see also [55] , [50] and were presented in [20] for SU(2g+1) σ-models.

7.6. We consider operators ϕ, ψ, U, V, W,... on an arbitrary Hilbert space H.

The general idea of the replacement of classical completely integrable equations by their operator analogs can be explained as follows. We mentioned in §6 that the most general two-dimensional completely integrable equations are represented in Zakharov-Mikhailov-Shabat scheme as the condition of the commutativity

$$\frac{\partial\psi}{\partial\xi} = A(\lambda)\psi \quad , \quad \frac{\partial\psi}{\partial\eta} = B(\lambda)\psi \tag{7.13}$$

of two linear problems, where $A(\lambda)$, $B(\lambda)$ are rational functions in the λ plane. In the classical case $A(\lambda)$, $B(\lambda)$ are $N \times N$ matrices and all such systems (7.13) are characterized by :
i) the positions of λ_j^o of poles of $A(\lambda)$, $B(\lambda)$ for fixed λ_j^o ;
ii) order of poles of $A(\lambda)$, $B(\lambda)$ at $\lambda = \lambda_j^o$.

In order to replace your classical system by an operator one you change $A(\lambda)$, $B(\lambda)$ to be operators keeping the same poles λ_j^o and orders of poles being fixed.

In general, in quantum case there appears also necessity to rearrange the order of terms in $A(\lambda)$, $B(\lambda)$ and in (7.13). Hence in each concrete case this must be done separately.

The first example is an operator non-linear Schrodinger equation (ONLS). This equation corresponds to the Lax representation

$$i\gamma \frac{dL}{dt} = [M,L] \tag{7.14}$$

with

$$L = \begin{pmatrix} I & 0 \\ 0 & -I \end{pmatrix} \frac{d}{dx} + \begin{pmatrix} 0 & A \\ B & 0 \end{pmatrix} \quad \text{and}$$

$$M = \begin{pmatrix} I & 0 \\ 0 & -I \end{pmatrix} \frac{d^2}{dx^2} + \begin{pmatrix} 0 & C \\ D & 0 \end{pmatrix} \frac{d}{dx} + \begin{pmatrix} U & V \\ W & Z \end{pmatrix}.$$

From (7.14) we obtain the following equivalent evolutionary operator equations

$$\begin{cases} 2i\,\gamma A_t = A_{xx} + 2ABA\ ; \\ -2i\gamma B_t = B_{xx} + 2BAB. \end{cases} \tag{7.15}$$

In the particular case $B = A^+$ we obtain from (7.15) operator nonlinear Schrödinger equation

$$i\,\psi_t = \psi_{xx} + \psi\psi^+\psi \tag{7.16}$$

Already the cases rank $(\psi) = 1$ or rank $(\psi) < \infty$ contain a lot of interesting examples. Let, e.g. H be $L^2(\Omega,\mu)$ and in H, $(a,b) = \int a(\omega)\, b(\omega)^* d\mu$. For any k we obtain as a particular case of (7.16) the following system of equations :

$$i\bar{\varphi}_{jt} = \bar{\varphi}_{jxx} + \sum_{i=1}^{k} c_i\, \bar{\varphi}_i(\bar{\varphi}_j,\bar{\varphi}_i) \quad : \quad j = 1,\dots,k \tag{7.17}$$

for $\bar{\varphi}_1(x,t),\dots,\bar{\varphi}_k(x,t)$ from H and constants $c_1,\dots,c_k$. In the particular case $k = 1$ we recover infinite-component nonlinear Schrödinger equation that was studied before. Now the equations (7.15)-(7.17) are "completely integrable" in the sense that the spectral representation (7.14) defines infinitely many conversation laws, inverse scattering procedure via Gelfand-Levitan equation of the solution of the Cauchy problem and natural action-angle variables if $\|A\|$, $\|B\|$, $\|\psi\|$ are rapidly decreasing on infinity $(|x| \to \infty)$.

7.7. We have natural operator generalization of the KdV equation $u_t = 6uu_x + u_{xxx}$. For this we consider Lax representation (7.14) where L is now a linear Schrodinger operator with "potential" ϕ :

$$L = -\frac{d^2}{dx^2} + \phi \quad , \tag{7.18}$$

and M is the operator of the third order in d/dx. However there is a big difference between operator KdV and other operator equations like ONLS : you cannot get from KdV any reasonable and non trivial multicomponent equations, when rank (ϕ) = 1. Because of this it is better to introduce operator KdV using operator modified KdV equation (OmKdV) that have multicomponent versions [9] .

The operator modified KdV equation arises from the Lax representation (7.14) where L is as in (7.14) and M is an operator of the third order in d/dx. This equation OmKdV naturally appears as the operator Hamiltonian flow commuting with the Hamiltonian flow for ONLS. We can write OmKdV as

$$\psi_t = \psi_{xxx} - 3(\psi_x\psi^2 + \psi^2\psi_x). \tag{7.19}$$

You can get in the case $H = L^2(\Omega,\mu)$ infinite-component mKdV as follows

$$\bar{\varphi}_t = \bar{\varphi}_{xxx} - 3\bar{\varphi}_x(\bar{\varphi},\bar{\varphi}) - 3\bar{\varphi}(\bar{\varphi}_x,\bar{\varphi}). \tag{7.20}$$

In order to get (7.20), see [9] , we can consider "two-component" generalization of (7.19). E.g. from (7.14) with L as in (7.14) we can get the following coupled system of operator equations

$$\begin{aligned} \phi_{1t} &= \phi_{1xxx} - 3\phi_{1x}\phi_2\phi_1 - 3\phi_1\phi_2\phi_{1x} ; \\ \phi_{2t} &= \phi_{2xxx} - 3\phi_{2x}\phi_1\phi_2 - 3\phi_2\phi_1\phi_{2x} . \end{aligned} \tag{7.21}$$

7.8. Our main result here is the establishing of the operator Miura transformation [52] relating OmKdV equation (7.19) and <u>operator KdV equation</u>

$$\phi_t = \phi_{xxx} + 3(\phi_x\phi + \phi\phi_x). \tag{7.22}$$

<u>Theorem 7.5.</u> For any ψ and $\varepsilon = \pm 1$ let us define

$$\phi_\varepsilon = \varepsilon\psi_x - \psi^2 \tag{7.23}$$

Then $\psi(x,t)$ is the solution of operator mKdV equation (7.19), iff both $\phi_\varepsilon : \varepsilon = \pm 1$ are the solutions of operator KdV equation (7.22).

In particular theorem 7.5 establishes Bäcklund transformations for both OmKdV and OKdV equations (7.19) and (7.22). As in classical situation, theorem 7.5 also reconstruct the reduction of the equation (7.22) to the Schrödinger linear operator (7.18). We put for this

$$\psi = -\varepsilon V_x V^{-1} \tag{7.24}$$

and making substitution $x \to x - 6\lambda t$, $t \to t$, $\phi_\varepsilon \to \phi_\varepsilon + \lambda$, we obtain from (7.22)-(7.24) that :

$$-V_{xx} - \phi V = \lambda V \tag{7.25}$$

where $\phi = \phi_\varepsilon$ is the solution of OKdV (7.22). Now using the scheme (7.24)-(7.25) and looking at the expansion of ψ in powers of λ - brings us : a) Hamiltonian structure for (7.22) or (7.19) ; b) infinitely many Hamiltonian flows commuting with those of (7.22) or (7.19) ; c) inverse scattering formalism. The most important quantity reflection coefficient $R(\lambda)$ is defined as usual brings us natural symplectic structure (c.f. [37]).

8. TWO DIMENSIONAL EQUATIONS OF PAINLEVE TYPE

Since the Riemann monodromy problem is closely connected with Painlevé transcendents, any multidimensional generalization must supply some kind of completely integrable systems of Painlevé type, i.e. with coefficients depending algebraically on the variables. Certain way for presenting these systems belongs to M. Sato et al. [62], [63] , who considered deformation theory of multidimensional Fuchsian systems. However from the point of view of generating truly multidimensional completely integrable systems this approach gives us nothing new. We obtain some non-linear equation on $u(t_1,\dots,t_k)$ where the behaviour of each of the variable is governed by commuting dynamical systems. This means that we are practically dealing with one system and it is first integral.

In order to get some genuinely multidimensional Painlevé systems we can look back on the way in which Painlevé transcendents arise from the deformation of linear Fuchsian differential equations. We find that Painlevé equations are equivalent to the commutativity condition of linear operators with additional spectral parameter λ, where the position of the singularities in the linear operator is not fixed λ_i but depends linearly on t. This suggests to us the way in which natural multidimensional generalizations can be obtained

We can consider for example Baker functions in the narrow and general sense, where the poles γ_i are also the functions of x_i. In this way we can obtain some of these systems considered at first by R. Fuchs [30] and then by Garnier [35].

In general the situation can be explained as follows : we consider the spectral parameter λ and functions (matrix or operator ones) $A_i(\bar{x},\lambda)$: $i = 1,\ldots,\ell$ being rational functions in λ with fixed orders of poles and the positions of poles being linear functions of variables $\bar{x} = (x_1,\ldots,x_\ell)$. E.g. we can assume the position of poles to be fixed. We consider the following systems of non-linear equations on the residues of poles $A_i(\bar{x},\lambda)$ in λ :

$$[A_i,A_j] = \frac{\partial A_i}{\partial x_j} - \frac{\partial A_j}{\partial x_i} \quad : \quad i,j = 1,\ldots,\ell.$$

We call this system of Painlevé type, if the poles of $A_i(\bar{x},\lambda)$ at λ are non-constant. If the positions of the poles of $A_i(x,\lambda)$ are fixed we obtain the usual system of Zakharov-Shabat type. In the one dimensional case there are quite a few of such Painlevé-type systems. These are all the Painlevé system [57] and all their generalizations found by Chazy [7] , Gambier [32] , Garnier [35] and others. These are also systems of equations satisfied by self similar (automodel) solutions of completely integrable systems in (x,t) dimensions.

In order to present non-trivial Painlevé systems in two dimensions we present the generalizations of chiral models.

Let us consider the following commutativity condition

$$i\psi_\xi = \frac{A}{\lambda + a\xi + b}\psi \quad ; \quad i\psi_\eta = -\frac{B}{\lambda + c\eta + d} \quad .(8.1)$$

The system (8.1) is equivalent to the following system of non-linear equations

$$A_\eta = -\frac{i}{c\eta - a\xi + d - b}[A,B] \; ; \tag{8.2}$$

$$B_\xi = \frac{i}{c\eta - a\xi + d - b}[A,B] \quad . \tag{8.3}$$

This system becomes familiar in a "g-form" of §6 as a field theory connected with a Lie group G.

In general using the idea of isomonodromy deformations combined with the differentiation with respect to spectral parameter equations one can obtain Painlevé type equations starting from any classical isospectral equations in (x,t) - or (x,y,t)-dimensions.

All these equations differ from the classical one as, say, the first Painlevé (I) differs from the elliptic function : there are additions of the non-autonomous terms and the orders of the equation is increased (usually by one).

It is impossible to apply directly the usual inverse scattering technique to solve these equations, though there always exists the reduction to the Riemann boundary value problem and a singular integral equation.

We present below certain examples of these equations. However here it should be mentioned that even for an ordinary Painlevé of types (I)-(VI) the method of solution of the equations with given asymptotics follows from the Riemann boundary value problem. Of course, as <u>this</u> equation is one-dimensional, you can relate it to a Gelfand-Levitan type of equation. However there will be no "normal" isospectral (or inverse scattering) interpretation of the solution of the Painlevé systems.

Examples of Painlevé-type systems in (x,t) dimensions.

1) Chiral models of the "Painlevé type". These examples in (x,t)-dimensions are connected with g-systems and σ-models. They correspond to the two-dimensional generalizations of "non-stationary Russian chains" ;

2) (x,t)-equations connected with Painlevé of types (I)-(II).

Some examples. The <u>simplest</u> :

$$v_t + v_{xxx} - 6v_x^2 = Cx, \quad C = \text{const.} \tag{8.4}$$

for $v_t \equiv 0$ and $v_x = w$ we get Painlevé (I). "Solitons" of (8.4) are solutions of the equation equivalent to Painlevé (I). Asymptotically they correspond to normal solitons of KdV or to 1-band elliptic solutions of KdV or to the Airy function (depending on asymptotic conditions on infinity).

One more complicated example connected with both Painlevé (I) and Painlevé (II).

$$u_{xt} + u_{xxxx} + u_x^2 + uu_{xx} = \mathfrak{P}[u] \tag{8.5}$$

for

$$\mathfrak{P}[u] = 8u - 3xu_x + 27tu_t + x^2u_{xx} + 6xtu_{xt} + 9t^2u_{tt}.$$

<u>Remark</u>. These equations are not "perturbed" classical equations. <u>E.g.</u> (8.4) is not an "inhomogeneous" KdV. It is possible to show using the Painlevé method that among the equations

$$u_t - 6u_x^2 + u_{xxx} = P(x)$$

the only equation without "movable" critical lines (or with meromorphic in (x,t) plane solutions) is the equation with

$$P(x) = Cx + D$$

Concluding remarks. We explained (see § 5) how to get from the Riemann problem under very special conditions all 1-dimensional completely integrable equations.

On the other hand general deformation (Schlesinger) equations describe equations of the type of the Painlevé transcendents (§3). These equations, as we know now 3.3, are also deduced from KdV type equations as self-similarity solutions. It is natural to refer to this case as 1.5-dimensional.

Two dimensional (x,t)-cases are already KdV or Zakharov-Shabat-Mikkailov systems (6.1 - 6.3).

The next, say 2,5 -dimensional case, that is still possible to consider, is the two-dimensional generalization of Painlevé transcendents. These equations appears, if you demand in the ZakHarov-Mikhailov scheme, to have poles in a λ-plane as functions of x and t (see (8.2)-(8.5)).

Summary of monodromy studies.

We presented the development of studies of the monodromy problems first formulated by Riemann. Our aim was to show how by using different solutions of the Riemann problem, it is possible to apply them to examine non-linear isomonodromy deformation equations.

First of all, for a one dimensional case we proved that this class of equations contains as a very particular limiting case isospectral deformation equations.

We tried to stress that the isomonodromy approach means new possibilities of deriving and studying multidimensional systems having exact solutions that are reduced to singular integral equations via the Riemann boundary problem. These classes of equations are very broad even in the case of two dimensional systems. The possibilities of four dimensional generalizations are not yet explored though we should mention again, that e.g. for self-dual Yang-Mills there is, due to Ward, [69] a certain mechanism for introduction of the Riemann boundary value problem. In this case it is also possible to build some analogs of the corresponding isomonodromy structure.

The possibility to treat the differential equations with respect to spectral parameter allows us to generalize the class of equations for which complete integrability can be rigorously proved. The variation of accessory parameters with respect to movable singularities is in fact the most general way for obtaining angle variables.

Detailed studies of the isomonodromy equations simplified or non-simplified are not finished at all and there are good possibilities to construct an adequate class of completely integrable systems in higher dimensions.

9. BAKER FUNCTION OR HOW TO SOLVE ISOSPECTRAL DEFORMATION EQUATIONS USING RIEMANN BOUNDARY VALUE PROBLEM ON RIEMANN SURFACE.

In this chapter we explain how to construct analytically the common eigenfunction of two commuting linear differential operators (so-called Baker function). This function is constructed using Riemann boundary value problem. In the Appendix the explicit expression is presented for hyperelliptic curves.

9.1. The reduction of certain system of non-linear equations to commutativity condition for linear operators does not solve, however, this system of equations. Practically, in order to solve this system of non linear equations we have to solve two linear problems simultaneously, or, what is easier, to find the common eigenfunction for two linear operators. The fact that a given function is the common eigenfunction for two different operators imposes very strong conditions on these eigenfunctions. Always these eigenfunctions have a certain precise algebrogeometrical sense.

In the one dimensional case it was found by Baker [3] and Burchnall-Chaundy [6] that for two linear scalar operators L_n and L_m of relatively prime orders m and n the common eigenfunction ψ (unique up to a multiplicative constant) is expressed in terms of the θ-function of an algebraic curve $\overline{Q}$ naturally associated with L_n and L_m. The curve $\overline{Q}$ is defined by the equation $Q(x,y) = 0$, where commuting operators L_m, L_n satisfy the algebraic relation $Q(L_n,L_m) = 0$.

In other words, in order to construct solutions of the system of non-linear equations $[L_n,L_m] = 0$ it is enough to find algebraically this common eigenfunction ψ called the Baker function [51] and then to find "quasipotentials", i.e. the coefficients of L_m, L_n from $L_n\psi = 0$, $L_m\psi = 0$.

This method easily gives us all the finite band potentials and in the same time the solutions of the KdV equation $u_t = 6uu_x - u_{xxx}$ and higher order KdV equations for periodic initial conditions $u(x) = u(x,0)$. We shall present the corresponding formulae due to [50],[3],[6] [55] [51].

In general, the Baker function corresponding to a given algebraic curve $\mathcal{T}$ is defined axiomatically in such way.

Let Γ be the Riemann surface of $\mathcal{T}$ of genus $g \geq 0$ with a fixed point P_o (often called $P_o = \infty$) and a local parameter $z = k^{-1}$ at $P = P_o$.

<u>Definition 9.1.</u> Let $x_1,\dots,x_g$ be independent variables and $\gamma_1,\dots,\gamma_g$ be g points on Γ. We call a function $\psi(\bar{x},\bar{x}_o,P)$ <u>Baker function</u> on Γ if :

1) $\psi(P)$ is a meromorphic function of P on $\Gamma \setminus P_o$ with g poles $\gamma_1,\dots,\gamma_g$ only (independent on $\bar{x} = (x_1,\dots,x_g)$) ;

2) at $P \to P_o$ or $k \to \infty$ the function $\psi(\bar{x},x_o,P)$ has the following asymptotics

$$\psi(\bar{x},\bar{x}_o,P) = \exp\{\sum_{i=1}^{g} k^i(x_i - x_{i_o})\} \times \{1 + \sum_{n=1}^{\infty} \psi_n(\bar{x}).k^{-n}\};$$

3) $\psi(\bar{x}_o,\bar{x}_o,P) \equiv 1$, i.e. $\psi_n(\bar{x}_o) = 0$.

For a divisor $D = (\gamma_1,\dots,\gamma_g)$ of a general position on Γ the function $\psi(\bar{x},\bar{x}_o,P)$ satisfying 1) - 3) exists and is unique [2] , [3] , [51] , [55] .

<u>Remark</u>. In fact the exponential part of $\psi(\bar{x},\bar{x}_o P)$ at $P \to P_o$ can be chosen in a more complicated way, e.g.

$$\psi(\bar{x};\bar{x}_o,P) = \exp\{kx + P(k)y + Q(k)t\} \times \{1 + \sum_{n=1}^{\infty} \psi_n(\bar{x})k^{-n}\}$$

for $\bar{x} = (x,y,t)$ and polynomials $P(k)$, $Q(k)$. Of course, when $g \geq \max\{d(P),d(Q)\}$ this representation is a particular case of the previous one. All such generalizations can be deduced from a more general definition of a Baker function given below.

The explicit expression for a Baker function $\psi(\bar{x},\bar{x}_o,P)$ can be given easily in terms of the ratio of θ-functions on Γ (cf.[55], [68]). Practically these formulae were known Picard [58] in his studies of uniform solutions of linear differential equations.

We present below in the Appendix the corresponding formulae in the hyperelliptic case (written by Baker) [2] , [3] , [55] , [50] .

It should be noted that these formulae are true not only for a finite g ; there are Baker's functions for an arbitrary (say hyperelliptic) Riemann surface of infinite genus (Mc Kean-Trubowitz [51]).

9.2. It is clear that function $\psi(\bar{x},\bar{x}_o,P)$ can be used as a common functions of L_n and L_m or $L_n - \frac{\partial}{\partial t}$, $L_m - \frac{\partial}{\partial y}$ for scalar differential operators only and only in the case when the rank r of the space of eigenfunctions φ common to L_n and L_m is one (then $\psi(\bar{x},\bar{x}_o,P)$ is a generator of this space).

For non-trivial and multidimensional systems there are many linearly independent common eigenfunctions and the leading term in the asymptotic expansion of $\psi(P)$ at $P \to P_o$ must have a more general form as well. We give a general definition of a Baker function that belongs to Novikov-Krichiver [56] .

We are working now with $m \times m$ matrix functions, though all our constructions are generalized for arbitrary operators.

The starting point for us is the existence of the solution $\psi(\bar{x},\bar{x}_o;k)$ of the system of equations

$$\frac{\partial \psi}{\partial x_j} = A_j(\bar{x},k)\psi \quad : \quad j = 1,\dots,\ell \tag{9.1}$$

for $A_i(\bar{x},k)$: $i = 1,\dots,\ell$ being rational functions in k independent of x_o and

$$\psi(\bar{x}_o,\bar{x}_o,k) \equiv I.$$

Here $\bar{x} = (x_1,\dots,x_\ell)$, $x_o = (x_{10},\dots,x_{\ell 0})$ and for $A_i(\bar{x},k)$: $i = 1,\dots,\ell$ the system (9.1) is equivalent to the following system of non-linear equations on the coefficient of $A_i(\bar{x},k)$: $i = 1,\dots,\ell$.

$$\frac{\partial A_j}{\partial x_i} - \frac{\partial A_i}{\partial x_j} = [A_i,A_j] \quad : \quad i,j = 1,\dots,\ell. \tag{9.2}$$

For $A_i(\bar{x},k)$: $i = 1,\dots,\ell$ being nontrivial functions in k the system (9.2) is non trivial.

<u>Definition 9.2.</u> (of a general Baker function). Let us on Γ have mg

points $(\gamma_1,\ldots,\gamma_{mg})$ and mg vectors $\bar{\alpha}_1,\ldots,\bar{\alpha}_{mg}$ of complex numbers of the size m-1 : $\bar{\alpha}_i = (\alpha_{i,1},\ldots,\alpha_{i,m-1})$: $i = 1,\ldots,mg$.

Let us suppose that we have an $m \times m$ matrix $\psi_o(\bar{x},\bar{x}_o,k)$ satisfying the system of equations.

$$\frac{\partial\psi_o}{\partial x_j} = A_j(\bar{x},k)\psi_o \quad : \quad j = 1,\ldots,\ell$$

for $A_j(\bar{x},k)$: $j = 1,\ldots,\ell$ rational functions in k (say, polynomials in k) and $\psi_o(\bar{x}_o,\bar{x}_o,k) = I$.

The $m \times m$ matrix function $\hat{\psi}(\bar{x},\bar{x}_o,P)$: $P \in \Gamma$ is called a general Baker function, corresponding to $(\gamma) = (\gamma_1,\ldots,\gamma_{mg})$, $(\alpha) = (\bar{\alpha}_1,\ldots,\bar{\alpha}_{mg})$ and $\psi_o(\bar{x},\bar{x}_o,P)$ if the following conditions are satisfied :

I) $\hat{\psi}(\bar{x},\bar{x}_o,P)$ as a function of P is meromorphic on Γ outside $P_o = \infty$;

II) the poles of $\hat{\psi}(\bar{x},\bar{x}_o,P)$ in P have an order 1 ; they do not depend on $\bar{x}$ and lie at points $\gamma_1,\ldots,\gamma_{mg}$ of Γ ;

III) For $\hat{\psi} = (\psi_{i,k})_{i,k = 1,\ldots,m}$ the residues of the functions $\psi_{i,k}(\bar{x},\bar{x}_o,P)$ at $P = \gamma_\lambda$: $\varphi_{i,k;\lambda}(\bar{x})$: $\lambda = 1,\ldots,mg$ satisfy the linear relations

$$\varphi_{i,k;\lambda}(\bar{x}) = \alpha_{\lambda,k}\,\varphi_{i,m;\lambda}(\bar{x}) \; : \; k = 1,\ldots,m-1,\; i = 1,\ldots,m \tag{9.3}$$

(where α_{j_1,j_2} are independent of $\bar{x}$) ; $\lambda = 1,\ldots,mg$

IV) In the neighbourhood of $P_o = \infty$ the function $\hat{\psi}(\bar{x},\bar{x}_o,P)$ is represented by the following asymptotic series

$$\hat{\psi}(\bar{x},\bar{x}_o,P) = (I + \sum_{n=1}^{\infty} \hat{\psi}_n(\bar{x},\bar{x}_o)k^{-n}) \quad \psi_o(\bar{x},\bar{x}_o,P), \tag{9.4}$$

$k = k(P)$.

V) $\hat{\psi}(\bar{x},\bar{x}_o,P) \equiv I$.

(the condition (9.3) means simply that for a given λ , $\lambda = 1,\ldots,mg$ among the function $\varphi_{i,j,\lambda}(\bar{x})$ for a given row i there is only one linearly independent.

9.3. In order to establish the existence of a general Baker function we use the solution of the boundary value problem on Riemann surface Γ in the form of Plemelj [59] and Koppelman [44] .

Roughly speaking, this problem means that we want to construct a single valued function W analytic on the Riemann surface with cuts C_ν the limit values of which after crossing the cuts C_ν satisfy boundary conditions

$$W^+ = m_\nu W^- + p_\nu ,$$

where m_ν, p_ν are given Hölder continuous functions on C_ν.

For our purposes we reduce the construct of $\hat{\psi}$ to the singular integral equation on a small circle of Γ around P_o with a Cauchy-type kernel. This equation is solved for each $\bar{x}$ separately and the condition (9.3) on the poles and residues of $\hat{\psi}$ uniquely determines the dependence of $\hat{\psi}$ on $\bar{x}$.

In order to construct the matrix $\hat{\psi}$ it is enough to construct each row separately. This means that all the conditions I)-IV) are satisfied, but (9.4) is replaced by

$$\varphi(\bar{x},\bar{x}_o,P) = (\bar{e}_i + \sum_{n=1}^{\infty} \varphi_n(\bar{x})k^{-n}) \psi_o(\bar{x},\bar{x}_o,k) , \qquad (9.4)'$$

for $\varphi = (\varphi_1,\ldots,\varphi_m)$-the i-th row of $\hat{\psi}$ and $\bar{e}_i = (0,\ldots,1,\ldots,0)$.

Let us now reduce the construction of the function $\varphi(\bar{x},P)$ satisfying I)-IV) ; (9.4)' to the Riemann boundary problem.

Let S be a boundary of a small neighbourhood of P_o at Γ. Let Γ^+ and Γ^- be external and internal domains on which S separates Γ.

The properties that are demanded from $\varphi(\bar{x},\bar{x}_o,P)$ are basically equivalent to the following ones ;

a) vector-functions $\varphi^+(\bar{x},\bar{x}_o,P) = \varphi(\bar{x},\bar{x}_o,P)$ for $P \in \Gamma^+$ and

$\varphi^-(\bar{x},\bar{x}_o,P) = \varphi_+(\bar{x},\bar{x}_o,P) \psi_o^{-1}(\bar{x},k(P))$ for $P \in \Gamma^-$ are meromorphic function in Γ^+ and Γ^- respectively ;

b) we have Riemann boundary problem

$$\varphi^+(\bar{x},\bar{x}_o,\zeta) = \varphi^-(\bar{x},\bar{x}_o,\zeta) \psi_o(\bar{x},\bar{x}_o k(\zeta)) \qquad (9.5)$$

for $\zeta \in S$;

c) we demand that the poles of all the coordinates $\varphi_i(P)$ of $\varphi(P)$ are lying in fixed points $(\gamma) = (\gamma_1,\ldots,\gamma_{mg})$ on Γ. Symbolically

$$(\varphi_i) + D \geq 0 \qquad (9.6)$$

for a divisor $-D = \gamma_1+\ldots+\gamma_{mg}$ with distinct γ_i.

d) As the solution of the problem (9.5)-(9.6) is, generally speaking, not unique, we have the set of complex vectors $(\bar{\alpha}) = (\bar{\alpha}_1,\ldots,\bar{\alpha}_{mg})$ satisfied for residues $\varphi_{i,j}(x)$ of φ_j at γ_i.

We reduce the boundary problem (9.5)-(9.6) further. We take on Γ a rational function f(P) with poles at points γ_i : $i = 1,\ldots,mg$ and with a zero of order $mg - g$ at P_o. Such a function f(P) exists and is unique up to a multiplicative constant.

We denote

$$\begin{aligned} \phi^+(\bar{x},\bar{x}_o,P) &= f(P)^{-1}\cdot \varphi^+(\bar{x},\bar{x}_o,P) ; \\ \phi^-(\bar{x},\bar{x}_o,P) &= k(P)^{-1}\cdot \varphi^-(\bar{x},\bar{x}_o,P). \end{aligned} \tag{9.7}$$

If $q_1,\ldots,q_g$ are the rest of zeroes of f(P) different from P_o and Δ is a divisor

$$\Delta = q_1 +\ldots+ q_g - P_o ,$$

then (9.5) - (9.6) in the notations (9.7) can reduced to

$$\begin{aligned} \phi^+(\bar{x},\bar{x}_o,\zeta) &= \phi^-(\bar{x},\bar{x}_o,\zeta)\, \psi_o(\bar{x},\bar{x}_o,k(\zeta))k(\zeta)f(\zeta)^{-1} , \\ (\phi_j) &\geq \Delta . \end{aligned} \tag{9.8}$$

The last condition in (9.8) means that the poles of the coordinates of ϕ are at points $q_1,\ldots,q_g$).

9.4. The problem (9.8) was solved by Koppelman [44] and Rodin (see [33]) using singular integral equation with Cauchy type kernel.

For the introduction of this meromorphic kernel A(p,q)dp on Γ we demand this kernel to be an abelian differential on P and as a function of q with the poles at $q_1,\ldots,q_g$ and a zero at P_o. When $p \to q$ we demand for singular part of A(p,q)dp to be

$$A(p,q)dp = \frac{dp}{p - q} .$$

Adopting Koppelman [44] notations we take on Γ 2g simple closed oriented curves $K_1,\ldots,K_{2g}$ which form a canonical homology basis : we assume that K_{2i-1} crosses K_{2i} once from left to right : $i = 1,\ldots,g$ and that no other crossing occurs.

We denote by $d\omega_i$: i = 1,...g the differentials of the first kind on Γ (i.e. analytic everywhere on Γ) that are normed by the conditions

$$\int_{K_{2i+1}} d\omega_j = \delta_{ji} \quad : \quad i,j = 1,\dots,g \quad ,$$

(here $d\omega_1,\dots,d\omega_g$ form a basis for the linear space of all regular differentials on Γ over $\mathbb{C}$).

For q_o, q_1 on Γ we denote $d\omega_{q_o,q_1}(p)$ the abelian differential of the third kind with two simple poles having residue -1 and +1 at $p = q_o$ and $p = q_1$ respectively with periods

$$\int_{K_{2i-1}} d\omega_{q_o,q_1}(p) = 0 \quad : \quad i = 1,\dots,g$$

Now we take g distinct points $q_1,\dots q_g$ on Γ (zeroes of f(P)). As $d\omega_i$: i = 1,...,g are linearly independent,

$$\det \left| \frac{d\omega_i(q_j)}{dq_j} \right| \quad i,j=1,\dots,g \quad \neq 0.$$

In other words we can solve a system of linear equations

$$\sum_{\mu=1}^{g} c_\mu(q) \frac{d\omega_\mu(q_i)}{dq_i} = - \frac{\partial \omega_{P_o,q}(q_i)}{\partial q_i} \; ; \; i = 1,\dots,g.$$

Then $c_\mu(q)$ depend analytically on q, vanish at $q = P_o$ and possess at most simple poles at the points $q = q_i$: i = 1...g. Now the kernel of the Cauchy type is defined by a Cramer rule :

$$A(p,q)dp = \frac{\begin{vmatrix} d\omega_{q,P_o}(p) & d\omega_1(p) \dots & d\omega_g(p) \\ d\omega_{q,P_o}(q_1) & d\omega_1(q_1) \dots & d\omega_g(q_1) \\ & \dots\dots & \dots\dots \\ d\omega_{q,P_o}(q_g) & d\omega_1(q_g) \dots & d\omega_g(q_g) \end{vmatrix}}{\begin{vmatrix} d\omega_1(q_1) \dots & d\omega_g(q_1) \\ \dots\dots & \dots\dots \\ d\omega_g(q_1) \dots & d\omega_g(q_g) \end{vmatrix}}$$

this is single-valued function with respect to q with simple poles at the points $q_1,\dots,q_g$ and zero at P_o.

Now let us use the Plemelj [59] jump formulae for this kernel

If $\phi(\bar{x},\bar{x}_o,P)$ is a solution of boundary problem (9.8), then we have

$$\begin{aligned} \phi(\bar{x},\bar{x}_o,P) &= \frac{1}{2\pi i}\int_S \rho(\bar{x},\bar{x}_o,\zeta)\, A(\zeta,P)d\zeta \ ; \\ \phi^+(\bar{x},\bar{x}_o,t) + \phi^-(\bar{x},\bar{x}_o,t) &= \\ &= \frac{1}{\pi i}\int_S \rho(\bar{x},\bar{x}_o,\zeta)\, A(\zeta,t)d\zeta \ ; \\ \phi^+(\bar{x},\bar{x}_o,t) - \phi^-(\bar{x},\bar{x}_o,t) &= \rho(\bar{x},\bar{x}_o,t). \end{aligned} \tag{9.9}$$

From the formulae (9.9) of Plemelj and from our boundary problem (9.8) it follows that $\rho(\bar{x},\bar{x}_o,t)$ satisfies the system of singular integral equations (see [53], [67], [33], [44]):

$$\rho(\bar{x},\bar{x}_o,t)\left[\frac{G+1}{2}\right] + \left[\frac{1}{2\pi i}\int_S \rho(\bar{x},\bar{x}_o,\zeta)\, A(\zeta,t)d\zeta\right] \times \frac{G-1}{2} = 0, \tag{9.10}$$

where by $G(\bar{x},\bar{x}_o,t)$ we denote

$$G(\bar{x},\bar{x}_o,t) = \psi_o(\bar{x},\bar{x}_o,t)f(t)^{-1}k(t). \tag{9.11}$$

In reverse, starting from any solution $\rho(\bar{x},\bar{x}_o,t)$ of (9.10) (9.11) we obtain the solution $\phi(\bar{x},\bar{x}_o,t)$ of (9.8). This solution is in fact unique.

Really, if $\phi_{1,2}(\bar{x},\bar{x}_o,P)$ are two solutions with the same jump function $\phi^+_\chi(\bar{x},\bar{x}_o,P) - \phi^-_\chi(\bar{x},\bar{x}_o,P) = \rho(\bar{x},\bar{x}_o,P)$: $\chi = 1,2$, then $\phi_1(\bar{x},\bar{x}_o,P) - \phi_2(\bar{x},\bar{x}_o,P)$ is already continuous on S, i.e. meromorphic on the whole Γ, every component of which has g poles and zero at P_o. Thus by Riemann-Roch theorem, every component is zero and $\phi_1(\bar{x},\bar{x}_o,P) = \phi_2(\bar{x},\bar{x}_o,P)$.

In other words, the number of linearly independent solutions of the boundary problem (9.8) is equal to the number of linearly independent solutions of the system (9.10)-(9.11). According to Koppelman [44], this number is equal to the index of (9.10) (9.11), i.e.

$$\kappa = [\arg \det G]_S$$

i.e. the change of the argument of G during travel along S in the positive direction.

Then (see [44], [53]) :

$$\kappa = m[\arg k(t) - \arg f(t)]_S = m(m-1)g + m.$$

Now the linear relations between the poles of $\varphi_j(P)$ at γ_i : $\varphi = (\varphi_1,\ldots,\varphi_m)$: $i = 1,\ldots,mg$ contained in $(9.3)^j$ give us the whole linearly independent family of the solutions of (9.8).

The last ambiguity in f(t) is liquidated by putting

$$\varphi(\bar{x}_o,\bar{x}_o,P) = (\sum_{n=0}^{\infty} \varphi_n(\bar{x}_o,\bar{x}_o)k^{-n})\psi_o(\bar{x}_o,\bar{x}_o,k(P)) = e_i$$

at $\bar{x} = \bar{x}_o$ when $P \to P_o$.

9.5. Conclusion. The Baker function $\psi(\bar{x},\bar{x}_o,P)$ (in the scalar case) can be always expressed in terms of the θ-function of a corresponding algebraic curve $\mathcal{T}$. In other words, the solutions of the corresponding Zakharov-Shabat system

$$L_i\psi = \frac{\partial}{\partial x_i}\psi \quad : \quad i = 1,\ldots,g$$

or

$$\frac{\partial L_i}{\partial x_j} - \frac{\partial L_j}{\partial x_i} + [L_i,L_j] = 0 \ ; \ i,j = 1,\ldots,g$$

are expressed in terms of Abelian integrals of the first and second kinds.

In general, however, Baker's function $\psi(\bar{x},\bar{x}_o,P)$ is a matrix and cannot be represented as combinations of θ-functions. The only way to construct the matrix Baker's function is to write a certain singular integral equation on Γ , whose solution will be ψ .

At the same time we are able to define a "nonstationary Baker's function", having movable poles on Γ that solves isomonodromy deformation equations of the Painlevé type (the corresponding definition is a natural generalization of 9.2).

9.6. Appendix to §9. Baker function and quasipotentials in the hyperelliptic case.

Let us present the description of the Baker's function (in the scalar case) on a hyperelliptic Riemann surface. It always happened that Baker's function ψ is, basically, a ratio of two θ-functions of a given Abelian variety. The corresponding "quasipotentials" $u_j(x_1,\ldots,x_k)$ - the coefficients of operators L_i such that $L_i\psi = k^i\psi$, $k = k(P)$, - are second logarithmic derivatives of θ-function of the same variety, restricted to a k-dimensional hyperplane. There is one extremely interesting feature of this incarnation of the Baker's function : the system of functions $\{u_j(x_1,\ldots,x_k), \psi(x_1,\ldots,x_k)\}$ satisfy an algebraic law of addition. All these features of the Baker's function are present already in the simplest case of an elliptic curve, where $\psi(x,k)$ is simply the solution of Lamé equation :

$$-\psi_{xx} + 2\mathcal{P}(x)\,\psi = k\,\psi,\ \psi(x,k) = \frac{\sigma(x \pm \alpha)}{\sigma(x)\sigma(x)}\,e^{\mp\zeta(\alpha)x},$$

$k = \mathcal{P}(\alpha)$.

Here $\mathcal{P}(x)$ - is Weirstrass elliptic function, uniformizing given elliptic curve $\mathcal{J}$ over $\mathbb{C}$ and $\psi(x,k)$ is so-called Hermite function. This example was treated in detail by Hermite, Picard, Painlevé [58] , [57] , (see [42] for references and further information).

Systems of functions satisfying an algebraic law of addition are always connected with Abelian varieties. In most transparent way this connection can be demonstrated using classical completely integrable systems naturally associated with Abelian varieties viewed as complex toruses.

Let us present briefly the Baker function and systems of functions satisfying an algebraic law of addition associated with a hyperelliptic curve [2], [50] , [55] , [51] .

Let Γ be the Riemann surface of the hyperelliptic curve $\mathcal{J}$: $y^2 = P_{2n+1}(x)$ of genus n. We take as the basis of circles on Γ :- $a_1,\ldots,a_n,b_1,\ldots,b_n$ where $a_i \circ a_j = b_i \circ b_j = 0$, $a_i \circ b_j = \delta_{ij}$. We take as the basis of holomorphic differentials on Γ

$$\Omega_k = \sum_{i=1}^{n} c_{ki}\,\frac{E^{i-1}dE}{P_{2n+1}(E)} \quad : \quad k = 1,\ldots,n$$

normed by the following conditions

$$\oint_{a_j} \Omega_{ki} = 2\pi i\,\delta_{jk}\,.$$

We obtain the matrix

$$B_{k\ell} = \oint_{b_\ell} \Omega_k$$

and θ-function

$$\theta(\eta_1,\dots,\eta_n) = \sum_{m_1,\dots,m_n} \exp\left\{\frac{1}{2}\sum_{j,k} B_{j,k}m_j m_k + \sum_k m_k\eta_k\right\}.$$

For a given set $(P_1^o,\dots,P_n^o)$ of points on Γ we define a potential $u(x)$:

$$u(x) = -2\frac{d^2}{dx^2}\log\ \theta(\eta_1^o + U_1(x-x_o) - K_1,\dots$$
$$\dots,\eta_n^o + U_n(x-x_o) - K_n),$$

where

$$K_j = \frac{1}{2}\sum_{\ell=1}^{n} B_{\ell j} - \pi i j \quad ; \quad \eta_j = \sum_{k=1}^{n}\int_{\infty}^{P_k^o}\Omega_j, \quad U_j = -\oint_{b_j}\Omega ,$$

and Ω is a differential such that

$$\Omega \sim i\frac{dz}{z^2} \quad \text{for} \quad E\to\infty \quad \text{and } z = \frac{1}{\sqrt{E}}$$

and $\oint_{a_j}\Omega = 0 \ : \ j = 1,\dots,n.$

Then $u(x)$ is a so-called n-band potential, satisfying an n th order stationary KdV (§5) equation of the order n. In other word $\mathcal{U} = \{u(x), u'(x),\dots,u^{(n-1)}(x)\}$ is a complete system of functions satisfying an algebraic law of addition. Now $u(x)$ depends on the choice of $(P_1^o,\dots,P_n^o)$. Different systems $\mathcal{U}$ are transformed into each other under the action of the higher KdV equations.

The Baker function corresponding to the same hyperelliptic curve $y^2 = \prod_{j=1}^{2n+1}(x - E_j)$ as the form

$$\psi(x,\bar{r}_o,P) = \exp\left\{\sum_{j=1}^{n}(i\omega(P))^j\right\} \times$$
$$\times \frac{\theta(A(P) + \bar{x}\,\vec{U} + \bar{r}_o).\ (\bar{r}_o)}{\theta(A(P) + \bar{r}_o).\ (x\,\vec{U} + \bar{r}_o)} ;$$

where

$$A(P)_k = \int_{\infty}^{P}\Omega_k \quad : \quad k = 1,\dots,n \ ;$$

$$\omega(P) = \int_{E_{2n+1}}^{P}\frac{z^n + b_1 z^{n-1} + \dots + b_n}{2\sqrt{\prod_{j=1}^{2n+1}(z - E_j)}}\,dz \ ,\oint_{a_k} d\omega(P) = 0 \ :$$

$k = 1,\dots,n.$

and

$$(U_j)_r = - \oint_{b_j} \bar{\Omega}_r$$

for a differential $\bar{\Omega}_r$ such that $\bar{\Omega}_r \sim i^r \frac{dz}{z^{2r}}$ for $P \to \infty$ and $z = 1/\sqrt{P}$ and $\oint_{a_j} \bar{\Omega}_r = 0 : j = 1,\ldots,n.$

Again $\{\psi(x,P),u(x),\ldots,u^{(n-1)}(x)\}$ satisfies the algebraic law of addition.

REFERENCES

1 Ablowitz, M.J., Segur, H. : 1977, Phys. Rev. Lett. 38, pp. 1103-1106.

2 Baker, H.F. : Abel's theorem and the allied theory including the theory of the theta functions, Cambridge, 1897.

3 Baker, H.F. : 1928, Proc. Royal. Soc. London, pp. 584-600.

4 Birkhoff, G.D. : 1909, Transaction of the AMS, pp. 436-470 ; 1910, Transaction of the AMS, pp. 199-202.

5 Boutroux, P. : 1913, Ann. de l'Ecole Norm. Sup. 30, pp. 255-375 ; 1914, Ann. de l'Ecole Norm. Sup. 31, pp. 99-159.

6 Burchnall, J.L., Chaundy, T.W. : 1922, Proc. London Math. Soc. 21, pp. 420-440 ; 1828, Proc. Soc. London Ser. A118, pp. 557-573.

7 Chazy, J. : 1911, Acta Math. 34, pp. 1-69 ; 1918, Acta Math. 41, pp. 29-69.

8 Chudnovsky, D.V.,: 1979, Nuovo Cimento B, (to appear) Service de Physique Théorique, Preprint DPh.T.79/38.

9 Chudnovsky, D.V.,: 1979, Phys. Rev. D. (to appear). Service de Physique Théorique, DPh.T. 79/80.

10 Chudnovsky, D.V. ; 1979, C.R. Acad. Sci. Paris, (to appear).

11 Chudnovsky, D.V. ; 1979, C.R. Acad. Sci. Paris, (to appear).

12 Chudnovsky, D.V. 1979 ; Service de Physique Théorique, (to appear) Preprint DPh.T.79/ .

13 Chudnovsky, D.V. : 1979, Proceedings of the International Conference "Non-Linear Evolution Equations and Dynamical Systems". Lecce, Italy, 1979, (to appear).

14 Chudnovsky, D.V. and Chudnovsky, G.V. : 1977, Nuovo Comento 40B, pp. 339-353.

15 Chudnovsky, D.V. and Chudnovsky, G.V. : (1977-1978) Seminaire sur les équations non-linéaires I, Centre de Mathématiques, Ecole Polytechnique, pp. 300,

16 Chudnovsky, D.V. and Chudnovsky G.V. : 1978, Lett. Nuovo Cimento, 22 pp. 31-36.

17 Chudnovsky, D.V. and Chudnovsky, G.V. : 1978, Lett. Nuovo Cimento, 22, pp. 47-51.

18 Chudnovsky, D.V. and Chudnovsky, G.V. : 1978, C.R. Acad. Sci. Paris 286A,pp. A1075 - A1078.
19 Chudnovsky, D.V. and Chudnovsky, G.V. : 1978, preprint IHES/M/78/236, pp. 1-12, to be published in Lectures Notes in Mathematics, Springer).
20 Chudnovsky, D.V. and Chudnovsky, G.V. : 1979, Phys. Lett. 72A, pp. 291-293 (Utrecht preprint).
21 Chudnovsky, D.V. and Chudnovsky, G.V. : 1979, Phys. Rev D (to appear, Service de Physique Theorique, Preprint DPh.T.79/110).
22 Chudnovsky, D.V. and Chudnovsky, G.V. : Service de Physique Theorique (to be published in Lecture Notes in Mathematics, Springer), Preprint DPh.T79/115.
23 Chudnovsky, D.V., and Chudnovsky, G.V. : 1979, Phys. Lett. A (to appear, IHES preprint).
24 Chudnovsky, D.V. and Chudnovsky G.V., 1979, Service de Physique Théorique, preprint
25 Chudnovsky, D.V. and Chudnovsky G.V., 1979, Seminar on Spectral Theory, Riemann Problem and Complete Integrability IHES, Paris.
26 Chudnovsky, G.V. : 1979, Cargese Lecture (see this volume).
27 Drach, J. : 1919, C.R. Acad. Sci. Paris 168, pp. 47-50 ; 1919, C.R. Acad. Sci. Paris 168, pp. 337-340.
28 Deift, P. and Trubowitz, E. : 1979, Comm. Pure Appl. Math. 32, pp. 121-200.
29 Erugin, N.P. : Linear Systems of Ordinary Differential Equations, Academic Press, 1966.
30 Fuchs, R. : 1905, C.R. Acad. Sci. Paris 141, pp. 555-558 ; 1907, Math. Ann. 63, pp. 301-321.
31 Fuchs, L. : Gesammelte Mathematische werke, bd 1-3, Berlin 1900-1906.
32 Gambier, B. : 1910, Acta Math. 33, pp. 1-55.
33 Gakhov, F.D. : Boundary Problems, Nauka, Moscow, 1977.
34 Gardner G., Greene, J.M., Kruskal, M.D., Miura, R.V. : 1974, Comm. Pure Appl. Math. 27, pp.97-120.
35 Garnier, R. : 1912, Ann. l'Ecole Norm. Sup. 48, pp. 1-126 ; 1945, C.R. Acad. Sci. Paris 221, pp. 276-278 ; 1956, Palermo Rend. 5, pp. 73-91.
36 Garnier, R. : 1919, Circolo Math. Palermo, 43, pp. 155-191.
37 Gelfand, J.M., Dikij, L.A. : 1975, Russian Math. Survey 30, pp. 37-100.
38 Gelfand, J.M., Dikij, L.A. : 1979, Funct. Anal. Appl. 13, pp. 10-20.
39 Gerard, R. and Ramis, J.P. : Equations différentielles et systèmes de Pfaff dans le champ complexe, Lecture Notes in Mathematics N°712, Springer, Berlin-New-York, 1979.
40 Golubev, V.V. : Lectures on the Analytic Theory of Differential equations, GTIL, Moscow-Leningrad, 1950.

41 Hilbert, D. : Grundzüge der Integralgleichungen, Leipzig-Berlin, 2-te Aufl., 1924.
42 Ince, E.L. : Ordinary differential equations, Dover, 1956.
43 Jost, R. and Newton, R.G. : 1955, Nuovo Cimento 1, pp. 590-622.
44 Koppelman, W. : 1961, J. Math. and Mech., pp. 247-277.
45 Kowalevski, S. : 1889, Acta Math. 12, pp. 177-232.
46 Lappo-Danilevsky, J.A. : Mémoires sur la Théorie des systèmes des Equations Différentielles Linéaires, Chelsea Publishing Company, 1953.
47 Lax, P. : 1968, Comm. Pure Appl. Math. 21, pp. 467-490.
48 Lax, P.D. : 1975, Comm. Pure Appl. Math. 28, pp. 141-200.
49 Mac Coy, B.M., Tracy, C.A. and Wu, T.T. : 1977, Phys. Lett. 61A, pp. 283-287.
50 Mc Kean, H.P., Van Moerbeke, P. : 1975, Invent. Math. 30, pp. 217-300.
51 Mc Kean, H.P., Trubowitz, E. : 1976, Comm. Pure Appl. Math. 29, pp. 143-226.
52 Miura, R.M. : 1976, SIAM Rev. 18, pp. 412-459.
53 Muskhelishvili, N.I. : Singular Integral Equations, P. Noordhoff, N.V., Groningen, Holland, 1953 ; Muskhelishvili, N.I.: Some basic problems of the mathematical theory of elasticity, P. Noordhoff N.V., Groningen, Holland, 1953.
54 Nogami, Y., Warke, C.S. : 1976, Phys. Lett. 59A, pp. 251-254.
55 Novikov, S.P., Dubrovin, B.A., Matveev,V.B. : 1976, Russian Math. Survey 31, pp. 59-100.
56 Novikov, S.P., Kritchiver, J.M. : 1978, Funct. Anal. Appl. 12, pp. 41-52.
57 Painlevé P. : Oeuvres Mathématiques, V 1-3, Paris, 1972.
58 Picard, E. : Leçons sur quelques types simples d'équations aux dérivées partielles, Paris, 1927.
59 Plemelj J. : Problems in the sense of Riemann and Klein, John Wiley, 1964.
60 Riemann, B., Oeuvres Mathémathiques, Albert Blanchard, Paris, 1968, pp. 353-363.
61 Röhrl, H. : 1957, Math. Ann. 133, pp. 1-25.
62 Sato, M., Miwa, T., and Jimbo, M. : 1977, Proc. Japan Acad. 53A, pp. 6-10, 147-152, 153-158, 183-185, 219-224 ; 1978 Proc. Japan Acad. 54A, pp. 1-5, 36-41, 309-312.
63 Sato, M., Miwa, T., and Jimbo, M. : Studies in Holonomic Quantum Fields, Preprints of the Research Institute for Mathematical Sciences, Kyoto University, Kyoto, Japan, 1977-1979.
64 Schlesinger L. : 1912, J. für Reine Angew. M. 141, pp. 96-145.
65 Schlesinger, L. : Einführung in die theorie der gewöhnlichen differentialgleichungen auf functionentheoretischer grundlage, 3aufl, Berlin-Leipzig, 1922.

66 Schesinger, L. Handbuch der theorie der linearen differentialgleichungen, Leipzig, bd. I-II, 1985-1898.
67 Vekua, I.N. : Singular Integral Equation, Fizmatgiz, Moscow 1950.
68 Verdier, J.L. : 1978, Seminaire Bourbaki.
69 Ward, R.S., Doctoral dissertation, Oxford, 1977.
70 Zakharov, V.E., Mikhailov, A.V. : 1978,JETP 74, pp. 1953-1970.
71 Zakharov, V.E., Shabat, A.B. : 1974, Funct. Anal. Appl. 8, pp. 43-57.

PADÉ APPROXIMANTION AND THE RIEMANN MONODROMY PROBLEM.

G.V. CHUDNOVSKY

CNRS, IHES, 35 route de Chartres, 91440 Bures-sur-Yvette, France
and
Dept. of Mathematics, Columbia University, New-York N.Y. 10027, USA

ABSTRACT

The general problem of Padé approximation to a system of functions satisfying linear differential equations is considered. We use the method of isomonodromy deformation to construct effectively the remainder function and Padé approximants in the case of N-point approximations of solutions of Fuchsian linear differential equations. Special attention is devoted to generalized hypergeometric functions ${}_qF_p(a_1,\ldots,a_q;b_1,\ldots,b_p;x)$. In various cases the asymptotics of the remainder function is presented. A separate section is devoted to applications of analytic methods to the problem of rational and diophantine approximations of the values at rational and algebraic points of functions satisfying linear differential equations. There is presented also an analytic method for investigation of the arithmetic nature of the constants, arising as values of hypergeometric functions, such as $L_2\ (1/q) : q \geq 14$, $\zeta(2)$, $\zeta(3)$ etc.. which occur in many physical situations.

C. Bardos and D. Bessis (eds.), Bifurcation Phenomena in Mathematical Physics and Related Topics, 449–510.

INTRODUCTION

The aim of this paper is to construct Padé approximations to the system of functions satisfying linear differential equations with rational function coefficients.

We consider Padé approximation problem as a general one following K. Mahler [20],[21] or [3][16] applied to the system of functions and one-or N-point. Such systems of Padé approximations considered for the first time by Hermite give rise to approximations of type I or type II.

Definition 0.1. Let $\{z_1,\ldots,z_N\}$ be N distinct points in $\mathbb{C}^1$ and $f_1(z),\ldots,f_n(z)$ be functions analytic in the neighborhood of $z = z_1,\ldots,z = z_N$. For n non-negative rational integers $m_1,\ldots,m_n$ we consider polynomials $A_i(z|f_1,\ldots,f_n;m_1,\ldots,m_n)$: of degrees $\leq m_i$ correspondingly : $i = 1,\ldots,n$ such that the function

$$R(z|f_1,\ldots,f_n;m_1,\ldots,m_n) \stackrel{\text{def}}{=} \sum_{i=1}^{n} A_i(z|f_1,\ldots,f_n;m_1,\ldots,m_n) \times f_i(z)$$

has zeroes at z_j : $j = 1,\ldots,N$:

$$\sum_{j=1}^{N} \text{ord}_{z_j}(R) \geq \left\{\sum_{i=1}^{n}(m_i + 1)\right\} - 1. \tag{0.1}$$

Then $R(z|f_1,\ldots,f_n;m_1,\ldots,m_n)$ is called the remainder function of N-point Padé approximation problem of type I to $(f_1(z),\ldots,f_n(z))$ with weights $(m_1,\ldots,m_n)$; $A_i(z|f_1,\ldots,f_n;m_1,\ldots,m_n)$ are called Padé approximants (polynomials of type I).

We call N-point Padé approximation perfect if always we have an equality in (0.1) and "almost perfect" if

$$\sum_{j=1}^{N} \text{ord}_{z_j}(R) - \sum_{i=1}^{n}(m_i + 1) + 1 \leq C(\bar{f})$$

for some absolute constant $C(\bar{f}) > 0$.

Usually only one-point Padé approximation problem is considered and in this case we just set $z_1 = 0$. We'll refer to this problem as Padé approximation problem [1][31][16].

The most interesting is the case of equal weights $m_i = N$: $i = 1,\ldots,n$ (the diagonal case) or almost equal weights $|m_i - N| \leq 1$: $i = 1,\ldots,n$ (the subdiagonal case). We can denote the corresponding remainder function in the diagonal case as $R_N(z)$ and polynomials of type I as $A_{i,N}(z)$.

There exists another definition of Padé approximation : the type II [16],[3] .

Definition 0.2. Let $f_1(z),\dots,f_m(z)$ be a formal power series, and $\rho_1,\dots,\rho_m$ be m positive integers. We say that $\{\mathcal{a}_1(z),\dots,\mathcal{a}_n(z)\}$ is the system of polynomials of type II corresponding to $(\rho_1,\dots,\rho_m)$, $(f_1(z),\dots,f_m(z))$, if $\{\mathcal{a}_i(z) : i = 1,\dots,m\}$ is a non-trivial system ;

1) $\deg \mathcal{a}_i(z) \leq \sigma - \rho_i$; $\sigma = \sum_{j=1}^{m} \rho_j$;

2) the order in $z = 0$ of

$$\mathcal{a}_\ell(z)\ f_k(z) - \mathcal{a}_k(z)\ f_\ell(z)$$

is at least $\sigma + 1$: $k,\ell = 1,\dots,m$.

The first example of the system of polynomials of type II was constructed in 1873 by Hermite in connection, of course, with the transcendence of e.

The polynomials N II and the usual Padé approximants $A_i(z|\rho_1,\dots,\rho_n)$: $i = 1,\dots,m$ to $f_1(z),\dots,f_m(z)$ (polynomials of type I) are connected, of course.

Under the conditions of perfectness and normality (see precise statements in the papers of Mahler, Jagier, Coates, de Brujn [3], [16], [20], [21]), for matrices

$$A(z|\rho_1,\dots,\rho_m) = (A_i(z|\rho_1 + \delta_{h1},\dots,\rho_n + \delta_{hm}))$$

$$\mathcal{a}(z|\rho_1,\dots,\rho_m) = (\mathcal{a}_i(z|\rho_1 - \delta_{h1},\dots,\rho_n - \delta_{hm})) :$$

$i,h = 1,\dots,m$ we have

$$A(z|\rho_1,\dots,\rho_m)\ \mathcal{a}^T(z|\rho_1,\dots,\rho_m) = \begin{pmatrix} c_1 z^\sigma & & 0 \\ & \ddots & \\ 0 & & c_n z^\sigma \end{pmatrix}$$

for some constants c_i.

In other words, the systems of polynomials of types I-II determine each other.

The main result of all our studies can be formulated as follows : we can *effectively* construct the system of Padé approximants at $z = 0$ (diagonal or subdiagonal) to any system of functions.

$$\{f_1(z),\dots,f_n(z)\} ,$$

satisfying a system of linear differential equations

$$f_i'(z) = \sum_{j=1}^{n} Q_{j,i}(z) : i = 1,\dots,n \qquad (0.2)$$

for rational functions $Q_{j,k}(z) : j = 1,\dots,n ; k = 1,\dots,n$ and being formal power series in $C[[z]]$*)

By *effective* construction we understand simultaneous solution of the following problems :

a) explicit expressions for differential equations that are satisfied by remainder function and Padé approximants ;

b) explicit determination of recurrent formulae that are satisfied by polynomial Padé approximants (and remainder function) at once ;

c) some "closed" (i.e. integral expression for remainder function $R_N(z)$ or $R(z; m_1,\dots,m_n)$ as function of variables N or $m_1,\dots m_n$;

d) determination of the asymptotics of remainder function $|R_N(z)|$ or $|R(z;m_1,\dots,m_n)|$ and asymptotics of polynomial Padé approximants $|A_{i,N}(z)|$ or $|A_i(z;m_1,\dots,m_n)|$ for fixed $z \neq 0$ and $N \to \infty$.

E.g. we should demand the determination of dependence on N in the recurrences b) in terms of some (known) function of N that can be written immediately starting from the equation (0.2). This differs from a usual theory of Padé approximation [1],[31],[30] because there famous three-term recurrence relations are expressed in terms of Hadamard determinants

$$\sigma_n = \begin{vmatrix} a_o \dots a_n \\ a_2 \dots a_{n+1} \\ a_n \dots a_{2n} \end{vmatrix}$$

for $f(z) = \sum_{i=0}^{\infty} a_i z^i$. We do not call this process effective.

Moreover it turns out that Hadamard determinants G_n can be effectively written as a function of n just in the same cases when we have effective Padé approximation satisfying a)-d), for example, in the case of f(z) satisfying the linear unhomogeneous differential equation of the first order, $f_1(z) = f(z)$, $f_2(z) = 1$.

So our solution of the problem of constructing the Padé approximation for the system of functions $\{f_1(z),\dots,f_n(z)\}$ satisfying linear differential equations (0.2) can be compared with Laguerre result [17] in the case of $n = 2$, $f_1(z) = f(z)$, $f_2(z) = 1$.

In general our result is based on a relationship existing between linear differential equations with irregular singularities and Fuchsian linear differential equations via generalized Borel transformation (e.g. Ramis [24], Ince [15], G. Birkhoff [19]).

General solution is based on isomonodromy deformation theory. We'll present below the exposition of explicit construction of Padé approximation in the case of Fuchsian linear differential equations only, i.e. in the case when (0.2) is a system of differential equations with regular singularities only.

We present in §§ 6-8 also arithmetic applications of our system of Padé approximation. In fact, all arithmetic applications are the development of physical attitude towards Padé approximation. Padé approximants are used to get good approximation (arithmetically speaking, diophantine approximations) to a given number α, say. If the corresponding approximations are very good, e.g.

$$0 < |\alpha - \frac{p}{q}| < |q|^{-\delta} \quad ; \quad \delta > 1$$

for $p,q \in \mathbb{Z}$, then α is irrational.

Of course, the system of Padé approximants to $\{f_1(x),\dots,f_n(x)\}$ can be used for the proof of linear independence of $f_1(\alpha),\dots,f_n(\alpha)$ over $\mathbb{Q}$ for $\alpha \in \mathbb{Q}$, $\alpha \neq 0$.

The text is organized as follows. In § 1 we follow original Riemann work [25] and present isomonodromy deformation (contiguous relations) for Fuchsian linear differential equation. The purpose of § 2 is to give effective contiguous relation between system of function of the same monodromy class. We compute degrees of the relations and also introduce certain polynomials that will appear as polynomial coefficients in Padé approximation problem. In § 3 we present one way to get Padé approximation formulae using multidimensional (Grothendieck) residue formula (this is a part of studies [5]). Then in § 3 we apply Global Residue Formula to construct Padé approximations to system of functions $\{1, {}_2F_1(1,\omega_j;\gamma;x) : j = 1,\dots,m\}$, $\{{}_2F_1(1,b;c;a_i x) : i = 1,\dots,n\}$ etc. These results (together with the asymptotics of the remainder function) generalize those of Padé [22],[23]. In § 4 we construct effectively Padé approximation system for system of generalized hypergeometric function ${}_{p+1}F_p\left(z\middle|\begin{matrix}a_1,\dots,a_{p+1}\\ b_1,\dots,b_p\end{matrix}\right)$ and p more functions contiguous to it.

Remainder function for Padé approximation again is expressed in a simple way in terms of generalized hypergeometric function. The asymptotics of the remainder function is computed and presented in § 4.

In §5 we give in terms of isomonodromy deformation explicit construction of the system of Padé approximations of type I for an arbitrary (fundamental) system of solutions of Fuchsian linear differential equation. This construction can be generalized for equations with irregular singularities (for discussion of these equations see Ince [15] or G. Birkhoff [19]).

In § 6 we turn to the arithmetic applications of our system of Padé approximations.

Together with the general discussion of the current situation we propose few general results about irrationality and transcendence of values of hypergeometric function ${}_2F_1(1,b;c;x)$ at rational or algebraic $x \neq 0$. In § 7 we define the most general class of Borel summable series (so-called C-functions) for which we can study the problem of arithmetic nature of the values of these functions at algebraic points. We consider different classes of such functions (C- and (G,C)-functions) and present Siegel's method of studying linear independence of values of these functions at algebraic points. We present both non-effective (Siegel's) method and an effective one, based on the knowledge of the Padé approximation. In § 8 we collect several results on measure of irrationality of values of generalized hypergeometric functions that we obtain using the effective Padé approximation.

1. RIEMANN'S RESULTS ON LINEAR DIFFERENTIAL EQUATIONS WITH REGULAR SINGULARITIES

As usual one should look in old books for new ideas. Of course, Riemann did everything.

This is indeed the case. The problems that we are treating now have roots in the Riemann-Hilbert monodromy problem, known as Hilbert problem N21. Riemann implicitly formulated this problem in his famous memoir "Zwei allgemeine Lehrsätze über lineäre Differentialgleichungen mit algebraischen Coefficienten" (1857), edited and published by H. Weber in 1876 [25].

Riemann's ideas are presented in a form that is very convenient for applications. It is worthwhile to present Riemann's considerations following his own notations (basically speaking).

Riemann starts with some finite set of points $a_1,\dots,a_n$ (or $a,b,\dots,g$) in the complex plane and some system of functions

$$y_1,\dots,y_n$$

regular in the whole plane except in $a_1,\dots,a_n$. If you move x around any of these branch points a_i, then functions y_i $(i=1,\dots,n)$ change into certain linear combinations of their previous values : this is the main assumption that Riemann made.

E.g. when you move around a, y_1 changes into

$$\sum_{j=1}^{m} A_j^{(1)} y_j, \dots, y_m \quad \text{into} \quad \sum_{j=1}^{m} A_j^{(m)} y_j, \dots .$$

In other words using modern (one hundred years old) terminology

$$\bar{y} \to A_i \bar{y}$$

when you move around the singularity a_i in a positive direction along the circle γ_i (that does not contain any other a_j inside). The group of linear substitutions, generated by A_i ($m \times m$ matrices) is called the group of the system $\bar{y}$ or monodromy group, $\mathfrak{M}$ say. Of course the matrices A_i : $i = 1,\dots,n$ are related by one equation

$$A_1 \dots A_n = I.$$

Any system $\bar{y}$ of m functions satisfying all the properties above is denoted by

$$Y\begin{pmatrix} a_1,\dots,a_n \\ A_1,\dots,A_n \end{pmatrix};x$$

(in the Riemann notations it was Q not Y). Now Riemann gives a definition that you cannot find in the modern literature : of a class of functions.

The system of functions $\bar{y}$ belongs to the class ($a_1,\dots,a_n$; $A_1,\dots,A_n$) if $\bar{y}$ has singularities at the a_i only, with the monodromy group $\mathfrak{M}$, generated by the matrices A_i : $i = 1,\dots,n$ for $A_1 \dots A_n = I$.

From now on, Riemann assumes, in addition, that the functions y_i : $i = 1,\dots,m$ do not have "singularities of infinite order". This means that y_j has regular behaviour at points a_i :

$$y_j(z) = (z - a_i)^{\rho} . \varphi_j(z)$$

where $\varphi_j(z)$ is analytic at a_i : $\varphi_j(a_i) \neq 0,\infty$.

Basing on these assumptions B. Riemann [25] stated his most important result :

<u>Theorem 1.1.</u> For any m+1 systems of functions $\bar{y}_j$: $j = 1,\dots,m+1$ belonging to the same *class* there exists a linear homogeneous relation with the coefficients being entire (polynomial) functions of x :

$$\sum_{j=1}^{m+1} A_j(x)\, \bar{y}_j(x) = 0.$$

In particular you can take derivatives of y_i with respect to x. In this way you obtain the system of linear differential equations with rational function coefficients that $\bar{y}$ satisfies, and Riemann stated :

Corollary 1.2. The system of functions $\bar{y} = (y_1,\ldots,y_m)$ belonging to a given class, satisfies a linear differential equation

$$\frac{d\bar{y}}{dx} = Q\,\bar{y} \qquad \text{or} \qquad L_m y_j = 0 : j = 1,\ldots,m$$

of order m. Here $Q = (Q_{ij})_{i,j = 1,\ldots,m}$ and $Q_{i,j}$ are rational functions of x and L_m is the differential operator of order m with rational function coefficients.

With respect to the famous contiguous functions, Riemann states that "every system y_1 belonging to the same class as $\bar{y}$ is a linear combination with rational coefficients of $\bar{y}$ and their derivatives up to the (m-1)-th order".

All these results are proved below in §2, where we present (along the lines of Riemann) effective way of finding these linear polynomial relations. As in [25] we restrict ourselves to the non-degenetate case, when A_i have distinct eigenvalues. See [15][18] for the general case.

However Riemann did not prove that for every class $(a_1,\ldots,a_n,A_1,\ldots,A_n)$ there exists a system $\bar{y}$ belonging to this class and this becomes the subject of the Riemann-Hilbert problem (see [14]). The Riemann-Hilbert problem has been solved in different ways, but the most important part of Riemann's ideas has not been developed since 1857. The possibility of using "contiguous functions" in order to build recurrent formulae and Padé approximations, this is the kernel of our pragmatic reading of Riemann.

Of course, the Riemann's ideas were developed and the theory of deformation of Fuchsian linear differential equations appeared.

Following L. Fuchs [13] and R. Fuchs, we consider functions

$$Y\begin{pmatrix} a_1,\ldots,a_n \\ A_1,\ldots,A_n \end{pmatrix}; x$$

to be functions of $a_1,\ldots,a_n$ also, assuming that the monodromy group $\mathcal{M}$ is fixed (and independent of $a_i : i = 1,\ldots,n$).

The existence of contiguous functions shows that there are linear recurrent relations between solutions of equations belonging to a class $(a_1,\ldots,a_n;A_1,\ldots,A_n)$. In particular there exists a natural system of polynomials associated with contiguous relations.

You can proceed in this way to get a Padé approximation to your system $\bar{y}$, what we'll do below.

As usual the best test is the hypergeometric function. We consider case of ${}_2F_1(1,b;c;x)$ and ${}_{p+1}F_p(x|\begin{smallmatrix}\cdots\\ \cdots\end{smallmatrix})$ functions as basic examples.

2. CONTIGUOUS RELATIONS BETWEEN SYSTEMS OF FUNCTIONS HAVING THE SAME MONODROMY GROUP.

In this chapter we write down relations between contiguous systems of functions existing according to Riemann theorem 1.1. We need to know degrees and explicit form of the polynomials entering these relations. This is done below using only the general structure of the systems of functions with a given monodromy group. The study of contiguous relations is only a part in a general study of isomonodromy deformation [4], but a significant part. E.g. Bäcklund transformation for all isomonodromy deformation equations like Painlevé I-VI [4] are deduced using contiguous relations for Fuchsian linear differential equations from which these nonlinear (isomonodromy deformation) equations arise.

We use standard Riemann notations. The complex numbers $a_1,a_2,\dots a_n$ denote the positions of (regular) singularities ; $A_1,\dots,A_n$ are $m\times m$ matrices such that

$$A_1 \dots A_n = 1.$$

As before we consider system of functions $\{f_1(z),\dots f_n(z)\}$ belonging to the class

$$Y\begin{pmatrix} z\ ; \begin{matrix} a_1,\dots,a_n \\ A_1,\dots,A_n \end{matrix}\end{pmatrix} ,$$

i.e. system of functions satisfying a system of Fuchsian linear differential equations with regular singularities at $a_1,\dots,a_n$ and with (fundamental) substitutions A_i at a_i : $i = 1,\dots,n$ and monodromy group $\mathfrak{M}$ generated by $\{A_1,\dots,A_n\}$. However there remains still ambiguity in the definition of $\{f_1(z),\dots f_n(z)\}$ since we did not specify local exponents at $a_1,\dots,a_n$.

The definition of local exponents of a regular point a_i is presented in lectures of D. Chudnovsky [4] §2 ; see also Ince [15] . If A_i have eigenvalues $\lambda_{i1},\dots,\lambda_{im}$: $i = 1,\dots,n$ then exponents ν_{ij} can be defined in such way that

$$\exp 2\pi i\nu_{kj} = \lambda_{kj} \quad : \quad k = 1,\dots,n \ , \ j = 1,\dots,m. \qquad (2.1)$$

In other words for the same λ_{ij} (the same A_i) numbers ν_{kj} may differ by rational integers. This difference in local exponents creates an ambiguity in the determination of $\{(f_1(z),\dots,f_m(z)\}$.

Now following B. Riemann by the following symbol

$$\begin{pmatrix} a_1 ; & a_2 ; & \dots ; & a_n \\ A_1 ; & A_2 ; & \dots ; & A_n \\ \nu_{11}; & \nu_{21}; & \dots ; & \nu_{n1} \\ \vdots & \vdots & & \vdots \\ \nu_{1m} & \nu_{2m} & & \nu_{nm} \end{pmatrix} \qquad (2.2)$$

we denote the system of functions $\{f_1(z),\dots f_m(z)\}$ that belongs to the class

$$Y\left(x ; \begin{matrix} a_1,\dots,a_n \\ A_1,\dots,A_n \end{matrix}\right)$$

and has at regular singularity a_i local exponents $\nu_{i1},\dots,\nu_{im}$: $i = 1,\dots,n$.

We denote that, because

$$\det(A_1)\dots\det(A_n) = 1 ,$$

we have always $\sum_{i=1}^{n} \sum_{j=1}^{m} \nu_{ij} = N$ - a rational integer. In order to normalize our class of functions we demand here and everywhere

$$\sum_{i=1}^{n} \sum_{j=1}^{m} \nu_{ij} = \frac{m(m-1)}{2} (\rho - 1)$$

for ρ-being n-1 if $\infty \in \{a_1,\dots,a_n\}$.

We call two systems (2.2) to be contiguous if $a_1,\dots,a_n$ and $A_1,\dots,A_n$ are the same and corresponding ν_{ij} and ν'_{ij} are different by rational integers.

One of the most important numbers connected with $(a_1,\dots,a_n ; A_1,\dots,A_n)$ is the number k of linearly independent over $\mathbb{C}[x]$ systems (2.2) of contiguous functions of the same class

$$Y\left(x; \begin{matrix} a_1,\dots,a_n \\ A_1,\dots,A_n \end{matrix}\right)$$

By B. Riemann, $k \leq m$ and, of course, the most interesting case for us is just $k = m$.

Remark You can easily construct a system (2.2) such that $k = 1 < m$. However for generalized hypergeometric functions *we always have* $k = m$. This is not at all trivial. Indeed you can always consider ${}_{p+1}F_p$-function as ${}_{q+1}F_q$-function for $q > p$ by identification

$$ {}_{p+1}F_p\left(x\middle|\begin{matrix}a_1,\dots,a_{p+1}\\ b_1,\dots,b_p\end{matrix}\right) = {}_{q+1}F_q\left(x\middle|\begin{matrix}a_1,\dots,a_{p+1},\omega,\dots,\omega\\ b_1,\dots,b_p,\ \omega\ ,\dots,\omega\end{matrix}\right) $$

for an arbitrary ω. However the families of functions contiguous to these two are entirely different. Precisely, for any arbitrary ${}_{p+1}F_p$-function the family of all functions contiguous to it has dimension $p+1$ over $\mathbb{C}[x]$.

Having the possibility of making projective transformations and assuming non-trivial situation of $n \geq 3$ we can put always

$$ a_1 = 0, \qquad a_2 = \infty \;, \; a_3 = 1. $$

In other words $\{0,\infty,1\} \subseteq \{a_1,\dots,a_n\}$.

We need also the following notations. Let $\vec{\lambda}_1,\dots,\vec{\lambda}_m$ be m vectors from $\mathbb{C}^m$ such that $\vec{\lambda}_i - \vec{\lambda}_j \in \mathbb{Z}^m$ for $i,j = 1,\dots,m$ ("contiguous" vectors). We define

$$ \varepsilon(\vec{\lambda}_1,\dots,\vec{\lambda}_m) $$

to be the following number. If

$$ \vec{\lambda}_i = (\lambda_{ij} : j = 1,\dots,m) \;, $$

then

$$ \varepsilon(\vec{\lambda}_1,\dots,\vec{\lambda}_m) = \min_{\pi \in S_m} \sum_{i=1}^{m} \lambda_{i\pi(i)} \;, $$

when the sum is over all permutations π of $\{1,\dots,m\}$.

We present the proof of the Riemann statement 1.1 together with the effective construction of the polynomial relation connected to the set of m+1 systems of the functions belonging to the same class.

Let's suppose that we have m + 1 systems of functions $\bar{f}_j(z) = (f_{1,j}(z);\ldots;f_{m,j}(z))$: $j = 1,\ldots,m+1$ belonging to the same class

$$Y\left(z\left|\begin{matrix}a_1,\ldots,a_n\\A_1,\ldots,A_n\end{matrix}\right.\right).$$

Then each $\bar{f}_j(z)$ is the fundamental system of solution of the linear differential equation of the m-th order

$$L_j[y] = 0 \quad : \quad y = f_{1,j}(z),\ldots,y = f_{m,j}(z) \quad (2.3)$$

having (non-apparent) regular singularities at $z = a_1,\ldots,z = a_n$ with corresponding monodromy substitutions $A_1,\ldots,A_n$.

For $A_1,\ldots,A_n$ not having equal eigenvalues we have non-singular matrices $U_1,\ldots,U_n$ such that

$$A_k = U_k\, B_k\, U_k^{-1} \quad (2.4)$$

for diagonal matrices

$$B_k = \mathrm{diag}\ (\lambda_{1,k};\ldots;\lambda_{m,k}) \ : \ k = 1,\ldots,n, \quad (2.5)$$

and $\vec{\lambda}_k = (\lambda_{1,k};\ldots;\lambda_{m,k})$: $k = 1,\ldots,n$. Now local exponents of $\bar{f}_j(z)$ at $z = a_k$ can be computed in terms of logarithms of elements of $\vec{\lambda}_k$: $k = 1,\ldots,n$. For complete discussion of the relation between global and local monodromies see [4] or [15].

For any $k = 1,\ldots,n$ we consider another fundamental system of solutions of (2.3) corresponding to local monodromy at $z = a_k$:

$$\bar{g}_j^{(k)}(z) \stackrel{\mathrm{def}}{=} U_k^{-1}\cdot\bar{f}_j(z) \stackrel{\mathrm{def}}{=} (g_{1,j}^{(k)}(z);\ldots;g_{m,j}^{(k)}(z)) \ : \quad (2.6)$$

$j = 1,\ldots,m+1$. Then by definition of A_k and (2.4) when we make a small circle γ_k around a_k starting at z (near a_k) and return to z, we have

$$\bar{g}_j^{(k)}(z)\Big|_{\gamma_k} = B_k\cdot\bar{g}_j^{(k)}(z) \quad ,$$

i.e. after returning at z,

$$g_{s,j}^{(k)}(z)\Big|_{\gamma_k} = \lambda_{s,k}\, g_{s,j}^{(k)}(z) \ : \ s = 1,\ldots,m.$$

Thus for k = 1,...,n and j = 1,...,m+1 we have

$$g^{(k)}_{s,j}(z) = (z - a_k)^{\nu^{(j)}_{s,k}} . h^{(k)}_{s,j}(z) \tag{2.7}$$

for $h^{(k)}_{s,j}(a_k) \neq 0,\infty$ and

$$e^{\nu^{(j)}_{s,k} . 2\pi i} = \lambda_{s,k} : s = 1,\dots,m. \tag{2.8}$$

In other words, $\nu^{(j)}_{s,k}$ differs from $\log \lambda_{s,k}$ by a rational integer.

Now in our notations the system $\bar{f}_j(z)$ belongs to the class

$$\begin{pmatrix} a_1 & \dots & a_n \\ A_1 & \dots & A_n \\ \nu^{(j)}_{1,1} & \dots & \nu^{(j)}_{1,n} \\ \vdots & & \vdots \\ \nu^{(j)}_{m,1} & \dots & \nu^{(j)}_{m,n} \end{pmatrix} \qquad j = 1,\dots,m+1.$$

We want to find explicitly relations between $\bar{f}_j(z): j = 1,\dots,m+1$.

For this we must solve the following system of equations :

$$\sum_{j=1}^{m+1} c_j . f_{i,j}(z) = 0 \quad : \quad i = 1,\dots,m, \tag{2.9}$$

for $c_j = c_j(z) : j = 1,\dots,m+1$. Cramer's rule gives the expression for c_j. We denote

$$\Delta_{j_o}(z) = \det(f_{i,j}(z))_{i=1,\dots,m \,;\, j=1,\dots,m+1;\, j\neq j_o} , \tag{2.10}$$

for $j_o = 1,\dots,m+1$. Then there exists the following solution of (2.9):

$$c_{j_o} = \Delta_{j_o}(z) \quad : \quad j_o = 1,\dots,m+1 . \tag{2.11}$$

We want to study now analytic properties of $\Delta_j(z)$. First of all,

$\Delta_j(z)$ may have (regular) singularities only at $z = a_1, \ldots, z = a_n$. At $z = a_k$ we can change to the basis $\bar{g}_j^{(k)}(z)$ by transformation

$$\bar{f}_j \to \bar{g}_j^{(k)} = U_k^{-1} \bar{f}_j \quad : \ j = 1, \ldots, m+1.$$

Then

$$\Delta_{j_o}(z) = \det(U_k) \det \left(g_{i,j}^{(k)} \right)_{j=1,\ldots,m+1, j \neq j_o}^{i=1,\ldots,m} ;$$

where $\det(U_k) \neq 0$ is constant.

We have $g_{i,j}^{(k)}(z) = (z - a_k)^{\nu_{i,k}^{(j)}} \cdot h_{i,j}^{(k)}(z)$, $h_{i,j}^{(k)}(a_k) \neq 0, \infty$.

If we assume non-degeneracy of A_k and linear independence of $g_{i,j}^{(k)}(z)$; i.e. $\Delta_{j_o}(z) \not\equiv 0$ for any $j_o = 1, \ldots, m+1$, then

$$\Delta_{j_o}(z) = (z - a_k)^{\vec{\varepsilon}_{j_o,k}} \cdot H_{j_o,k}(z) \tag{2.12}$$

for $H_{j_o,k}(z) \neq 0, \infty$ and

$$\vec{\varepsilon}_{j_o,k} = \vec{\varepsilon} \left(\vec{\nu}_k^{(1)}, \ldots, \vec{\nu}_k^{(j_o - 1)}, \vec{\nu}_k^{(j_o + 1)}, \ldots, \vec{\nu}_k^{(m+1)} \right)$$

for

$$\vec{\nu}_k^{(j)} = \left(\nu_{1,k}^{(j)}, \ldots, \nu_{m,k}^{(j)} \right) : j = 1, \ldots, m+1 \quad .$$

We can now conclude that the function

$$\Delta_{j_o}(z) . (z - a_1)^{-\vec{\varepsilon}_{j_o,1}} \cdot \ldots .(z - a_k)^{-\vec{\varepsilon}_{j_o k}} \cdot \ldots .(z - a_n)^{-\vec{\varepsilon}_{j_o,n}} \tag{2.13}$$

is a function analytic at complex plane. If in the term corresponding to the point on infinity is excluded, then

$$P_{j_o}(z) = \Delta_{j_o}(z) \prod_{a_k \neq \infty} (z - a_k)^{-\vec{\varepsilon}_{j_o,k}} \tag{2.14}$$

is a polynomial of degree

$$\sum_{k=1}^{n} \{-\vec{\varepsilon}_{j_o,k}\} . \tag{2.15}$$

All systems $\bar{f}_j(z)$ correspond to the same class. In other words for any $k=1,\ldots,n$, $i = 1,\ldots,m$

$$\nu_{i,k}^{(j_1)} - \nu_{i,k}^{(j_2)} \quad \text{- is an integer : } j_1,j_2 = 1,\ldots,m+1.$$

As a consequence

$$\vec{\varepsilon}_{j_1,k} - \vec{\varepsilon}_{j_2,k} \quad \text{is an integer : } j_1,j_2 = 1,\ldots,m+1 .$$

Let

$$\delta_k = \min\{\vec{\varepsilon}_{j,k} \ : \ j = 1,\ldots,m+1\} .$$

We have by (2.9)

$$\sum_{j=1}^{m+1} f_j(z)\ \Delta_j(z) = 0 . \tag{2.16}$$

Now taking into account (2.14) we obtain

$$\sum_{j=1}^{m+1} f_j(z)\ R_j(z) = 0 \tag{2.17}$$

for polynomials $R_j(z)$,

$$R_j(z) = \Delta_j(z) \prod_{k=1,a_k\neq\infty}^{n} (z - a_k)^{-\delta_k}$$

of degree

$$\sum_{k=1,a_k\neq\infty}^{n} \{-\delta_k\} -\vec{\varepsilon}_{j,2} \quad \text{for} \quad a_2 = \infty .$$

This way we obtain a polynomial relation (2.17) connecting $f_1(z),\ldots,f_{m+1}(z)$. Explicit construction of polynomials $R_j(z)$ can be used for the effective construction of Padé approximation.

3. GROTHENDIECK RESIDUE FORMULA AND ITS APPLICATIONS TO PADÉ APPROXIMATION SYSTEMS FOR HYPERGEOMETRIC FUNCTIONS ${}_2F_1(1,b_{ij};c_{ij};x_i)$.

We can use contiguous relations straightforward in order to get Padé approximations (of type I) for a given system of solutions of Fuchsian linear differential equations. This will be done in §§ 4-5. Here we prefer to use the method of integral representation in the complex plane in order to obtain several different formulae of Padé approximations to systems of ${}_2F_1(1,b;c;x)$-functions. We use contour integrals and Cauchy residue in $\mathbb{C}^1$ but for further generalizations we present contour integral method and residue formula in the case of $\mathbb{C}^n$.

3.1. Let U be the ball $\{z \in \mathbb{C}^n : |z| < \varepsilon\}$ and $f_1,\dots,f_n \in \theta(\bar{U})$ functions holomorphic in a neighborhood of the closure $\bar{U}$ of U. We assume that

$$D_i = (f_i) = \text{divisors of } f_i : i = 1,\dots,n$$

have the origin as their set-theoretic intersection,

$$f^{-1}(0) = \{\bar{0}\}$$

for

$$f = (f_1,\dots,f_n) : U^* = U \setminus \{\bar{0}\} \to C^n \setminus \{\bar{0}\} .$$

We are interested in residues associated to a meromorphic n-form

$$\omega = \frac{g(z)\, dz_1 \wedge \dots \wedge dz_n}{f_1(z) \dots f_n(z)} \quad : g \in \theta(\bar{U})$$

having polar divisor

$$D = D_1 + \dots + D_n.$$

In order to define the Grothendieck residue symbol we take the cycle of integration

$$\Gamma = \{z : |f_i(z)| = \varepsilon\}$$

with the orientation, say, $d(\arg f_1)\wedge\dots\wedge d(\arg f_n) \geq 0$). Then the *residue* of ω at $\bar{0}$ is

$$\mathrm{Res}_{\{\bar{0}\}}\, \omega = \left(\frac{1}{2\pi i}\right)^n \int_\Gamma \omega \quad . \tag{3.1}$$

First of all, $\mathrm{Res}_{\{\bar{0}\}}\, \omega$ possesses all "normal" local properties.

Lemma 3.1. In the generic case, when the D_i are smooth and meet transversely, i.e. Jacobian of f

$$J_f(\bar{0}) = \frac{\partial(f_1,\dots,f_n)}{\partial(z_1,\dots,z_n)}(\bar{0}) \neq 0, \qquad (3.2)$$

then

$$\mathrm{Res}_{\{\bar{0}\}}\omega = g(\bar{0})/J_f(\bar{0}) . \qquad (3.3)$$

The most interesting for us is the Global Residue formula proposed by Griffiths for $\mathbb{P}^n$. We assume that $D_1,\dots,D_n$ are hypersurfaces of respective degrees $d_1,\dots,d_n$ with intersections at isolated points P_ν and this intersection we present as a zero cycle

$$D_1\dots D_n = \sum_\nu m_\nu P_\nu ,$$

where the Bezout theorem

$$\sum_\nu m_\nu = d_1\dots d_n .$$

We'll assume below simply that all P_ν lie in $\mathbb{C}^n \subset \mathbb{P}^n$ and that D_i is defined by

$$f_i(x_1,\dots,x_n) = 0$$

for polynomials f_i of degree d_i.

The most general meromorphic n-form on $\mathbb{P}^n$ with polar divisor $D = D_1 + \dots + D_n$ has in $\mathbb{C}^n$ an expression

$$\omega = \frac{g(x)dx_1\wedge\dots\wedge dx_n}{f_1(x)\dots f_n(x)}$$

for a polynomial $g(x)$. Here ω does not have the hyperplane at infinity as a component of its polar divisor when the degree of $g(x)$ satisfies :

$$\deg(g) \leq (d_1+\dots+d_n) - (n+1).$$

Thus the Global Residue formula is *Generalized Jacobi-Kronecker formula.*

$$\sum_\nu \mathrm{Res}_{P_\nu}\left(\frac{g(x)dx_1\wedge\dots\wedge dx_n}{f_1(x)\dots f_n(x)}\right) = 0 \qquad (3.4)$$

Kronecker himself proved the following important identity

$$\sum_{\nu} \frac{g(P_\nu)}{J_f(P_\nu)} = 0 , \tag{3.5}$$

if $\deg(g) \leq \sum_{i=1}^{n} d_i - (n+1)$ and D_i meet transversely at $d_1 \dots d_n$,

$$J_f = \frac{\partial(f_1,\dots,f_n)}{\partial(z_1,\dots,z_n)}$$

is the Jacobian of f.

Of course (3.5) follows from (3.4) and (3.3).

In particular, from these formulae we obtain an interesting multidimensional generalization of Hermite interpolation formula.

Corollary 3.3. Let f(x) be holomorphic in $\bar{U}$ and we define in the neighborhood of $\bar{0}$ a new function $F(\bar{x})$ by

$$F(\bar{x}) = \left(\frac{1}{2\pi i}\right)^n \int_\Gamma \frac{f(\overline{xz})\, dz_1 \wedge \dots \wedge dz_n}{f_1(z)\dots f_n(z)}$$

Then $F(\bar{x})$ vanishes at $\bar{x} = \bar{0}$ of order

$$\geq \sum_{i=1}^{n} d(f_i) - n.$$

We can write an explicit expression for $F(\bar{x})$ in terms of $\partial_z^k f$ in certain cases, e.g. when $f_1,\dots,f_n$ have only simple intersection or f_i' are powers of such f_i, etc..., or in any case when the singularities of the intersections are known. In each of these cases $F(\bar{x})$ is a linear form from partial derivatives

$$\partial_z^{k_1,\dots,k_n} f(\overline{xz})\Big|_{\bar{z}=P_\nu} .$$

For $\mathbb{C}^1$ the corresponding expression is presented in lemma 3.4.

3.2. The possibilities of applying the Global Residue Formula to the construction of Padé approximations are unbelievably rich. All *explicitly known* Padé approximations can be represented in a nice short contour integral form. This does not mean that the Global Residue formula is the only source for Padé approximation tables, but it is one of the best.

We shall illustrate the power of the contour integral method by the *simplest* example, the hypergeometric function

$$_2F_1(1,b;c;x)$$

where b, c are complex numbers, c is not a non-positive integer,

$$_2F_1(1,b;c;x) = \sum_{n=0}^{\infty} \frac{(b)_n}{(c)_n} x^n ,$$

where, as usual, for a number a, and $n \geq 0$, $(a)_o = 1$, $(a)_n = a(a+1)\dots(a+n-1)$.

This functions satisfies, as all $_2F_1$-hypergeometric functions do, an ordinary hypergeometric equation

$$\left[z(1-z) \frac{d^2}{dz^2} + [c - bz] \frac{d}{dz} - b\right] f = 0 ,$$

but satisfies also an equation of the *first* order

$$(z(1-z) \frac{d}{dz} + [c - bz]) f = c .$$

The hypergeometric equation is Fuchsian and the local exponents for this equation in the case of $_2F_1(1,b;c;x)$ are

$$P \begin{pmatrix} 0 & \infty & 1 \\ 0 & 1 & 0 \\ 1-c & b & c-b-1 \end{pmatrix}$$

in the Riemann's sense. Two principal branches at x = 0 are as usual

$$P^{(\alpha)} = {}_2F_1(1,b;c;x) \quad \text{and}$$

$$P^{(\alpha')} = x^{1-c} {}_2F_1(2-c,b-c+1;2-c;x) = x^{1-c}(1-x)^{c-b-1} .$$

One of the most important features of hypergeometric functions (as, generally, of all Fuchsian systems) is the existence of contiguous relations.

In our case this means that any function

$$_2F_1(n,b+m;c+k;x)$$

for integers n,m,k can be expressed as a linear combination of

$$1 = {}_2F_1(0,b;c;x) \quad \text{and}$$

$$_2F_1(1,b;c;x)$$

with the coefficients being rational functions in b,c,x.

Examples

$$(1-z)^{\omega} = {}_2F_1(1,-\omega;1;z) \quad ;$$

$$\ln(1-z) = z\,{}_2F_1(1,1;2,z) \; ;$$

$$\ln\frac{1+z}{1-z} = 2z\,{}_2F_1(1,\tfrac{1}{2};\tfrac{3}{2};z^2) \; ;$$

$$\operatorname{arctg} z = z\,{}_2F_1(1,\tfrac{1}{2};\tfrac{3}{2};-z^2) \; .$$

Historically, Padé approximants to $_2F_1(1,b;c;x)$ were constructed by Padé in 1909. However in this case much was done before. Already Euler wrote a continued fraction expansion for the solution of Riccati's equation, i.e. for the ratio of two solutions of the linear differential equation of the second order. Then Gauss in 1812 found the explicit continued fraction for

$$\frac{{}_2F_1(a+1,b;c+1;z)}{{}_2F_1(a,b;c;z)} = \cfrac{1}{1-\cfrac{u_1 z}{1-\cfrac{v_1 z}{1-\cfrac{u_2 z}{1-\cfrac{v_2 z}{\ddots}}}}}$$

where

$$u_n = \frac{(b+n-1)(c-a+n-1)}{(c+2n-2)(c+2n-1)} \;;\; v_n = \frac{(a+n)(c-b+n)}{(c+2n-1)(c+2n)} \quad ;$$

$n = 1,2,3\ldots.$

In particular for $a=0$, $_2F_1(0,b';c';z) = 1$, one obtains the continued fraction expansion for $_2F_1(1,b;c;z)$. The partial fractions of this expansion were carefully investigated by Riemann (1857) [25] who even found the asymptotics of the remainder term and the coefficients for [n/n] and [n/n-1] Padé approximants to ${}_2F_1(1,b;c;z)$ as $n \to \infty$.

3.3. We present here three new classes of Padé approximation connected with ${}_2F_1(1,b;c;x)$.

1) Padé approximation to any system of functions

$$ {}_2F_1(1;\omega_j;c;x) \; : \; j = 0,1,\ldots,m, $$

where $\omega_i - \omega_j \notin Z$ for $i \neq j$ (part 3.4).

2) Padé approximation to any system of functions

$$ {}_2F_1(1,b;c;a_i\, x) \; : \; i = 0,\ldots,m $$

with distinct $a_o,\ldots,a_m$ (part 3.5) :

3) some N-point approximations associated with ${}_2F_1(1,x;c;z_i)$ (part 3.6).

The idea of all the applications of the Global Residue formulae is very simple. You consider some function

$$ F(x,z), $$

analytic in z and a polynomial P(z) of degree n and for a contour C containing all zeroes of P(z) you define

$$ R(x) = \frac{1}{2\pi i} \oint_C \frac{F(x,z)dz}{P(z)} $$

If at $x \cong x_o$, $F(x,z) = (x-x_o)^M F_o(x,z) + F_1(x,z)$ and $F_1(x,z)$ is polynomial in z of degree < n, then

$$ \operatorname{ord}_{x=x_o} R(x) \geq M $$

In particular cases it is convenient to choose the kernel F(x,z) to be F(xz).

We want to present the explicit expression for R(x) in terms of F(x,z), $\partial_z^k F(x,z)$ and P(z) only. This expression is contained in the following version of Hermite interpolation formula:

Lemma 3.4 (Hermite interpolation formula). Let f(z) be regular in the interior of the closed curve C ; where the interior of C contains the origin z = 0 and n points $z_1,\ldots,z_n$. We define

$$R = \frac{1}{2\pi i} \oint_C \frac{f(z)}{P_n(z)} dz ,$$

for $P_n(z) = \prod_{i=1}^{n} (z - z_i)$.

Then we have a closed expression for R in terms of $f(z_i)$ and its derivatives at z_i. Let the set $\{z_1,\ldots,z_n\}$ consist of m different points $\omega_1,\ldots,\omega_m$; where ω_j is repeated $n_j + 1$ times :

$j = 1,\ldots,m$, $\sum_{j=1}^{m} (n_j + 1) = n$

$$P_n(z) = \prod_{i=1}^{n} (z-z_i) = \prod_{j=1}^{m} (z-\omega_j)^{n_j+1}$$

Then we have

$$R = \sum_{j=1}^{n} \frac{1}{n_j!} \left(\frac{\partial}{\partial z}\right)^{n_j} \left\{\frac{f(z).(z-\omega_j)^{n_j+1}}{P_n(z)}\right\}\Bigg|_{z=\omega_j}.$$

3.4. We can write down the expression for the remainder function for the Padé approximation to the system of functions

$$1, {}_2F_1(1,\omega_i;\gamma;x) \quad : \quad i = 1,\ldots,m$$

for complex numbers $\omega_1,\ldots,\omega_m$ such that $\omega_i - \omega_j$ are not integers for $i \neq j$.

We put

$$R_o\left(x;\begin{matrix}\omega_1,\ldots,\omega_m\\ K_1,\ldots,K_m\end{matrix}\right) = \frac{1}{2\pi i} \oint_{C_o} \frac{{}_2F_1(1,s;\gamma;x)ds}{\phi_o(s)}$$

for

$$\phi_o(s) = \prod_{i=1}^{m} \prod_{K=0}^{K_i} (s + K - \omega_i) \prod_{K=0}^{K_o} (s+K)$$

where C_o is a contour in the s-plane containing all the zeroes of $\phi_o(s)$.

Then for $K_o \geq \max(K_1,\ldots,K_m)$ we have

$$R_o\left(x;\begin{matrix}\omega_1,\ldots,\omega_m\\ K_1,\ldots,K_m\end{matrix}\right) = \sum_{i=0}^{m} P_i(x)\, f_i(x) ,$$

$f_o(x) = 1$, $f_i(x) = {}_2F_1(1,\omega_i;\gamma;x) \quad i = 1,\ldots,m.$

Here $P_i(x)$ are polynomials in x of degree K_i : $i = 0,1,\dots,m$.

In the particular case $m = 1$ we reconstruct the famous Padé formulae for the Padé approximations for ${}_2F_1(1,b;c;x)$ in terms of Jacobi polynomials [22],[23].

For Padé approximation to ${}_2F_1(1,\omega_j;c;x)$ we know the asymptotics of remainder function and polynomial coefficients.

Let $\omega_1,\dots,\omega_m$ be non-integers such that $\omega_i - \omega_j$ is not an integer : $i \neq j$. We consider the remainder function to be normalized in the following way

$$R\left(x;\begin{matrix}\omega_1,\dots,\omega_m\\ K_o,\dots,K_m\end{matrix}\right) = \frac{K_o!K_1!\dots K_m!}{2\pi i}\oint_{C_o}\frac{{}_2F_1(1,s;c;x)ds}{\phi_o(s)}$$

for $\phi_o(s) = \prod_{i=1}^{m}\prod_{k=0}^{K_i}(s+k-\omega_i)\prod_{k=0}^{K_o}(s+k)$ and C_o containing all zeroes of $\phi_o(s)$.

Then

$$R\left(x;\begin{matrix}\omega_1,\dots,\omega_m\\ K_o,\dots,K_m\end{matrix}\right) = \sum_{i=0}^{m} A_i\left(x;\begin{matrix}\omega_1,\dots,\omega_m\\ K_o,\dots,K_m\end{matrix}\right) f_i(x) \quad ,$$

$f_o(x) = 1$, $f_i(x) = {}_2F_1(1,\omega_i;c;x)$: $i = 1,\dots,m$ and

$\deg(A_i) \leq K_i$: $i = 0,1,\dots,m$. We have the following

Theorem. Let $K_i = N + \rho_i$: $i = 0,\dots,m$ and $x,\rho_o,\dots,\rho_m$ be fixed. Then for $N \to \infty$,

$$\frac{1}{N}\log\left|R\left(x;\begin{matrix}\omega_1,\dots,\omega_m\\ K_o,\dots,K_m\end{matrix}\right)\right| \to \log\left|1 - \sqrt[m]{1-x}\right|^m \quad ,$$

$$\frac{1}{N}\log\left|A_i\left(x;\begin{matrix}\omega_1,\dots,\omega_m\\ K_o,\dots,K_m\end{matrix}\right)\right| \to \log\left|1 + \sqrt[m]{1-x}\right|^m$$

with $i = 0,\dots,m$.

3.5 Padé approximation to

$${}_2F_1(1,b;c;a_i x) \ : \ i = 0,1,\dots,m$$

is really the simplest possible. You define

$$P(x) = \prod_{i=0}^{m} (x - a_i)^n$$

where $n \geq 0$ and $a_o, a_1, \ldots, a_m$ are distinct numbers. Now we have the following contour integral

$$R(x; a_o, \ldots, a_m, n) = \frac{1}{2\pi i} \oint_C \frac{{}_2F_1(1, b+n, c; xz)dz}{P(z)},$$

where C contains all a_i. Then $R(x;\ldots)$ is the remainder function of the Padé approximants to

$${}_2F_1(1, b; c; a_i x)$$

corresponding to the case of equal weights $n_i = n : i = 0, 1, \ldots, m$.

3.6. Another N-point Padé approximation system.

Let $\zeta_1, \ldots, \zeta_k$ be a sequence of distinct numbers, different from 0 and 1. We define

$$P(z) = \prod_{i=1}^{k} (z - \zeta_i)^{n_i}$$

for positive integers $n_i : i = 1, \ldots, k$.

We put for a contour C containing all zeroes of $P(z)$,

$$F(x) = \frac{1}{2\pi i} \oint_C \frac{{}_2F_1(1, x; c; z)dz}{P(z)} .$$

Then $F(x)$ is the remainder function of the Padé approximation system to

$$f_i(x) = {}_2F_1(1, x; c; \zeta_i) \quad : \quad i = 1, \ldots, k$$

and $$f_o(x) = 1$$

corresponding to N-point Padé approximation at simple zeroes

$$-x : x = 0, \ldots, \sum_{i=1}^{k} n_i + k - 2.$$

4. EXPLICIT PADÉ APPROXIMATION OF TYPE I TO THE SYSTEM OF GENERALIZED HYPERGEOMETRIC FUNCTIONS. ASYMPTOTICS.

4.1. We present a construction of the system of Padé approximations to the system of arbitrary generalized hypergeometric functions. For this we find necessary to formulate the classical Riemann theory of P-functions (hypergeometric functions of ${}_2F_1$-type).

Following Riemann, we denote by

$$P\begin{pmatrix} a & b & c & \\ \alpha & \beta & \gamma & x \\ \alpha' & \beta' & \gamma' & \end{pmatrix}$$

any branch of a certain many-valued analytic function of x with the following properties.

(i) Every branch is finite and holomorphic, except at the three singular points $x = a,b,c$.

(ii) Any three branches are linearly dependent.

(iii) At $x = a$ there are two principal branches $(P^{(\alpha)},P^{(\alpha')})$ which are "regular" and belong to the exponents (local multiplicities) (α,α'). Similarly there are two regular branches $(P^{(\beta)},P^{(\beta')})$ belonging to the exponents (β,β') at $x = b$ and $(P^{(\gamma)},P^{(\gamma')})$ belonging to the exponents (γ,γ') at $x = c$.

(iv) The exponent-differences $(\alpha' - \alpha)$, $(\beta' - \beta)$, $(\gamma' - \gamma)$ are not integers ; and the six exponents are always connected by the relation

$$\alpha + \alpha' + \beta + \beta' + \gamma + \gamma' = 1$$

Of course, the condition (iv) can be omitted if one considers the case of integral $\alpha' - \alpha$, $\beta' - \beta$, $\gamma' - \gamma$ as the limit of the corresponding "nondegenerate" branches.

We can assume without loss of generality that $a = 0$, $b = \infty$, $c = 1$. For this we can use arbitrary projective transformation. Then P-function of Riemann will be denoted simply

$$P\begin{pmatrix} \alpha & \beta & \gamma & x \\ \alpha' & \beta' & \gamma' & \end{pmatrix}$$

and for $\alpha = 0, \gamma = 0$ we obtain Gauss function ${}_2F_1(\beta,\beta';1-\alpha';x)$.

We want to specify further contiguous relations from §1-2 for the case of P-functions.

Let us consider three P-functions

$$P_i\begin{pmatrix}\alpha_i & \beta_i & \gamma_i \\ \alpha'_i, & \beta'_i, & \gamma'_i, x\end{pmatrix} \quad (\sum \nu_i = 1) \; : \; i = 1,2,3$$

where the corresponding exponents (α_i and α_j, β_i and β_j etc...) are congruent modulo 1. In other words (the case of 2×2 matrices, with 3 singularities) we consider 3 systems of functions with the same monodromy group, whose exponents differ by integers.

These systems of functions are contiguous.

Again by theorem 1.1. these three functions (more precisely, systems of functions) are connected by a polynomial homogeneous relation.

The method of Riemann (§2) enables us to find explicitly the degrees of this polynomial relation and even to find the polynomials in a closed form.

Of course, we can use the very simple substitution

$$x^{\delta}(1 - x)^{\varepsilon}\, P\begin{pmatrix}\alpha & \beta & \gamma \\ \alpha' & \beta' & \gamma'\end{pmatrix}x = P\begin{pmatrix}\alpha + \delta & \beta-\delta-\varepsilon & \gamma + \varepsilon \\ \alpha' + \delta & \beta'-\delta'-\varepsilon & \gamma' + \varepsilon\end{pmatrix}x$$

and so we may restrict ourselves to a reduced (hypergeometric) form :

$$P_i\begin{pmatrix} 0 & \frac{1}{2}(1 - \lambda_i - \mu_i - \nu_i) & 0 & \\ & & & x \\ \lambda_i & \frac{1}{2}(1 - \lambda_i + \mu_i - \nu_i) & \nu_i & \end{pmatrix} \; : \; i = 1,2,3$$

for $\lambda_i = \alpha_i - \alpha'_i$, $\mu_i = \beta_i - \beta'_i$, $\nu_i = \gamma_i - \gamma'_i$ – the exponent-differences.

Now let us define the following functions of x :

$$\begin{aligned} S_{ij} &= (P_i^{(\alpha)} P_j^{(\alpha')} - P_i^{(\alpha')} P_j^{(\alpha)} = \\ &= (\alpha_\beta \alpha'_{\beta'} - \alpha'_\beta \alpha_{\beta'})(P_i^{(\beta_i)} P_j^{(\beta'_j)} - P_i^{(\beta'_i)} P_j^{(\beta_j)}) = \\ &= (\alpha_\gamma \alpha'_{\gamma'} - \alpha'_\gamma \alpha_{\gamma'})(P_i^{(\gamma_i)} P_j^{(\gamma'_j)} - P_i^{(\gamma'_i)} P_j^{(\gamma_j)}). \end{aligned}$$

Now S_{ij} belongs to the exponent

$$\alpha_{ij} = \min(\alpha_i + \alpha_j', \alpha_i' + \alpha_j)$$

at $x = 0$, to

$$\beta_{ij} = \min(\beta_i + \beta_j', \beta_i' + \beta_j)$$

at $x = \infty$, and to

$$\gamma_{ij} = \min(\gamma_i + \gamma_j' , \gamma_i' + \gamma_j)$$

at $x = 1$.

In other notations

$$\alpha_{ij} = \frac{1}{2} [\alpha_i + \alpha_j + \alpha_i' + \alpha_j' - |\lambda_i - \lambda_j|] \quad ;$$

$$\beta_{ij} = \frac{1}{2} [\beta_i + \beta_j + \beta_i' + \beta_j' - |\mu_i - \mu_j|] \quad ;$$

$$\gamma_{ij} = \frac{1}{2} [\gamma_i + \gamma_j + \gamma_i' + \gamma_j' - |\nu_i - \nu_j|] \quad .$$

<u>Lemma 4.1</u>. With the notations as above we have

$$P_1 S_{23} + P_2 S_{31} + P_3 S_{12} = 0$$

where the function

$$\chi_{ij} = x^{-\alpha_{ij}} (x-1)^{-\gamma_{ij}} S_{ij}$$

belongs at the infinity to the exponent

$$(\alpha_{ij} + \beta_{ij} + \gamma_{ij})$$

and so is a polynomial of degree

$$-(\alpha_{ij} + \beta_{ij} + \gamma_{ij}) = \frac{1}{2}[|\lambda_i - \lambda_j| + |\mu_i - \mu_j| + |\nu_i - \nu_j| - 2]$$

This result is a very natural basis for the whole construction of Padé approximations for hypergeometric functions.

Moreover this result as well as its generalizations in §2 are a natural basis for the proof that wide class of G-functions satisfying a system of Fuchsian linear differential equations belong also to (G,C) - class (see §7 for corresponding definitions).

Let us present the remainder function for the generalized hypergeometric function for the Padé approximation problem and polynomials of type I.

As usual we work with ${}_qF_p$-hypergeometric functions. For complex numbers $a_1,\ldots,a_q;b_1,\ldots,b_p$ such that $b_1,\ldots,b_p$ are different from non-positive integers we define

$$ {}_qF_p\left(x\,\middle|\,\begin{matrix}a_1,\ldots,a_q\\ b_1,\ldots,b_p\end{matrix}\right) = \sum_{n=0}^{\infty} \frac{(a_1)_n\cdots(a_q)_n}{(b_1)_n\cdots(b_p)_n\, n!}\, x^n. $$

This function is of special interest to us for $q = p+1$, when it is analytic with radius of convergence one (though an arbitrary p and q are easily treated).

Remark 4.2. The function

$$ {}_qF_p\left(x\,\middle|\,\begin{matrix}a_1,\ldots,a_q\\ b_1,\ldots,b_p\end{matrix}\right) $$

is an entire function of the parameters $a_1,\ldots,a_q$. Moreover, the following function

$$ \frac{{}_qF_p\left(x\,\middle|\,\begin{matrix}a_1,\ldots,a_q\\ b_1,\ldots,b_p\end{matrix}\right)}{\Gamma(b_1)\cdots\Gamma(b_p)} $$

is an entire function of all the variable parameters

$$ a_1,\ldots,a_q,b_1,\ldots,b_p. $$

Remark 4.3. The function

$$ {}_{p+1}F_p\left(x\,\middle|\,\begin{matrix}a_1,\ldots,a_{p+1}\\ b_1,\ldots,b_p\end{matrix}\right) \qquad (\text{note } q = p+1) $$

satisfies a natural Fuchsian linear differential equation of order p+1. This is one of the few cases when the monodromy group has been computed. Here the singularities at $\{0,1,\infty\}$ only, are regular, and the corresponding local multiplicities are

$$\left(\begin{matrix} \underline{0} & \underline{\infty} & \underline{1} \\ 0 & a_1 & 0 \\ 1-b_1 & a_2 & 1 \\ \vdots & \vdots & \vdots \\ 1-b_p & a_{p+1} & p-1 \\ & & d \end{matrix} \quad x\right)$$

for $d = \sum_{i=1}^{p} b_i - \sum_{j=1}^{p+1} a_j$.

Here as usual, the total sum over all local multiplicities is

$$\sum \mu_\nu = \frac{n(n-1)(\rho-1)}{2} ,$$

where $n(=p+1)$ is the order of the equation and $\rho(= 2)$ is the number of singularities in the finite part of the plane.

Problem 4.4. To find all classes of functions having regular singularities only at

$$\{0,1,\infty\}$$

and satisfying Fuchsian equations of order n.

A) Can we reduce all these functions to some combinations of Appel's functions

$$F(\bar{\alpha};\bar{\beta};x_1,\ldots,x_K)$$

restricted to lines ?

B) What kind of multiple integral formulae correspond to these functions ?

Now we are ready to present the formulae for the remainder function in the Padé approximation problem of type I. The corresponding Padé approximants of type II have been constructed before in [8], [9] and they are even more convenient for applications.

Let us introduce some notations to make things look simpler. We put

$$\bar{a} = (a_1,\ldots,a_{p+1}) \in \mathbb{C}^{p+1} \; ; \quad \bar{b} = (b_1,\ldots,b_p) \in \mathbb{C}^p$$

and define

$$f(x|\bar{a};\bar{b}) = {}_{p+1}F_p\left(x\middle|\begin{matrix}a_1,\dots,a_{p+1}\\ b_1,\dots,b_p\end{matrix}\right)$$

Instead of working with $f(x|\bar{a};\bar{b})$ and its derivatives we prefer, as usual in the theory of hypergeometric functions, to work with the system of "canonical" contiguous functions of the form

$$f(x|\bar{a}+\bar{e}\ ;\ \bar{b}+\bar{f})$$

for $\bar{e} \in (\mathbb{N} \cup \{0\})^{p+1}$, $\bar{f} \in (\mathbb{N} \cup \{0\})^p$ of ℓ_1 - norm one. For such a system of $p+1$ functions we can construct all diagonal, lower and upper diagonal approximations.

We fix parameters

$$a_1,\dots,a_{p+1} \quad \text{and} \quad b_1,\dots,b_p$$

Now we have our function

$$f_o(x) \overset{\text{def}}{=} {}_{p+1}F_p\left(x\middle|\begin{matrix}a_1,\dots,a_{p+1}\\ b_1,\dots,b_p\end{matrix}\right)$$

and define p more functions in the following way (for example):

$$f_i(x) \overset{\text{def}}{=} {}_{p+1}F_p\left(x\middle|\begin{matrix}a_1,\dots,a_i+i,\dots,a_{p+1}\\ b_1,\dots,b_i+i,\dots,b_p\end{matrix}\right) :$$

$i = 1,\dots,p.$

i.e. we add to a_i, b_i in the i-th function an integer $i : i = 1,\dots,p$.

We construct a system of Padé approximants to the functions

$$\{f_o(x), f_1(x), \dots, f_p(x)\}$$

in the following way.

Let $N \geq 1$ be an integer. We want to find a (remainder) function $R_N(x)$ such that $R_N(0) \neq 1$ and

$$x^{N(p+1)+p}\, R_N(x) = \sum_{i=0}^{p} P_{N,i}(x)\, f_i(x)$$

with polynomials $P_{N,i}$ of degree N (the so-called diagonal case).

Theorem 4.5. We define for $N \geq 1$

$$\mathcal{R}_N(x) = {}_{p+1}F_p\left(x\left|\begin{matrix} a_1+Np,\dots,a_{p+1}+Np \\ b_1+(p+1)N,\ \dots,b_p+(p+1)N\end{matrix}\right.\right). \quad (4.1)$$

Then we have

$$x^{(p+1)N-1}\ \mathcal{R}_N(x) = \sum_{i=0}^{p} f_i(x).P_{N,i}(x)\ , \quad (4.2)$$

where $P_{N,i}(x)$ are polynomials of degrees $N-1$: $i = 0,1,\dots,p$.

Here, of course, $\mathcal{R}_N(0) \neq 0$.

The rule here is extremely simple : in order to get

$$\underbrace{[N-1,\dots,N-1]}_{p+1}$$

Padé approximants of type I to

$$\{f_0(x),\dots,f_p(x)\}$$

you change a_i and b_i in the definition of $f_0(x)$ to $a_i + Np$, $b_i + (p+1)N$, respectively, and multiply by x to the power of the maximal order of zero. If you want to get

$$[N-1,\dots,\ N-1-j,\dots,\ N-1]$$

j-th place

Padé approximants to

$$\{f_0(x),\dots,f_j(x),\dots,f_p(x)\}$$

j-th place

you add Np to the upper parameters in $f_j(x)$ and $N(p+1)$ to the lower parameters in $f_i(x)$; then you multiply by $x^{(p+1)N-1-j}$.

4.2. One of the most difficult problem that exists in Padé approximations (cf. Nutall [32]) : the determination of the convergence of Padé approximants is completely solved for our effective Padé approximation to generalized hypergeometric functions.

Theorem 4.6. Let

$$f_1(z) = {}_{p+1}F_p\left(z\left|\begin{matrix}a_1,\dots,a_{p+1}\\ b_1,\dots,b_p\end{matrix}\right.\right)$$

by any ${}_{p+1}F_p$ generalized hypergeometric function and

$$f_i(z) = {}_{p+1}F_p\left(z\left|\begin{matrix}a_1+k_{1,i},\dots,a_{p+1}+k_{p+1,i}\\ b_1+m_{1,i},\dots,b_p+m_{p,i}\end{matrix}\right.\right) : \quad (4.3)$$

$i = 2,\dots,p+1$

be a system of functions contiguous with $f_1(z)$ (i.e. $k_{j,i}$, $m_{\ell,i}$ be rational integers ; $j = 1,\dots,p+1, \ell = 1,\dots,p,\ i = 2,\dots,p+1$), linearly independent together with $f_1(z)$ over $\mathbb{C}[z]$. Let

$$R(z|f_1,\dots,f_{p+1};\ n_1,\dots,n_{p+1})$$

be a remainder function of Padé approximation system of type I corresponding to the system of functions $(f_1(z),\dots,f_{p+1}(z))$ and weights $(n_1,\dots,n_{p+1})$ at $z = 0$:

$$R(z|f_1,\dots,f_{p+1};n_1,\dots,n_{p+1}) = \sum_{i=1}^{p+1} f_1(z)\ A_i(z|f_1,\dots,f_{p+1};n_1,\dots,n_{p+1}) \quad , \quad (4.4)$$

where $A_i(z|f_1,\dots,f_{p+1};n_1,\dots,n_{p+1})$ be polynomials of degree n_i : $i = 1,\dots,p+1$. We assume $R(z|f_1,\dots,f_{p+1};n_1,\dots,n_{p+1})$ to be normalized in the following way :

$$R(z|f_1,\dots,f_{p+1};n_1,\dots,n_{p+1}) = z^M R_o(z), \quad (4.5)$$

where $R_o(0) = 1$ and $M \geq \left\{\sum_{i=1}^{p+1}(n_i + 1)\right\} - 1$ by the definition of Padé approximants.

Suppose that z, $z\neq 0$, $|z| < 1$ is fixed and $n_i = N + \rho_i$ where ρ_i are fixed : $i = 1,\dots,p+1$. Then the asymptotics of the remainder function $|R(z \mid f_1,\dots,f_{p+1};n_1,\dots,n_{p+1})|$ as $N \to \infty$ is described by the following formula :

$$\frac{1}{N}\log|R(z|f_1,\dots,f_{p+1};n_1,\dots,n_{p+1})|$$
$$\to \log|z^{p+1}\ (1+\tfrac{1}{p})^{p(p+1)}\ t^*(z)^{p(p+1)}| \qquad (4.6)$$

as $N\to\infty$. Here $t = t^*(z)$ is a root of the polynomial

$$zt^{p+1} - (p+1)t + p = 0 \qquad (4.7)$$

such that $t^*(0) = 1 + \frac{1}{p}$.

Remark 4.7. We want to explain why M in (4.5) is not defined as $\left\{\sum_{i=1}^{p+1}(n_i+1)\right\} - 1$. This is connected with arbitrariness of $f_2(z),\dots,f_{p+1}(z)$ in (4.3). E.g. the defect $M+1 - \sum_{i=1}^{p+1}(n_i+1)$ may be non zero ; but this defect will be always bounded by a constant depending only on p. If the choice of $f_2(z),\dots,f_{p+1}(z)$ in (4.3) is a canonical one, like in Theorem 4.5 with

$$f_i(z) = {}_{p+1}F_p\left(z\left|\begin{matrix}a_1,\dots,a_{i-1}+i-1,\dots,a_{p+1}\\ b_1,\dots,b_{i-1}+i-1,\dots,b_p\end{matrix}\right.\right) \quad i = 2,\dots,p$$

then in (4.5), $M = \left\{\sum_{i=1}^{p+1}(n_i+1) - 1\right\}$.

Remark 4.8. The statement of theorem can be refined in the sense that for $N\to\infty$,

$$N^{-C_1}|z|^{(p+1)N}(1+\tfrac{1}{p})^{p(p+1)N}|t^*(z)|^{p(p+1)}$$
$$\le |R(z|f_1,\dots,f_{p+1};n_1,\dots,n_{p+1})|$$
$$\le N^{C_2}|z|^{(p+1)N}(1+\tfrac{1}{p})^{p(p+1)N}|t^*(z)|^{p(p+1)}$$

for two constants $C_1, C_2 > 0$ depending on $f_1,\dots,f_{p+1}$, z and $\rho_1,\dots,\rho_{p+1}$.

Remark 4.9. The choice of the root $t = t^*(z)$ of $zt^{p+1} - (p+1)t + p = 0$ means that $t^*(z) \sim \frac{p}{p+1} + O(z)$, $|z|\to 0$

Remark 4.10. We would like to point the attention on the normalization in (4.5): $R_o(0) = 1$. For applications it is necessary to change this normalization often, multiplying $R(z|f_1,\dots,f_{p+1}; n_1,\dots,n_{p+1})$ by, say ,

$$\left\{ \frac{(n_1+\dots+n_{p+1})!^p}{\prod_{i=1}^{p+1} (\sum_{j\neq i} n_j)!} \right\}^{-1} .$$

Proof of the theorem 4.6. According to the previous construction we can write $R(z|f_1,\dots,f_{p+1};n_1,\dots,n_{p+1})$ as

$$R(z|f_1,\dots,f_{p+1};n_1,\dots,n_{p+1}) = z^{-\sum_{i=1}^{p+1}(n_i+1)-1} P_o(z).F(z),$$

where $P_o(z)$ is a polynomial, depending only on $f_1(z),\dots,f_{p+1}(z)$ and $F(z)$ is a generalized ${}_{p+1}F_p$-function. In the case of equal weight $n_1,\dots,n_{p+1}$ or weights n_i of the form $n_i = N + \rho_i$: $i = 1,\dots,p+1$ with fixed $\rho_1,\dots,\rho_{p+1}$ the function $F(z)$ can be represented in such a form :

$$F(z) = {}_{p+1}F_p\left(z\left|\begin{matrix} c_1+Np,\dots,c_{p+1}+Np \\ d_1+N(p+1),\dots,d_p+N(p+1)\end{matrix}\right.\right) \qquad (4.8)$$

for $c_1,\dots,c_{p+1};d_1,\dots,d_p$ depending only on $f_1(z),\dots,f_{p+1}(z)$ (or on parameters $a_\ell, b_j, a_\ell+k_{\ell,i}, b_j+m_{\ell,i}$ defining $f_i(z)$ in (4.3)) and on $\rho_1,\dots,\rho_{p+1}$. In the case of equal weights n_i the corresponding result was presented before ; the case of $n_i = N+\rho_i$ with $N\to\infty$ and fixed ρ_i : $i = 1,\dots,p+1$ does not differ from equal weight case. Now for $F(z)$ in (4.8) we can use general Pochammer p-fold integral representation of ${}_{p+1}F_p$-generalized hypergeometric function [29] . We have for an arbitrary system $\alpha_1,\dots,\alpha_{p+1};\beta_1,\dots,\beta_p$ of complex parameters with $Re(\beta_i) > Re(\alpha_i)$: $i = 1,\dots,p$.

$$\begin{aligned} {}_{p+1}F_p\left(z\left|\begin{matrix}\alpha_1,\dots,\alpha_{p+1}\\ \beta_1,\dots,\beta_p\end{matrix}\right.\right) &= \prod_{i=1}^{p} \frac{\Gamma(\beta_i)}{\Gamma(\alpha_i)\,\Gamma(\beta_i-\alpha_i)} \\ &\times \int_0^1 dt_1\dots\int_0^1 dt_p \,.\, \prod_{i=1}^{p}(1-t_i)^{\beta_i-\alpha_i-1}\, t_i^{\alpha_i-1}\,(1-t_1\dots t_p\, z)^{-\alpha_{p+1}} \end{aligned} \qquad (4.9)$$

The formula (4.9) is used now for $F(z)$ from (4.8)with $N\to\infty$. The asymptotics of Γ-factors is determined by Stirling's formula

$$\frac{1}{N}\log\left\{\prod_{i=1}^{p}\frac{\Gamma(Np+N+d_i)}{\Gamma(Np+c_i)\Gamma(N+d_i-c_i)}\right\}$$
$$\to \log\left\{(p+1)^{p(p+1)}.p^{-p^2}\right\} \quad \text{as} \quad N\to\infty \quad . \tag{4.10}$$

The asymptotics of the integrand of the p-fold integral in (4.9) is determined by the maximum of the

$$\varphi(t,z) = \frac{t^p(1-t)}{(1-t^pz)}$$

for fixed z, $0 < |z| < 1$, and $0 \le t \le 1$. By solving $\varphi_t'(t,z) = 0$ we come to the equation $zt^{p+1} - (p+1)t + p = 0$. The maximum of the integrand of the integral in (4.9) is of the form

$$N^{C_3}\,|\varphi(t^*(z),z)|^{pN} . \tag{4.11}$$

Then the asymptotics of $R(z|f_1,\dots,f_{p+1};n_1,\dots,n_{p+1})$ is described by the combination of (4.10) and (4.11). We come to the formula (4.6).

One of the most interesting consequence of this theorem is the independence of the leading term of the asymptotics of the remainder function on parameters a_i, b_i in (4.3). The same is true for the asymptotics of the coefficients $|A_i(z|f_1,\dots,f_{p+1};n_1,\dots,n_{p+1})|$. Before in [8],[9] we gave an asymptotics of remainder function and polynomial coefficients in Padé approximation problem of type II for generalized hypergeometric function.

5. EFFECTIVE CONSTRUCTION OF THE PADÉ APPROXIMATIONS TO THE SOLUTIONS OF FUCHSIAN LINEAR DIFFERENTIAL EQUATIONS.

In this chapter we construct one-point Padé approximations to the solutions of Fuchsian linear differential equations. The same construction based on results of §§ 1-3 gives us N-point Padé approximations to the solutions of arbitrary system of linear differential equations. The corresponding N-point approximation will be presented elsewhere.

One of the important aspects of isomonodromy deformation and the corresponding contiguous relations is its effectiveness.

We can use this effectiveness to get explicitly Padé approximations (and even N-point Padé approximations) to the arbitrary solutions of the Fuchsian linear differential equations.

We start with the Theorem 5.1.generalizing the Gauss (or even the Euler) construction of the continued fraction expansion for

$$\frac{F(a,b+1;c+1;z)}{F(a,b;c;z)},$$

or Padé approximations to $\frac{d}{dz}\log F(a,b;c;z)$.

We consider the class of functions $\bar{y}$:

$$Y\left(x;\begin{matrix}a_1,\ldots,a_n\\A_1,\ldots,A_n\end{matrix}\right)$$

in Riemann's sense §§1-2 . If $\lambda_{j,k}$: $j=1,\ldots,m$ are eigenvalues of A_k : $k=1,\ldots,n$; then possible local exponents of $\bar{y}(x)$ at $x=a_k-\mu_{j,k}$: $j=1,\ldots,m$ are connected with eigenvalues $\lambda_{i,k}$: $j=1,\ldots,m$ by

$$e^{2\pi i\mu_{j,k}}=\lambda_{j,k} : k=1,\ldots,n ;\quad j=1,\ldots,m.$$

We say that $\bar{y}(x)=(y_1(x),\ldots,y_m(x))$ belongs to the class

$$\begin{pmatrix}a_1,\ldots,a_n\\ \overline{}\quad\overline{}\\ A_1,\ldots,A_n\\ \mu_{1,1},\ldots,\mu_{1,n}\\ \vdots\qquad\vdots\\ \mu_{m,1},\ldots,\mu_{m,n}\end{pmatrix}$$

if $\bar{y}$ belongs to the class $Y\begin{pmatrix}a_1,\ldots,a_n\\A_1,\ldots,A_n\end{pmatrix}$ and has local exponents at $x=a_K$ which are $\mu_{j,k}=j=1,\ldots,m$, $k=1,\ldots,n$.

Of course, local exponents $\mu_{j,k}$ for all $\bar{y}(x)$ belonging to the same class $Y\begin{pmatrix}a_1,\ldots,a_n\\A_1,\ldots,A_n\end{pmatrix}$ differ only by integers. These systems of functions are called contiguous.

According to Riemann' theorem there are at most m systems of functions from $Y\left(x;\begin{matrix}a_1,\ldots,a_n\\A_1,\ldots,A_n\end{matrix}\right)$, which are linearly independent over $\mathbb{C}[z]$. In the non-degenerate case, when all eigenvalues of A_k are distinct : $k=1,\ldots,n$, we have exactly m systems of functions belonging to the class $Y\begin{pmatrix} x; & a_1,\dots,a_n \\ & A_1,\dots,A_n \end{pmatrix}$.

We may assume, without loss of generality, that

$$a_1 = 0, \quad a_2 = \infty, \quad a_3 = 1.$$

Now we want to construct a Padé approximation to the system of m contiguous functions or, what is the same, to m solutions of a Fuchsian linear differential equation

$$L_m y = 0 ,$$

of order m. It is convenient to choose one of the branches of $\bar{y}$ to be non-singular at x = 0.

For this we use the following simple transformation :

If $Y(x)$ corresponds to the exponents $(\mu_{1,k},\dots,\mu_{m,k})$ at $x = a_k$; then $x^{\delta}(1-x)^{\varepsilon}Y(x)$ corresponds to the exponents $(\mu_{1,1}+\delta,\dots,\mu_{m,1}+\delta)$ at $x = a_1 = 0$, to $(\mu_{1,2}-\delta-\varepsilon,\dots,\mu_{m,2}-\delta-\varepsilon)$ at $x = a_2 = \infty$, to $(\mu_{1,3}+\varepsilon,\dots,\mu_{m,3}+\varepsilon)$ at $x = a_3 = 1$ and to $(\mu_{1,k},\dots,\mu_{m,k})$ at $x = a_k$ for $k > 3$.

Thus we can always assume that at $x = a_1 = 0$, $\mu_{1,1} = 0$ and we have one branch of $Y(x)$ at $x = 0$, $y_1(x)$, say, such that

$$y_1(x) \text{ is analytic at } x = 0, \ y_1(0) \neq 0 .$$

Assume now that we have m arbitrary contiguous systems of functions, linearly independent over $\mathbb{C}[x]$, belonging to classes

$$Y^{(j)}(x) = \begin{pmatrix} a_1 = 0, & a_2 = \infty, & \dots, & a_n \\ \overline{A_1} & \overline{A_2} & \dots, & \overline{A_n} \\ t\bar{\mu}_1^{(j)}, & t\bar{\mu}_2^{(j)}, & \dots, & t\bar{\mu}_3^{(j)} \\ & & \dots, & \end{pmatrix} \quad : j = 1,\dots,m \tag{5.1}$$

where $\bar{\mu}_k^{(j)} \in \mathbb{C}^m$, and

$$\bar{\mu}_k^{(j)} - \bar{\mu}_k^{(j')} \in \mathbb{Z}^m \quad \text{for } j,j' = 1,\dots,m, \quad k = 1,\dots,n.$$

We choose branches $y_1^{(j)}(x) \overset{\text{def}}{=} f_j(x) : j = 1,\dots,m$ from the system $Y^{(j)}(x)$ such that $f_j(x)$ is analytic at $x = 0$, $f_j(0) \neq 0$.

Now we want to contruct a Padé approximation system to functions

$$\{f_1(x),\dots,f_m(x)\}$$

corresponding to the weights $\{n_1,\dots,n_m\}$. We consider now only the diagonal case $n_1 = \dots = n_m = N$ or the subdiagonal case $|n_i - N| \leq 1 : i = 1,\dots,m$.

In these cases the expression for the remainder function can be obtained by considering a new, (m+1)-th system of contiguous functions of the form

$$Y_N(x) = \begin{pmatrix} \dfrac{0}{A_1} & \dfrac{\infty}{A_2} & & \dfrac{a_n}{A_n} \\ \nu_{1,1}+mN & \nu_{1,2}-N & & \nu_{1,n} \\ \nu_{2,1} & \nu_{2,2}-N & \dots & \nu_{2,n} \\ \vdots & & & \vdots \\ \nu_{m,1} & \nu_{m,2}-N & & \nu_{m,n} \end{pmatrix} \qquad (5.2)$$

where $\nu_{i,j}$ depend only on $\bar{\mu}_k^{(i)}$ and $|n_i - N| : i = 1,\dots,m$, $k = 1,\dots,n$.

If we now take the branch $f_n(x)$ of $Y_N(x)$ at $x = 0$, corresponding to the exponent $\nu_{1,1}+mN$ at $x = 0$, we obtain by results of §§1-3 the remainder function for the Padé approximation problem for $\{f_1(x),\dots,f_m(x)\}$:

$$f_N(x) = \sum_{i=1}^{m} P_i(x|f_1,\dots,f_m ; n_1,\dots,n_m)\, f_i(x) \qquad (5.3)$$

where $P_i(x|f_1,\dots,f_m;n_1,\dots,n_m)$ are polynomials of degree $\leq n_i$: $i = 1,\dots,m$ and

$$\mathrm{ord}_{x=0}\, f_N(x) \geq \sum_{i=1}^{m} \{n_i + 1\} - 1 \quad .$$

It is easy to show that for a certain system of n contiguous functions $Y^{(i)}(x)$: $i = 1,\dots,m$ it may happen that

$$\mathrm{ord}_{x=0}\, f_N(x) > \sum_{i=1}^{m} \{n_i + 1\} - 1.$$

However,

$$\left|\mathrm{ord}_{x=0}\, f_N(x) - \sum_{i=1}^{m} \{n_i + 1\} + 1\right|$$

is always bounded by an absolute constant (depending only on the exponents $\bar{\mu}_k^{(i)}$).

For the case of the canonical system of contiguous functions, we have a perfect system of Padé approximations of type I.

Theorem 5.1. Let us consider the following canonical system of contiguous functions :

$$Y^{(i)}(x) = \begin{pmatrix} \frac{0}{A_1} & \frac{\infty}{A_2} & \frac{1}{A_3} & \cdots & \frac{a_n}{A_n} \\ \mu_{1,1}^{(i)} & \mu_{1,2}^{(i)} & \mu_{1,3} & & \mu_{1,n} \\ \vdots & \vdots & \vdots & \cdots & \vdots \\ \mu_{m,1}^{(i)} & \mu_{m,2}^{(i)} & \mu_{m,3} & & \mu_{m,n} \end{pmatrix}$$

with $\bar{\mu}_k^{(i)} = \bar{\mu}_k : k > 2$ and $\mu_{1,1}^{(i)} = 0$,

$$\mu_{\ell,1}^{(i)} = \mu_{\ell,1} - \delta_{\ell,i} \cdot (i-1) : \; i = 1,\ldots,m \;\; ; \; \ell = 1,\ldots,m$$

$$\mu_{\ell,2}^{(i)} = \mu_{\ell,2} + \delta_{\ell,i} \cdot (i-1) : \; i = 1,\ldots,m \;\; ; \; \ell = 1,\ldots,m.$$

We take the branch $f_i(x)$ of $Y^{(i)}(x)$ at $x = 0$ such that $f_i(0) \neq 0,\infty$.

Let $N \geq 1$ and let us consider a new system of contiguous functions

$$Y_N(x) = \begin{pmatrix} \frac{0}{A_1} & \frac{\infty}{A_2} & \cdots & \frac{a_k}{A_k} & \cdots & \frac{a_n}{A_n} \\ \nu_{1,1} & \nu_{1,2} & & \nu_{1,k} & & \nu_{1,n} \\ \vdots & \vdots & & \vdots & & \vdots \\ \nu_{m,1} & \nu_{m,2} & \cdots & \nu_{m,k} & \cdots & \nu_{m,n} \end{pmatrix}$$

where $\nu_{1,1} = 0$, $\nu_{j,1} = \mu_{j,1} - mN$; $j = 2,\ldots,m$; $\nu_{j,2} = \mu_{j,2} + (m-1)N$; $j = 1,\ldots,m$. Let $f_N(x)$ be a branch of $Y_N(x)$ corresponding to a local exponent $\nu_{1,1} = 0$ at $x = 0$.

Then

$$x^{(mN)-1} \cdot f_N(x) = \sum_{i=0}^{m} P_{i,N}(x)\, f_i(x) \qquad (5.5)$$

where $P_{i,N}(x)$ are polynomials of degree $\leq N-1$ and $f_N(0) \neq 0$.

In the nondegenerate case, all $P_{i,N}(x)$ are of the exact degree $N-1$.

In the proof of theorem 5.1 we use simply the computations of degrees of polynomial relations between systems $Y^{(i)}(x)$: $i = 1,\dots,n$, $Y_N(x)$ of contiguous functions. The formula (5.5) is simple consequence of theorem 1.1. together with degree estimates (2.13)-(2.17).

6. ARITHMETIC APPLICATIONS OF PADÉ APPROXIMATIONS OF CLASSICAL ANALYTIC FUNCTIONS.

In this chapter we discuss arithmetic applications of Padé approximation to the solutions of linear differential equations.

In the appendix to § 6 we give asymptotics of polynomial coefficients of Padé approximants of type I to generalized hypergeometric functions. These results together with asymptotics of remainder function in Theorem 4.6 complete the studies of Padé approximation to generalized hypergeometric functions started in [8],[9].

6.1. The possibility of further progress in the investigation of the arithmetical nature of systems of analytic functions, satisfying a system of linear differential equations can be explained because we have an effective construction of Padé approximations.

More precisely, let $f_i(x)$: $i = 1,\dots,m$, $\bar{f} = (f_1,\dots,f_m)$ be a system of functions satisfying a linear differential equation

$$\bar{f}_x = \hat{Q}\,\bar{f},$$

where $\hat{Q} = \hat{Q}(x)$ is an $m \times m$ matrix with elements that are rational functions of x. Let us assume that $\bar{f}$ is nonsingular at $x = 0$ (or is at most regularly singular at $x = 0$). Then Padé approximants to $(f_1(x),\dots,f_m(x))$ of types I or II can be constructed effectively in terms of $\bar{f}(x)$, $\hat{Q}(x)$ only.

E.g. polynomials $R_j(x;N)$ of degrees $\leq N$: $j = 1,\dots,m$ can be constructed such that

1) $R_j(x;N)\, f_i(x) - R_i(x;N)\, f_j(x)$: $i \neq j$

has order at $x = 0$ at least $m[\frac{N}{m-1}] + 1$ and

2) $R_j(x;N)$ satisfy the linear recurrent system in N of order bounded by a constant, depending only on m and $\deg(\hat{Q})$.

Analogously, we can construct polynomials

$$P_1(x;\mathcal{N}),\dots,P_m(x;\mathcal{N}) \tag{6.2}$$

of degrees $\leq \mathcal{N}$ such that

$$\mathrm{ord}_{x=0} \sum_{i=1}^{m} P_i(x;\mathcal{N})\ f_i(x) \geq \mathcal{N}m + O(1) \tag{6.3}$$

and the sequence

$$\bar{P}(x;\mathcal{N}) = (P_1(x;N),\dots,P_m(x;N)) \tag{6.4}$$

satisfies a linear recurrent relation of order O(1).

Here, in (6.3) and (6.4) the constants in O(1) depend on m and deg(Q) only.

In particular we can obtain an *effective* proof of Shidlovski's lemma or what K. Mahler [3],[16],[21] "almost perfectness" of systems of functions satisfying linear differential equations.

Only the case

$$m = 2, \quad f_1 = 1, \quad f_2(x) = f(x), \tag{6.5}$$

i.e. in (1)

$$f' = Q_1 f + Q_0 \tag{6.5'}$$

for Q_0, $Q_1 \in \mathbb{C}(x)$, has been considered before. We obtain in this case from our general formulae the construction of M. Laguerre [17].

Moreover, wa can improve Laguerre's result and find the equation in $\partial/\partial t$ which is satisfied by the generating function

$$\sum_{n=1}^{\infty} P_n(x)\ t^n$$

for fractions

$$f(x) - \frac{Q_n(x)}{P_n(x)}$$

in the continued fraction expansion of f(x).

From the analytic point of view our construction of Padé approximations solves all the problems. However two main questions are still unresolved :

A1) What is the arithmetic nature of the coefficients of the polynomials $P_i(x;\mathcal{N})$ or $R_j(x;\mathcal{N})$?

A2) What is the asymptotic nature of the remainder function and polynomial coefficients

$$R(x;\mathcal{N}) = \sum_{i=1}^{m} P_i(x;\mathcal{N})\, f_i(x)$$

in (6.3) when $\mathcal{N} \to \infty$ and x is fixed ?

We are able to answer these questions in many important cases.

First of all *it is possible* to investigate the case of *entire* functions $f_i(x)$: $i = 1,\dots,m$ with $\hat{Q} \in GL(m, \bar{Q}(x))$ but which are *not* E-functions [28]. The simplest situation where we have now an algebraic independence result is the case of the generalized hypergeometric equation

$$\{x\, P(\delta) - Q(\delta)\} \quad f = 0$$

for $P(x)$, $Q(x) \in Q[x]$, $\delta = x\frac{d}{dx}$ and $q = d(Q) > d(P) = p$, where

$$f = {}_pF_{q-1}\left(x\middle|\begin{matrix}a_1,\dots,a_p\\ b_1,\dots,b_{q-1}\end{matrix}\right)$$

Another important case is the case of G-functions [29]. We recall that we still do not know , for an arbitrary system $\bar{f}(x)$ of G-functions satisfying (6.1) (i.e. $\hat{Q}(x) \in GL(m, \bar{Q}(x))$), will it be the case that in

$$T^k(x)\,\frac{\bar{f}^{(k)}(x)}{K!} = \hat{R}_k(x)\,\bar{f}(x)$$

for $\hat{R}_K(x) \in GL(m, \bar{Q}\,x\,)$, that the denominators of the coefficients of the polynomials in $\hat{R}_K(x)$ are bounded by C^K : $K \to \infty$. This is the so-called (G,C)-function condition. We have; of course, E. Bombieri and Dwork's recent great success [12] in this direction but the problem is still open (at least as far as I know the recent Princeton achievements).

Nevertheless, for general classes of functions both problems A1), A2) can be solved. These problems have already been solved by the author for an *arbitrary* system of hypergeometric ${}_{p+1}F_p$ or simply ${}_qF_p$-functions [8],[9] or §§ 3-4.

New results for linear independence have been obtained in this direction. Now as a new step forward one can treat even the case of an arbitrary product of hypergeometric functions.

As an example we would like to present the following result.

Theorem 6.1. Let c and b be rational numbers, $b \notin \mathbb{Z}$, $c \notin \mathbb{Z}$, $b-c \notin \mathbb{Z}$, and $f(x) = {}_2F_1(1,b;c;x)$ be a transcendental function. Then for any $D \geq 1$ there exists a constant

$$C(\bar{b},\bar{c},D) > 0,$$

depending only on D and the denominators $\bar{b}$ of b and $\bar{c}$ of c, such that for

$$x \in \bar{Q}\ ,\quad x \neq 0 \text{ and } d(x) \leq D\ ,\quad |x| < C(\bar{b},\bar{c},D))$$

the number ${}_2F_1(1,b:c:x)$ is not of degree $\leq D$.

The main achievement in this theorem is *no dependence* of $C(\bar{b},\bar{c},D)$ on H(x). Moreover we have a bound for $|P(f(x))|$ with $P(z) \in \mathbb{Z}[z]$. E.g. under the same assumption as in theorem A, for $P(z) \in \mathbb{Z}[z]$, $P \not\equiv 0$,

$$|P(f(x))| > H(P)^{-c_2(x,f,D)}$$

assuming $d(P) \leq D$ and $H \geq c_3(x,f,D)$.

APPENDIX TO § 6

6.2. The asymptotics of the polynomial coefficients of Padé approximants of type I to generalized hypergeometric functions.

Here we solve the problem A2) in the case of generalized hypergeometric functions. In §4 we gave already the asymptotics for remainder term ; now we give the asymptotics of polynomial coefficients.

We can present now the asymptotics for the polynomial coefficients of Padé approximants of type I in the case of generalized hypergeometric functions. This bound together with the investigations of the arithmetical properties of the coefficients are necessary for the proof of linear independence of values of hypergeometric functions over Q (see §.8 for applications).

First of all, as in §.4 we assume that we have p + 1 generalized contiguous functions

$$f_i(x) = {}_{p+1}F_p\left(x\left|\begin{matrix} a_1+k_{1,i},\dots,a_{p+1}+k_{p+1,i} \\ b_1+m_{1,i},\dots,b_{p+1}+m_{p,i}\end{matrix}\right.\right)$$

$i = 1,\dots,p+1$ linearly independent over $C[x]$ ($k_{j;i}$, $m_{\ell,i}$ are rational integers : $j = 1,\dots,p+1$, $\ell = 1,\dots,p$, $i = 1,\dots,p+1$) and the remainder function

$$R(z|f_1,\dots,f_{p+1};\ n_1,\dots,n_{p+1})$$

of the Padé approximations of type I corresponding to $(f_1(x),\dots,f_{p+1}(x))$ and weights $(n_1,\dots,n_{p+1})$ at $x = 0$.

Theorem 6.2. Let, as before, $f_1(x),\dots,f_{p+1}(x)$ be p+1 contiguous functions of ${}_{p+1}F_p$-generalized hypergeometric functions, linearly independent over $\mathbb{C}[x]$. Let $R(x|f_1,\dots,f_{p+1};\ n_1,\dots,n_{p+1})$ be the remainder function of Padé approximation of type I to $(f_1(x),\dots,f_{p+1}(x))$ with weights $(n_1,\dots,n_{p+1})$:

$$R(x|f_1,\dots,f_{p+1};\ n_1,\dots,n_{p+1}) =$$

$$= \sum_{j=1}^{p+1} f_j(x)\ A_j(x|f_1,\dots,f_{p+1};\ n_1,\dots,n_{p+1}).$$

Here $A_j(x|f_1,\dots,f_{p+1};\ n_1,\dots,n_{p+1})$ are polynomials of degree $\le n_j$: $j = 1,\dots,p+1$ and

$$R(x|f_1,\dots,f_{p+1};n_1,\dots,n_{p+1}) = x^M R_o(x) ,$$

where $M \ge \{\sum_{j=1}^{p+1} (n_j + 1)\} - 1$ and $R_o(0) = 1$.

Suppose that $n_j = N + \rho_j$: $j = 1,\dots,p+1$ where $\rho_1,\dots,\rho_{p+1}$ are fixed. Then for fixed x, $0 < x < 1$, we have

$$\frac{1}{N} \log|A_j(x|f_1,\dots,f_{p+1}\ ;n_1,\dots,n_{p+1})| \to$$

$$\to \log \left\{(p+1)^{p(p+1)}\ p^{-p(p+1)} \left(\frac{p+1 - t_*(x)p}{t_*(x)}\right)^p\ x\right\}$$

where $t_*(x) = t$ is a minimal (by modulus) root of the polynomial

$$p\ t^{p+1} - (p+1)\ t^p + x = 0.$$

Proof. We can use theorem 4.5-4.6 (see also (4.8)) to get explicit expressions for coefficients $A_j(x|f_1,\dots,f_{p+1};n_1,\dots,n_{p+1})$ in terms of hypergeometric functions. However for our purposes it is enough to find the leading (exponential) term of the asymptotics of $A_j(x|f_1,\dots,f_{p+1};\ n_1,\dots,n_{p+1})$ with fixed x. Thus we notice that from §.4 it follows that the leading term of the asymptotics of $A_j(x|f_1,\dots,f_{p+1};\ n_1,\dots,n_{p+1})$ for $n_i = N + \rho_i$: $i = 1,\dots,p+1$ is given by the hypergeometric polynomial of degree N in (6.7).

First of all, as it was suggested in Remark 4.10, we multiply $R(z|f_1,\dots,f_{p+1};\ n_1,\dots,n_{p+1})$ and so all $A_j(z|\ f_1,\dots,f_{p+1};\ n_1,\dots,n_{p+1})$ by

$$\frac{\prod_{i=1}^{p+1} (\sum_{j\neq i} n_j)!}{(n_1 + \ldots + n_{p+1})!^p}$$

In this way we see that the leading exponential term of the asymptotics of $A_j(x|f_1,\ldots,f_{p+1};n_1,\ldots,n_{p+1})$ for $0 < x < 1$ is given by

$$\frac{(n_1 + \ldots + n_{p+1})!^p}{\prod_{i=1}^{p+1} (\sum_{j\neq i} n_j)!} \cdot \frac{(Np)!}{(N!)^p} . x^N . {}_{p+1}F_p\left(x^{-1} \middle| \begin{matrix} Np,-N,\ldots,-N \\ \nu_1,\ldots,\nu_p \end{matrix}\right) . \tag{6.7}$$

Since ρ_i and ν_j are fixed, we find that for $N \to \infty$ the leading exponential term of the asymptotics of $A_j(x|f_1,\ldots,f_{p+1};n_1,\ldots,n_{p+1})$ is given by

$$\frac{(p+1)^{p(p+1)N}}{p^{p^2 N}} . x^N . {}_{p+1}F_p\left(x^{-1} \middle| \begin{matrix} Np,-N,\ldots,-N \\ 1,\ldots,1 \end{matrix}\right) .$$

Now we can use integral representation of generalized hypergeometric function. This time integral representation is different from (4.9) and we must use another Pochammer integral (cf. [2]) :

$$\begin{aligned} {}_{p+1}F_p\left(z \middle| \begin{matrix} \alpha_1,\ldots,\alpha_{p+1} \\ \beta_1,\ldots,\beta_p \end{matrix}\right) &= \prod_{i=1}^{p} \frac{-\sqrt{-1}\ \Gamma(\beta_i)\exp(-\sqrt{-1}\pi\alpha_i)}{\Gamma(\alpha_i)\Gamma(\beta_i - \alpha_i) 2 \sin\pi\alpha_i} \times \\ &\times \int_1^{(0+)} dt_1 \ldots \int_1^{(0+)} dt_p \prod_{i=1}^{p} t^{\alpha_i - 1} (1-t)^{\beta_i - \alpha_i - 1} (1-t_1\ldots t_p z)^{-\alpha_{p+1}} \end{aligned} \tag{6.8}$$

valid for $\mathrm{Re}\beta_i > \mathrm{Re}\alpha_i$, $|\mathrm{ard}(-z)| < \pi$.

Using functional equation for Γ-function we find the following expression for leading exponential term ($z = x^{-1}$)

$$x^N \frac{(p+1)^{p(p+1)N}}{p^{p^2 N}} \int_1^{(0+)} \ldots \int_1^{(0+1)} (1 - t_1\ldots t_p z)^{-Np} \prod_{i=1}^{p} \frac{(1-t_i)^N}{t_i^{-N}} dt_i .$$

We study the maximum of

$$\varphi(t,z) = \frac{(1-t)}{t(1-t^p z)}, \quad z^{-1} = x,$$

and we obtain the equation :

$$pt^{p+1} z - (p+1)t^p z + 1 = 0 \ : \ z = x^{-1}.$$

We take the smallest (by modulus) root $t = t_*(x)$ of this equation and

$$t_*(x) \sim (p+1)^{-1/p} . x^{1/p} \quad \text{when } x \to 0.$$

Then

$$\varphi(t_*,z) = \frac{(p+1) - t_* .p}{p.t_*} .$$

This together with the representation (6.7)-(6.8) completes the proof of theorem 6.2.

In the case of $p = 1$ we obtain classical Riemann's result [25].

7. C-AND (G,C)-FUNCTIONS. THEIR ANALYTIC AND ARITHMETIC PROPERTIES. SIEGEL'S METHOD.

We consider analytic properties of solutions of linear differential equations. We classify them according to the order of growth of coefficients and then use generalized Borel transform to study arithmetic properties of coefficients. In this way we arrive to C- and (G,C)-classes of functions generalizing in a natural way E- and G-functions of C. Siegel [27],[28]. We describe then Padé and Siegel's methods of proof of linear independence of values of C-functions and present several particular results.

7.1. According to Ramis' results [24] we can classify the order of growth of solutions of linear differential equations in a very simple way.

E.g. Let $f(x)$ be a formal power series from $\mathbb{C}[[x]]$ and $f(x)$ satisfy a linear differential equation

$$D[x]\ f(x) = 0$$

where $D[x]$ is a linear differential operator in $\frac{d}{dx}$ with coefficients from $\mathbb{C}(x)$:

$$D[x] = \sum_{i=0}^{n} a_i(x) \frac{d^i}{dx^i}, \ a_i(x) \in \mathbb{C}(x) : i = 0,\ldots,n.$$

Then, unless $f(x) \in \mathbb{C}[x]$ (this case must be excluded), there exists a number $k = k(f)$ such that k is a rational number, and from

$$f(x) = \sum_{n=0}^{\infty} a_n x^n$$

it follows that the function

$$B^k f(x) = \sum_{n=0}^{\infty} a_n (n!)^k x^n$$

has a non zero, but finite radius of convergence.

The number $k = k(f) \in \mathbb{Q}$ will be called the *index (of growth)* of $f(x)$.

7.2. The general definition of C- and (G,C)-functions.

We start with the definitions, generalizing the definitions of E- and G- functions in Siegel's sense.

Definition of the C-function 7.1. Let $f(x) \in \bar{\mathbb{Q}}[[x]]$ and $f(x)$ be a solution of a linear differential equation in d/dx with the coefficients from $\bar{\mathbb{Q}}[x]$. Let

$$F(x) = \sum_{n=0}^{\infty} a_n x^n$$

and $f(x)$ have the index (of growth) $k = k(f)$. We say that $f(x)$ is a C-function (Symbolically $f \in C[k]$) if the common denominator of

$$\{a_i (i!)^k \quad : \quad i = 0,1,\ldots,n\}$$

is bounded by C^n (by geometric progression).

In other words there are integers $A_n \in \mathbb{Z}$ such that

$$A_n a_i (i!)^k \quad \text{are algebraic integers for all}$$

$$i = 0,1,\ldots,n \text{ and for } C > 1, \ |A_n| \le C^n \text{ or}$$

$$|A_n| \underset{\varepsilon}{<<} C^{n(1+\varepsilon)} \quad \text{for any } \varepsilon > 0 \text{ and } n \to \infty.$$

By definition we include all polynomials from $\bar{\mathbb{Q}}[x]$ in the class of C-functions. The index is undefined for polynomials.

Remark 7.2. The number $(n!)^k$ is not an integer if $k \notin \mathbb{Z}$, but this is not important for the definition of the C-function. Indeed we can always write

$$k = \frac{A}{B}, \quad (A,B) = 1, \quad A,B\text{-integers}$$

and $B > 0$ and

$$n! = (n_B)^B . m_B(n) ,$$

where n_B and $m_B(n)$ are integers and

$$m_B(n) \ll_{\delta,B} e^{n(1+\delta)} \quad \text{for any } \delta > 0 \text{ and } n \to \infty.$$

In other words we can always estimate the common denominator of algebraic numbers

$$\{a_m . m_B^A \quad : \quad m = 0,1,\ldots,n\}$$

where $k = A/B$.

Remark 7.3. If $f(x) \in \bar{\mathbb{Q}}[[x]]$ is a C-function and

$$f(x) = \sum_{n=0}^{\infty} a_n x^n ,$$

then by the definition of $k = k(f)$ we have

$$(n!)^k \, \overline{|a_n|} \leq R^n$$

for some $R \geq 1$.

Here for an algebraic α and its conjugates $\alpha^{(2)},\ldots,\alpha^{(d)}$,

$$|\bar{\alpha}| = \max\{|\alpha|, |\alpha^{(2)}|, \ldots, |\alpha^{(d)}|\}$$

If we now consider functions of exponential growth, i.e. $k(f) = 1$, then C-functions are exactly E-functions in the Siegel sense [28].

If we consider, however, functions analytic at the origin but with finite radius of the convergence, i.e. $k(f) = 0$, then the class of C-functions coincides with the class of G-functions, in the sense of [27] .

If we speak about the system $\bar{f}(x) = (f_1(x),\ldots,f_K(x))$ of functions being the system of C-functions, this usually means that the constant $\underline{C}$ (in the definition of the C-class) *is the same* for all functions $f_1(x),\ldots,f_K(x)$.

The next definition was formulated for the G-function case for the first time by Galochkin (1974) [33] who tried to reprove Siegel's unpublished (1929) [27] G-function results.

Definition of the (G,C)-function 7.4. Let $f(x) \in \bar{\mathbb{Q}}[[x]]$ and $f(x)$ satisfy a linear differential equation

$$\text{(E)} \qquad Df(x) = 0$$

of order $n: D = \sum_{j=0}^{n} a_j(x) \frac{d^j}{dx^j}$, $a_j(x) \in \bar{\mathbb{Q}}[x]: j = 0,1,\ldots,n.$

Let $k = k(f)$ be the index (of growth) of $f(x)$. We iterate equation (E) and we obtain for any $m \geq 0$, the following representation

$$(E^+) \qquad f^m(x)(m!)^{k-1} = \sum_{j=0}^{n-1} B_{j,m}(x)\, f^{(j)}(x)$$

with $B_{j,m}(x) \in \bar{\mathbb{Q}}(x)$. There exists a polynomial $T(x) \in \bar{\mathbb{Q}}[x]$ such that

$$(E^{++}) \qquad T^m(x)\, f^{(m)}(x)(m!)^{k-1} = \sum_{j=0}^{n-1} A_{j,m}(x) f^{(j)}(x)$$

where now $A_{j,m}(x) \in \bar{\mathbb{Q}}[x]$: $j = 0,1,\ldots,n-1$; $m \geq 0$. We say that the function $f(x)$ belongs to the (G,C)-class, if a sequence of integers D_m : $m \geq 0$ exists such that the polynomials

$$D_m \cdot A_{j,k}(x) \quad : \quad j = 0,1,\ldots,n-1, \quad k = 0,1,\ldots,m$$

all have algebraic *integer* coefficients and

$$|D_m| \leq C^m \quad : \quad m \geq 0 \qquad \text{or}$$

$$|D_m| \ll_{\varepsilon} C^{(1+\varepsilon)m} \quad \text{for any} \quad \varepsilon > 0 \quad \text{and } m \to \infty.$$

This definition is more restrictive than the definition of the C-function since it requires knowledge of p-adic properties of $f(x)$ outside of $x = 0$.

The class of (G,C)-functions is, however, the only class of functions for which Siegel's method [29] is applicable.

It is very interesting to note that Siegel's method can be applied both in $\mathbb{C}$ and $\mathbb{C}_p$, but in an archemedian situation only the case $k > 0$ is an easy one. In a p-adic situation only the case $k = 0$ is an easy one.

7.3. Unfortunately the class of C-functions (and of (G,C)-functions as well) *cannot* cover all the functions satisfying *even the simplest* linear differential equations over $\bar{\mathbb{Q}}[x]$ or even over $\mathbb{Q}[x]$.

E.g. any generalized hypergeometric function

$$f(x) = {}_qF_p\left(x\left|\begin{matrix}a_1,\dots,a_q\\ b_1,\dots,b_q\end{matrix}\right.\right) = \sum_{n=0}^{\infty}\frac{(a_1)_n\cdots(a_q)_n}{(b_1)_n\cdots(b_p)_n n!}x^n$$

is a C-function if $a_i, b_j \in \mathbb{Q}$. But if a_i, b_j, $a_i - b_j$ are different from integers (this is assumed for the sake of simplicity)**) and *at least one* of a_i, b_j *is irrational*, then $f(x)$ is not a C-function, though $f(x)$ may satisfy a linear differential equation even on $\mathbb{Q}(x)$.

In order to treat this class of functions we must expand our definitions and allow denominators to grow faster than a geometric progression.

<u>Definition 7.5</u>. Let $(f_1(x),\dots,f_n(x)) = \bar{f}(x)$ be a system of functions from $(\bar{\mathbb{Q}}[[x]])^n$ satisfying a linear differential equation

$$\text{(E)} \qquad \frac{d}{dx}\bar{f} = \hat{Q}_1(x).\bar{f}$$

with $n\times n$ matrix $\hat{Q}_1(x)$ having elements from $\bar{\mathbb{Q}}(x)$. Let

$$f_i(x) = \sum_{m=0}^{\infty} a_{i,m}.x^m \quad : \quad i = 1,\dots,n.$$

We say that $(f_i(x) : i = 1,\dots,n) = \bar{f}(x)$ belongs to the class $\mathcal{F}[k;s]$ if

1) $\overline{|a_{i,m}(m!)^k|} \le R_1^m \quad : \quad m \to \infty\ ,\ i = 1,\dots,n$ and

2) the common denominator $\mathcal{D}_m$ of

$$\mathcal{D}_m = \text{den}\,\{a_{i,\ell}(\ell!)^k : \ell = 0,1,\dots,m\ ;\ i = 1,\dots,n\} \text{ is}$$

bounded by $(m!)^s$:

$$\overline{|\mathcal{D}_m|} \ll_{\varepsilon} (m!)^{s(1+\varepsilon)} \quad \text{for any } \varepsilon > 0.$$

Let now for $m \ge 0$,

$$(E_m) \quad (m!)^{k-1}.T^m(x).\frac{d^m}{dx^m}\bar{f} = \hat{Q}_m(x).\bar{f}(x),$$

where $T(x) \in \bar{\mathbb{Q}}[x]$ and elements of $\hat{Q}_m(x)$ are polynomials from $\bar{\mathbb{Q}}[x]$. We say that $\bar{f}(x)$ belongs to the class $\mathcal{E}[k;r]$ if condition 1) is satisfied (i.e. $\bar{f}(x)$ has index k) and there exists a number

$\mathcal{B}_m$ $(m \geq 0)$ - an algebraic integer from the field $\mathbb{Q}[a_{i,m} :$ $i = 1,\dots,n,\ m \geq 0]$ such that

3) all the polynomials in the matrices

$$\hat{Q}_\ell \cdot \mathcal{B}_m \quad : \ell = 0,1,\dots,m$$

have algebraic *integer* coefficients and

$$0 < |\overline{\mathcal{B}_m}| \underset{\varepsilon}{<<} (m!)^{r(1+\varepsilon)} \quad \text{for any} \quad \varepsilon > 0.$$

Remark. There are actually simple functions (e.g. hypergeometric) with $k=0$ (e.g. of type ${}_{p+1}F_p\ (x|\begin{smallmatrix}\cdots\\ \cdots\end{smallmatrix}))$ and from the class $\mathfrak{F}[0,s]$, for *any* given large $s > 0$, that do not belong to $\mathfrak{F}[0,s']$ for smaller (integer) $s' < s$. It is clear, however, that having fixed n we can bound both s and r. But for applications we need something better than a trivial

$$s \leq n.$$

We need $s \leq n-1$.

Warning : If $x=0$ is a singularity of the system (E), which may easily happen, the classes $\mathfrak{F}[k;s]$ and $\mathfrak{E}[k;s]$ become absolutely different ! Moreover in this case in (E_m) it is better to multiply

$$\frac{d^m}{dx^m}\bar{f} \quad \text{not by} \quad (m!)^{k-1}$$

but by the quantity $(m!)^{k'-1}$ with $k' \leq k$ and r in $\mathfrak{E}[k;r]$ may be even zero.

Example 7.6. 1) e^x is a (G,C)-function, $x = 0$ is not a singular point of $y' = y$, so

$$e^x \in \mathfrak{F}[1;0] \cap \mathfrak{E}[1;0].$$

2) Let $\theta(x) = x^m + \dots + \theta_o \in \overline{\mathbb{Q}}[x]$ and

$$y = y_\theta(x) = 1 + \sum_{n=1}^{\infty} \frac{x^n}{\prod_{j=1}^{n} \theta(j)}$$

satisfy the equation

$$\theta(x\frac{d}{dx})\, y = xy + \theta_o.$$

Now $x=0$ is a singularity of the equation. If all zeros of $\theta(x)$ are rational numbers, then y_θ is a $\bar{C}$-function and a (G,C)-function.

However for at least one non-rational root of $\theta(x)$, the function $y_\theta(x)$ is not a C-function any more.

Nevertheless $y_\theta(x)$ is always a (G,C)-function !

Moreover, we can replace the factor $(m!)^{k-1}$ in front of $\frac{d^m}{dx^m} y_\theta(x)$ simply by $(m!)^0 = 1$. Here $k = \deg(\theta)$, of course.

Lemma 7.7. Let $\theta(x) \in \mathbb{K}[x]$ and $\theta(x) = \prod_{j=1}^{m} (x - \lambda_j)$, where λ_j are not integers and

$$\tau = m - \sum_{j=1}^{m} \frac{1}{\deg(\lambda_j)} .$$

Then $y_\theta(x),\ldots,y_\theta^{(m-1)}(x)$ belong to the class $\mathcal{H}[m; [\mathbb{K} : Q]\tau]$.

The addition of $[\mathbb{K}:Q]$ in $\mathcal{H}[m;.]$ is clear ; if $\mathcal{D}_m$ is the denominator of $a_{i,\ell}$, in 2), then $\mathcal{D}_m \in \mathbb{K}$, and

$$\text{Norm } \mathbb{K}/Q\ (\mathcal{D}_m) \leq (m!)^{\tau.[\mathbb{K}:Q]} .$$

As we mentioned before, $y_\theta(x),\ldots,y_\theta^{(m-1)}(x)$ are always (G,C)-functions.

7.4. Siegel's method of linear forms

Before proceeding to some effective results it is important to explain Siegel's non-constructive method (1929).

Siegel managed to avoid the use of explicit Padé approximations. The scheme of the proof of, say, the linear independence of the numbers

$$f_1(\alpha),\ldots,f_n(\alpha)$$

for the system of functions $f_1(x),\ldots,f_n(x)$ must, in general, be the following.

We assume that the system of functions $f_1(x),\ldots,f_n(x)$ is analytic in the domain D (complex or p-adic) containing the point 0, and that at $x = 0$ *all* functions $f_1(x),\ldots,f_n(x)$ have *algebraic* Taylor coefficients from a given algebraic number field $\mathbb{K}$. Usually some additional information on the denominators of the coefficients is required.

This will lead us to the class of C-functions.

Now we can construct an "approximation form" to the system of functions

$$f_1(x),\ldots,f_n(x).$$

This means, basically, that we take some positive integers $m_1,\ldots,m_n$ and try to find polynomials

$$P_1(x),\ldots,P_n(x)$$

from $\mathbb{K}[x]$ of degrees $m_1,\ldots,m_n$ respectively such that "the remainder function"

$$R(x) = P_1(x)\, f_1(x) + \ldots + P_n(x)\, f_n(x)$$

has at $x = 0$ a zero of "high" order relative to m_j.

Trivial linear algebra shows that it is possible to find

$$P_j^o(x) \in \mathbb{K}[x]\ ,\quad \deg(P_j^o) \le m_j \quad : \quad j = 1,\ldots,n$$

such that

$$\mathrm{ord}_{x=0}\, R(x) = \mathrm{ord}_{x=0}\, \Big(\sum_{j=1}^{n} P_j^o(x)\, f_j(x)\Big)$$

is at least $\sum_{j=1}^{n} (m_j + 1) - 1$.

This system of polynomials was called in § 0 the system of Padé approximants of type I to the system of functions $\{f_1(x),\ldots,f_n(x)\}$.

However there is no *actual* way to construct these Padé approximants, unless we can evaluate $(n_1 + \ldots + n_m + m-1)\times(n_1 + \ldots + n_m + m-1)$ determinants built from the coefficients of $f_j(x)$.

We assume that by some trick, good luck or Siegel's non-constructive method one *did* construct "approximation forms" $\{P_1(x),\ldots,P_n(x)\}$ such that

$$P_i(x) \in \mathbb{K}[x],\ \deg(P_i) \le m_i \quad : i = 1,\ldots,n$$

$$\mathrm{ord}_{x=0}\, R(x) = \mathrm{ord}_{x=0}\, \Big(\sum_{i=1}^{n} P_i(x)\, f_i(x)\Big) \ge \mathcal{N}$$

(where usually at least $\mathcal{N} \ge \sum_{j=1}^{n} \frac{m_j+1}{2}$ is satisfied) and *the sizes* of the polynomials $P_i(x)$ *are also evaluated* :

$$\max_{i=1,\ldots,n} \overline{\|P_i\|} \le H$$

(here $\overline{\|P\|}$ is the maximum of the sizes and denominators of the coefficients of P).

Having this *system* of "approximation forms" constructed, we can try to prove linear independence over $\bar{\mathbb{Q}}$ for the system of numbers

$$\{f_1(\alpha),\dots,f_n(\alpha)\}$$

where α is an algebraic number, $\alpha \neq 0$, from the domain of analyticity D of $\bar{f}(x) = (f_1(x),\dots,f_n(x))$. If our system of approximation forms is the Padé system, then roughly speaking, we simply substitute α in $R(x)$.

In order to prove something using Siegel's method we must, unfortunately, take derivatives. This means that the remainder function

$$R(x) = P_1(x)\, f_1(x) + \dots + P_n(x)\, f_n(x)$$

must be differentiable and, as we are interested in linear relations (for the moment), the system $f_1(x),\dots,f_n(x)$ must satisfy a linear differential equation with coefficients from $\bar{\mathbb{Q}}(x)$.

However the operation of differentiation (innocent from the analytic point of view) is very dangerous for arithmetic applications.

Indeed, we have

$$\left(\frac{d}{dx}\right)^m R(x) = \mathcal{R}_{1,m}(x)\, f_1(x) + \dots + \mathcal{R}_{n,m}(x)\, f_n(x)$$

for rational functions $\mathcal{R}_{m,i}(x) \in \bar{\mathbb{Q}}(x)$. We knew before the sizes of the polynomials $P_i(x)$, but what will the sizes and especially the denominators of the coefficients of $\mathcal{R}_{j,m}(x)$ be ?

This problem is not very important for E-function results (or for any C-function with $k > 0$), but becomes crucial for G-functions. This forces us to introduce the class of (G,C)-functions.

7.5. Let us look at Siegel's actual proof. The basic ingredient is Siegel's lemma :

Siegel's lemma 7.8. Let $\mathbb{K}$ be an algebraic number field of degree d over $\mathbb{Q}$. Let a_{ij} $(1 \leq i \leq n,\ 1 \leq j \leq m)$ be algebraic integers from $\mathbb{K}$ and

$$A \geq \max_{1 \leq j \leq m} \sum_{i=1}^{n} \overline{|a_{i,j}|} .$$

If $n > m$, then there exists a non-trivial solution $(x_1,\dots,x_n)$ (in the integers from $\mathbb{K}$) of the system

$$\sum_{i=1}^{n} a_{i,j}\, x_i = 0 \quad : \quad j = 1,\dots,m$$

and, moreover,

$$\max_{i=1,\ldots,n} |\overline{x_i}| \leq (C_1 A)^{(m/n-m)} .$$

So, when, $m \to n$, the sizes of the solutions increase essentially.

Here $C_1 = C_1(\mathbb{K}) > 0$, more precisely $C_1 = C_1(d) > 0$.

We apply Siegel's lemma to the system of functions from $\bar{\mathbb{Q}}[[x]]$

$$f(x) = (f_1(x),\ldots,f_m(x))$$

satisfying a linear differential equation over $\bar{\mathbb{Q}}[x]$and having exponent k.

We take an auxiliary function in the following form :

n is some parameter (a sufficiently large integer), and all other numbers are constants ;

$$F_n(z) = (n!)^k . P(z,\bar{f}(z)), \tag{7.1}$$

where

$$P(z,\bar{f}(z)) = \sum_{j=1}^{m} P_j(z)\ f_j(z) \tag{7.2}$$

and $P_j(z)$ are polynomials from $\mathbb{K}[z]$ of degrees $\leq n : j = 1,\ldots,m$.
We put $\mathbb{K} = Q(a_{j,\ell} : j = 1,\ldots,m,\ \ell \geq 0)$ for

$$f_j(z) = \sum_{\ell=0}^{\infty} a_{j,\ell}\ z^{\ell} \quad : j = 1,\ldots,m . \tag{7.3}$$

If we write

$$P_j(z) = \sum_{i=0}^{n} \frac{C_{i,j}}{(i!)^k} . z^i \quad : j = 1,\ldots,m \tag{7.4}$$

then we can represent the condition

$$\mathrm{ord}_{z=0}\ F_n(z) \geq (m-\delta).n \quad , \qquad \delta > 0 \tag{7.5}$$

as the system of very simple $(m-\delta)n$ conditions on $m(n+1)$ quantities $C_{i,j} : i \leq n,\ j = 1,\ldots,m$.

Now we apply Siegel's lemma directly.

Lemma 7.9. Let $\bar{f}(x) = (f_1(x),\ldots,f_m(x))$ be the system of C-functions, where C - the constant in the definition of the "C-function class" - is the same for all $f_i(x) : i = 1,\ldots,m$. Then there exists a non-trivial set of polynomials $P_1(x),\ldots,P_m(x)$ from

$\mathbb{K}[x]$ with algebraic integer coefficients $C_{i,j}$ of the form (7.4) such that for

$$F_n(z) = (n!)^k \sum_{j=1}^{m} P_j(z)\, f_j(z)$$

the condition (7.5) is satisfied :

$$\mathrm{ord}_{z=0}\, F_n(z) \geq (m-\delta)n \quad , \quad m > \delta > 0$$

and

$$\max_{i \leq n, j=1,\ldots,m} |\overline{C_{i,j}}| \leq C^{n.\frac{(m-\delta)^2}{\delta}} . R_1^{n\frac{(m-\delta)^2}{\delta}} \qquad ,(7.6)$$

here R_1 depends only on k.

If however $\bar{f}(z)$ belongs to the class $\mathfrak{H}[k;s]$ with $s > 0$; then instead of (7.6) we can prove only the following bound

$$\max_{i=0,\ldots,n,\ j=1,\ldots,m} |\overline{C_{i,j}}| \leq$$

$$\leq (n!)^{\frac{(m-\delta)^2}{\delta} s} . R_2^{n.\frac{(m-\delta)^2}{\delta}} \quad , R_2 = R_2(\bar{f}) > 0. \quad . \quad (7.7)$$

The bound (7.6) for $k > 0$ is enough to finish the proof independently, whether $\bar{f}$ is a (G,C)-function or not.

Really, the main contribution to sizes of the coefficients of $F_n(z)$ comes from $(n!)^k$;

$$\mathrm{size}\ (F_n) \sim (n!)^k \quad ; \qquad (7.8)$$

also for $k > 0$ we have such an analytic bound for the value of $F_n(z)$; following (7.1)-(7.6) :

$$|F_n(z)| \leq \{[(m-\delta)n]!\}^{-k} (n!)^k R_3(f)^n . |z|^{(m-\delta)n}$$

uniformly in z, or

$$|F_n(z)| \leq n^{-n(m-\delta-1)^k} . R_4^n . |z|^{(m-\delta)n} \quad . \qquad (7.9)$$

It is easy to show (Siegel - ... ...kind of arguments) that in order to get m linearly independent forms it is enough to differentiate $F_n(z)$ at $z = \alpha$, for $\alpha \neq 0$, at most $\delta n + O(1)$ times. This will not spoil both analytic and arithmetic bounds (7.8) - (7.9) and we obtain Siegel's type of result for $\delta \to 0$.

Let us present corollaries of this "non-effective" approach along the lines of Siegel's paper [27] :

General Assumption. Functions $f_1(x),\ldots,f_m(x)$ are algebraically independent over $\mathbb{C}[x]$.

Corollary 7.10. Let $f_1(x),\ldots,f_m(x)$ be the system of C-functions corresponding to the case $k > 0$. If $\alpha \neq 0$ is an algebraic number different from the singularities of the system (E) of linear differential equations satisfied by $\bar{f} = (f_1(x),\ldots,f_m(x))$, then $f_1(\alpha),\ldots,f_m(\alpha)$ are algebraically independent over Q.

This result is, in fact, equivalent to E-function results [27], [28].

Corollary 7.11. Let $f_1(x),\ldots,f_m(x)$ be the system of G-functions (i.e. the system of C-functions and $k=0$) being simultaneously the system of (G,C)-functions. Let $\varepsilon > 0$, $D \geq 0$ and $\alpha \neq 0$ be algebraic numbers such that

$$|\alpha| < \exp\left(-C(\bar{f},D,\varepsilon)\,\{\log H(\alpha)\}^{\frac{1}{2}+\varepsilon}\right)$$

for $d(\alpha) \leq D$ and certain constant $c(\bar{f},D,\varepsilon) > 0$ depending on $\bar{f} = (f_1,\ldots,f_m)$, D and ε. Then for any polynomial $P(x_1,\ldots,x_m) \in \mathbb{Z}[x_1,\ldots,x_m]$ of degree $\leq D$,

$$P(f_1(\alpha),\ldots,f_m(\alpha)) \neq 0$$

(the numbers $f_1(\alpha),\ldots,f_m(\alpha)$ are not connected by any algebraic relation of degree $\leq D$).

This type of result was essentially considered in [33] (cf. recent Bombieri results).

Also our methods are applicable to entire functions $\mathfrak{H}[k,s]$ with $k > 0$, $s > 0$. We'll present corrésponding results separately.

8. MEASURES OF IRRATIONALITY OF NUMBERS CONNECTED WITH GENERALIZED HYPERGEOMETRIC FUNCTIONS.

Here we apply effective construction of Padé approximation to generalized hypergeometric functions in order to get results on linear independence of values of hypergeometric functions at rational points. We use methods of §§ 6-7 together with asymptotics for Padé approximations from theorems 4.6 and 6.2.

Let us present one rough result, which gives explicit constants in the linear independence for the values of hypergeometric functions.

Theorem 8.1. Let $a_1,\dots,a_q$ and $b_1,\dots,b_p$, $q = p+1$ be rational numbers (and b_j be different from negative integers) and

$$a_i = \frac{\alpha_i}{\beta_i}, \quad b_j = \frac{\gamma_j}{\delta_j},$$

where $\alpha_i,\beta_i,\gamma_j,\delta_j$ are integers, $(\alpha_i,\beta_i) = 1, (\gamma_j,\delta_j) = 1$: $i = 1,\dots,q$, $j = 1,\dots,p$. Let $f(x)$ be a ${}_qF_p$-hypergeometric function :

$$f(x) = {}_qF_p\left(x\left|\begin{matrix} a_1,\dots,a_q \\ b_1,\dots,b_p \end{matrix}\right.\right)$$

and $f(x;\bar{n},\bar{m})$ be the functions, contiguous to $f(x)$:

$$f(x;\bar{n},\bar{m}) = {}_qF_p\left(x\left|\begin{matrix} a_1+n_1,\dots,a_q+n_q \\ b_1+m_1,\dots,b_p+m_p \end{matrix}\right.\right)$$

for integers n_i, m_i. Let $\ell \geq 0$ and $f_1(x),\dots,f_\ell(x)$ be ℓ functions among $f(x;\bar{n},\bar{m})$ such that $1, f_1(x),\dots,f_\ell(x)$ are linearly independent over $\mathbb{C}[x]$.

Let P,Q are integers, $Q > P \geq 1$. If

$$Q^{1/p} > P^{(1+(1/p))}.e^p. \prod_{i=1}^{q} \beta_i \prod_{j=1}^{p} \delta_j \prod_{i=1}^{q} \prod_{\gamma|\beta_j} \gamma^{(1/\gamma-1)}$$

$$\times \prod_{j=1}^{p} \prod_{\gamma|\delta_j} \gamma^{(1/\gamma-1)} \tag{8.1}$$

(γ denotes a prime number), then

$$1, f_1(\tfrac{P}{Q}),\dots,f_\ell(\tfrac{P}{Q})$$

are linearly independent over $\mathbb{Q}$.

E.g. in the general situation, when neither a_i, nor $a_i - b_i$ are integers, $\ell = p+1 = q$ and under the condition (8.1), the numbers

$$1, f(\tfrac{P}{Q}),\ f'(\tfrac{P}{Q}),\dots,f^{(p)}(\tfrac{P}{Q})$$

are linearly independent over $\mathbb{Q}$.

8.2. As one of the applications of our Padé approximation to periods elliptic curves we can mention :

Theorem 8.2. For any integer $q \geq 14$, the numbers

$$1, F(\tfrac{1}{2},\tfrac{1}{2};1;\tfrac{1}{q}) \quad \text{and} \quad F(\tfrac{1}{2},-\tfrac{1}{2};1;\tfrac{1}{q})$$

are linearly independent over $\mathbb{Q}$. Moreover for integers n_o, n_1, n_2

$$|n_o + n_1 F(\tfrac{1}{2},\tfrac{1}{2};1;\tfrac{1}{q}) + n_2 F(\tfrac{1}{2},-\tfrac{1}{2};1;\tfrac{1}{q})| > H^{-c(q)}$$

for $H \geq H_o(q)$. Here $H = \max(|n_o|,|n_1|,|n_2|)$ and $c(q) \to 2+0$ as $q \to \infty$.

For example, for an elliptic curve $y^2 = (1-x^2)(1-k^2x^2)$ and rational $k^2 = 4a/b$ where $b \geq 14a^2$, we have for the corresponding period ω and quasi-period η(for definitions see [2],[34]) :

$$|n_o\pi + n_1\omega + n_2\eta| > H^{-c_1(k^2)} \quad ;$$

$$H = \max(|n_o|,|n_1|,|n_2|) \geq H_1(k^2) \ ,$$

where $c_1(k^2) \to 2$ as $\log b/\log a \to \infty$.

We can use also the notations of the Legendre integrals. E.g.

$$|n_o\pi + n_1 K(k) + n_2 E(k)| > H^{-c_1(k^2)} \ ,$$

here $\quad c_1(k^2) = 2 + O(\frac{1}{k})$.

Of course we have similar results for the incomplete Legendre elliptic integrals [2] $E(\varphi, k)$, $F(\varphi,k)$ of the first and the second kind. These results are effective as they are related to the solutions of the linear differential equation of the first order with rational coefficients. For a solution of such an equation there exists an effective scheme of the construction of the Padé approximations due to Laguerre.

E.g.

Proposition 8.3. If k^2 is a rational number, $0 < k^2 < 1$ and $\sin\varphi = a/b$, where $b > c_2.a^2$ for $c_2 = c_2(k^2)$ then for integers n_o,n_1,n_2, $H = \max(|n_o|,|n_1|,|n_2|) \geq H_2(k^2,b)$,

$$|n_o + n_1 F(\varphi,k) + n_2 E(\varphi,k)| > H^{-c_3(k^2,\varphi)}$$

where $c_3(k^2,\varphi) \to 2+0$ as $\frac{\log b}{\log a} \to \infty$.

While for the linear independence of the periods we had previous results [34] produced by non-effective methods, there are *no* non-effective methods to treat *quadratic* relations among the periods.

The possibility to prove new results for "quadratic independence" is based on the existence of the relations between products of the generalized hypergeometric functions and solutions of other Fuchsian equations (sometimes again of hypergeometric type).

E.g. we have[29] Bailey (1935) identities, from which it follow

$${}_2F_1(\tfrac{1}{2}, \tfrac{1}{2}; 1; z)^2 = {}_3F_2\left(4z(1-z)\,\middle|\,\begin{matrix}1/2, & 1/2, & 1/2\\ 1, & 1\end{matrix}\right) .$$

or

$${}_2F_1(\tfrac{1}{2}, \tfrac{1}{2}; 1; z)^2 = (1-z)^{-1/2}\ {}_3F_2\left(\frac{z^2}{4(1-z)}\,\middle|\,\begin{matrix}1/2,1/2,1/2\\ 1, \quad 1\end{matrix}\right)$$

These identities can be used to get lower bound of the linear forms in

$$\omega^2,\ \omega\eta\ ,\ \eta^2,\ \pi^2$$

for small rational k^2, $0 < |k^2| < 1$.

We present only one example :

<u>Corollary 8.4</u>. If $k^2 = \frac{a}{b} \in Q$ and $b > 25a^2$, then for the periods ω and η corresponding to the modulus k^2 we have

$$|n_o\pi^2 + n_1\omega^2 + n_2\omega\eta + n_3\eta^2| > H^{-c_4(k^2)}$$

where $n_o,\dots,n_3$ are integers, $H = \max(|n_o|,\dots,|n_3|) \geq H_2(k^2)$ and

$$c_4(k^2) \to 3+0 \quad \text{as} \quad \log b/\log a \to \infty.$$

<u>8.3.</u> Padé approximation for polylogarithmic functions.

We consider the general polylogarithmic functions

$$L_k(x) = \sum_{n=1}^{\infty} \frac{x^n}{n^k} \ : \ k = 1,2,3,\dots \qquad .$$

See [5],[18] for their role in a solution of the Riemann monodromy problem. Different physical and mathematical problems [5], [11],[18] lead to investigations of linear independence of values of $L_k(\alpha)$ at algebraic $\alpha \neq 0$. We can notice only that $L_k(1) = \zeta(k)$ for $k \geq 2$. The problem of the arithmetic nature of values of $L_k(x)$ at rational (or algebraic) x, $0 < |x| < 1$ can be treated successfully using Padé approximations. In [11] we used some simple explicit construction and got the irrationality of $L_2(1/q)$ for a rational integer $q \geq 2^4.e^2$ together with more general but less effective results on $L_k(\alpha)$ for the small algebraic number α , $0 < |\alpha| < 1$. In [8],[9] the effective Padé approximants of type II were constructed for an arbitrary generalized hypergeometric function and functions contiguous to it.

For polylogarithms results of [8],[9] or our results of §§ 4,6 implies.

Corollary 8.5. There exists an absolute constant $C_1 > 0$ such that for rational integer q, if

$$q \geq c_1 \frac{k^k}{(k+1)^{k+1}} e^{k^2} ,$$

then

$$1, L_1(1/q) = -\log(1 - \frac{1}{q}), \ldots, L_k(\frac{1}{q})$$

are linear independence over $\mathbb{Q}$.

Also it follows from[8],[9] that $L_2(1/q)$ is an irrational number for any rational integer $q \geq 14$. The system of rational approximations to $L_k(x)$ designed for the purpose of evaluating certain Feynman integrals was presented in [5].

REFERENCES

[1] Baker Jr., G.A. : Essentials of Padé approximants, Academic Press, 1975.
[2] Bateman, H. and Erdélyi, A. : Higher transcendental functions, Mc Graw-Hill, 3V, 1953.
[3] de Bruin, M.G.: Generalized c-fractions and a multidimensional Padé table, Amsterdam, 1974.
[4] Chudnovsky, D.V. : Cargèse Lectures (see this volume) 1979.
[5] Chudnovsky, D.V. and Chudnovsky G.V.: Service de Physique Théorique, Preprint DPh.T79/115 (to be published in Lectures Notes in Mathematics, Springer).
[6] Chudnovsky, G.V. ; C.R. Acad. Sci. Paris, 288A, pp. A-439 - A-440.

[7] Chudnovsky G.V.: C.R. Acad. Sci. Paris, 288A, pp. A-965-A-967.
[8] Chudnovsky G.V.: C.R. Acad. Sci. Paris, 288A, pp. A-1001 - A-1004.
[9] Chudnovsky G.V.: J. Math. Pure Appl. (to appear) 1979.
[10] Chudnovsky G.V.: C.R. Acad. Sci. Paris, 288A, pp. A-607 - A-609.
[11] Chudnovsky G.V.: 1979 Preprint IHES/M/79/26 (to be published in Lecture Notes in Mathematics, Springer).
[12] Dwork B. : 1979 (to be published).
[13] Fuchs L.: Gesammelte Mathematische werke, Bd.1-3, Berlin 1900-1906.
[14] Hilbert D.: Grundzüge der Integralgleichungen, Leipzig-Berlin, 2-te Aufl., 1924.
[15] Ince I.L. : Ordinary Differential Equations, Dover, 1959.
[16] Jager H.: A Multidimensional Generalization of the Padé Table, Drukkerij Holland N.V., Amsterdam, 1964.
[17] Laguerre E. : J. de Math., 1, p. 135-165, 1885.
[18] Lappo-Danilevsky, J.A.: Mémoires sur la Théorie des Systèmes des Equations Différentielles Linéaires. Chelsea Publishing Company, 1953.
[19] Birkhoff G.D.: Transcation of the AMS, pp. 436-470, 1909.
[20] Mahler K.: Mathemat. Ann. 168, pp. 200-227, 1976.
[21] Mahler K.: Phil. Trans. Royal Soc. London, 245 A, pp.371-398, 1953.
[22] Padé H.: Ann. Ecole Norm. Sup. (3) 16, pp. 156-259, 1899.
[23] Padé H.: Ann. Ecole Norm. Sup. 24, pp. 341-400, 1907.
[24] Ramis J.P.: Asterisque 59-60, pp. 173-204, 1978.
[25] Riemann B., Oeuvres Mathématiques, Albert Blanchard, Paris pp. 353-363, 1968.
[26] Rossum H. van : A Theory of Orthogonal Polynomials Based on Padé Table, Utrecht, 1953.
[27] Siegel C.L., Abh. Preussichen Akad Wissen. Phys. Math. Cl Classe, N° 1.
[28] Siegel C.L.: Transcendental Numbers, Princeton University Press, 1949.
[29] Slater, L.J.: Generalized Hypergeometric Functions, Cambridge Univ. Press. 1966.
[30] Szëgo G.: Orthogonal Polynomials, Providence, 1939.
[31] Wall H.S.: The analytic Theory of Continued Fractions, Chealsea, 1948.
[32] Nuttall J. : Cargèse Lectures (see this volume) 1979.
[33] Galotchkin G. : Math. Skornic, 95, pp. 396-417, 1974.
[34] Chudnovsky G.V.: Proceedings ICM, Helsinki, pp. 169-177, 1978.

(*) We do not demand in general for the system (0.2) to be non-singular at $z = 0$.

(**) However if one of a_{i_0} is a negative integer, then $f(x)$ is a polynomial.

V

NONLINEAR PARTIAL DIFFERENTIAL EQUATIONS AND APPLICATIONS

NON LINEAR EVOLUTION EQUATIONS OF SCHRODINGER TYPE

Haïm BREZIS

Université PARIS VI

We shall report on a joint work with T. Gallouet (see [2] for details) concerning the equation

$$(1) \quad \begin{cases} i \dfrac{\partial u}{\partial t} - \Delta u + k|u|^2 u = 0 & \text{in } \Omega \times [\,0,\infty\,] \\ u(x,t) = 0 & \text{in } \Gamma \times [\,0,\infty\,] \\ u(x,0) = u_o(x) & \text{in } \Omega \end{cases}$$

where Ω is a domain in $\mathbb{R}^2$ with compact smooth boundary Γ (in fact, the compactness does not play an essential role ; Ω could be for example a half space). Here k denotes a real given constant and $u(x,t)$ is a complex valued function defined on $\Omega \times [\,0, +\infty\,]$. This kind of equation occurs in nonlinear optics (see e.g. the lecture of Prof. Bullough at this meeting) especially in the study of the propagation of laser beams. The question of existence and uniqueness has been extensively studied when $\Omega = \mathbb{R}^2$ (see Ginibre-Velo [5], Lin-Strauss [7], Baillon-Cazenave-Figueira [1]). The earlier proofs do not apply when $\Omega \neq \mathbb{R}^2$. Our purpose is to present a proof which is valid when $\Omega = \mathbb{R}^2$ (and when $\Omega = \mathbb{R}^2$ as well).
Our main result is the following :

C. Bardos and D. Bessis (eds.), Bifurcation Phenomena in Mathematical Physics and Related Topics, 513–520.

THEOREM 1.

Assume $u_o \in H^2(\Omega) \cap H_o^1(\Omega)$ [1]

Assume either $k \geqslant 0$

$$\text{or} \quad k < 0 \quad \text{and} \quad |k| \int_\Omega |u_o(x)|^2 dx < 4 .$$

Then there exists a unique global solution of (1) which is " almost " classical ; more precisely

$$u \in C([0,\infty) ; H^2(\Omega)) \cap C^1([0,\infty] ; L^2(\Omega)) .$$

REMARKS.

1) When $k < 0$ and $|k| \int_\Omega |u_o(x)|^2 dx \geqslant 4$ it may occur for <u>some</u> initial datas and <u>some</u> domains Ω that the solution u of (1) - which exists for small time - <u>blows up in finite time</u> (this is an easy modification of the argument in [6])

2) When $\Omega = \mathbb{R}^2$, it is known (see [4]) that $\| u(;t)\|_{L^\infty} = 0(\frac{\log t}{t})$ as $t \to \infty$. Such a decay can not hold in bounded domains since $\| u(.,t)\|_{L^2} = \| u_o \|_{L^2}$. A natural question is whether $\| u(.,t)\|_{L^\infty}$ remains bounded as $t \to \infty$. (the proof belows shows only that $\| u(.,t)\|_{L^\infty}$ is bounded by $Ce^{\alpha t}$ for some constants C and α).

The proof of Theorem 1 relies on an inequality of Sobolev type which we describe in details since it could be useful for other purposes

LEMMA 2.

We have $\| u \|_{L^\infty} \leqslant C [1+(\log(1+ \| u \|_{H^2}))^{1/2}]$

$$\forall u \in H^2 \text{ with } \| u \|_{H^1} \leqslant 1$$

where C depends only on Ω .

---

(1) That is, u_o together with all its derivatives of order equal or less than two belong to $L^2(\Omega)$; in addition $u_o(x)$ vanishes on Γ .

REMARKS.

1) It is well known that a function $u \in H^1(\Omega)$ is not necessarily bounded. Lemma 2 deals with the critical case of the Sobolev embedding theorem. It is clear that $\|u\|_{L^\infty} \leq C\|u\|_{H^2}$ $\forall\, u \in H^2$. However, such an estimate is far from optimal if we have the additional information that $\|u\|_{H^1} \leq 1$. Lemma 2 asserts that if we keep $\|u\|_{H^1} \leq 1$, then $\|u\|_{L^\infty}$ could blow up, but at most like $\sqrt{\log \|u\|_{H^2}}$ as $\|u\|_{H^2} \to \infty$.

2) Lemma 2 has been extended in [3] to $\Omega \subset R^N$.
In full generality we have

$$\|u\|_{L^\infty} \leq C\,[\,1+(\log(1+\|u\|_{W^{\ell,q}}))^{1-\frac{1}{p}}\,]$$

$$\forall\, u \in W^{\ell,q} \quad \text{with} \quad \|u\|_{W^{k,p}} \leq 1$$

provided $kp = N$ and $\ell q > N$.

3) Lemma 2 seems "somewhat" related to Trudinger's inequality (see [11]) which asserts that

$$\int_\Omega e^{k|u|^2}\,dx \leq C \qquad \forall\, u \in H^1 \quad \text{with} \quad \|u\|_{H^1} \leq 1$$

for some constants $k > 0$ and C (assuming $\Omega \subset R^2$ is bounded). However the exact relationship is not clear. Trudinger's inequality implies (see e.g. [3]) that for $\Omega \subset R^2$ and $u \in H^2$

$$|u(x)-u(y)| \leq C\,|x-y|\sqrt{1+|\log|x-y||} \quad \forall\, x,y \in \Omega\,.$$

It has been pointed out by C. Bardos that logarithmic estimates occur frequently in limity cases. For example $\Delta u \in L^\infty$ implies $|\nabla u(x)-\nabla u(y)| \leq C|x-y|(1+|\log|x-y||$; however such an inequality can not be improved and does not seem to be connected with the previous estimates.

PROOF OF LEMMA 2.

Using standard extension techniques we can always assume that $\Omega = \mathbb{R}^2$. We denote by $\hat{u}$ the Fourier transform of u so that

$$\|u\|_{L^\infty} \leq \|\hat{u}\|_{L^1}\,.$$

On the other hand, for fixed $R > 0$ (to be chosen later) we have

$$\| \hat{u} \|_{L^1} = \int_{|\xi|<R} (1+|\xi|)|\hat{u}(\xi)| \frac{1}{1+|\xi|} d\xi + \int_{|\xi|>R} (1+|\xi|^2)|\hat{u}(\xi)| \frac{1}{1+|\xi|^2} d\xi$$

$$\leqslant \| u \|_{H^1} [\int_{|\xi|<R} \frac{d\xi}{(1+|\xi|)^2}]^{1/2} + \| u \|_{H^2} [\int_{|\xi|>R} \frac{d\xi}{(1+|\xi|^2)^2}]^{1/2}$$

$$\leqslant [\log(1+R)]^{1/2} + \| u \|_{H^2} \frac{C}{1+R} .$$

We obtain the conclusion of Lemma 2 by choosing $R = \| u \|_{H^2}$.

The proof of the existence of a local solution of (1) relies on the following Lemma essentially due to Segal [10].

LEMMA 3.

Assume X is a Banach space and $A : D(A) \subset X \to X$ is a linear densely defined operator which generates a contraction semigroup $S(t)$. Assume $F : D(A) \to D(A)$ is a non linear mapping which is Lipschitz on every bounded subset of $D(A)$ (for the graph norm). Then for every $u_o \in D(A)$ there exists a unique solution $u(t)$ defined on $[0,T_{max})$ such that

$$u \in C([0,T_{max}) ; D(A)) \cap C^1([0,T_{max}) ; X) \quad \text{and}$$

$$(2) \qquad \begin{cases} \dfrac{du}{dt} + Au = Fu & \text{on} \quad [0,T_{max}) \\ u(0) = u_o \end{cases}$$

In addition we have the following alternative :
<u>either</u> $T_{max} = \infty$

<u>or</u> $T_{max} < \infty$ and $\lim_{t \uparrow T_{max}} (\| u(t)\| + \| Au(t)\|) = \infty$

PROOF OF LEMMA 3

It follows from Segal's Theorem (see [10])-applied in the space $D(A)$ with the graph norm- that there exists a unique function $u \in C([0,T_{max}) ; D(A))$ satisfying

$$(3) \qquad u(t) = S(t)u_o + \int_0^t S(t-s)F(u(s))ds \quad \text{for} \quad t < T_{max}$$

and <u>either</u> $T_{max} = \infty$ <u>or</u> $T_{max} < \infty$ and $\lim_{t \uparrow T_{max}} \| u(t)\| + \| Au(t)\| = \infty$

Finally we use for example Corollary 2.4 in [9] to conclude that $u \in C^1([0,T_{max}) ; X)$ and that u satisfies (2).

PROOF OF THEOREM 1.

We apply Lemma 3 with $H = L^2(\Omega ; \mathbb{C})$
$Au = i\Delta u$, $D(A) = H^2 \cap \overset{o}{H}{}^1$ and $Fu = ik|u|^2 u$. It is easy to verify that F is lipschitz on every bounded subset of $D(A)$. Hence we obtain a solution of (1) defined on $[0,T_{max})$ with the alternative :

<u>either</u> $T_{max} = \infty$ <u>or</u> $T_{max} < \infty$ and $\lim\limits_{t \uparrow T_{max}} \| u(\cdot,t)\|_{H^2} = \infty$.

Next, we shall establish a bound for $\| u(\cdot,t)\|_{H^2}$ of the form $e^{\alpha e^{\beta t}}$ for some constants α,β . This will imply that $T_{max} = \infty$, thereby proving Theorem 1.
We shall divide the proof into 3 steps

<u>Step 1.</u> $\| u(\cdot,t)\|_{L^2} = \| u_o \|_{L^2} \qquad \forall\ t < T_{max}$.

Indeed, multiplying (1) by $\bar{u}$ and considering the imaginary part leads to $\frac{1}{2} \frac{d}{dt} \int_\Omega |u(x,t)|^2 dx = 0$.

<u>Step 2.</u> $\| u(\cdot,t)\|_{H^1} \leqslant C \qquad \forall\ t < T_{max}$,

where C depends only on u_o .

Indeed multiplying (1) by $\frac{\partial u}{\partial t}$ and considering the real part leads to
$$\frac{1}{2}\frac{d}{dt} \int_\Omega |Du(x,t)|^2 dx + \frac{k}{4}\frac{d}{dt} \int |u(x,t)|^4 dx = 0 .$$
Thus
$$\frac{1}{2} \int_\Omega |Du(x,t)|^2 dx + \frac{k}{4} \int_\Omega |u(x,t)|^4 dx \equiv E_o ,$$
where
$$E_o = \frac{1}{2} \int_\Omega |Du_o(x)|^2 dx + \frac{k}{4} \int |u_o(x)|^4 dx .$$
If $k \geqslant 0$, the conclusion is obvious. So we assume $k < 0$.

We have
$$\int_\Omega |Du(x,t)|^2 dx \leqslant \frac{|k|}{2} \int_\Omega |u(x,t)|^4 dx + 2E_o .$$

On the other hand an inequality of Gagliardo and Nirenberg (see [8]) asserts that
$$\int_\Omega |Du(x,t)|^2 dx \leqslant \frac{|k|}{2} \int_\Omega |u(x,t)|^4 dx + 2E_o .$$
On the other hand an inequality of Gagliardo and Nirenberg

(see [8]) asserts that

$$\int |u|^4 dx \leqslant \frac{1}{2}\left(\int_\Omega |u|^2 dx\right)\left(\int_\Omega |Du|^2 dx\right)$$

and the conclusion follows easily provided $|k|\int |u_o|^2 < 4$.

Next we deduce from (3) that

$$Au(t) = S(t)Au_o + \int_0^t S(t-s)AFu(s)ds$$

and therefore

(4) $\| u(.,t)\|_{H^2} \leqslant C \| u_o\|_{H^2} + C \int_0 \| Fu(.,s)\|_{H^2} ds$

Finally we shall use the following

LEMMA 4.

$$\| Fu \|_{H^2} \leqslant C \| u \|^2_{L^\infty} \| u \|_{H^2} \qquad \forall\ u \in H^2 \cap H^1_o$$

PROOF OF LEMMA 4.

It is clear that

$$\| Fu \|_{H^2} \leqslant C \| u \|^2_{L^\infty} \| u \|_{H^2} + C \| u \|_{L^\infty} \| Du \|^2$$

But

$$\| Du \|^4_{L^4} = \int (Du)^3 Du\ dx = -3 \int u(Du)^2 (D^2 u) dx$$
$$\leqslant 3 \| u \|_{L^\infty} \| Du \|^2_{L^4} \| D^2 u \|_{L^2}$$

and so

$$\| Du \|_{L^4} \leqslant 3 \| u \|_{L^\infty} \| D^2 u \|_{L^2} \leqslant 3 \| u \|_{L^\infty} \| u \|_{H^2} .$$

PROOF OF THEOREM 1. CONCLUDED

We deduce from Lemma 2, Lemma 4 and Step 2 that

(5) $\| Fu(.,s)\|_{H^2} \leqslant C \| u(.,s)\|_{H^2} (1+\log(1+\| u(.,s)\|_{H^2}))$.

Set $\varphi(t) = \| u(.,t)\|_{H^2}$. It follows from (4) and (5) that

(6) $\varphi(t) \leqslant C + C \int_0^t \varphi(s)(1+\log(1+\varphi(s)))ds$

Such an inequality - which is almost of Gronwall type - can be rewritten as

$$G'(t) \leqslant CG(t)(1 + \log(1+G(t))$$

where $G(t)$ denotes the right hand side in (6).
We obtain

$$\frac{d}{dt} \log [1+\log(1+G(t))] \leq C$$

and then $G(t) \leq e^{\alpha e^{\beta t}}$ for some constants α and β.
The same estimate holds for $\|u(.,t)\|_{H^2}$.

REFERENCES

[1] J.B. BAILLON, T CAZENAVE, M. FIGUEIRA,
Equation de Schrödinger nonlinéaire, C.R. Acad. Sc. 284 (1977) p. 869-872.

[2] H. BREZIS, T. GALLOUET,
Non linear Schrödinger evolution equations, J. Nonlinear Analysis (to appear).

[3] H. BREZIS, S. WAINGER,
A note on limiting cases of Sobolev embeddings and convolution inequalities (to appear).

[4] T.CAZENAVE,
Equations de Schrödinger non linéaires (to appear).

[5] J.GINIBRE et G. VELO,
On a class of nonlinear Schrödinger equations (to appear)

[6] R.T. GLASSEY,
On the blowing up of solutions to the Cauchy problem for the nonlinear Schrödinger equation, J. Math. Phys. 18 (1977) p. 1794-1799.

[7] J. LIN and W. STRAUSS,
Decay and scattering of solutions of a nonlinear Schrödinger equation, J. Funct. Anal. 30 (1978) p.245-263

[8] L. NIRENBERG,
On elliptic partial differential equations, Ann. Sc. Norm. Sup. Pisa 13 (1959), p.115-162.

[9] A. PAZY
Semigroups of linear operators and applications to partial differential equations, Lecture Notes, Univ. of Maryland (1974).

[10] I. SEGAL,
Nonlinear semi groups, Ann. of Math. 78 (1963)p.339-364.

[11] N. TRUDINGER
On imbeddings into Orlicz spaces and some applications, J. Math. Mech. 17 (1967) p.473-483.

ASYMPTOTIC BEHAVIOR OF SOLUTIONS OF HYPERBOLIC BALANCE LAWS

C. M. Dafermos

Lefschetz Center for Dynamical Systems
Division of Applied Mathematics
Brown University
Providence, Rhode Island 02912 USA

We survey the asymptotic behavior of solutions to hyperbolic conservation laws and balance laws, as well as systems of such laws with the property that each characteristic field is either genuinely nonlinear or linearly degenerate.

1. INTRODUCTION

We discuss the asymptotic behavior, as $t \to \infty$, of solutions to the problem for systems of strictly hyperbolic balance laws

$$u_t + f(u)_x + g(u,x,t) = 0 \tag{1.1}$$

in one space variable. The vector field $u(x,t)$ takes values in R^n and the functions $f\colon R^n \to R^n$ and $g\colon R^n \times R \times R^+ \to R^n$ are assumed smooth. Strict hyperbolicity means that for each $u \in R^n$ the $n \times n$ matrix $\nabla f(u)$ has n real distinct eigenvalues (characteristic speeds) $\lambda_1(u),\dots,\lambda_n(u)$.

When the supply term g vanishes, (1.1) reduces to the system of hyperbolic conservation laws

$$u_t + f(u)_x = 0. \tag{1.2}$$

Solutions of (1.2) vanishing at $x = \pm\infty$ conserve the quantity

$$\int_{-\infty}^{\infty} u(x,t)dx.$$

C. Bardos and D. Bessis (eds.), Bifurcation Phenomena in Mathematical Physics and Related Topics, 521–533.

Even so, L^p norms of $u(x,t)$, $p > 1$, may decay, as $t \to \infty$, provided that the solution is "spreading out". When $f(u)$ is linear, characteristic speeds are constant so that each mode travels with its fixed speed and does not decay. When $f(u)$ is nonlinear, the dependence of characteristic speeds upon u causes the convergence of characteristics so that, even if the initial data $u(x,0)$ are very smooth, smooth solutions generally break down in a finite time and shock waves develop. Shocks and simple waves interact and cancel each other out thus inducing decay. Another indication of decay is provided by "entropy inequalities" [1]

$$\eta(u)_t + q(u)_x \leq 0, \tag{1.3}$$

satisfied by admissible weak solutions of certain systems (1.2), as abstract forms of the second law of thermodynamics. The entropy decay mechanism is particularly effective in the presence of strong shocks.

Since in the linear case there is no dissipation, one should expect that solutions of (1.2) will decay faster when $f(u)$ is "very nonlinear". It turns out that the relevant condition on $f(u)$ is <u>genuine nonlinearity</u> [2],

$$\nabla\lambda_k(u)\cdot r_k(u) \neq 0, \qquad u \in R^n, \tag{1.4}$$

where $r_k(u)$ is the right eigenvector of $\nabla f(u)$ associated with $\lambda_k(u)$. Condition (1.4) is not always satisfied by all characteristic fields in systems of conservation laws arising in continuum physics. As a matter of fact, there are classical examples (e.g. the system of conservation laws of gas dynamics) where certain characteristic fields are <u>linearly degenerate</u>:

$$\nabla\lambda_j(u)\cdot r_j(u) \equiv 0, \qquad u \in R^n. \tag{1.5}$$

One should not expect decay of components of solutions corresponding to such fields.

Returning to general balance laws (1.1) we remark that whenever the supply term g is dissipative it collaborates with the dissipative mechanism of (1.2) to speed up decay. On the other hand, when g is not dissipative, there is competition with the dissipative mechanism of (1.2) the outcome of which depends on relative strength.

In the case of a single genuinely nonlinear conservation law very precise information is available which is described in Section 2. In Sections 3 and 4 we discuss asymptotic behavior for the single genuinely nonlinear balance law and the general single conservation law. Known results for systems are surveyed in Sections 5 and 6.

2. SINGLE CONSERVATION LAW. THE CONVEX CASE

In this section we discuss the asymptotic behavior of solutions to the single conservation law

$$u_t + f(u)_x = 0, \quad f(u) \text{ strictly convex.} \tag{2.1}$$

In the genuinely nonlinear case, $f''(u) > 0$, Lax [2] discovered an explicit solution for the initial value problem of (2.1) which yields precise information on the asymptotic behavior of solutions. A different approach, which also applies when $f''(u)$ may vanish at isolated points, employs the concept of generalized characteristics [3].

The following proposition [3] exhibits the influence on decay of (a) the "strength of convexity" of $f(u)$ and (b) the deployment of initial data.

<u>2.1. Theorem.</u> Assume that

$$c|u|^p \le f''(0) \le C|u|^p, \qquad p \ge 0, \quad 0 < c < C, \tag{2.2}$$

for u in some neighborhood of 0. Let $u(x,t)$ be an admissible solution of (2.1) on $(-\infty,\infty) \times [0,\infty)$ with

$$\int_x^{x+L} u(y,0)dy = O(L^s), \quad \text{as} \quad L \to \infty, \tag{2.3}$$

uniformly in $x \in (-\infty,\infty)$, for some $s \in [0,1)$. Then

$$u(x,t) = O(t^{\frac{s-1}{p(1-s)+2-s}}), \quad \text{as} \quad t \to \infty, \tag{2.4}$$

uniformly in $x \in (-\infty,\infty)$.

<u>2.2. Corollary.</u> Assume that $f(u)$ satisfies (2.2) in some neighborhood of 0 and let $u(x,t)$ be an admissible solution of (2.1) on $(-\infty,\infty) \times [0,\infty)$ with $u(x,0) \in L^q(-\infty,\infty)$, $1 \le q < \infty$. Then

$$u(x,t) = O(t^{-\frac{1}{p+q+1}}), \quad \text{as} \quad t \to \infty, \tag{2.5}$$

uniformly in $x \in (-\infty,\infty)$.

The variation of a certain function of the solution also decays, namely,

2.3. Theorem. Let $u(x,t)$ be an admissible solution of (2.1) on $(-\infty,\infty) \times [0,\infty)$ with initial data $u(x,0) \in L^1(-\infty,\infty)$. Then, for any $t \in (0,\infty)$,

$$\operatorname{Var}_x\{f(u(x,t)) - u(x,t)f'(u(x,t))\} \le \frac{2}{t}\int_{-\infty}^{\infty} |u(x,0)|dx. \tag{2.6}$$

In connection to the above theorem, see the discussion following Theorem 4.3, in Section 4 below.

When the initial data are periodic, one has information on the asymptotic shape of the solution:

2.4. Theorem. Let $u(x,t)$ be an admissible solution of (2.1) on $(-\infty,\infty) \times [0,\infty)$ with initial data $u(x,0)$ periodic of mean $\bar{u}$. Assume $f''(\bar{u}) \neq 0$. Then, as $t \to \infty$, $u(x,t)$ is asymptotic to order $o(t^{-1})$ to a periodic sawtooth function. The number of "teeth" per period equals the number of points in a period interval in which the function

$$\int_0^x [u(y,0) - \bar{u}]dx$$

attains its minimum.

Interesting asymptotic shapes also emerge when initial data have compact support. In that case solutions behave asymptotically as N-waves.

2.5. Theorem. Assume $f''(0) > 0$. Let $u(x,t)$ be an admissible solution of (2.1) on $(-\infty,\infty) \times [0,\infty)$ with initial data of compact support. Then

$$u(x,t) = \begin{cases} 0 & x \le \eta_-(t) \\ \frac{1}{f''(0)}\left[\frac{x}{t} - f'(0)\right] + o(\frac{1}{t}) & \eta_-(t) < x \le \eta_+(t) \\ 0 & \eta_+(t) < x \end{cases} \tag{2.7}$$

where

$$\eta_{\mp}(t) = tf'(0) \mp [2I_{\mp}tf''(0)]^{1/2} + o(1) \tag{2.8}$$

with I_-, I_+ constants (invariants of the solution) depending upon the initial data.

When $f''(0)$ vanishes, the N-wave is deformed, as the following example indicates.

2.6. Theorem. Let $u(x,t)$ be an admissible solution on $(-\infty,\infty) \times [0,\infty)$ of the equation

$$u_t + (u^{2k})_x = 0, \qquad k = 1,2,\ldots, \tag{2.9}$$

with initial data of compact support. Then

$$u(x,t) = \begin{cases} 0 & x \leq \eta_-(t) \\ \left(\frac{1}{2k}\frac{x}{t}\right)^{\frac{1}{2k-1}} + O(t^{-\frac{1}{k}}) & \eta_-(t) < x \leq \eta_+(t) \\ 0 & \eta_+(t) < x \end{cases} \tag{2.10}$$

where

$$\eta_{\mp}(t) = \mp 2k\left(\frac{1}{2k-1} I_{\mp}\right)^{\frac{2k-1}{2k}} t^{\frac{1}{2k}} + O(1) \tag{2.11}$$

with I_-, I_+ the same invariants appearing in (2.8).

There are also results [3,4] on the asymptotic behavior of solutions of (2.1) when the initial data approach different constants u_- and u_+ as $x \to -\infty$ and $x \to \infty$. It turns out that the solution approaches the wave fan that solves the Riemann Problem [2] corresponding to (u_-,u_+).

3. SINGLE BALANCE LAW. THE CONVEX CASE

We now consider the single balance law

$$u_t + f(u)_x + g(u,x,t) = 0, \quad f(u) \text{ strictly convex.} \tag{3.1}$$

Although no analog of Lax's explicit solution of (2.1) is known for (3.1), the method of generalized characteristics [5] applies and yields precise information.

As a first example we consider the case where g is dissipative and we generalize Corollary 2.2.

3.1. Theorem. Assume that $f(u)$ satisfies (2.2) in some neighborhood of 0 and that

$$ug(u,x,t) \geq 0, \quad u \in (-\infty,\infty), \; x \in (-\infty,\infty), \; t \in [0,\infty). \tag{3.2}$$

Let $u(x,t)$ be an admissible solution of (3.1) on $(-\infty,\infty) \times [0,\infty)$ with $u(x,0) \in L^q(-\infty,\infty)$, $1 \leq q < \infty$. Then

$$u(x,t) = O(t^{-\frac{1}{p+q+1}}), \quad \text{as} \quad t \to \infty, \tag{3.3}$$

uniformly in $x \in (-\infty,\infty)$.

In the next example g is not necessarily dissipative but it decays at infinity so that the dissipative mechanism of the conservation law prevails.

3.2. Theorem. Assume that

$$|g(u,x,t)| \leq a(x)b(t), \; u \in (-\infty,\infty), \; x \in (-\infty,\infty), \; t \in [0,\infty), \tag{3.4}$$

where $a(x)$ is a bounded function such that $a(x) \to 0$ as $|x| \to \infty$ while $b(t) \in L^1(0,\infty)$. Let $u(x,t)$ be an admissible solution of (3.1) on $(-\infty,\infty) \in [0,\infty)$ with $u(x,0) \in L^1_{loc}(-\infty,\infty)$ such that $u(x,0) \to 0$ as $|x| \to \infty$. Then

$$u(x,t) = o(1), \quad \text{as} \quad t \to \infty, \tag{3.5}$$

uniformly in $x \in (-\infty,\infty)$.

It is also possible to establish decay of the variation of solutions. Here is a typical result for the periodic case.

3.3. Theorem. Consider the balance law

$$u_t + f(u)_x + g(u) = 0 \tag{3.6}$$

with $f''(u) \geq a > 0$, $g(0) = 0$, and $g'(u) \geq 0$. Let $u(x,t)$ be an admissible solution of (3.6) on $(-\infty,\infty) \times [0,\infty)$ with initial data periodic of period T. Then

$$\operatorname*{Var}_{[0,T]} u(x,t) \leq \frac{2T}{at}, \quad t \in (0,\infty). \tag{3.7}$$

When the initial data have compact support, the solution attains an asymptotic profile analogous to the N-wave of the conservation law (Theorem 2.5). We present the result in the context of a simple concrete example.

3.4. Theorem. Let $u(x,t)$ be an admissible solution on $(-\infty,\infty) \times [0,\infty)$ of the balance law

$$u_t + \left(\frac{u^2}{2}\right)_x + u^3 = 0 \tag{3.8}$$

with initial data of compact support. Then

$$u(x,t) = \begin{cases} 0 & x \leq \eta_-(t) \\ \dfrac{2x}{x^2+2t} + O\left(\dfrac{1}{t}\right) & \eta_-(t) < x \leq \eta_+(t) \\ 0 & \eta_+(t) < x. \end{cases} \tag{3.9}$$

The fronts $\eta_-(t), \eta_+(t)$ of the wave are determined asymptotically as solutions of an equation which is too complicated to be illuminating.

4. SINGLE CONSERVATION LAW. THE GENERAL CASE

Next we consider the single conservation law

$$u_t + f(u)_x = 0 \tag{4.1}$$

without any convexity restrictions on $f(u)$. No explicit representation of solutions is known. The method of generalized characteristics works but not as effectively as in the convex case.

In general, decay is to be expected when $f''(u)$ vanishes only at isolated points. The rate of decay will depend on the "flattness" of $f(u)$ at points of inflexion. The following result, established [6,7] by methods of topological dynamics, starts from very weak assumptions on $f(u)$ but, in return, yields no information whatsoever on the rate of decay.

4.1. Theorem. Assume that the set of points on which $f''(u)$ vanishes has no (finite) accumulation point on the real axis. Let $u(x,t)$ be an admissible solution of (4.1) on $(-\infty,\infty) \times [0,\infty)$ with initial data that are L^1-almost periodic on $(-\infty,\infty)$ of mean 0. Then $u(x,t)$ decays to 0, as $t \to \infty$, in $L^1_{loc}(-\infty,\infty)$.

In the test case where $f(u)$ has only one inflexion point, one may derive the asymptotic behavior of solutions from the convex case (Corollary 2.2) by employing the ordering property of admissible solutions of (4.1), i.e., that $u(x,0) \le \bar{u}(x,0)$, $x \in (-\infty,\infty)$, implies $u(x,t) \le \bar{u}(x,t)$, $x \in (-\infty,\infty)$, $t \in [0,\infty)$.

4.2. Theorem. Assume that $f''(u) > 0$ for $u > 0$, $f''(u) < 0$ for $u < 0$ and

$$c|u|^p \le |f''(u)| \le C|u|^p, \quad p > 0,\ 0 < c < C, \tag{4.2}$$

for u in some neighborhood of 0. Let $u(x,t)$ be an admissible solution of (4.1) on $(-\infty,\infty) \times [0,\infty)$ with $u(x,0) \in L^q(-\infty,\infty)$, $1 \le q < \infty$. Then

$$u(x,t) = O(t^{-\frac{1}{p+q+1}}), \quad \text{as} \quad t \to \infty, \tag{4.3}$$

uniformly in $x \in (-\infty,\infty)$.

Using a clever scaling argument, Benilan and Crandall [8] obtain a decay estimate for orbits of Lipschitz continuous semigroups generated by homogeneous generators. Since $-f(u)_x$ generates a contraction semigroup on $L^1(-\infty,\infty)$, the above result applies to (4.1), when $f(u)$ is homogeneous, and gives

4.3. Theorem. Assume that for some $\alpha \ge 2$

$$f(\lambda u) = \lambda^\alpha f(u), \quad u \in (-\infty,\infty), \quad \lambda \in [0,\infty). \tag{4.4}$$

Let $u(x,t)$ be an admissible solution of (4.1) on $(-\infty,\infty) \times [0,\infty)$ with $u(x,0) \in L^1(-\infty,\infty)$. Then

$$\operatorname*{Var}_x f(u(x,t)) \le \frac{2}{(\alpha-1)t}\int_{-\infty}^{\infty} |u(x,0)|dx, \quad t \in (0,\infty). \tag{4.5}$$

In particular, (4.5) implies $|u(x,t)|^\alpha = O(t^{-1})$ which is in agreement with (2.5), (4.3). Note that (4.5) also yields information on the decay of $u_t(x,t)$. In the convex case, $f(u) = |u|^\alpha$, (4.5) reduces to (2.6). It is not known whether (2.6), or any analog of it, holds for general conservation laws.

The decay of solutions with periodic initial data was invesitgated by Greenberg and Tong [9] who obtain the following

4.4. Theorem. Assume that $f(u)$ is as in Theorem 4.2 and let $u(x,t)$ be an admissible solution of (4.1) on $(-\infty,\infty) \times [0,\infty)$ with initial data periodic of mean 0. Then there is a sequence

$\{t_n\}$, with $t_n \to \infty$, such that

$$u(x,t_n) = O(t_n^{-\frac{1}{p+1}}), \quad n \to \infty, \tag{4.6}$$

uniformly in $x \in (-\infty,\infty)$.

The asymptotic shape of periodic solutions, under the assumptions of Theorem 4.4, has also been studied [9,10].

Our understanding of the asymptotic behavior of solutions to the general conservation law (4.1) is still imperfect. Additional research will be required for the completion of the program.

5. SYSTEMS OF CONSERVATION LAWS

All available information on the asymptotic behavior of solutions to systems of conservation laws has been obtained through explicit studies of wave interactions, within the framework of the construction scheme of Glimm [11], and is thus restricted to solutions of small oscillation or variation.

We discuss first the case where the system is genuinely nonlinear, i.e., (1.4) is satisfied for $k = 1,\ldots,n$. The crucial observation is that weak waves propagate and interact as waves of a single conservation law, modulo an "error" which is at most of second order in the strength of interacting waves. The general strategy is to show that the cumulative effect of these errors is insignificant and thus decay of waves in systems is governed by precisely the same laws that rule decay of waves in a single conservation law. Systems of two equations are special in that, in Riemann invariants coordinates, interaction errors are actually of third order and this makes the study of decay easier. The earliest result is due to Glimm and Lax [12]:

5.1. Theorem. When the initial data $u(x,0)$ have small oscillation, there is a solution $u(x,t)$ on $(-\infty,\infty) \times [0,\infty)$ to the strictly hyperbolic, genuninely nonlinear system (1.2) of two equations. Furthermore, if $u(x,0)$ is periodic of mean 0,

$$u(x,t) = O(\tfrac{1}{t}), \quad \text{as} \quad t \to \infty, \tag{5.1}$$

while if $u(x,0)$ has compact support,

$$\operatorname{Var}_x u(x,t) = O(t^{-\frac{1}{2}}), \quad \text{as} \quad t \to \infty. \tag{5.2}$$

Starting from the above result, DiPerna [13] was able to show that in systems of two equations the two characteristic fields decouple at a rate $O(t^{-3/2})$ and each one approaches, at a rate $O(t^{-1/6})$, an individual N-wave of the type exhibited in Theorem 2.5. After a careful analysis, Liu [14] observed that in a genuninely nonlinear system of n equations interactions between waves of different families, which produce errors of second order, are rapidly completed, since these fields propagate with distinct characteristic speeds. On the other hand, interactions between waves of the same family, which persist longer, produce errors of the third order. He thus establishes the following

5.2. Theorem. When the initial data $u(x,0)$ have compact support and small total variation, there is a solution $u(x,t)$ on $(-\infty,\infty) \times [0,\infty)$ to the strictly hyperbolic, genuinely nonlinear system (1.2) of n equations and

$$\operatorname*{Var}_{x} u(x,t) = O(t^{-\frac{1}{2}}), \quad \text{as} \quad t \to \infty. \tag{5.3}$$

Furthermore, as $t \to \infty$, $u(x,t)$ approaches in L^1 at a rate $O(t^{-1/6})$, a system of n N-waves propagating with speeds $\lambda_1(0),\ldots,\lambda_n(0)$.

It is not known whether the rate $O(t^{-1/6})$ in the above theorem is optimal.

Next, we consider systems with the property that each characteristic field is either genuninely nonlinear or linearly degenerate. DiPerna [15] studied the asymptotic behavior of solutions to systems in the above class that are endowed with an entropy. Subsequently, Liu [16] established more precise results based exclusively upon considerations on wave interactions.

5.3. Theorem. Consider a strictly hyperbolic system of n equations each characteristic field of which is either genuinely nonlinear or linearly degenerate. When the initial data have compact support and small total variation, a global solution to the Cauchy problem exists and, as $t \to \infty$, each genuinely nonlinear characteristic field approaches in L^1, at a rate $O(t^{-1/6})$, an N-wave, while every linearly degenerate characteristic field approaches in L^1, at a rate $O(t^{-1/2})$, a travelling wave (generalized contact).

Liu also considers [16,17,18] initial data which approach different constant states u_- and u_+, as $x \to -\infty$ and $x \to \infty$, and shows that the resulting solution approaches the wave fan that solves the Riemann problem associated with (u_-, u_+). Thus, the investigation of asymptotic behavior within the framework of small solutions and Glimm's method is virtually complete. In contrast, very little is known about the asymptotic behavior of solutions with large initial data or of solutions to systems that are not genuinely nonlinear.

6. SYSTEMS OF BALANCE LAWS

The investigation of the asymptotic behavior of solutions to systems of balance laws (1.1) has only begun recently. Using an adaptation of the method of Glimm, Liu [19] studies systems of balance laws

$$u_t + f(u)_x + g(u,x) = 0 \tag{6.1}$$

under the assumption that g decays rapidly as $|x| \to \infty$ and no characteristic speed is 0. Liu's scheme is based upon an approximate resolution of discontinuities by (6.1), analogous to the Riemann problem solution for (1.2) which is the building block of Glimm's scheme. A typical result is that when the initial data $u(x,0)$ have small total variation and approach constant states u_- and u_+ as $x \to -\infty$ and $x \to +\infty$, there is a solution $u(x,t)$ of (6.1) on $(-\infty,\infty) \times [0,\infty)$ which approaches, as $t \to \infty$, a fan of waves, which do not produce interactions and cancellations and essentially consist of simple and shock waves of the associated conservation law, together with a steady state solution

$$\frac{df(u)}{dx} + g(u,x) = 0 \tag{6.2}$$

of (6.1) which occupies the center.

It is not known what happens in the "resonance" situation where one of the characteristic speeds is 0. Clearly, a lot of work is still needed in order to understand completely asymptotic behavior in systems of balance laws.

REFERENCES

[1] Lax, P.D., Shock waves and entropy. "Contributions to Functional Analysis", ed. E.A. Zarantonello. New York: Academic Press 1971, pp. 603-634.

[2] Lax, P.D., Hyperbolic systems of conservation laws II. 1957, Comm. Pure Appl. Math. 10, pp. 537-566.

[3] Dafermos, C.M., Characteristics in hyperbolic conservation laws. "Nonlinear Analysis and Mechanics", Vol. I. ed. R.J. Knops. London: Pitman 1977, pp. 1-58.

[4] Liu, T.-P., Invariants and asymptotic behavior of solutions of a conservation law. 1978, Proc. Am. Math. Soc. 71, pp. 227-231.

[5] Dafermos, C.M., Generalized characteristics and the structure of solutions of hyperbolic conservation laws. 1977, Ind. U. Math. J. 26, pp. 1097-1119.

[6] Dafermos, C.M., Asymptotic behavior of solutions of a hyperbolic conservation law. 1972, J. Diff. Eqs. 11, pp. 416-424.

[7] Dafermos, C.M., Asymptotic behavior of solutions of evolution equations. "Nonlinear Evolution Equations", ed. M.G. Crandall. New York: Academic Press 1978, pp. 103-124.

[8] Benilan, Ph. and Crandall, M.G., Personal communication.

[9] Greenberg, J.M. and Tong, D.D.M., Decay of periodic solutions of $\partial u/\partial t + \partial f(u)/\partial x = 0$. 1973, J. Math. Anal. Appl. 43, pp. 56-71.

[10] Conlon, J.G., Asymptotic behavior for a hyperbolic conservation law with periodic initial data. 1979, Comm. Pure Appl. Math. 32, pp. 99-112.

[11] Glimm, J., Solutions in the large for nonlinear hyperbolic systems of equations. 1965, Comm. Pure Appl. Math. 18, pp. 697-715.

[12] Glimm, J. and Lax, P.D., Decay of solutions of systems of nonlinear hyperbolic conservation laws. 1970, Mem. Am. Math. Soc. 101, pp. 1-112.

[13] DiPerna, R.J., Decay and asymptotic behavior of solutions to nonlinear hyperbolic systems of conservation laws. 1975, Ind. U. Math. J. 24, pp. 1047-1071.

[14] Liu, T.-P., Decay to N-waves of solutions of general systems of nonlinear hyperbolic conservation laws. 1977, Comm. Pure Appl. Math. 30, pp. 585-610.

[15] DiPerna, R.J., Decay of solutions of hyperbolic systems of conservation laws with a convex extension. 1977, Arch. Rat. Mech. Anal. 64, pp. 1-46.

[16] Liu, T.-P., Linear and nonlinear large time behavior of solutions of general systems of hyperbolic conservation laws. 1977, Comm. Pure Appl. Math. 30, pp. 767-796.

[17] Liu, T.-P., Asymptotic behavior of solutions of general systems of nonlinear hyperbolic conservation laws. 1978, Ind. U. Math. J. 27, pp. 211-253.

[18] Liu, T.-P., Large time behavior of solutions of initial and initial-boundary value problems of a general system of hyperbolic conservation laws. 1977, Comm. Math. Phys. 55, pp. 163-177.

[19] Liu, T.-P., Quasilinear hyperbolic systems (to appear).

ACKNOWLEDGEMENT

This research was supported in part by the National Science Foundation under MCS 79-05774, in part by the Office of Naval Research under N0014-76-C-0278-P0002 and in part by the United States Army under ARO-DAAG27-76-G-0294.

NONLINEAR PROBLEMS ARISING IN THE STUDY OF NEMATIC LIQUID CRYSTALS

João-Paulo DIAS
Department of Mathematics, University of Lisbon

ABSTRACT

Following [2] and [1], we deduce, from Leslie's model (cf. [9]), a weak formulation for the time dependent coupled equations of an incompressible nematic liquid crystal submitted to a strong homogeneous magnetic field. We also give some results about the existence, uniqueness, regularity and asymptotic behavior of weak solutions (cf. [2], [4],[1] and [5]).

1. THE EQUATIONS AND THE VARIATIONAL APPROXIMATED MODEL

Let Ω be a smooth open subset of $\mathbb{R}^N$, $N = 2$ or 3, with boundary Γ, where lies an incompressible nematic liquid crystal which is submitted during the interval of time $[0,T]$, $0 < T < +\infty$, to a strong homogeneous magnetic field $h \in C([0,T] ; \mathbb{R}^N)$.

Research supported by a grant of the Gulbenkian's Foundation.

C. Bardos and D. Bessis (eds.), Bifurcation Phenomena in Mathematical Physics and Related Topics, 535–550.

Let $u = u(x,t)$, $v = v(x,t)$, $\pi = \pi(x,t)$ be, respectively, the director field, the velocity field and the pressure, and consider the Leslie's equations (cf. [9],[11]), where we have neglected the inertia of molecular orientation (we use the summation convention and we put $u_{i,k} = \frac{\partial u_i}{\partial x_k}$, $A_{ij}(u) = \frac{1}{2}(u_{i,j} + u_{j,i})$, $B_{ij}(u) = \frac{1}{2}(u_{j,i} - u_{i,j})$) :

$$\gamma_1 \left[\frac{\partial u_i}{\partial t} + v_k\, u_{i,k} + B_{ik}(v)\, u_k\right] + \gamma_2\, A_{ik}(v)\, u_k - \left(\frac{\partial f}{\partial u_{i,j}}\right)_{,j} + \frac{\partial f}{\partial u_i} - \lambda u_i = 0,$$

$$|u| = (u \,.\, u)^{1/2} = 1,$$

$$\rho\left(\frac{\partial v_i}{\partial t} + v_k v_{i,k}\right) = -\pi_{,i} + \Big\{-\frac{\partial f}{\partial u_{k,j}}\, u_{k,i} + \alpha_1 A_{kp}(v) u_k u_p u_i u_j + \alpha_2\left[\frac{\partial u_i}{\partial t} + v_k u_{i,k} + B_{ik}(v) u_k\right] u_j + \alpha_3\left[\frac{\partial u_j}{\partial t} + v_k u_{j,k} + B_{jk}(v) u_k\right] u_i + \alpha_4 A_{ji}(v) + \alpha_5 A_{ik}(v) u_j u_k + \alpha_6 A_{jk}(v) u_i u_k\Big\}_{,j},$$

$$\operatorname{div} v = 0,$$

where α_i, $i = 1,\dots,6$, are the characteristic viscosities, $\gamma_1 = \alpha_3 - \alpha_2 > 0$, $\gamma_2 = \alpha_6 - \alpha_5 = \alpha_2 + \alpha_3$, $\rho > 0$ is the usual density,

$$f = \frac{1}{2}\left[k_1 (\operatorname{div} u)^2 + k_2 (u.\operatorname{rot} u)^2 + k_3 |u \times \operatorname{rot} u|^2 - \chi (u \,.\, h)^2\right]$$

is the density of free energy, $k_i > 0$, $i = 1,2,3$, $\chi > 0$ and

$\lambda = \lambda(x,t)$ is the orientation stress.

If we assume that $k_1 = k_2 = k_3$, we obtain, since $|u| = 1$,

$$f = \frac{1}{2}\left[k_1 (\text{div } u)^2 + k_1 \, |\text{rot } u|^2 - \chi (u \,.\, h)^2\right] ,$$

and so $\left(\dfrac{\partial f}{\partial u_{i,j}}\right)_{,j} = k_1 \, \Delta \, u_i$.

We deduce from the first equation (eliminating λ),

$$\gamma_1 \frac{\partial u_i}{\partial t} - k_1 \, \Delta \, u_i + \gamma_1 v_k u_{i,k} + \chi(u \,.\, h)^2 u_i - \chi(u \,.\, h) h_i + \\ + \theta_i - k_1 u_{k,p} u_{k,p} u_i = 0, \qquad (1.1)$$

where $\theta_i = \gamma_1 B_{ik}(v) u_k + \gamma_2 A_{ik}(v) u_k - \gamma_2 A_{kp}(v) u_k u_p u_i$.

We do not consider the two last terms of the first member of (1.1) and so we lost the property $|u| = 1$.

We take, as boundary conditions,

$$u \,.\, \nu = 0, \quad \frac{\partial u}{\partial \nu} \times \nu = 0 \text{ on } \Gamma,$$

where ν the unitary outward normal to Γ.

We obtain

$$\gamma_1 \int_\Omega \frac{\partial u}{\partial t} \,.\, m \, dx + k_1 \int_\Omega u_{i,k} m_{i,k} dx + \gamma_1 \int_\Omega v_k u_{i,k} m_i \, dx + \\ + \chi \int_\Omega (u \,.\, h)^2 (u \,.\, m) dx - \chi \int_\Omega (u \,.\, h)(h \,.\, m) dx = 0 \, , \forall \, m \in W,$$

where

$W = \{u \in (H^1(\Omega))^N \mid u \,.\, \nu = 0 \text{ on } \Gamma\}$, with the norm

$\| u \| = (\int_\Omega u_{i,k} u_{i,k} dx)^{1/2}$ which is equivalent (in W) to the usual $(H^1(\Omega))^N$ norm (cf. [3]).

Now, let us introduce the spaces (cf. [10], chap. 1)

$V = \{v \in (H_0^1(\Omega))^N \mid \operatorname{div} v = 0 \text{ in } \Omega\}$, with the same norm $\| \cdot \|$,

$H = \{v \in (L^2(\Omega))^N \mid \operatorname{div} v = 0 \text{ in } \Omega \text{ and } v \cdot \nu = 0 \text{ on } \Gamma\}$, with the usual $(L^2(\Omega))^N$ norm, and, for $u \in (L^\infty(\Omega))^N$, consider the continuous bilinear form over $V \times V$ which is defined by

$$a(u ; v , w) = \alpha_4 \int_\Omega A_{ij}(v) A_{ij}(w) dx +$$
$$+ (\alpha_1 - \gamma_2) \int_\Omega A_{kp}(v) u_k u_p A_{ij}(w) u_i u_j dx + 2 \alpha_6 \int_\Omega A_{ik}(v) u_k A_{ij}(w) u_j dx +$$
$$+ 2 \alpha_3 \int_\Omega B_{ik}(v) u_k A_{ij}(w) u_j dx, \quad \forall\, v, w \in V.$$

If we also work with the equations of the velocity field and if we made some simplifications (in particular, we do not consider the terms containing the derivatives of the director field) (see [2] for details), we obtain the following weak formulation (cf. [2], [1]) (we put $\mu = \frac{k_1}{\gamma_1}$ and we replace $(\frac{\chi}{\gamma_1})^{1/2} h$ by h) :

$$\begin{cases} u \in L^2(0,T;W) \cap L^\infty(0,T;(L^2(\Omega))^N), \quad (u \cdot h)u \in L^2(0,T;(L^2(\Omega))^N), \\ \frac{d}{dt} \int_\Omega u \cdot m \, dx + \mu \int_\Omega u_{i,k} m_{i,k} dx + \int_\Omega (u \cdot h)^2 (u \cdot m) dx - \\ \qquad - \int_\Omega (u \cdot h)(h \cdot m) dx + \int_\Omega v_k u_{i,k} m_i dx = 0, \quad \forall\, m \in W, \\ u(0) = u_0 \in (L^\infty(\Omega))^N, \quad \text{with } |u_0|_\infty \le 1, \end{cases} \tag{1.2}$$

$$\left\{\begin{array}{l} v \in L^2(0,T\ ;\ V) \cap L^\infty(0,T\ ;\ H), \\ \rho \dfrac{d}{dt}\displaystyle\int_\Omega v\ .\ w\,dx + \rho \int_\Omega v_k v_{i,k} w_i\,dx + a(u_o\ ;\ v,w) + \\ \qquad\qquad + b_1(u_o,v_o\ ;\ u - u_o,w) = 0\ ,\ \forall\ w \in V, \\ v(0) = v_o \in H \cap (W^{1,p}(\Omega))^N, \qquad \text{with } p > N, \end{array}\right. \tag{1.3}$$

where

$$b_1(u_o,v_o\ ;\ m,w) = \int_\Omega \frac{\partial}{\partial u_q}\left[(\alpha_1 - \gamma_2)A_{kp}(v_o)u_k u_p A_{ij}(w)u_i u_j + \right.$$

$$\left. + 2\alpha_6 A_{ik}(v_o)u_k A_{ij}(w)u_j + 2\,\alpha_3 B_{ik}(v_o)u_k A_{ij}(w)u_j\right]_{u=u_o} m_q\,dx \quad ,$$

is a continuous bilinear form over $(L^r(\Omega))^N \times V$, where $\frac{1}{r} = \frac{1}{2} - \frac{1}{p}$ (notice that $W \hookrightarrow (L^r(\Omega))^N$).

<u>Remark 1</u> : If N = 2 we have (cf. [2])

$$u \in L^2(0,T\ ;\ W) \cap L^\infty(0,T\ ;\ (L^2(\Omega))^2) \hookrightarrow L^4(0,T\ ;\ (L^4(\Omega))^2)\ ,$$

and so $(u\ .\ h)u \in L^2(0,T)\ ;\ (L^2(\Omega))^2)$. ∎

<u>Remark 2</u> : We can deduce from (1.2), (1.3) that we have $u' \in L^1(0,T\ ;\ W')$, $v' \in L^1(0,T\ ;\ V')$. This gives a sense to the initial conditions $u(0) = u_o$, $v(0) = v_o$. ∎

We will assume that

$$\alpha_1 - \gamma_2 \geq 0\quad ,\quad \alpha_3 \leq 0\quad ,\quad \alpha_6 \leq 0 \tag{1.4}$$

This is a reasonable hypothesis, which is satisfied in the usual cases (cf.[11]).

Lemma 1 (cf. [2]) : suppose that $N = 2$, $u_o \in (L^\infty(\Omega))^2$ and $|u_o|_\infty \leq 1$. Then, we have

$$a(u_o\ ;\ v,v) \geq \alpha \| v \|^2 , \qquad \forall\ v \in V, \tag{1.5}$$

where $\alpha = \frac{1}{2}(\alpha_3 + \alpha_4 + \alpha_6) > 0$ is the Miesowicz's viscosity (cf. [11]).

Proof of lemma 1 : It is easy to see that

$$\int_\Omega A_{ij}(v)\ A_{ij}(v)\ dx = \frac{1}{2} \int_\Omega v_{i,j}\ v_{i,j}\ dx \quad , \forall\ v \in V. \tag{1.6}$$

Now, take $v \in V$. We have $A_{12} = A_{21}$, $A_{11} + A_{22} = 0$, where $A_{ij} = A_{ij}(v)$, and

$$2 \int_\Omega A_{ik}\ u_{ok}\ A_{ij}\ u_{oj}\ dx = 2 \int_\Omega [A_{11}^2(u_{o1}^2 + u_{o2}^2) + A_{12}^2(u_{o1}^2 + u_{o2}^2)]dx \leq$$
$$\leq 2 \int_\Omega (A_{11}^2 + A_{12}^2)\ dx = \int_\Omega A_{ij}\ A_{ij}\ dx \tag{1.7}$$

It can also be proved that

$$2 \int_\Omega B_{ik}(v)\ u_{ok}\ A_{ij}(v)\ u_{oj}\ dx \leq \frac{1}{2} \int_\Omega v_{i,j}\ v_{i,j}\ dx\ . \tag{1.8}$$

Lemma 1 is now an easy consequence of (1.4), (1.6), (1.7) and (1.8). ∎

If $N = 3$ the situation is not so clear, but we can give some algebraic conditions for u_o in order to satisfy (1.5) (for example, if $|u_o|_\infty \leq \frac{1}{2}$).

We also need to introduce the space

$W_2 = \{u \in (H^2(\Omega))^N \mid u \cdot \nu = 0 \quad , \quad \frac{\partial u}{\partial \nu} \times \nu = 0 \text{ on } \Gamma\}$,

with the usual $(H^2(\Omega))^N$ norm.

The space W_2 is dense in W (cf. [1]).

2. GENERAL RESULTS

The three following results (cf. [2] and [1]) can be proved with the methods introduced in [8], [7] and [10]:

Theorem 1 : If N = 2, or if N = 3 and u_o satisfies (1.5), then there exists one solution $\{u,v\}$ of the system (1.2), (1.3). If N = 2 this solution is unique.

Theorem 2 : If N = 2, $h \in C^1([0,T] ; \mathbb{R}^2)$, $u_o \in W_2$ and $v_o \in V \cap (H^2(\Omega))^2$, then the (unique) solution $\{u,v\}$ of (1.2), (1.3) satisfies $u \in C([0,T] ; W)$, $v \in C([0,T] ; V)$, $u' \in L^2(0,T ; W) \cap L^\infty(0,T ; (L^2(\Omega))^2)$, $v' \in L^2(0,T ; V) \cap L^\infty(0,T ; H)$.

Theorem 3 : If N = 3, $h \in C^1([0,T] ; \mathbb{R}^3)$, $u_o \in W_2$ and satisfies (1.5) and $v_o \in V \cap (H^2(\Omega))^3$, then there exists $T_1 \in]0,T]$ such that, in the interval $]0,T_1[$, the system (1.2), (1.3) has a unique solution $\{u,v\}$ satisfying $u \in C([0,T_1] ; W)$, $v \in C([0,T_1] ; V)$, $u' \in L^2(0,T_1 ; W) \cap L^\infty(0,T_1 ; (L^2(\Omega))^3)$, $v' \in L^2(0,T_1 ; V) \cap L^\infty(0,T_1 ; H)$.

The following results will be proved in the next section :

Theorem 4 (cf. [1]) : Under the hypothesis of theorem 2 (respectively, of theorem 3), we have

$$u(t) \in W_2 , \quad \text{a.e. on }]0,T[\quad (\text{respectively, on }]0,T_1[\ ),$$

and $|u(t)|_\infty \leq 1$.

Theorem 5 (cf. [4]) : Assume that $h \in C^1([0,T] ; \mathbb{R}^N)$ ($N = 2$ or 3), $u_o \in W_2$ and $v_o = 0$. Then, there exists a unique function

$$u \in C([0,T] ; W_2) \cap C^1([0,T] ;(L^2(\Omega))^N), \tag{2.1}$$

such that

$$u' - \mu \Delta u + (u \cdot h)^2 u - (u \cdot h) h = 0 \text{ in } [0,T], \tag{2.2}$$

$$u(0) = u_o \tag{2.3}$$

Remark 3 : The function u, which satisfies (2.1), (2.2), (2.3), will be called a strong solution of (2.2), (2.3). ∎

3. PROOFS OF THEOREMS 4 AND 5

Proof of theorem 4 : We have, for all $m \in W$ and a.e. on $]0,T_1[$ (with $T_1 = T$ if $N = 2$) :

$$\int_\Omega u'.m\, dx + \mu \int_\Omega u_{i,k} m_{i,k} dx + \int_\Omega (u \cdot h)^2 (u \cdot m) dx - $$
$$- \int_\Omega (u \cdot h)(h \cdot m) dx + \int_\Omega v_k u_{i,k} m_i dx = 0 \tag{3.1}$$

If we apply the regularity results of [12], we easily deduce that, a.e. on $]0,T_1[$,

$$u(t) \in W_2,\ u' - \mu \Delta u + (u \,.\, h)^2 u - (u \,.\, h)h + v_k u_{,k} = 0\ ,$$

a.e. in Ω.

Now, take $m = |u|^{2p} u$, $p = 0,1,2,\ldots$ We have $m(t) \in W$, a.e. on $]0,T_1[$. We obtain, from (3.1),

$$\frac{1}{2(p+1)} \frac{d}{dt} |u|^{2(p+1)}_{2(p+1)} + \int_\Omega (u \,.\, h)^2 |u|^{2(p+1)} dx \leq \int_\Omega (u \,.\, h)^2 |u|^{2p} dx,$$

and so, for $\tau \in]0,T_1]$,

$$\int_0^\tau \int_\Omega (u.h)^2 |u|^{2(p+1)} dx\ dt \leq \frac{1}{2(p+1)} |u_0|^{2(p+1)}_{2(p+1)} +$$
$$\int_0^\tau \int_\Omega (u.h)^2 |u|^{2p} dx\ dt \leq |\Omega| \left[\frac{1}{2(p+1)} + \frac{1}{2p} + \ldots + \frac{1}{2}\right] +$$
$$+ \int_0^\tau \int_\Omega (u.h)^2 dx\ dt, \text{ with } |\Omega| = \text{measure of } \Omega.$$

Hence, with $A_{k,\tau} = \{(x,t) \in \Omega \times]0,\tau[\ |\ |u(x,t)| \geq k\}$, $k > 1$,

$$\int_{A_{k,\tau}} (u.h)^2 dx\ dt \leq |\Omega| k^{-2(p+1)} \left[\frac{1}{2(p+1)} + \frac{1}{2p} + \ldots + \frac{1}{2}\right] +$$
$$+ k^{-2(p+1)} \int_0^\tau \int_\Omega (u.h)^2 dx\ dt \underset{p \to +\infty}{\to 0},$$

and so we obtain

$$u' - \mu \Delta u + v_k u_{,k} = 0,$$
$$\text{a.e on } A_2 = \{(x,t) \in \Omega \times]0,\tau[\ |\ |u(x,t)| > 1\}.$$

Now, consider $\varphi = |u|^2$. We have

$$\varphi(t) \in H^2(\Omega), \text{ a.e. on }]0,\tau[,\ \frac{\partial \varphi}{\partial \nu} = 2\, u \,.\, \frac{\partial u}{\partial \nu} = 0 \text{ on } \Gamma,$$

$$\varphi' - \mu\,\Delta\varphi + v_k\,\varphi_{,k} \le 0, \quad \text{a.e. on } A_\tau .$$

We deduce, with $\varphi_1 = \varphi - \min\{\varphi, 1\}$,

$$\int_0^\tau\int_\Omega(\varphi' - \mu\,\Delta\,\varphi + v_k\varphi_{,k})\varphi_1 dx\ dt = \int_{A_\tau}(\varphi' - \mu\,\Delta\,\varphi + v_k\varphi_{,k})\ _1 dx\ dt \le 0.$$

Furthermore, since $|u_o|_\infty \le 1$, $\frac{\partial\varphi}{\partial\nu} = 0$ on Γ and $v_{k,k} = 0$ in Ω, we obtain

$$\int_0^\tau\int_\Omega(\varphi' - \mu\,\Delta\,\varphi + v_k\varphi_{,k})dx\ dt = \frac{1}{2}\,|\varphi_1(\tau)|_2^2 + \mu\int_0^\tau\int_\Omega|\nabla\varphi_1|^2 dx\ dt,$$

and so $|\varphi_1(\tau)|_2 \le 0$. This completes the proof of theorem 4. ∎

<u>Proof of theorem 5</u> : We can consider the strongly continuous contraction semi-group in $(L^2(\Omega))^N$ generated by $-\mu\,\Delta$ with domain W_2. By applying a result of Segal (cf. [13]) it is not difficult to prove the existence of a local strong solution for the equation (see [4] for details). Now, we must show that, if u is a strong solution,

$$|u(t)|_{W_2} + |u'(t)|_2 \le c(t), \text{with } c \in C([0,T]\ ;\ \mathbb{R}_+).$$

As in the proof of theorem 4 we obtain

$$|u(t)|_\infty \le 1, \quad \forall\ t \in [0,T] .$$

Now, put $f = -(u\ .\ h)^2 u + (u\ .\ h)h$. We have

$$f \in C^1([0,T]\ ;\ (L^2(\Omega))^N),\ |f'(t)|_2 \le c(t)(1 + |u'(t)|_2),$$
$$\forall\ t \in [0,T] \tag{3.2}$$

Let us consider the continuous linear operator $A: W \to W'$ defined by

$$< A\, m,w >_{W'\times W} = \mu \int_\Omega m_{i,k} w_{i,k} dx, \quad \forall\, m,w \in W,$$

and let $v \in L^2(0,T\ ;\ W)$ be the solution of

$$\begin{cases} v' + A\, v = f'\ , \text{ with } v' \in L^2(0,T\ ;\ W'), \\ v(0) = u'(0) \in (L^2(\Omega))^N. \end{cases}$$

It is easy to see that $v = u'$. Hence, we obtain, a.e. on $]0,T[$,

$$< u'',u' >_{W'\times W} + < A\, u',u' >_{W'\times W} = \int_\Omega f' \,.\, u' dx,$$

$$\frac{1}{2}\frac{d}{dt} |u'(t)|_2^2 \leq |f'(t)|_2\, |u'(t)|_2 \,. \qquad (3.3)$$

From (3.2), (3.3), we infer

$$\frac{d}{dt} |u'(t)|_2^2 \leq c(t)(1 + |u'(t)|_2^2)\ ,$$

and so, by Gronwall's inequality (notice that $u'(0) = \mu\, \Delta\, u(0) + f(0)$),

$$|u'(t)|_2 \leq c(t).$$

Hence, $|\mu\, \Delta\, u(t)|_2 \leq |u'(t)|_2 + |f(t)|_2 \leq c(t)$.
But $|u(t)|_{W_2} \leq c\ |\mu\, \Delta\, u(t)|_2$. This completes the proof of theorem 5. ∎

4. SOME REMARKS ABOUT THE ASYMPTOTIC BEHAVIOR

Let h_o be in $\mathbb{R}^N$ and u be a solution of the following stationary variational problem :

$$u \in W,\ \mu \int_\Omega u_{i,k} m_{i,k} dx + \int_\Omega (u.h_o)^2 (u.m)dx - \int_\Omega (u.h_o)(h_o.m)dx, \quad \forall\ m \in W \qquad (4.1)$$

We easily deduce from (4.1), applying the results of [12] (notice that $W \subset (L^6(\Omega))^N$) :

$$u \in W_2 \cap (C^2(\Omega))^N, \quad -\mu \Delta u + (u.h_o)^2 u - (u.h_o)h_o = 0 \text{ in } \Omega \quad (4.2)$$

Introducing the constant

$$c_o = \sup_{u \in W,\, u \neq 0} (|u|_2 / \| u \|), \qquad (4.3)$$

we obtain the following results :

Proposition 1 (cf.[4]) : We have $|u(x)| \leq 1$, $\forall\ x \in \Omega$. Furthermore, if $c_o |h_o| < \mu^{1/2}$, then $u = 0$.

Theorem 6 (cf. [4]) : Let h be in $C^1([0,+\infty[,\mathbb{R}^N)$, u_o in W_2 and let u be the strong solution of (2.2),(2.3), considered in the interval $[0,+\infty[$. Assume that there exists h_o in $\mathbb{R}^3$ such that $c_o\ |h_o| < \mu^{1/2}$, where c_o is defined by (4.3), and $\int_t^{t+1} |h(s) - h_o|^2 ds \underset{t \to +\infty}{\to} 0$. Then, we have

$$|u(t)|_2 \underset{t \to +\infty}{\to} 0 \quad \text{and} \quad \int_t^{t+1} \| u(s) \|^2 ds \underset{t \to +\infty}{\to} 0 .$$

Theorem 7 (cf. [5]) : Under the hypothesis of theorem 6 , assume that $\int_t^{t+1} |h'(s)|^2 ds \le c, \ \forall\, t \in \mathbb{R}_+$. Then, we have

$$|u(t)|_{W_2} \underset{t \to +\infty}{\to} 0 .$$

Proof of theorem 6 : We have (cf. the proof of theorem 5),

$$|u(t)| \le 1, \qquad \forall\, t \in [0, +\infty[.$$

Now, we obtain from (2.2), multiplying by u and integrating over Ω,

$$\frac{1}{2}\frac{d}{dt}|u|_2^2 + \mu \| u \|^2 + \int_\Omega (u.h)^2 \, |u|^2 dx = \int_\Omega (u.h)^2 dx \qquad (4.4)$$

Furthermore, we have,

$$(u.h)^2 = (u.h)(u.h-h_o) + (u.h-h_o)(u.h_o) + (u.h_o)^2 \qquad (4.5)$$

and, with c_o defined by (4.3),

$$\left| \int_\Omega (u.h)(u.h-h_o)dx \right| \le \int_\Omega |(u.h)u| \, |h-h_o| dx \le$$

$$\le \frac{1}{2}\int_\Omega (u.h)^2 \, |u|^2 dx + \frac{1}{2}|\Omega| \, |h-h_o|^2, \qquad (4.6)$$

$$\left| \int_\Omega (u.h-h_o)(u.h_o)dx \right| \le \varepsilon \int_\Omega |h_o|^2 |u|^2 dx + c(\varepsilon) \int_\Omega |h-h_o|^2 |u|^2 dx \le$$

$$\le \varepsilon |h_o|^2 \, c_o^2 \| u \|^2 + c(\varepsilon) \, |\Omega| \, |h-h_o|^2, \qquad (4.7)$$

$$\int_\Omega (u.h_o)^2 dx \le |h_o|^2 \int_\Omega |u|^2 dx \le |h_o|^2 \, c_o^2 \, \| u \|^2 \qquad (4.8)$$

Now, by the hypothesis, take $\varepsilon > 0$ such that $\rho = (1+\varepsilon)|h_o|^2 c_o^2 < \mu$. We deduce, from (4.4), (4.5), (4.6), (4.7) and (4.8)

$$\frac{1}{2}\frac{d}{dt}|u|_2^2 + (\mu - \rho)\| u \|^2 \le |\Omega| \, (\tfrac{1}{2} + c(\varepsilon)) \, |h-h_o|^2 \qquad (4.9)$$

and so, with $c_1 = |\Omega| \ (\frac{1}{2} + c(\varepsilon))$, we obtain

$$\frac{1}{2} \frac{d}{dt} |u|_2^2 + (\mu - \rho) c_o^2 |u|_2^2 \leq c_1 |h - h_o|^2 ,$$

which implies, since $\int_t^{t+1} |h(s) - h_o|^2 ds \underset{t \to +\infty}{\to} 0$,

$$|u(t)|_2^2 \underset{t \to +\infty}{\to} 0 .$$

Now, we derive, from (4.9),

$$\int_t^{t+1} \| u(s) \|^2 ds \leq \frac{1}{2(\mu-\rho)} |u(t)|_2^2 + \frac{c_1}{\mu-\rho} \int_t^{t+1} |h(s)-h_o|^2 ds \underset{t \to +\infty}{\to} 0$$

■

Now, consider the strip $\Omega = \mathbb{R} \times]0,1[\subset \mathbb{R}^2$ and take $h_o = (h_{o1}, 0) \in \mathbb{R}^2$. Let $u = (u_1, u_2)$ be a solution of problem (4.2) and put $h = h_o/\mu^{1/2} = (h_1, h_2)$. We obtain

$$\begin{cases} - \Delta u + (u.h)^2 u - (u.h)h = 0 \text{ in } \Omega, \\ u \cdot \nu = 0 \text{ and } \frac{\partial u}{\partial \nu} \times \nu = 0 \text{ on } \Gamma. \end{cases} \tag{4.10}$$

This can be written as follows :

$$\begin{cases} - \Delta u_1 + h_1^2 u_1^3 - h_1^2 u_1 = 0 & \text{in } \Omega \\ \frac{\partial u_1}{\partial \nu} = 0 & \text{on } \Gamma, \end{cases} \tag{4.11}$$

$$\begin{cases} - \Delta u_2 + h_1^2 u_1^2 u_2 = 0 & \text{in } \Omega, \\ u_2 = 0 & \text{on } \Gamma. \end{cases} \tag{4.12}$$

We deduce from (4.12) that $u_2 = 0$. Furthermore, if we put $v = h_1 u_1$, $\lambda = h_1^2$, we obtain from (4.11) :

$$\begin{cases} -\Delta v + v^3 - \lambda v = 0 & \text{in } \Omega , \\ \dfrac{\partial v}{\partial \nu} = 0 & \text{on } \Gamma. \end{cases} \tag{4.13}$$

The study of nontrivial solutions of (4.13) and of their stability will be made in [5] (see [6] for the study of this problem with Dirichlet's boundary conditions).

REFERENCES

[1] D. CIORANESCU and J.P. DIAS, A time dependent coupled system related to the tridimensional equations of a nematic liquid crystal, J. of Math. Anal. and Appli., to appear.

[2] J.P. DIAS, A simplified variational model for the bidimensional coupled evolution equations of a nematic liquid crystal J. of Math. Anal. and Appli. 67 (1979), pp. 525-541.

[3] J.P. DIAS, Un problème aux limites pour un système d'équations non linéaires tridimensionnel, Boll.Unio.Mat.Ital. 16-B (1979), to appear.

[4] J.P. DIAS, On the existence of a strong solution for a nonlinear evolution system, J. of Nonlinear Analysis, to appear.

[5] J.P. DIAS and A. HARAUX, Smoothing effect and asymptotic behavior for the solutions of a nonlinear time dependent system, to appear.

[6] K. KIRCHGÄSSNER and J. SCHEURLE, On the bounded solutions of a semilinear elliptic equation in a strip, J. of Diff. Eq. 32 (1979), pp. 119-148.

[7] O.A. LADYZHENSKAYA, The mathematical theory of viscous incompressible flow, Gordon and Breach, New-York, 1963.

[8] J. LERAY, Etude de diverses équations intégrales non linéaires et de quelques problèmes que pose l'hydrodynamique, J. de Math. Pures et Appli. 12 (1933), pp. 1-82.

[9] F.M. LESLIE, Some constitutive equations for liquid crystals Arch. Rat. Mech. and Anal. 28 (1968), pp. 265-283 .

[10] J.L. LIONS, Quelques méthodes de résolution des problèmes aux limites non linéaires, Dunod et Gauthier-Villars, Paris, 1969.

[11] A.F. MARTINS, Contribution à l'étude de la dynamique moléculaire dans les phases nématique et isotrope des cristaux liquides, Portugaliae Phys. 8 (1972), pp. 1-166.

[12] M. SCHOENAUER, Quelques résultats de régularité pour la solution d'un système elliptique avec des conditions aux limites couplées, to appear.

[13] I. SEGAL, Nonlinear semi-groups, Annals. of Math. 78 (1963), pp. 339-364.

SOME PROPERTIES OF FUNCTIONAL INVARIANT SETS FOR NAVIER-STOKES EQUATIONS

R.TEMAM

Departement de Mathématiques
Université Paris-Sud
91405 - Orsay, France

INTRODUCTION

Our aim in this article is to describe a few properties of

- the set of stationary solutions to the Navier-Stokes equations for a viscous incompressible fluid in a bounded domain Ω of $\mathbb{R}^n$, n=2 or 3,
- any functional invariant set for the time-dependent Navier-Stokes equations, under the same conditions, the dimension of space being n=2.

We start in Section 1 by recalling the functional setting of Navier-Stokes following[8]. In Section 2 we describe the results contained in [1-2-6] concerning the structure of the set of stationary solutions, In section 3 we present the result in [3] concerning any functional invariant set and another remark concerning the time periodic solutions.

1. FUNCTIONAL SETTING OF NAVIER-STOKES EQUATIONS

Let Ω denotes an open bounded set in $\mathbb{R}^n$, n=2 or 3. Let $L^2(\Omega)$ denotes the space of real square integrable functions on Ω, and $H_0^1(\Omega)$ denotes the Sobolev space :

$$H_0^1(\Omega)= \left\{u\in L^2(\Omega),\ \frac{\partial u}{\partial x_i} \in L^2(\Omega),\quad i=1,\dots,n,\ u=0 \text{ on } \partial\Omega \right\}$$

We set $\mathbb{L}^2(\Omega)=L^2(\Omega)^n$, and $\mathbb{H}_0^1(\Omega)=H_0^1(\Omega)^n$. These space are equiped with the usual scalar products which we denote by

$$(u,v) = \int_\Omega u(x)v(x)dx$$

C. Bardos and D. Bessis (eds.), Bifurcation Phenomena in Mathematical Physics and Related Topics, 551–554.

for $u,v \in L^2(\Omega)$ or $\mathbb{L}^2(\Omega)$ and by

$$((u,v)) = \sum_{i=1}^{n} \left(\frac{\partial u}{\partial x_i}, \frac{\partial v}{\partial x_i}\right)$$

for $u,v \in H_o^1(\Omega)$ or $\mathbb{H}_o^1(\Omega)$.

The space $\mathbb{L}^2(\Omega)$ is the direct sum of the two orthogonal spaces:

$$H = \left\{u \in \mathbb{L}^2(\Omega),\ \text{div } u=0,\ u.\underset{\approx}{n}=0 \text{ on } \partial\Omega\right\}, \tag{1.1}$$

$\underset{\approx}{n}$ the unit outward normal on $\partial\Omega$, and,

$$G = \left\{u \in \mathbb{L}^2(\Omega),\ u=\text{grad } p,\ p \in H^1(\Omega)\right\} \tag{1.2}$$

(cf. [8]).

Let P_H denotes the projector in $\mathbb{L}^2(\Omega)$ onto H. We define the unbounded operator A in H the domain of which is

$$D(A) = \left\{u \in \mathbb{H}_o^1(\Omega) \cap \mathbb{H}^2(\Omega),\ \text{div } u=0\right\},$$

and $Au=-P_H\Delta u$, $\forall u \in D(A)$. One can prove that A is an *isomorphism* from D(A) onto H. For $u,v \in D(A)$ we set also.

$$B(u,v) = P_H\left(\sum_{i=1}^{n} u_i \frac{\partial v}{\partial x_i}\right),$$

and

$$B(u) = B(u,u)$$

Given $\nu > 0$, $u_o \in H$, and a function f from [0,T] into H, the resolution of the time-dependent Navier-Stokes equations on the interval of time [0,T], amounts to solving the functional evolution equation :

$$\left.\begin{aligned} &u_i[0,T] \to D(A) \\ &\frac{du}{dt}(t) + \nu Au(t) + B(u(t)) = f(t) \text{ on } [0,T] \\ &u(0) = u_o \end{aligned}\right\} \tag{1.3}$$

For the details cf. [4-8] .

2. STRUCTURE OF THE SET OF STATIONARY SOLUTIONS

The stationary Navier-Stokes equations takes the form

$$\left.\begin{array}{l} u \in D(A) \\ \nu Au + Bu = f \end{array}\right\} , \qquad (2.1)$$

where f is given in H.

The set $S(f,\nu)$ of solutions of (2.1) is not empty (existence of solutions) and reduceed to a single point if $|f|/\nu^2$ is sufficiently small (uniqueness). Otherwise it was shown in [1-2] that

THEOREM 1
$S(f,\nu)$ is a real analytic compact set of finite dimension m ; $m \leq m_*$, where the number m_* can be estimated in term of ν and $|f|$.

THEOREM 2
For every $\nu > 0$, there exists an open dense set $0 \subset H$, and for every $f \in H$, the set $S(f,\nu)$ is finite, the number of elements of $S(f,\nu)$ is odd and constant an every connected component of 0 .

If the boundary data of the problem are different from 0 ($u = \phi$ on $\partial\Omega$), then the equation (2.1) is replaced (cf.[1]) by an equation

$$\left.\begin{array}{l} u \in D(A) \\ \nu Au + B(u+\phi) = f \end{array}\right\} \qquad (2.2)$$

where ϕ belongs to an appropriate subspace X of $\mathbb{H}^2(\Omega)$. Then we have results similar to Theorem 1 and 2 (cf.[6]) :

THEOREM 1'
$S(f,\phi,\nu)$ is a real analytic compact set of finite dimension m , $m \leq m_*$, where m_* depends on ν , $|f|$ and the norm of ϕ in $\mathbb{H}^2(\Omega)$.

THEOREM 3
For every $\nu > 0$ and f fixed, there exists an open dense set 0' in X , and for every ϕ in 0' , the set $S(f,\phi,\nu)$ is finite, the number of elements of $S(f,\phi,\nu)$ is odd and constant an every connected component of 0'.

3. FUNCTIONAL INVARIANT SET

The dimension of space is now n=2. A functional invariant set is a subset Y of

$$V = \left\{ u \in \mathbb{H}^1_o(\ ) \ , \ \text{div} \ u=0 \right\}$$

such that $S(t)u_o$ is defined for every u_o in Y and $t > 0$ and $S(t)Y = Y$, where $S(t)u_o = u(t)$, u being the (unique) solution of (1.3). Such a set may be a stationary solution, or the trajectory of a periodic solution, or an attractor. It is known that any functional invariant set is bounded in V(n=2). We proved in [3]:

THEOREM 4

The Hausdorff dimension of any functional invariant set Y *is bounded, and less that an expression* $d_*(\nu,|f|,M)$, *where* $M = \sup_{u \in Y} \|u\|$.

It is proved in [3] that, for a periodic solution, there exists a finite dimensional projector which is periodic on Y. Cf. also [7-6] for some other remarks concerning time periodic solutions.

REFERENCES

[1] C.Foias, R.Temam, *Structure of the set of stationary solutions of the Navier-Stokes equations*, Comm. Pure Appl.Math. Vol.XXX (1977) p.149-164.

[2] C.Foias, R.Temam, *Remarques sur les équations de Navier-Stokes stationnaires et les phénomènes successifs de bifurcation*, Annali Scuola Norm.Sup.di Pisa, Serie IV, Vol.5 (1978) p.29-63 et volume dédié à J.Leray.

[3] C.Foias, R.Temam, *Some analytic and geometric properties of the solutions of the evolution Navier-Stokes equations*, J.Math.Pures Appl., 58 (1979) p.339-368.

[4] J.L.Lions, *Quelques méthodes de résolution des problèmes aux limites non linéaires*, Dunod-Gauthier-Villars, Paris 1969.

[5] J.C.Saut, R.Temam, *Generic properties of nonlinear boundary value problems*, Communications in P.D.E., vol.4 n°3, 1979, p.293-319.

[6] J.C.Saut, R.Temam, *Generic properties of Navier-Stokes equations : genericity with respect to the boundary values*, Indiana Math.Journal, to appear.

[7] R.Temam, *Une propriété générique de l'ensemble des solutions stationnaires ou périodiques des équations de Navier-Stokes*, Symposium Franco-Japonais, Septembre 1976, à paraître.

[8] R.Temam, *Navier-Stokes equations, theory and numerical analysis*, North-Holland, Amsterdam 1977, second edition 1979.

AN INTRODUCTION TO THE TIME-DEPENDENT HARTREE-FOCK THEORY IN NUCLEAR PHYSICS

B.G. Giraud

DPh.T. CEN-Saclay - 91190 Gif-sur-Yvette, France

ABSTRACT

The emphasis is put on the fact that a system with many degrees of freedom can actually show behaviours where only a few degrees, of a collective nature, are excited. Such behaviours relate to the solitonic solutions of TDHF. The lectures first introduce the physical ideas which govern the theory, then provide derivations of TDHF and its properties, and finally show how the solitonic solutions can be recognized by a criterion of geometrical nature in the fiber bundle of the Slater determinants. The collective degrees of freedom are then derived from first principles, as the algebra which governs the TDHF evolution of the solitons.

C. Bardos and D. Bessis (eds.), Bifurcation Phenomena in Mathematical Physics and Related Topics, 555–583.

1. BASIC CONCEPTS

We consider a set of N fermions (nucleons) with coordinates $\underset{\sim}{r}_i$, $i = 1,\ldots,N$. For the sake of simplicity protons and neutrons will not be distinguished and their spins will also be disregarded. Typically N = 100, and there is a range of phenomena (low energy nuclear physics) where the coupling of nucleons to pions and other "elementary" fields or hadronic excited states need not be explicit. Also in this range one can neglect relativistic corrections. The system is then described at each time t by a wave function $\Psi(\underset{\sim}{r}_1,\ldots,\underset{\sim}{r}_N,t)$. In principle one should let Ψ also depend on other degrees of freedom, such as those of the electromagnetic field, for radiative processes may occur. However the dependence of Ψ upon 301 variables may be considered as a problem of sufficient complexity from the outset.

We shall admit that the time evolution of Ψ is given by the linear time-dependent Schrödinger equation (TDSE),

$$i\,\frac{\partial\Psi}{\partial t} = \mathcal{H}\,\Psi \equiv \Big[-\frac{\hbar^2}{2m}\sum_{i=1}^{N}\Delta_i + \sum_{i>j=1}^{N} v(|\underset{\sim}{r}_i - \underset{\sim}{r}_j|)\Big]\,\Psi, \qquad (1.1)$$

where $\hbar$ is the Planck constant, m the nucleon mass, Δ_i the Laplacian with respect to $\underset{\sim}{r}_i$. From now on we take $\hbar = 2m = 1$ in a suitable system of units. The pairwise interaction between nucleons is in general more complicated than a central potential $v(|\underset{\sim}{r}_i - \underset{\sim}{r}_j|)$. In fact, it may depend on spins, isospins, velocities, densities, etc, thus converting eq. (1.1) into a system of coupled integro-differential equations. Three-body forces are even possible. As it stands, the complexity of eq. (1.1) will be sufficient for the present discussion.

Although linear, the TDSE, eq. (1.1), is impossible to solve with present computers. There are just too many variables. Among the recent developments of nuclear theory one may thus notice three approaches, all of which reduce in practice the number of independent degrees of freedom, namely i) the collective model, ii) the shell model and iii) more recently, the time-dependent Hartree-Fock approximation (TDHF).

The collective model replaces $\underset{\sim}{r}_1,\ldots,\underset{\sim}{r}_N$ by a much smaller number of degrees of freedom $Q_1,Q_2\ldots Q_{N'}$, $N' \ll 3N$, which are guessed from experimental evidence. For example, the liquid drop model of the nucleus can incorporate phenomena of compression of nuclear matter, and this can be measured by the "collective" degree of freedom $Q = \Sigma_{i=1}^{N} r_i^2$. One has then to postulate a Schrödinger equation for a collective wave function $\Psi_{coll}(Q_1,\ldots,Q_{N'})$ by means of an empirical collective Hamiltonian $\mathcal{H}_{coll}$. Despite the great successes obtained by Bohr and Mottelson and their collaborators with this model, it is clear that its connection with the more fundamental eq. (1.1) is not obvious.

Rather than reducing the number of degrees of freedom, the shell model and TDHF approaches make restrictions on the wave functions ψ. In the case of TDHF, the function Ψ is approximated, at each time, by a Slater determinant Φ. While Φ then remains a function of $\underset{\sim}{r}_1,\ldots,\underset{\sim}{r}_N$, the evolution equation, eq. (1.1), is replaced by a non-linear equation

$$i \frac{\partial\Phi}{\partial t} = H\,\Phi \equiv \left[-\sum_{i=1}^{N} \Delta_i + \sum_{i=1}^{N} u(\underset{\sim}{r}_i,t)\right]\Phi - W\Phi \quad , \qquad (1.2)$$

where u and W depend on Φ, quadratically in general, and W is an integral operator. More details on the definition of u and W will be given later in these lectures. It is enough to realize at this stage that the TDHF equation, eq. (1.2), is cubic in general. Paradoxically, it is easier to solve on a computer than eq. (1.1).

As regards the shell model, the function Ψ is approximated by a linear combination of a finite number of Slater determinants, namely by projection of eq. (1.1) in a subspace of finite dimension. The choice of this subspace is usually based upon a specific solution of eq. (1.2) of the form

$$\Phi(\underset{\sim}{r}_1,\ldots,\underset{\sim}{r}_N,t) = \exp\,(-i\,\varepsilon_o\,t)\;\Phi_o^s(\underset{\sim}{r}_1,\ldots,\underset{\sim}{r}_N) \quad , \qquad (1.3)$$

where ε_o is (algebrically) as small as possible. Actually Φ_o^s is the ground state of the static Hartree-Fock equation

$$\left[-\sum_{i=1}^{N} \Delta_i + \sum_{i=1}^{N} u(\underset{\sim}{r}_i)\right]\Phi - W\,\Phi = \varepsilon_o\Phi \quad , \qquad (1.4)$$

and one can define from Φ_o^s a denumerable set of other Slater determinants, the various "particle-hole excitations", which correspond to a natural hierarchy in the nuclear dynamics. The shell-model subspaces are obtained by truncations of this hierarchy.

It turns out that some wave functions of interest in the shell model, or in TDHF, are of the form $\Psi^* = Q\;\Phi_o^s$, where Q is among the "collective" operators of the collective model. It is then clear that a certain amount of unification is available between the three approaches mentioned above. Keeping in mind the motivation of this work, namely the search for a reasonable procedure of reduction of degrees of freedom, we shall divide these lectures into the following points i) an investigation of the properties of Slater determinants and their associated algebra of one-body operators, which are our main theoretical tools (section 2), ii) a derivation of TDHF and consideration of its main properties (conservation laws, solitonic behaviours, possible bifurcations, etc), often from numerical evidence instead of detailed mathematical proofs (section 3) and finally, iii) an identification of necessary, and hopefully sufficient, conditions for collective behaviour of the nucleons (section 4).

Many of the considerations which follow are plain definitions, or can be found in standard textbooks. Unnecessary (and sometimes tedious) proofs will thus be omitted. Other mathematical statements are less traditional and will therefore receive some justification. Most of the results described in these lectures, however, are already published or under press. The justifications will therefore be kept to the strict minimum necessary for qualitative understanding. References are provided *in fine*.

2. THE MANIFOLD OF SLATER DETERMINANTS

2.1. Single-particle basis

Wave functions $\varphi(\underset{\sim}{r})$ for one nucleon are usually taken as complex, with a square integrable modulus, and twice differentiable. Although the Hermitian scalar product is complex, it is convenient, for physical reasons, to select as an orthonormal basis of their vector space a set of real valued functions $\psi_\alpha(\underset{\sim}{r})$, $\alpha = 1\ldots\infty$. This is because the average momentum of a nucleon in such a real state identically vanishes

$$\langle\alpha|\underset{\sim}{p}|\alpha\rangle \equiv \int d\underset{\sim}{r}\ \psi_\alpha(\underset{\sim}{r})\ \frac{1}{i}\ \underset{\sim}{\nabla}_r\ \psi_\alpha(\underset{\sim}{r}) = 0 \quad , \tag{2.1}$$

with the usual momentum operator $\underset{\sim}{p} = -i\underset{\sim}{\nabla}$. Imaginary values are then easily related with nuclear velocities.

Any other orthonormal basis $\{|\lambda\rangle\}$ can be deduced from the basis $\{|\alpha\rangle\}$ by a unitary transformation

$$\varphi_\lambda(\underset{\sim}{r}) = \sum_{\alpha=1}^{\infty} c_{\lambda\alpha}\ \psi_\alpha(\underset{\sim}{r}) \quad , \ \lambda = 1,\ldots\infty, \tag{2.2}$$

where $(c_{\lambda\alpha})$ is a unitary matrix. For reasons which will be given in the next subsections, it is convenient from now on to call the states $|\lambda\rangle$ hole states or particle states, respectively, whether $\lambda \leq N$ or $\lambda > N$.

2.2. Slater determinants and their corresponding particle-hole basis

The Slater determinant $\Phi_{\lambda_1 \dots \lambda_N}$, or in short notation $|\lambda_1 \dots \lambda_N\rangle$, is just the determinant of single-particle wave functions

$$\Phi_{\lambda_1 \dots \lambda_N}(\underset{\sim}{r}_1, \dots, \underset{\sim}{r}_N) = (N!)^{-1/2} \sum_P (-)^P P \, \varphi_{\lambda_1}(\underset{\sim}{r}_1) \dots \varphi_{\lambda_N}(\underset{\sim}{r}_N), \tag{2.3}$$

where P is any permutation of the nucleons. The following theorem is trivial,

<u>Theorem 1.</u> The set of all (distinct) Slater determinants, $\{|\lambda_1 \dots \lambda_N\rangle\}$, built upon a complete, orthonormal basis of single particle states, $\{|\lambda\rangle\}$, makes a complete, orthonormal basis of the space of wave functions of N fermions.

It is traditional in nuclear physics to call "reference vacuum" the determinant obtained from the choice $\lambda_1 = 1, \dots \lambda_N = N$,

$$|0\rangle \equiv |1, \dots, N\rangle \quad . \tag{2.4}$$

It is then possible to replace one of the hole states by a particle state. This defines a "1-particle - 1 hole" determinant

$$|ph\rangle \equiv |1, \dots, h-1, p, h+1, \dots N\rangle \; . \tag{2.5}$$

Two substitutions generate "2 particle - 2 hole" states

$$|pp' \, hh'\rangle \equiv |1, \dots, h-1, p, h+1, \dots h'-1, p', h'+1, \dots N\rangle \; , \tag{2.6}$$

and so on, up to "N particle - N hole" states. Under this relabelling, Theorem 1 reads alternatively

<u>Theorem 1'.</u> The set $\{|0\rangle, |ph\rangle, \dots |Np, Nh\rangle\}$ of determinants of the particle-hole hierarchy makes a complete, orthonormal basis for the states of N fermions.

2.3. One- and two-body operators

By definition a one-body operator is a sum of identical operators which act upon one nucleon only at a time

$$A \equiv \sum_{i=1}^{N} a(\underset{\sim}{r}_i, \underset{\sim}{\nabla}_i). \tag{2.7}$$

Two-body operators are sums of operators which act upon two nucleons

$$B \equiv \sum_{i>j=1}^{N} b(\underset{\sim}{r}_i, \underset{\sim}{r}_j, \underset{\sim}{\nabla}_i, \underset{\sim}{\nabla}_j) . \tag{2.8}$$

As an example, consider the Hamiltonian $\mathcal{H}$ of eq. (1.1). It is the sum of a one-body operator, the kinetic energy $T \equiv \Sigma(-\Delta_i)$, and a two-body operator, the potential energy $V = \Sigma\, v_{ij}$. The operator H of eq. (1.2) is a one-body operator, $H = \Sigma(-\Delta_i + u_i - w_i)$.

There are many reasons which make one-body operators specially interesting. For instance, many "collective" operators, such as the centre-of-mass coordinate $\underset{\sim}{R} \equiv N^{-1} \Sigma_i\, \underset{\sim}{r}_i$, the total momentum $\underset{\sim}{P} = -i\, \Sigma_i\, \underset{\sim}{\nabla}_i$, the average square radius $Q_o = N^{-1} \Sigma_i\, \underset{\sim}{r}_i^2$, the total angular momentum $\vec{J} = \Sigma_i\, (\underset{\sim}{r}_i \times \underset{\sim}{p}_i)$, and so on, are typically of a one-body nature. Furthermore, calculations with one-body operators are obviously simpler than with two-body operators. In particular the following theorem applies,

Theorem 2. Consider a Slater determinant $|0\rangle$ and its associated 1p - 1h, 2p - 2h,... basis. Consider an arbitrary one-body operator A. Then all matrix elements $\langle Mp\ Mh|\,A\,|0\rangle$ vanish if $M \geq 2$. In the case of a two-body operator B, one needs $M \geq 3$ to cancel identically the matrix elements $\langle Mp\ Mh|\,B\,|0\rangle$.

Last but not least, while the TDSE evolution operator $\exp(-i\mathcal{H}t)$ is usually impossible to calculate, or even to approximate reasonably, the exponential of a one-body operator may be easy to evaluate. It is therefore interesting to look for cases where $\exp(-i\mathcal{H}t)$ can be approximated by the exponential of a one-body operator.

2.4. Non-linearity of the manifold. Natural metric.

It is well known that the sum of two determinants does not make a determinant in general. The set of Slater determinants, although imbedded in the linear space of wave functions, is thus a curved manifold. The tangent manifold must be defined. It is given by the Thouless theorem,

Theorem 3. An infinitesimal variation of a Slater determinant can be expanded in just a sum of 1 particle-1 hole states relating to that determinant (and also a component on the determinant itself).

$$|\Phi\rangle \equiv |0\rangle \rightarrow |\Phi'\rangle \equiv |\Phi + \delta\Phi\rangle \quad ,$$

$$|\delta\Phi\rangle = \delta c_o|0\rangle + \sum_{ph} \delta c_{ph}|1p1h\rangle \quad . \tag{2.9}$$

For instance, to vary the two-nucleon determinant $\varphi_1(\underset{\sim}{r}_1)\varphi_2(\underset{\sim}{r}_2) - \varphi_1(\underset{\sim}{r}_2)\varphi_2(\underset{\sim}{r}_1)$, one can modify φ_1 into $\varphi_1 + \delta c\,\varphi_3$ and φ_2 into $\varphi_2 + \delta c'\,\varphi_4$, and the first order variation reads $\delta c|32\rangle - \delta c'|14\rangle$. It is trivial to generalize this example into a proof of the theorem. The $|ph\rangle$ space can thus be visualized as the tangent space at point $|0\rangle$.

It is now interesting to define a metric on this manifold. Because of the Hilbert space in which it is usually embedded, the corresponding scalar product provides automatically the length element

$$d\ell^2 = \langle d\Phi|d\Phi\rangle = \sum_{ph} |dc_{ph}|^2 \quad , \tag{2.10}$$

where it is assumed that dc_o, physically uninteresting, vanishes.

As an application of this metric consider the simplest case of a one nucleon determinant (a single-particle wave-function) which is taken as Gaussian and evolves by translation from average position $\underset{\sim}{s}_1$ to average position $\underset{\sim}{s}_2$,

$$\Phi_{\underset{\sim}{s}}(\underset{\sim}{r}) \equiv \pi^{-3/4}\,\sigma^{-3/2}\,\exp\left[-\frac{(\underset{\sim}{r}-\underset{\sim}{s})^2}{2\sigma^2}\right] . \tag{2.11}$$

The distance is

$$L = \int_0^1 d\lambda \left\langle \frac{d\Phi_{\underset{\sim}{s}}}{d\lambda} \middle| \frac{d\Phi_{\underset{\sim}{s}}}{d\lambda} \right\rangle^{1/2} \quad , \quad \underset{\sim}{s} = \underset{\sim}{s}_1 + \lambda(\underset{\sim}{s}_2 - \underset{\sim}{s}_1)$$

$$= |\underset{\sim}{s}_2 - \underset{\sim}{s}_1|\,(\sigma\sqrt{2})^{-1} \quad . \tag{2.12}$$

The width of the wave packet is seen to govern the scale of the metric. This is physically significant : a narrow wave packet is more modified by a displacement than a flat wave packet by the same displacement.

2.5. A little physics

A simple model of the low lying spectrum of a nucleus consists in describing i) the ground state by the Slater determinant Φ^s solution of the static Hartree-Fock problem, eq. (1.4), then ii) each low-lying excited state by a pure 1p-1h configuration built upon Φ_o^s, then iii) higher-lying excited states by 2p-2h configurations and so on.

An actual nuclear spectrum strongly deviates from such a single picture. Nevertheless, such a simple model proves to be of great help, at least for some nuclei, to classify and/or label the eigenstates. There are nuclear states, however, which definitely differ totally from a more or less pure particle-hole excitation. They are "collective" states, of the form $\Psi^* \simeq Q\,\Phi_o^s$, where Q is at least a one-body operator for which none of the matrix elements $<ph|\,Q\,|\Phi_o^s>$ vanishes a priori. In other words the collective states read

$$|\Psi^*> \;=\; \sum_{ph} Q_{ph}\;|ph> \quad , \tag{2.13}$$

where all the numbers Q_{ph} may be of the same order of magnitude. This can be considered as an intermediate case between i) quasi-pure one-particle, one-hole states, for which all numbers Q_{ph} are small except one, and ii) very mixed states, which contain 2p-2h, 3p-3h (and even more) components together with 1p-1h components. All these cases are found in actual nuclei.

Consider the situation where, at time t = 0, the nucleus is slightly excited from its ground state, with a small admixture δc of a collective state. Assume that the ground state can be approximated by Φ_o^s and the collective state by eq. (2.13). Then at all times the wave function of the nucleus is

$$\Psi(t) \;=\; \exp(-iE_o t)\left[\,|\Phi_o^s> + \exp(-iE^* t)\;\delta c\;\sum_{ph} Q_{ph}|ph>\right], \tag{2.14}$$

where E_o is the energy of the ground state and E^* the excitation energy of the collective state. From eq. (2.9) one recognizes that Ψ is a Slater determinant with global phase $\exp(-iE_o t)$ and

$$\delta c_{ph} \;=\; \exp(-iE^* t)\;Q_{ph}\;\delta c \quad . \tag{2.15}$$

Therefore Ψ can be described by TDHF.

This is only a limit case (δc small) and it will be seen later that TDHF can also describe large scale evolutions, in the sense of the metric established in subsection 2.4. The search for interesting modes of nuclear dynamics can therefore proceed through the selection of some TDHF trajectories $\Phi(t)$, simple enough so that one can recognize from them the influence of an operator Q, later identified as "collective".

3. DERIVATIONS AND PROPERTIES OF TDHF

3.1. TDSE and the principle of least action (PLA)

Consider a time-dependent wave-function $\Psi(t)$ (not a Slater determinant, but a general N-nucleon wave-function) and the action functional

$$\mathcal{S} \equiv \int_{t_1}^{t_2} \langle\Psi(t)|(\mathcal{H} - i\frac{\partial}{\partial t})|\Psi(t)\rangle \quad , \tag{3.1}$$

where the symbol $\langle\ |\ \rangle$ means integration with respect to all variables but time. It is trivial to relate TDSE to a stationarity condition of $\mathcal{S}$ with respect to the trajectory $\Psi(t)$. Indeed, schematically, the condition

$$\delta\mathcal{S} = \int_{t_1}^{t_2} \langle\delta\Psi(t)|(\mathcal{H} - i\frac{\partial}{\partial t})|\Psi(t)\rangle = 0, \ \forall\ \delta\Psi(t), \tag{3.2}$$

reduces to $(\mathcal{H} - i\frac{\partial}{\partial t})|\Psi\rangle = 0$ since $\delta\Psi$ is arbitrary.

3.2. TDSE and Hamilton equations

If $\Psi(t)$ is parametrized by its expansion on a complete basis $\{\Phi_k\}$ in the N-body space

$$\Psi(t) = \sum_{k=1} c_k(t)\ \Phi_k \quad , \tag{3.3}$$

the action functional is governed by the energy functional

$$\mathcal{E} \equiv \langle\Psi|\mathcal{H}|\Psi\rangle = \sum_{\ell k} c_k^* \mathcal{H}_{k\ell}\, c_\ell \quad , \tag{3.4}$$

$$\mathcal{H}_{k\ell} \equiv \langle\Phi_k|\mathcal{H}|\Phi_\ell\rangle$$

and TDSE then reads

$$i\frac{dc_k}{dt} = \frac{\partial\mathcal{E}}{\partial c_k^*} \quad . \tag{3.5}$$

Define a "coordinate"

$$q_k = \sqrt{2}\ \mathrm{Re}\ c_k \tag{3.6}$$

and a "momentum"

$$p_k = \sqrt{2}\ \mathrm{Im}\ c_k \quad . \tag{3.7}$$

One obtains from eq. (3.5)

$$\frac{dq_k}{dt} = \frac{\partial \mathcal{E}}{\partial p_k} , \quad \frac{dp_k}{dt} = - \frac{\partial \mathcal{E}}{\partial q_k} , \tag{3.8}$$

which expresses that TDSE is equivalent to an infinite set of linear Hamilton equations.

3.3. TDHF and PLA.

When Ψ is restricted to a Slater determinant, the variation of the action becomes

$$\delta \mathcal{S} = \int_{t_1}^{t_2} \langle \delta\Phi(t) | (\mathcal{H} - i \frac{\partial}{\partial t}) | \Phi(t) \rangle \tag{3.9}$$

and we notice again that $\delta\Phi$ is an arbitrary linear combination of the particle-hole determinants derived from $\Phi(t)$. This leads to a first formulation of TDHF

$$i \langle ph | \frac{d\Phi}{dt} \rangle = \mathcal{H}_{ph} , \quad \forall\ ph \quad , \tag{3.10}$$

$$\mathcal{H}_{ph} \equiv \langle ph | \mathcal{H} | \Phi \rangle \quad .$$

3.4. TDHF and Hamilton equations

The determinant $\Phi(t)$ can be parametrized by its hole states $\varphi_\lambda(\underset{\sim}{r},t)$, $\lambda = 1...N$, and more precisely by their components on a fixed single-particle basis, as defined in eq.(2.2). The action integral is then governed by the energy functional

$$\mathcal{F} \equiv \langle \Phi | \mathcal{H} | \Phi \rangle = \sum_{\alpha\beta} t_{\alpha\beta}\, \rho_{\alpha\beta} + \frac{1}{2} \sum_{\alpha\beta\gamma\delta} (u_{\alpha\beta\gamma\delta} - w_{\alpha\beta\gamma\delta})\, \rho_{\alpha\gamma} \rho_{\beta\delta} ,$$

$$\alpha,\beta,\gamma,\delta = 1...\infty \tag{3.11}$$

with

$$\rho_{\alpha\beta} = \sum_{\lambda=1}^{N} c^*_{\lambda\alpha} c_{\lambda\beta} \quad , \tag{3.12}$$

$$t_{\alpha\beta} = - \int d\underset{\sim}{r} \;\; \psi_\alpha(\underset{\sim}{r}) \; \Delta\psi_\beta(\underset{\sim}{r}) \quad , \tag{3.13}$$

$$u_{\alpha\beta\gamma\delta} = \int d\underset{\sim}{r}\, d\underset{\sim}{r}' \; \psi_\alpha(\underset{\sim}{r}) \; \psi_\beta(\underset{\sim}{r}')v(|\underset{\sim}{r}-\underset{\sim}{r}'|) \; \psi_\gamma(\underset{\sim}{r}) \; \psi_\delta(\underset{\sim}{r}') \; , \tag{3.14}$$

$$w_{\alpha\beta\gamma\delta} = \int d\underset{\sim}{r}\, d\underset{\sim}{r}' \; \psi_\alpha(\underset{\sim}{r}) \; \psi_\beta(\underset{\sim}{r}')v(|\underset{\sim}{r}-\underset{\sim}{r}'|) \; \psi_\gamma(\underset{\sim}{r}')\psi_\delta(\underset{\sim}{r}) \; . \tag{3.15}$$

The proof of eqs. (3.11) to (3.15) relies only on the fact that Φ is a Slater determinant. The calculations are slightly tedious but straightforward. They can be found in standard textbooks[1] of quantum mechanics.

The functional derivative of the action with respect to $c^*_{\lambda\alpha}(t)$ then gives an alternate formulation of TDHF,

$$i \frac{dc_{\lambda\alpha}}{dt} = \frac{\partial \mathcal{F}}{\partial c^*_{\lambda\alpha}} \quad , \alpha = 1\ldots\infty \; , \lambda = 1\ldots.N \quad . \tag{3.16}$$

The coordinates $q_{\lambda\alpha} \equiv \sqrt{2}$ Re $c_{\lambda\alpha}$ and momenta $p_{\lambda\alpha} \equiv \sqrt{2}$ Im $c_{\lambda\alpha}$ thus obey an infinite set of Hamilton equations, with $\mathcal{F}$ as Hamiltonian. It can be noticed that $\mathcal{F}$ is a polynomial of degree 4 in q's and p's, and the equations of motion are thus of degree 3 (the degree becomes 5 if there are three-body forces).

3.5. TDHF in coordinate space.

There is a simpler way to derive TDHF equations than in the above subsections, provided one uses the coordinate rather than the particle-hole representation. The same energy functional, eq. (3.11), can be written as

$$\mathcal{F} = - \sum_\lambda \int d\underset{\sim}{r} \;\; \varphi^*_\lambda(\underset{\sim}{r},t) \;\; \Delta \; \varphi_\lambda(\underset{\sim}{r},t)$$

$$+ \frac{1}{2} \sum_{\lambda,\mu=1}^{N} \int d\underset{\sim}{r}\; d\underset{\sim}{r}' \;\; \varphi^*_\lambda(\underset{\sim}{r},t)\varphi^*_\mu(\underset{\sim}{r}',t) \;\; v(|\underset{\sim}{r}-\underset{\sim}{r}'|)\varphi_\lambda(\underset{\sim}{r},t)\varphi_\mu(\underset{\sim}{r}',t)$$

$$- \frac{1}{2} \sum_{\lambda,\mu=1}^{N} \int d\underset{\sim}{r}\; d\underset{\sim}{r}' \;\; \varphi^*_\lambda(\underset{\sim}{r},t)\varphi^*_\mu(\underset{\sim}{r}',t)v(|\underset{\sim}{r}-\underset{\sim}{r}'|)\varphi_\mu(\underset{\sim}{r},t) \;\; \varphi_\lambda(\underset{\sim}{r}',t) \; . \tag{3.17}$$

The functional derivative $\delta\mathcal{F}/\delta\varphi^*_\lambda(r,t)$ yields, as a condition for stationary action, a third (and still equivalent) formulation of TDHF,

$$i\frac{\partial\varphi_\lambda}{\partial t} = -\Delta\varphi_\lambda + u(\tilde{r},t)\varphi_\lambda - \int d\tilde{r}'\, w(\tilde{r},\tilde{r}',t)\,\varphi_\lambda(\tilde{r}',t)\,,\qquad \lambda = 1,\ldots N \tag{3.18}$$

with

$$u(\tilde{r},t) = \sum_{\mu=1}^{N}\int d\tilde{r}'\, v(|\tilde{r}-\tilde{r}'|)\,|\varphi_\lambda(\tilde{r}',t)|^2, \tag{3.19}$$

$$w(\tilde{r},\tilde{r}',t) = \sum_{\mu=1}^{N}\varphi_\mu(\tilde{r},t)\, v(|\tilde{r}-\tilde{r}'|)\,\varphi^*_\mu(\tilde{r}',t)\quad . \tag{3.20}$$

This is nothing but eq. (1.2), where $W = \sum_{i=1}^{N} w(\tilde{r}_i,\tilde{r}'_i,t)$. It is seen that eq. (1.2) has been reduced to a set of N coupled equations, one for each single-particle wave function φ_λ .

3.6. Numerical resolution schemes.

The typical radius of a nucleus is a few times 10^{-13}cm, say 5 fm. A mesh of 10 steps of 1 fm, in three dimensions (1000 points) is thus in order. This is only an order of magnitude : a parallelepipedic mesh is better, for instance, if two nuclei collide. One might also consider a spherical mesh if implosion phenomena are studied.

A typical time step Δt for the motion of a single nucleon is known to be 10^{-23} s. Collective motion is slower, with a typical time of order 10^{-21} s. A TDHF calculation usually involves one or two hundred time steps, in order to go from 10^{-23} to 10^{-21} s.

The most trivial algorithm to solve eqs. (3.18) is to calculate φ_λ at time t + Δt as

$$\varphi_\lambda(t+\Delta t) = \left[1 + i\,\Delta t/2(\Delta - u(t) + w(t))\right]\varphi_\lambda(t) \tag{3.21}$$

where the Laplacian is approximated by a finite difference formula and u and w are calculated at time t from eqs. (3.19) and (3.20). A more refined (implicit) algorithm reads

$$\varphi_\lambda(t+\Delta t) = \left[1 + i\,\Delta t/2(\Delta - u(t) + w(t))\right]\left[1 - i\,\Delta t/2(\Delta - u(t) + w(t))\right]^{-1}\varphi_\lambda(t)\,. \tag{3.22}$$

This formula, eq (3.22) has the advantage of an explicit unitary transformation of φ_λ from t to t + Δt, but the disadvantage of demanding a matrix inversion.

In any case, these algorithms remain costly. A typical TDHF computation for the collision of two middle-weight nuclei may take 15 minutes on a CDC 7600. Approximations to eqs. (3.18) to (3.22), and other algorithms, have been necessary. More details can be found in the references.

3.7. Initial conditions

One specifies the single particle wave functions $\varphi_\lambda(\underset{\sim}{r},t)$ at time $t = 0$. If one wants to study the behaviour of one nucleus, all these wave functions are wave packets centered around the center of the integration mesh (one-center problem).

If one wants to study the collision of two nuclei, one usually defines a two-center problem in the following way. Let Φ^s_a be the (real) solution of the static Hartree-Fock equation (1.4) for the projectile a, with single particle orbitals $\psi^a_\lambda(\underset{\sim}{r})$, $\lambda = 1,\ldots,N_a$, which are localized around the center of the mesh. The wave functions ψ^a_λ will now be translated and "boosted", defining

$$\varphi^a_\lambda(\underset{\sim}{r},t{=}0) \;=\; \exp(i\,\underset{\sim}{k}_a\cdot\underset{\sim}{r})\;\psi^a_\lambda(\underset{\sim}{r}-\underset{\sim}{r}_a)\quad, \tag{3.23}$$

where $\underset{\sim}{r}_a$ is the new mesh point selected as center for nucleus a and $\underset{\sim}{k}_a$ is the momentum given to a. Indeed it is trivial to check that, in connection with eq. (2.1),

$$<\varphi^a_\lambda|\underset{\sim}{p}|\varphi^a_\lambda> \;=\; \underset{\sim}{k}_a \quad . \tag{3.24}$$

In the same way, the static Hartree-Fock orbitals ψ^A_λ, $\lambda = N_a+1,\ldots,N_a+N_A=N$, of the target nucleus A are shifted and boosted by a different amount, defining

$$\varphi^A_\lambda(\underset{\sim}{r},t{=}0) \;=\; \exp(i\,\underset{\sim}{k}_A\cdot\underset{\sim}{r})\;\psi^A_\lambda(\underset{\sim}{r}-\underset{\sim}{r}_A). \tag{3.25}$$

The set of orbitals $\{\varphi^a_\lambda,\varphi^A_\lambda\}$ can be used to build a Slater determinant $\Phi(t{=}0)$, describing two nuclei at initial distance $\underset{\sim}{r}_a - \underset{\sim}{r}_A$, and with momenta $\underset{\sim}{k}_a$, $\underset{\sim}{k}_A$. Central collision occurs if $\underset{\sim}{k}_a$ is parallel and opposite to $\underset{\sim}{r}_a - \underset{\sim}{r}_A$ and $\underset{\sim}{k}_A$ parallel and oriented like $\underset{\sim}{r}_a - \underset{\sim}{r}_A$.

A one-center problem can be defined, however, for collisions provided one considers implosion or explosion phenomena. A typical single particle wave function for such a situation is

$$\varphi_\lambda(\underset{\sim}{r}) = \exp(\mp i\, k_\lambda |\underset{\sim}{r}|) \quad \psi_\lambda(|\underset{\sim}{r}| - |\underset{\sim}{r}_\lambda|), \qquad (3.26)$$

where ψ_λ is real and one must notice the dependence on the radial coordinate $|\underset{\sim}{r}|$ only. It would be interesting to reintroduce a slight dependence on the orientation of $\underset{\sim}{r}$ in order to test stability with respect to a deviation of sphericity.

3.7. Conservation laws

Theorem 4. The scalar product of single-particle wave-functions which evolve under TDHF is conserved.

Indeed, consider eqs. (3.18) to (3.20). The operator $-\Delta + u - w$ is obviously Hermitian and thus generates a unitary evolution.

Theorem 5. The energy is conserved by the TDHF evolution. Indeed the time derivative of the energy is

$$\frac{d}{dt} \langle\Phi(t)|\mathcal{H}|\Phi(t)\rangle = i\langle\Phi(t)|(\mathcal{H} H - H\mathcal{H})|\Phi(t)\rangle\,, \qquad (3.27)$$

where H is that one-body operator defined by eq. (1.2). By virtue of Theorem 2, only the 1p - 1h matrix elements of H can contribute to the right-hand side of eq. (3.27). One finds

$$i\langle\Phi(t)|(\mathcal{H} H - H\mathcal{H})|\Phi(t)\rangle = i \sum_{ph} \left(\begin{array}{c} \langle\Phi(t)|\mathcal{H}|ph\rangle\langle ph|H|\Phi(t)\rangle \\ \langle\Phi(t)|H|ph\rangle\langle ph|\mathcal{H}|\Phi(t)\rangle \end{array} \right) \qquad (3.28)$$

which vanishes identically since, by eqs. (2.1) and (3.10), the particle-hole matrix elements of H and $\mathcal{H}$ are identical.

Theorem 6. A conservation law valid for TDSE is also valid for TDHF if the conserved quantity is the expectation value of a one-body operator A fixed in time.

Indeed the time derivative of this quantity is

$$\begin{aligned} \frac{d}{dt} \langle\Phi(t)|A|\Phi(t)\rangle &= i\langle\Phi(t)|(AH - HA)|\Phi(t)\rangle \\ &= i \sum_{ph} (\langle\Phi(t)|A|ph\rangle\langle ph|H|\Phi(t)\rangle - \langle\Phi(t)|H|ph\rangle\langle ph|A|\Phi(t)\rangle) \\ &= i \sum_{ph} (\langle\Phi(t)|A|ph\rangle\langle ph|\mathcal{H}|\Phi(t)\rangle - \langle\Phi(t)|\mathcal{H}|ph\rangle\langle ph|A|\Phi(t)\rangle) \\ &= i \langle\Phi(t)|(A\mathcal{H} - \mathcal{H}A)|\Phi(t)\rangle\,, \end{aligned} \qquad (3.29)$$

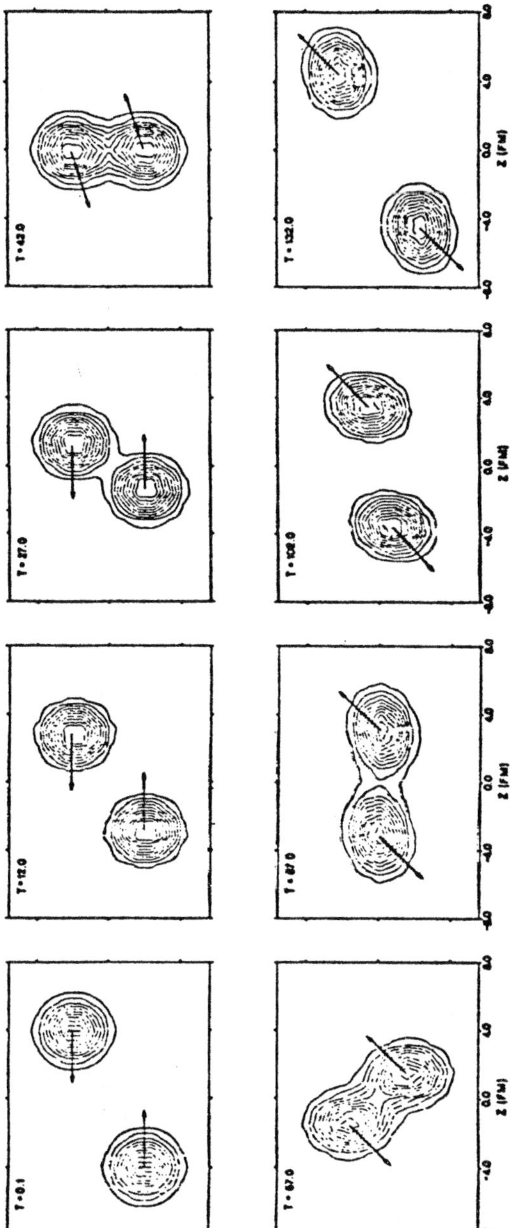

Figure 1 - Density plot for α-α scattering

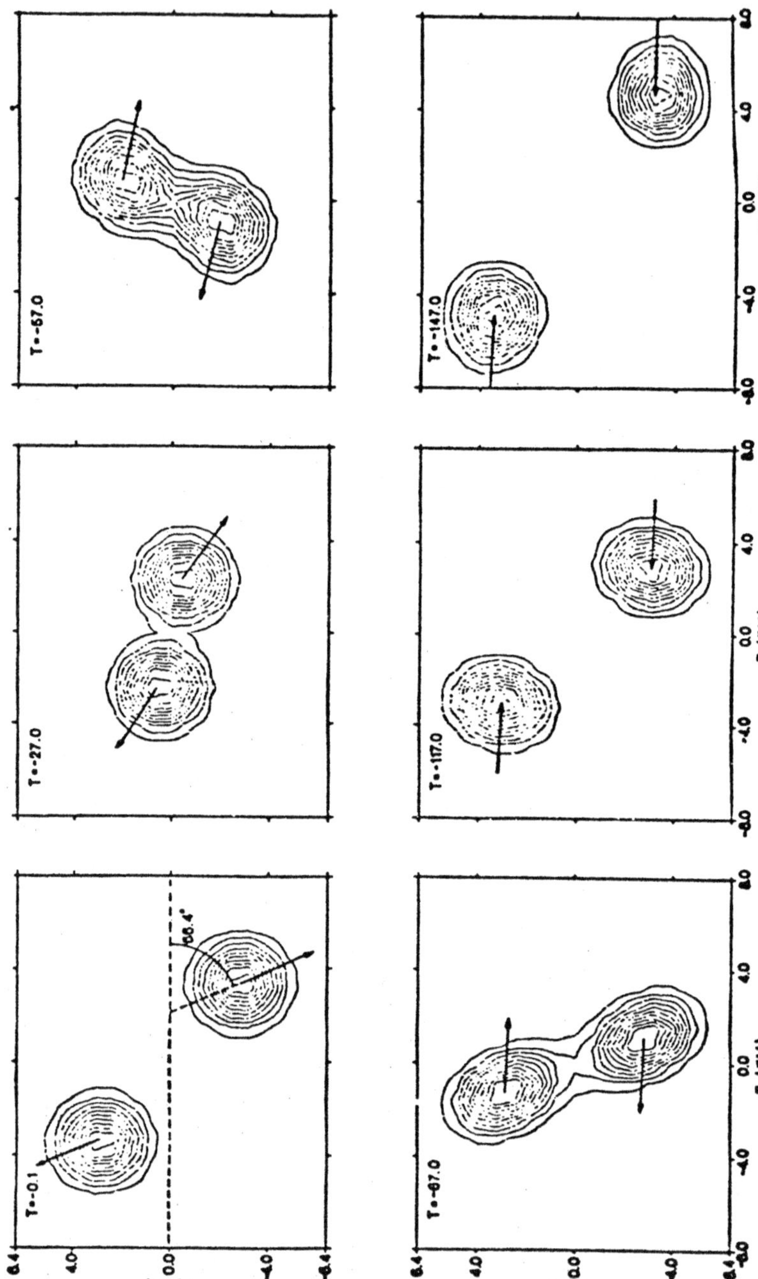

Figure 2 - Density plot for α-α scattering (time-reversed).

which vanishes, since A, the conserved operator, commutes by definition with $\mathcal{H}$.

All these conservation laws are of special interest to test the accuracy of a numerical algorithm, or even to stabilize the algorithm.

3.8. Solitonic solutions.

The meaning of "soliton" in the present discussion is the following. While the conservation laws of subsection 3.7 are valid for any solution of TDHF, specific initial conditions may lead to specific solutions which have additional conservation laws. For instance, all the geometrical parameters which describe a nuclear shape might remain constant, except one of them. This shape rigidity reminds of a soliton, and the only degree of freedom which remains moving as a function of time might then be interpreted as a collective coordinate. This collective coordinate looks like a normal mode, decoupled from the other degrees of freedom. The reason why the other degrees of freedom are not excited is either that their coupling is small, or alternately that they can move so fast that, in a Born-Oppenheimer sense, they are locked to the slower, heavier, "collective" degree of freedom. In any case, TDHF solutions with rigidity properties are of special interest, for they represent disentangled degrees of freedom.

A first solitonic solution is provided by following theorem :

Theorem 7. A static Hartree-Fock solution boosted in the translation mode goes on evolving by pure translation without spreading.

Indeed compare the particle-hole matrix elements $\langle p_o h_o | \mathcal{H} | \Phi_o^s \rangle$ and $\langle ph | \mathcal{H} | \Phi(t) \rangle$ referring to the static HF solution and the boosted, translated state deduced from Φ_o^s. Since by ansatz one considers

$$\Phi(t) = \exp(-i\ \underset{\sim}{P}.\underset{\sim}{v}t)\ \exp(i\ \underset{\sim}{k}.\underset{\sim}{R})\ \Phi_o^s \quad , \tag{3.30}$$

where $\underset{\sim}{P} \equiv \sum_{i=1}^{N} \underset{\sim}{p}_i$ and $R \equiv N^{-1} \sum_{i=1}^{N} \underset{\sim}{r}_i$ are the (one-body) generators for displacement and boost, respectively, and where $\underset{\sim}{v}$ and $\underset{\sim}{k}$ are the velocity and momentum, the particle-hole basis for $\Phi(t)$ is deduced from that of Φ_o^s by the same transformation

$$|ph\rangle = \exp(-i\ \underset{\sim}{P}.\underset{\sim}{v}t)\ \exp(i\ \underset{\sim}{k}.\underset{\sim}{R})\ |p_o h_o\rangle \ . \tag{3.31}$$

One thus finds

$$\langle ph | \mathcal{H} | \Phi(t) \rangle = \langle p_o h_o | \exp(-i\ \underset{\sim}{k}.\underset{\sim}{R})\ \mathcal{H}\ \exp(i\ \underset{\sim}{k}.\underset{\sim}{R}) | \Phi_o^s \rangle \quad , \tag{3.32}$$

because $\mathcal{H}$ commutes with translations. Furthermore the only term

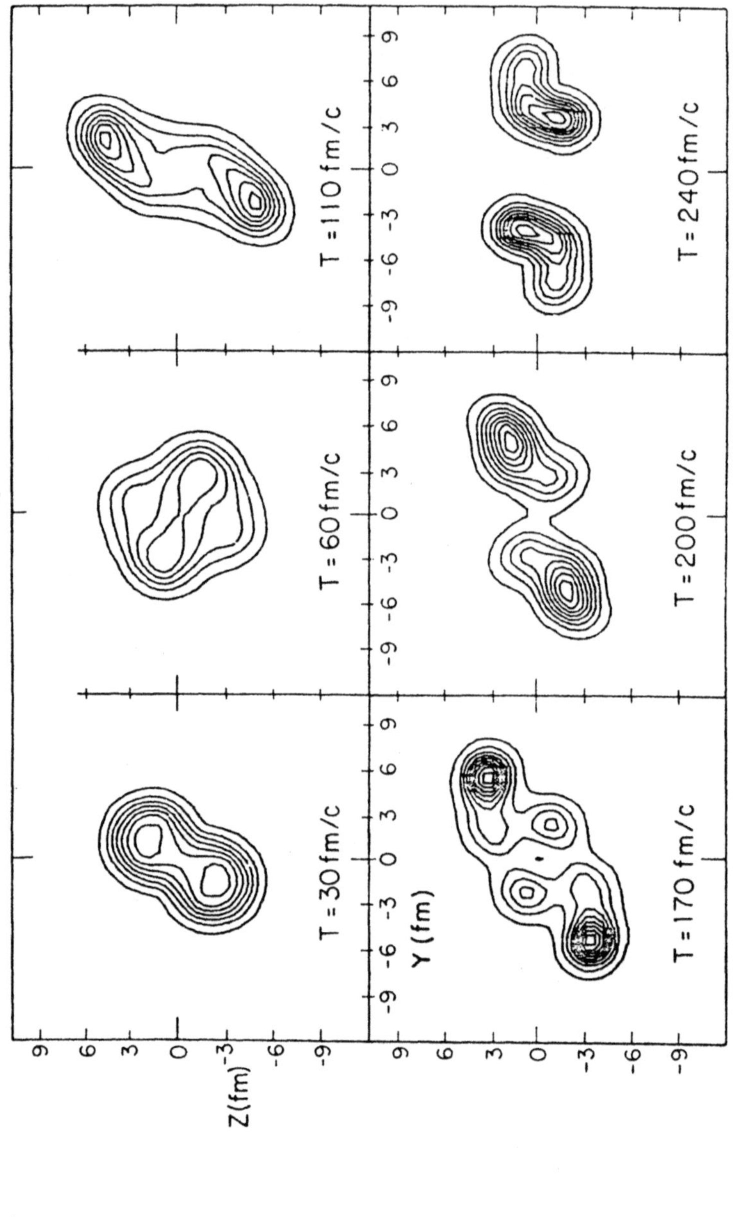

Figure 3 - Density plot for ^{16}O - ^{16}O scattering.

in $\mathcal{H}$ which does not commute with boosting is the center-of-mass kinetic energy $N^{-1} P^2$. One finds then

$$<ph|\mathcal{H}|\Phi(t)> = <p_o h_o|N^{-1}(\underset{\sim}{P} + \underset{\sim}{k})^2 + \mathcal{H} - N^{-1}P^2|\Phi_o^s> \quad , (3.33)$$

and finally

$$<ph|\mathcal{H}|\Phi(t)> = 2N^{-1} \underset{\sim}{k}. <p_o h_o|\underset{\sim}{P}|\Phi_o^s> \quad , \qquad (3.34)$$

since $|p_o h_o>$ and $|\Phi_o^s>$ are orthogonal and the matrix elements $<p_o h_o|\mathcal{H}|\Phi_o^s>$ vanish for the static HF solution Φ_o^s. It is seen from eq. (3.30) and the condition $\underset{\sim}{v} = 2N^{-1} \underset{\sim}{k}$ that eq. (3.34) reads

$$<ph|\mathcal{H}|\Phi(t)> \;=\; i <ph|\frac{d\Phi}{dt}> \quad ,$$

which is nothing but eq. (3.10). This proves that $\Phi(t)$, as given by the ansatz, eq. (3.40), is a TDHF solution.

The behaviour expressed by eq. (3.40) is purely solitonic.
The wave - packet propagates by translation, with all other degrees of freedom frozen. This is the first evidence of collective behaviour in a specific TDHF solution.

3.9. Quasi-solitonic solutions

One sees on Figs. 1 and 2 taken from a calculation of $\alpha - \alpha$ scattering by Devi, Irvine and Strayer [2] that the density plots of the He nuclei do not show major deformations of the nuclear densities. Mainly translations and rotations seem to occur, with a certain amount of polarisation at closest approach. The nuclei seem largely undisturbed after collision. This rigidity reminds of solitonic behaviour.

It may be pointed out that Fig. 2 is a time- reversed TDHF calculation ($i \to -i$ in eq. (3.18) for instance). Time-reversed evolution is an interesting problem by itself and anyhow provides TDHF computer codes with accuracy checks.

3.10. Non-solitonic solutions

It is clear from Fig. 3 , taken from a calculation of $^{16}O - {}^{16}O$ scattering by Cusson, Smith and Maruhn [3] , that the nuclear density plots exhibit a large amount of deformations, which vary strongly when time goes on. This is therefore probably not a solitonic solution in the sense discussed above (rigidity). One may notice, however, that the nuclear wave packets do not seem to spread when the collision is finished. This apparent confinement might be a residue of solitonic behaviour.

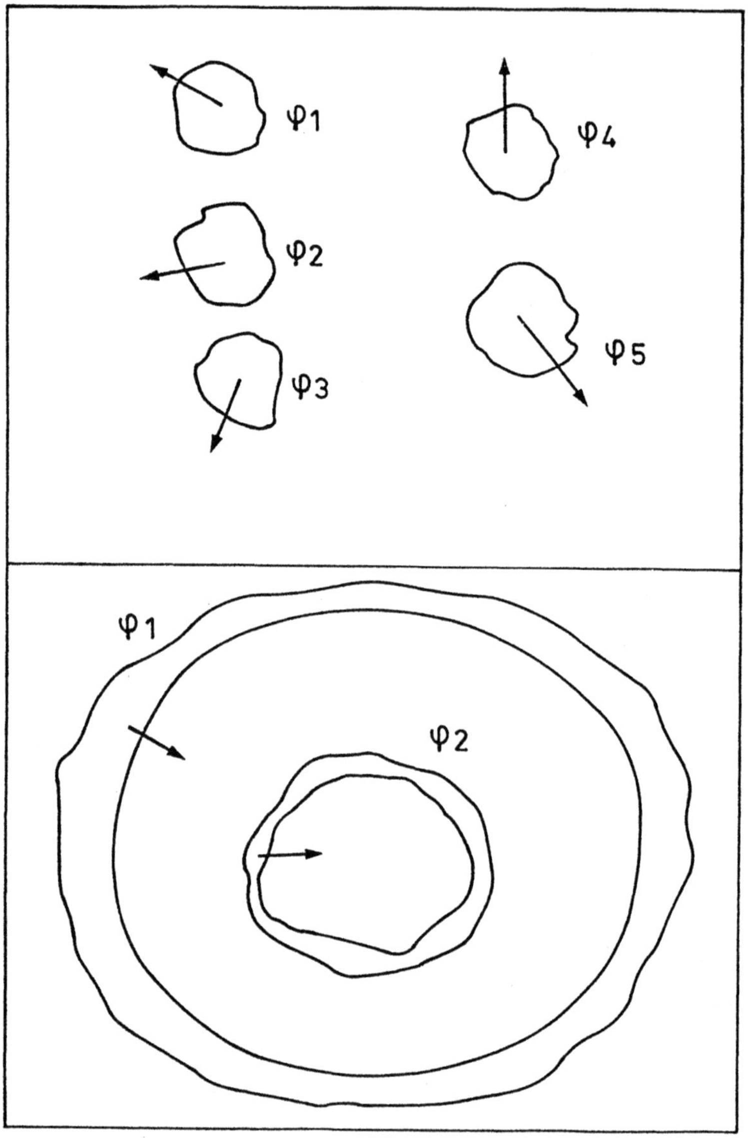

Figure 4 - Initial condition for spreading (top) or implosion (bottom).

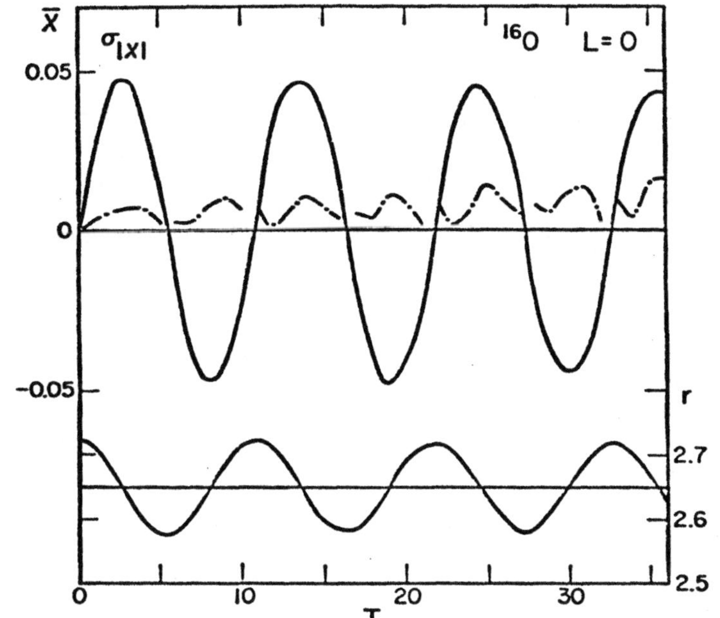

Figure 5 - Oscillation of mean square radius of ^{16}O.

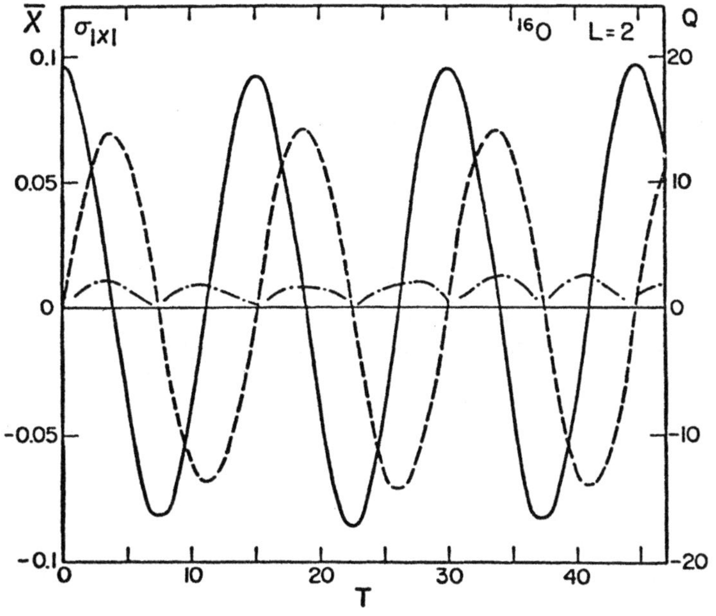

Figure 6 - Oscillation of quadrupole moment of ^{16}O.

An initial condition such as that shown by Fig. 4, which consists in separate wave packets for all nucleons, flying apart, gives an example of a strongly non-collective behaviour. For the action of the potentials u and w upon the wave functions φ_λ vanish for such an initial condition. Indeed, as shown by eqs. (3.19) and (3.20) the domains of u and w cannot exceed the domains of the nuclear density by more than a fringe equal to the range of v. The non linearity of TDHF is thus switched off, and H reduces to $\Sigma_i(-\Delta_i)$. Wave packets spread in that limit case, where TDSE and TDHF become equivalent.

3.11. Periodic solutions

A first case of periodic solutions relates to the solutions of the static Hartree-Fock problem, see eqs. (1.3) and (1.4). As it only involves the global phase of $\phi(t)$, it probably carries less information than a "truly" periodic solution which would let the nuclear density oscillate as a function of time.

Numerical evidence of such oscillatory behaviour has been found by Bonche, Doubre and Quentin [4] for the monopole and quadrupole vibrations of ^{16}O. The data plotted on Figs. 5 and 6 seem to behave harmonically. Further studies indicate that these solutions could be of the form shown by eq. (2.14), where Q would be closely related to the average square radius or the quadrupole moment. All other degrees of freedom being apparently frozen, this kind of solution therefore looks like solitonic and/or collective. There remains to check whether the same conclusions hold for large amplitude vibrations which are more interesting than small amplitude vibrations. As a matter of fact, the latter case can be proved to give automatically an harmonic limit.

3.12 The case of fusion.

As shown by Fig. 7, taken from a calculation by Grammaticos [5] there are cases where the two nuclei do not seem to separate after a few times 10^{-21} s (maybe they separate after 10^{-20} s, but it is too expensive to run the computer for that span of time !). The point is, there seems to be a clear cut separation between the two extreme cases i) separation after a collision time of order a few times 10^{-22} s and ii) no separation if the interaction lasts more than 10^{-21} s ("fusion").

This reminds of a bifurcation situation. Assume for instance that the nuclear interaction v depends on a parameter ν (global strength, range...). Consider a fixed initial condition described by eqs. (3.23) and (3.25). Most numerical calculations available up to now show that, when the parameter ν is varied, the nuclei either collide or "fuse", with a sharp transition between the two regimes.

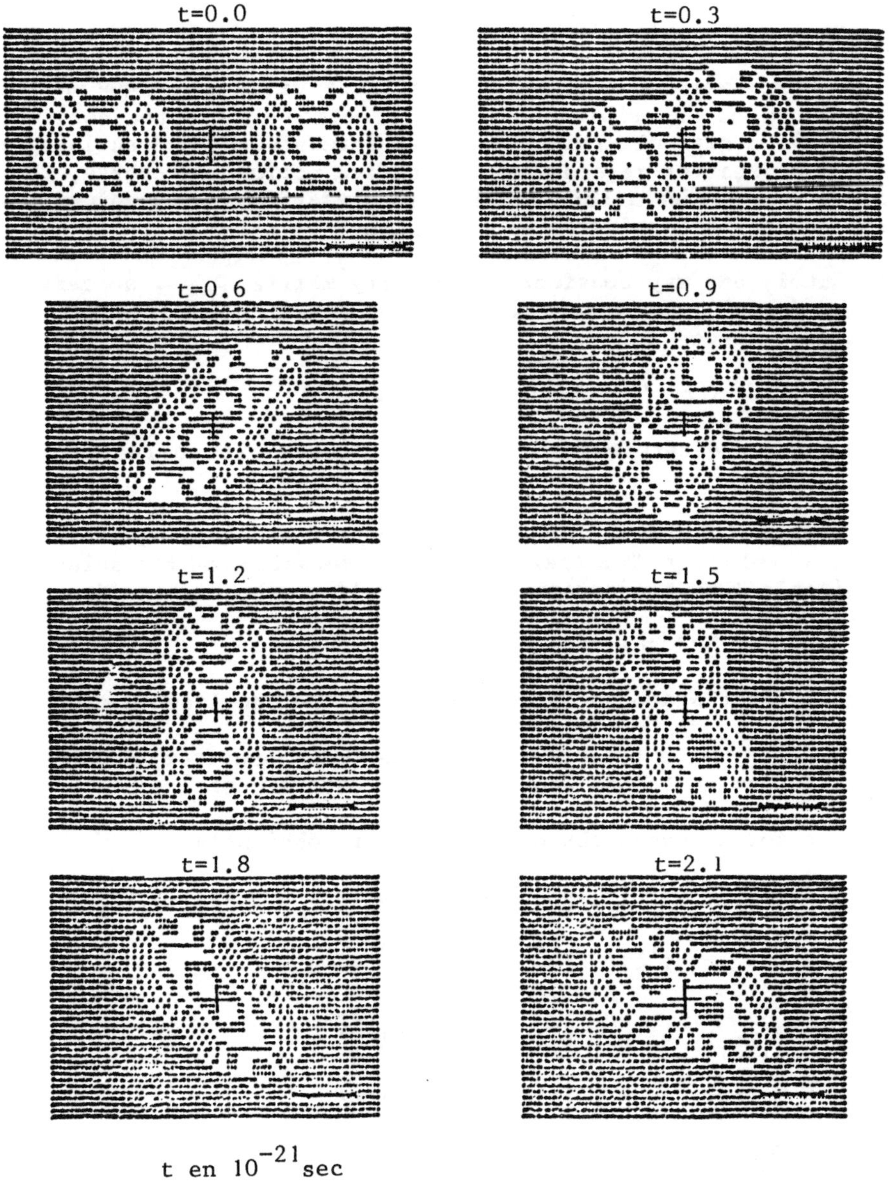

Figure 7 - Collision leading apparently to fusion.

4. SEARCH FOR COLLECTIVE MOTIONS

4.1. Real and complex determinants

In the same way as real single-particle wave-functions may be interpreted as describing no coherent motion of the nucleon, see eq. (2.1), a real determinant will be interpreted as describing a nucleus at rest as regards any collective motion. Of course, this demands that Slater determinants which differ from one another by a global phase be considered as equivalent.

Alternately one may consider the density matrix $\rho_{\alpha\beta}$, as defined by eq. (3.12) on a real basis, or the same density matrix in coordinate space

$$\rho(\underset{\sim}{r},\underset{\sim}{r}') \equiv \sum_{\lambda=1}^{N} \varphi_\lambda^*(\underset{\sim}{r})\, \varphi_\lambda(\underset{\sim}{r}') \quad . \tag{4.1}$$

A motionless determinant is then associated with a real ρ and vice-versa.

Any complex Φ , or ρ, will be interpreted as carrying a certain amount of velocity. The analysis of this velocity, and the selection of suitable TDHF trajectories $\Phi(t)$, will now be discussed.

4.2. Boosting and shifting operators

We shall first state an integral form of the Thouless theorem

<u>Theorem 3'</u>. The exponential of a one-body operator converts a Slater determinant into another Slater determinants.

Consider for instance the case N=2, where obviously

$$\exp\left[\hat{a}(\underset{\sim}{r}_1,\underset{\sim}{p}_1) + \hat{a}(\underset{\sim}{r}_2,\underset{\sim}{p}_2)\right]\left[\varphi_1(\underset{\sim}{r}_1)\,\varphi_2(\underset{\sim}{r}_2) - \varphi_2(\underset{\sim}{r}_1)\,\varphi_1(\underset{\sim}{r}_2)\right]$$
$$= \tilde{\varphi}_1(\underset{\sim}{r}_1)\,\tilde{\varphi}_2(\underset{\sim}{r}_2) - \tilde{\varphi}_2(\underset{\sim}{r}_1)\tilde{\varphi}_1(\underset{\sim}{r}_2) \quad , \tag{4.2}$$

where

$$\tilde{\varphi}_1(\underset{\sim}{r}_1) = \exp\,[\hat{a}(\underset{\sim}{r}_1,\underset{\sim}{p}_1)]\,\varphi_1(\underset{\sim}{r}_1) \quad , \tag{4.3a}$$

$$\tilde{\varphi}_2(\underset{\sim}{r}_2) = \exp\,[\hat{a}(\underset{\sim}{r}_2,\underset{\sim}{p}_2)]\,\varphi_2(\underset{\sim}{r}_2) \quad , \tag{4.3b}$$

$$\tilde{\varphi}_2(\underset{\sim}{r}_1) = \exp\,[\hat{a}(\underset{\sim}{r}_1,\underset{\sim}{p}_1)]\,\varphi_2(\underset{\sim}{r}_1) \quad , \tag{4.3c}$$

$$\tilde{\varphi}_1(\underset{\sim}{r}_2) = \exp\,[\hat{a}(\underset{\sim}{r}_2,\underset{\sim}{p}_2)]\,\varphi_1(\underset{\sim}{r}_2) \quad . \tag{4.3d}$$

What is striking in eqs (4.3) is that the argument $\underset{\sim}{r}_i$ is mute and that actually

$$\tilde{\varphi}_\lambda = \exp[\hat{a}]\varphi_\lambda \quad . \tag{4.4}$$

The generalization to N > 2 is trivial.

Assume now that $\hat{a}$ is real and consider $\mathcal{A} \equiv \sum_{i=1}^{N} \hat{a}_i$. When acting on a real determinant, $\exp \mathcal{A}$ converts it into another real determinant. This transformation corresponds to a modification of "positions", without any velocities. It can be understood as a shifting operation. It is actually convenient to assume that $\mathcal{A}$ is anti-Hermitian, in other words that $\mathcal{P} \equiv i\mathcal{A}$ is Hermitian and purely imaginary. Thus the proposition :

Prop. 1. Purely imaginary and Hermitian one-body operators $\mathcal{P}$ are generators of displacements $\exp(-i\mathcal{P})$ in the algebra of TDHF.

If $\hat{a}$ in eq. (4.4) is purely imaginary, the transformation of φ_λ into $\tilde{\varphi}_\lambda$ makes (almost always !) $\tilde{\varphi}_\lambda$ complex if φ_λ is real. It is again convenient to assume $\hat{a}$ anti-Hermitian and therefore $\mathcal{Q} \equiv i\mathcal{A}$ is Hermitian and real. The transformation $\exp(-i\,\mathcal{Q})$ is then interpreted as converting a real determinant into a complex one, with the same "positions", but now with non-vanishing velocities. Hence the proposition :

Prop. 2. Purely real and Hermitian one-body operators $\mathcal{Q}$ are generators of accelerations (boosting operators) in the algebra of TDHF.

Special cases are the center-of-mass momentum $\underset{\sim}{P}$ and coordinate $\underset{\sim}{Q}$ which were used in the ansatz of a special solution, eq. (3.30). Our task is now to look for TDHF solution of the form

$$\Phi(t) = \exp[-i\,\tau(t)\mathcal{Q}]\;\exp[-i\,\sigma(t)\mathcal{P}]\quad \Phi(0) \quad , \tag{4.5}$$

in order to interpret σ as a generalized position, τ as a generalized momentum and $\mathcal{P}$ and $\mathcal{Q}$ as the corresponding momentum and position operators.

4.3. Adiabaticity. Fibration

A formula like eq. (4.5) also reads

$$\Phi(t) = \exp[-i\;\tau\mathcal{Q}]\;\Phi_r(t) \quad , \tag{4.6}$$

where $\Phi_r(t)$ is that real determinant derived from $\Phi(0)$ by the transformation $\exp(-i\;\sigma\mathcal{P})$ if $\Phi(0)$ is also a real determinant Φ_{ro}. The problem is to derive $\mathcal{Q}$ and Φ_r from $\Phi(t)$, for the TDHF calculation only gives $\Phi(t)$.

A theorem by Baranger and Veneroni [6] establishes that such a decomposition, eq. (4.6), is unique, provided some conditions are imposed upon $\tau\mathcal{Q}$. In particular the spectrum of $\tau\mathcal{Q}$ must be strongly bounded. This means that the velocities brought by $\tau\mathcal{Q}$ are restricted to be small enough, and thus the decomposition is connected with adiabaticity conditions.

Another interpretation of this decomposition is to consider the manifold of Slater determinants as a fiber bundle. The basis would be the manifold of real determinants Φ_r. The fiber would be the group of exponentials $\exp(i\mathcal{Q})$ of purely imaginary and anti-Hermitian one-body operators $i\mathcal{Q}$.

In any case, there is a serious difficulty brought by this kind of decomposition. For one finds, for any TDHF trajectory,

$$\Phi(t) = \exp[-i\ \tau(t)\mathcal{Q}(t)]\exp[-i\ \sigma(t)\mathcal{S}(t)]\ \Phi_{r_o}, \qquad (4.7)$$

where $\mathcal{Q}$ and $\mathcal{S}$ depend on time. This is not satisfactory for our search of a fixed algebra of collective operators $\mathcal{S}$, $\mathcal{Q}$.

4.4. Stop and go analysis

Let us nonetheless proceed with the decomposition, eq. (4.6) in the adiabatic case. In particular, consider a TDHF calculation initiated at time $t = 0$ from a (motionless) real determinant Φ_o. The corresponding Hartree-Fock Hamiltonian $H_o \equiv H(\Phi_o)$, as defined by eqs. (1.2), (3.18) to (3.20), is obviously real. Therefore the TDHF evolution

$$\Phi(dt) = (1 - i\ dt\ H_o)\ \Phi_o \qquad (4.8)$$

expresses that Φ has received a slight imaginary modification. It is then obvious that H_o is a boost generator. In other word, H_o is the candidate for the rôle of "collective coordinate" operator $\mathcal{Q}$.

The next time step then reads

$$\Phi(2dt) = \left[1 - i\ dt\ H(\Phi(dt))\right]\Phi(dt)\ , \qquad (4.9)$$

where $H(\Phi(dt))$ is that Hartree-Fock Hamiltonian derived from $\Phi(dt)$ through eqs. (3.19) and (3.20). Since $\Phi(dt)$ and Φ_o differ from each other by the imaginary component $-idt\ H_o\Phi_o$, it is trivial to check from these eqs. (3.19) and (3.20) that $H(\Phi(dt))$ and H_o differ from each other by a purely imaginary (and Hermitian) operator H_1 (at order dt)

$$H(\Phi(dt)) = H_o + H_1\ . \qquad (4.10)$$

Then eq. (4.9) reads

$$\Phi(2dt) = \left[1 - idt\ (H_o + H_1)\right]\Phi(dt)\ . \qquad (4.11)$$

In the same way as H_o, a real, Hermitian, one-body operator, behaves like a boosting operator, one may interpret H_1, an imaginary, Hermitian, one-body operator as a candidate for shifting operator (collective momentum $\mathcal{P}$).

For any Φ_o it is thus possible to calculate, numerically and sometimes analytically, the operators H_o and H_1. They are the only acceptable candidates for collective coordinate and momentum, since they are derived from the dynamical equations (4.8) and (4.9). It is seen they result from a dynamical analysis of collective motion starting from rest at any point Φ_o. This is why this analysis can be called "stop and go".

The problem to be faced again is that H_o and H_1 depend strongly on H_o in general. The search for collective dynamics thus amounts to the search of a set of real determinants Φ_o which induce a set of operators H_o and H_1 weakly varying in time in some sense. This difficult question is investigated in the next subsection.

4.5. Curvature analysis (qualitative)

On a purely intuitive basis, it is clear that the particle-hole matrix elements of H_o define the acceleration vector of the TDHF trajectory as viewed from the tangent plane to the Slater manifold at Φ_o. In the same way the particle-hole matrix elements of H_1 define the velocity vector in that plane. The algebra made of H_o and H_1 thus remains (qualitatively !) constant if the trajectory is "straight" on the manifold. A study of geodesics is thus in order. It is also tempting to make the acceleration and the velocity parallel, in order to minimize "centrifugal" forces. The equation

$$< ph \,|(H_o - \Lambda H_1)|\Phi_o> \;=\; 0 \quad , \tag{4.12}$$

where Λ is a proportionality constant, thus expresses this parallelism.

Since H_o and H_1 are defined by Φ_o, the condition, eq. (4.12) is thus an equation for Φ_o. It can be checked that it amounts to a constrained Hartree-Fock equation for H_o, of degree 5 if there are only two-body forces. The exact properties of this equation have not yet been investigated in detail. It is clear, however, that the solution Φ_o is parametrized by Λ. One thus gets a collective path $\Phi_o(\Lambda)$ and the corresponding generators $H_o(\Lambda)$ and $H_1(\Lambda)$ make an algebra whose stability as a function of Λ can be investigated.

Although intuitive rather than rigorous, the condition, eq. (4.12) appears like a first answer [7] to the major problem raised at the beginning : derive collective algebras and paths from TDHF.

4.6. Curvature analysis (more rigorous).

We shall now find a necessary, although non sufficient, condition for collective dynamics. We are looking for TDHF trajectories of the form

$$\Phi(t) = \exp [i\, \alpha(t)\, Q]\, \Phi(0) \quad , \tag{4.13}$$

where Q is a fixed Hermitian one-body operator and must later be identified as "collective". It is known from Theorem 3' that, whatever $\alpha(t)$ real, this transformation, eq. (4.13), is a unitary transformation among Slater determinants. The following theorem can now be proved,

Theorem 8. According to the natural metric defined in subsection 2.4 the curvature of the trajectories defined by eq. (4.13) is comprised between 1 and $(3-2N^{-1})^{1/2}$.

In practice, since the particle number N is large, the maximum reaches $\sqrt{3}$. The proof of this theorem goes through the definition of the unit tangent vector $|T>$,

$$|T> \equiv <\Phi(t)|Q^2|\Phi(t)>^{-1/2}\ i\, Q|\Phi(t)> , \tag{4.14}$$

and the unit normal vector $|N>$

$$C|N> \equiv -<\Phi(t)|Q^2|\Phi(t)>^{-1}\ Q^2|\Phi(t)> \quad , \tag{4.15}$$

where C is the curvature. The detailed proof is a bit technical. It has been published elsewhere [8].

The physical point of interest is that all collective operators Q deduced from experimental evidence (translation, rotation, vibration...) correspond to the maximum curvature. It can be checked in particular that all such operators have particle-hole matrix elements Q_{ph} of the same order of magnitude. On the contrary, those operators Q such that $Q_{ph} = 0$ except for one matrix element $Q_{p_o h_o} = 1$ are recognized from experimental evidence as non collective, and they induce a minimal curvature.Thus the proposition

Prop. 3. Collective TDHF trajectories can be recognized from the fact that their curvature is as close as possible to $\sqrt{3}$.

In particular, the normal vector of a trajectory initiated from a real determinant Φ_o is given by

$$C|N> = -<\Phi_o|H_o^2|\Phi_o>^{-2}$$
$$\left[<\Phi_o|H_o^2|\Phi_o>(H_o^2 + i\, H_1) - \frac{1}{2}\, i <\Phi_o|H_1H_o + H_oH_1|\Phi_o>H_o\right]|\Phi_o> \tag{4.16}$$

and the equation

$$c^2 = 3 - 2 N^{-1} \tag{4.17}$$

is, like eq. (4.12), an equation for Φ_o. It is an open question whether eqs. (4.12) and (4.17) have common solutions. In any case, it must be understood that eq. (4.17), although more rigorous that eq. (4.12), is only more or less necessary (curvatures deviating slightly from $\sqrt{3-2N^{-1}}$ would be acceptable) but not sufficient.

5. CONCLUSION

Some progress has been achieved towards the extraction of a reduced number of dominant degrees of freedom in a complicated system. On one hand, TDHF, although an approximation to TDSE, does provide a calculable trajectory starting from any starting point. On the other hand, consideration of the algebra of displacement and acceleration along the trajectory provides candidates for the degrees of freedom searched for. Last but not least, one has found a beginning of selection rules (proportionality, curvature) in order to identify initial conditions for trajectories of interest, namely quasi-solitonic dynamics.

REFERENCES

General reference : "Proceedings of the Orsay-Saclay workshop on the TDHF method", to be ordered at editions de Physique, BP. n°112, 91402 - Orsay, France.

1. Messiah A., "Quantum Mechanics", Chapter XVIII.
   Thouless D.J., "Quantum Mechanics of Many-Body Systems", Chapter III. The key point of the argument is the fact that the initial Hamiltonian and the Hartree-Fock Hamiltonian have equal one-particle one-hole matrix elements, $\langle ph|\mathcal{H}|\Phi\rangle = \langle ph|H|\Phi\rangle$.
2. Sandhya Devi K.R., Strayer M.R. and Irvine J.M.;(1979) J. Phys. G 5, 281.
3. Cusson R.Y., Smith R.K. and Maruhn J.A., (1976)Phys. Rev. Lett. 36, 1166.
4. Bonche P., Doubre H., and Quentin P., (1979) Phys. Lett. 85B, 5
5. Grammaticos B., Ph-D Thesis, University of Paris XI, 1977.
6. Baranger M. and Vénéroni M., (1978) Ann. Phys. 114, 123.
7. Giraud B.G. and Le Tourneux J., "Geometrical derivation of collective operators and paths in TDHF", Saclay preprint DPh.T. 78/57.
8. Giraud B.G. and Rowe D.J., (1979) J. de Physique Lettres 8, 177.

LIST OF PARTICIPANTS

- ABRAHAM Douglas — Oxford University
  1, South Parks Road
  OXFORD
  (England)

- AUBRY Serge — Laboratoire Léon Brillouin
  Orme des Merisiers
  BP n°2
  91190 GIF-sur-YVETTE
  (France)

- BAKER George — T 11
  Los Alamos Laboratory
  LOS ALAMOS, New Mexico
  (U.S.A.)

- BALL J.M. — Department of Mathematics
  University of California
  BERKELEY, California 94720
  (U.S.A.)

- BANAI Nozar — Université Libre de Bruxelles
  Faculté des Sciences
  Campus de la Plaine
  C.P. 231
  Bld. du Triomphe
  B-1050 BRUXELLES
  (Belgique)

- BARDOS Claude — Université Paris-Nord
  Département de Mathématiques
  Av. J.B. Clément
  93430 VILLETANEUSE
  (France)

- BARNSLEY Michael — Service de Physique Théorique
  CEN-Saclay
  BP n°2
  91190 GIF-sur-YVETTE
  (France)

- BENACHOUR Saïd — Université des Sciences et Techniques d'Alger
  Institut de Mathématiques
  BP 9 DAR EL BEIDA
  ALGER
  (Algérie)

- BERESTYCKI Henri	Université Paris VI Analyse Numérique Tour 55-65 4 place Jussieu 75230 PARIS CEDEX 05 (France)
- BESSIS Daniel	Service de Physique Théorique CEN-Saclay BP n°2 91190 GIF-sur-YVETTE (France)
- BIVAR-WEINHOLTZ Antonio	R. Castilho n°73 4° esq. 1200 LISBONNE (Portugal)
- BOCCARDO Lucio	Universita di Roma Istituto Matematico "Guido Castelnuovo" 00100 ROMA (Italia)
- BRAUNER Claude	Ecole Centrale de Lyon Laboratoire de Mathématiques et d'Informatique 36, route de Dardilly 69130 ECULLY (France)
- BREZIS Haim	Université Paris VI Analyse Numérique Tour 55-65 4 place Jussieu 75230 PARIS CEDEX 05 (France)
- BRUSCHI Mario	Istituto "G. Marconi" Universita degli Studi ROMA (Italia)
- BUFFET Emmanuel	Laboratoire de Physique Théorique Ecole Polytechnique Fédérale de Lausanne P.O. Box 1024 CH-1001 LAUSANNE (Suisse)
- BULLOUGH Robin	NORDITA Blegdamsvej 17 DK-2100 COPENHAGEN Ø (Denmark)

- BUZANO Ernesto — Istituto di Analisi Matematica
  Universita di Torino
  Via Carlo Alberto 10
  10123 TORINO
  (Italia)

- CALOGERO Francesco — Istituto di Fisica
  Universita di Roma
  Piazzale delle Scienze 5
  00185 ROMA
  (Italia)

- CANUTO Claudio — Istituto di Analisi Matematica
  Universita di Torino
  Via Carlo Alberto 10
  10123 TORINO
  (Italia)

- CHADAM John Martin — RR # 3
  CALEDONIA, Ontario
  (Canada)

- CHUDNOVSKY Gregory — I.H.E.S.
  35 route de Chartres
  91440 BURES-sur-YVETTE
  (France)

- CHUDNOVSKY David — Service de Physique Théorique
  CEN-Saclay
  BP n°2
  91190 GIF-sur-YVETTE
  (France)

- CRANDALL Michael — Mathematics Research Center
  University of Wisconsin
  610 Walnut Street
  MADISON, WI 53706
  (U.S.A.)

- DAFERMOS Constantin — Division of Applied Mathematics
  Brown University
  PROVIDENCE, R.I. 02912
  (U.S.A.)

- DALLA CASA Valerio — Universita degli Studi di Parma
Istituto di Fisica
Via Massino d'Azeglio 85
43100 PARMA
(Italia)

- DAUDPOTA Q. Isa — University of Edinburgh
Department of Physics
James Clerk Maxwell Building
Mayfield Road
EDINBURGH EH9 3JZ
(Great Britain)

- DA VEIGA Hugo — Universita di Trento
Facolta di Scienze
38050 POVO (Trento)
(Italia)

- DELLA DORA Jean — Mathématiques Appliquées
Université Scientifique
et Médicale de Grenoble
B.P. 53X
38041 GRENOBLE CEDEX
(France)

- DERRIDA Bernard — Ecole Normale Supérieure
Laboratoire de Physique du Solide
24 rue Lhomond
75231 PARIS CEDEX 05
(France)

- DIAS Joao Paulo — Université Pierre et Marie Curie
Analyse Numérique
Tour 55-65
4, place Jussieu
75230 PARIS CEDEX 05
(France)

- DI PERNA Ronald — Mathematics Research Center
University of Wisconsin - Madison
610 Walnut Street
MADISON, WI 53706
(U.S.A.)

- ECKMANN J.P. — Département de Physique Théorique
Université de Genève
32, bld. d'Yvoy
CH-1211 GENEVE 4
(Suisse)

- ERNEUX Thomas	Université Libre de Bruxelles Faculté des Sciences Campus de la Plaine Bld. du Triomphe C.P. 231 B-1050 BRUXELLES (Belgique)
- FEINGENBAUM Mitchell	T-9 Los Alamos Scientific Laboratory P.O. Box 1663 LOS ALAMOS, New Mexico 87545 (U.S.A.)
- FERRINI Federico	Istituto di Fisica Piazza Torricelli, 2 56100 PISA (Italia)
- FISHER Michael	Department of Chemistry Baker Laboratory ITHACA, N.Y. 14853 (U.S.A)
- FRITZSCH Günter	Max-Planck Institut für Biophysik Kennedy Allee 70 D-6000 FRANKFURT a.M. 70 (R.F.A.)
- GAL-CHEN Tzvi	Department of Physics University of Ontario TORONTO, Ontario Canada M5S 1A7
- GAMMEL John	Physics Department Washington University SAINT-LOUIS, MO 63130 (U.S.A)
- GHIDOUCHE Hamid	Université Paris-Nord Département de Mathématiques Av. J.B. Clément 93430 VILLETANEUSE (France)
- GHIL Michael	Courant Institute of Mathematical Sciences New York University 251 Mercer Street NEW YORK, N.Y. 10012 (U.S.A.)

- GIDAS Basilis	The Rockefeller University 1230 York Avenue NEW YORK, N.Y. 10021 (U.S.A.)
- GIRAUD Bertrand	Service de Physique Théorique CEN-Saclay B.P. n°2 91190 GIF-SUR-YVETTE (France)
- GLASSEY Robert	Indiana University BLOOMINGTON, Indiana 47405 (U.S.A.)
- GRABERT Hermann	Institut für Theoretische Physik Pfaffenwaldring 57 7000 STUTTGART 80 (R.F.A.)
- GRUNDLAND Alfred	Universytet Warszawski Institut Fizyki Teorezycznej U.W. Ul. Hoza 69 00-681 WARSZĄWA (Pologne)
- HAHN Harro	Theoretische Physik "A", T.U. Mandelssohnstr. 1a D-3300 BRAUNSCHWEIG (R.F.A.)
- HANYGA Andrzej	Polish Academy of Sciences Institute of Geophysics 00-973 WARSAW Ul. Pasteura 3 P.O. Box 155 (Poland)
- HERNANDEZ Jesus	Departamento de Matemàticas Universidad Autonoma MADRID 34 (Espagne)
- KERNEVEZ J.P.	U.T.C. BP. 233 60206 COMPIEGNE (France)

- KOMOROWSKI Jacek
Institute of Math. Methods in Physics
Warsaw University
Ul. Hoza 74
00682 WARSAW
(Poland)

- KUO PEN YU
Collège de France
Département de Mathématiques
11 place Marcelin Berthelot
75231 PARIS CEDEX 05
(France)

- LABBE Jacques
Laboratoire P.M.T.M.
Université Paris-Nord
Av. J.B. Clément
93430 VILLETANEUSE
(France)

- LICHNEWSKY
Université Paris-Sud
Lab. de Mathématiques
Bât. 425
91405 ORSAY
(France)

- LIONS J.L.
Collège de France
Département de Mathématiques
11 place Marcelin Berthelot
75231 PARIS CEDEX 05
(France)

- LIONS Pierre-Louis
Université Paris VI
Analyse Numérique
Tour 55-65
4 place Jussieu
75230 PARIS CEDEX 05
(France)

- LI TA TSIEN
Collège de France
Département de Mathématiques
11 place Marcelin Berthelot
75231 PARIS CEDEX 05
(France)

- MALLACH Lawrence
University of California
Department of Anatomy
School of Medicine
LOS ANGELES, California 90024
(U.S.A.)

- MANCINI Giovani	Universita di Bologna Istituto Matematico "Salvatore Pincherle" Piazza di Porta S. Donato, 5 BOLOGNA (Italia)
- MARKOWSKI Boyan	Institute for Nuclear Research and Nuclear Engineering 1113 SOFIA (Bulgaria)
- MEHTA Madan Lal	Service de Physique Théorique CEN-Saclay BP n°2 91190 GIF-sur-YVETTE (France)
- MERCIER Bertrand	MA/EM Centre d'Etudes de Limeil BP n°27 94190 VILLENEUVE St-GEORGES (France)
- MIGNOT Fulbert	Université Paris VI Tour 55 SE 9, quai Saint-Bernard 75005 PARIS (France)
- MIGUENS LOURO Bento J.	Faculdade de Ciencias de Lisboa Rua da Escola Politecnica LISBOA 2 (Portugal)
- MINTCHEV Mihail	Scuola Normale Superiore Piazza dei Cavalieri 7 56100 PISA (Italia)
- MULLIN John	Mathematics Institute University of Oxford 23-27 St-Giles OXFORD Oxon (England)
- NICOLAENKO Basil	University of California Los Alamos Scientific Laboratory Mathematical Analysis Group T-7 LOS ALAMOS, New Mexico 87545 (U.S.A.)

- NIGGEMANN Michael — Universität Stuttgart
Mathematisches Institut A
1 Lehrstuhl
Postfach 801 140
D-700 STUTTGART 80
(R.F.A.)

- NUTTALL John — University of Western Ontario
Department of Physics
LONDON
Ontario
Canada N6A 3K7

- PENEL Patrick — Université de Toulon et du Var
Chateau St-Michel
83130 LA GARDE
(France)

- POMEAU Yves — Service le Physique Théorique
CEN-Saclay
BP n°2
91190 GIF-sur-YVETTE
(France)

- PUEL Jean-Pierre — Université Pierre et Marie Curie
Analyse Numérique, Tour 55-65
4 place Jussieu
75230 PARIS CEDEX 05
(France)

- RABINOWITZ Paul — Mathematics Research Center
University of Wisconsin
610 Walnut Street
MADISON, Wisconsin 53706
(U.S.A.)

- RAGNISCO Orlando — Istituto "G. Marconi"
Università degli Studi
ROMA
(Italia)

- SANCHEZ Luis — Centro de Matematica e Aplicaçoes Fundamentais
Av. Prof. Gama Pinto, 2
LISBOA 4
(Portugal)

- SANTINI Paolo Maria — Istituto "G. Marconi"
Università degli Studi
ROMA
(Italia)

- SARMA Gobalakichena
Service de Physique du Solide et de Résonance Magnétique
CEN-Saclay
BP n°2
91190 GIF-sur-YVETTE
(France)

- SCHATZMAN Michelle
Université Paris-Nord
Département de Mathématiques
Av. J-B. Clément
93430 VILLETANEUSE
(France)

- SCHEURER Bruno
MA/EM
Centre d'Etudes de Limeil
BP n°27
94190 VILLENEUVE St-GEORGES
(France)

- SCHEURLE Jurgen
Universität Stuttgart
Mathematisches Institut A
1 Lehrstuhl,
Postfach 801 140
D-700 STUTTGART 80
(R.F.A.)

- SIJBRAND J.
Rijksuniversiteit Utrecht
Mathematisch Instituut
Budapestlaan 6
Post Bus 80 010
3508 TA UTRECHT
(The Netherlands)

- SMOLLER Joel
Mathematics Department
University of Michigan
ANN ARBOR, MI 48109
(U.S.A.)

- SPARROW Collin
Department of Pure Mathematics and Mathematical Statistics
University of Cambridge
16, Mill Lane
CAMBRIDGE , CB2 1SB
(England)

- SCHONBEK Maria
Mathematics Department
Purdue University
LAFAYETTE, IN 47907
(U.S.A.)

- STORA Raymond
Division Théorique
C E R N
1211 GENEVE 23
(Suisse)

- STRAUSS W.
Mathematics Department
Brown University
PROVIDENCE, Rhode Island 02912
(U.S.A.)

- SULEM Pierre-Louis
Observatoire de Nice
BP 252
06007 NICE CEDEX
(France)

- SULEM Catherine
Université Paris-Nord
Département de Mathématiques
Av. J-B. Clément
94190 VILLETANEUSE
(France)

- TARTAR Luc
Département de Mathématiques
Université Paris-Sud
Bât.425
91405 ORSAY
(France)

- TEMAM R.
Département de Mathématiques
Université Paris-Sud
Bât.425
91405 ORSAY
(France)

- TRUE Hans
The Technical University of Denmark
Lab. of Applied Mathematical Physics
Bldg. 303 B
DK-2800 LYNGBY
(Denmark)

- TURCHETTI Giorgio
Istituto di Fisica "A. Righi"
Via Irnerio 46
40126 BOLOGNA
(Italia)

- VALLI Alberto
Università di Trento
Facoltà di Scienze
38050 POVO (Trento)
(Italia)

- VAZQUEZ Luis
Universidad Complutense de Madrid
Facultad de Ciencias Fisicas
Departamento de Fisica Teorica
MADRID-3
(Espagne)

- VOROS André
Service de Physique Théorique
CEN-Saclay
BP n°2
91190 GIF-sur-YVETTE
(France)

- WU T. T.
Physics Department
Harvard University
CAMBRIDGE, MA 02138
(U.S.A.)

GPSR Compliance

The European Union's (EU) General Product Safety Regulation (GPSR) is a set of rules that requires consumer products to be safe and our obligations to ensure this.

If you have any concerns about our products, you can contact us on ProductSafety@springernature.com

In case Publisher is established outside the EU, the EU authorized representative is:

Springer Nature Customer Service Center GmbH
Europaplatz 3
69115 Heidelberg, Germany

Batch number: 10816188

Printed by Printforce, the Netherlands